SOLUTIONS TO BLACK EXERCISES

Roxy Wilson
University of Illinois, Urbana-Champaign

CHEMISTRY

THE CENTRAL SCIENCE

13TH EDITION

BROWN | LeMAY
BURSTEN | MURPHY
WOODWARD | STOLTZFUS

PEARSON

Boston   Columbus   Indianapolis   New York   San Francisco   Upper Saddle River
Amsterdam   Cape Town   Dubai   London   Madrid   Milan   Munich   Paris   Montréal   Toronto
Delhi   Mexico City   São Paulo   Sydney   Hong Kong   Seoul   Singapore   Taipei   Tokyo

Editor in Chief: Adam Jaworski

Senior Acquisitions Editor: Terry Haugen

Acquisitions Editor: Chris Hess, Ph.D.

Executive Marketing Manager: Jonathan Cottrell

Associate Team Lead, Program Management, Chemistry and Geosciences: Jessica Moro

Team Lead, Project Management, Chemistry and Geosciences: Gina M. Cheselka

Project Manager: Erin Kneuer

Full-Service Project Management/Composition: PreMediaGlobal

Operations Specialist: Christy Hall

Supplement Cover Designer: Seventeenth Street Studios

Cover Image Credit: "Metal-Organic Frameworks" by Omar M. Yaghi, University of California, Berkeley

Credits and acknowledgments borrowed from other sources and reproduced, with permission, in this textbook appear on the appropriate page within the text.

www.pearsonhighered.com

1 2 3 4 5 6 7 8 9 10—EBM—18 17 16 15 14
ISBN-10: 0-321-94927-7; ISBN-13: 978-0-321-94927-1

Contents

Introduction

Chemistry: The Central Science, 13th edition, contains more than 2600 end-of-chapter exercises. Considerable attention has been given to these exercises because one of the best ways for students to master chemistry is by solving problems. Grouping the exercises according to subject matter is intended to aid the student in selecting and recognizing particular types of problems. Within each subject matter group, similar problems are arranged in pairs. This provides the student with an opportunity to reinforce a particular kind of problem. There are also a substantial number of general exercises in each chapter to supplement those grouped by topic. Visualizing Concepts, general exercises which require students to analyze visual data in order to formulate conclusions about chemical concepts, and Integrative Exercises, which require students to integrate concepts from several chapters, are continuing features of the 13th edition. Answers to the odd numbered topical exercises plus selected general exercises, about 1100 in all, are provided in the text. These appendix answers help to make the text a useful self-contained vehicle for learning.

This manual, **Solutions to Black Exercises** in **Chemistry: The Central Science, 13th edition**, was written to enhance the end-of-chapter exercises by providing documented solutions for those 1600 problems _not_ answered in the appendix of the text. The manual assists the instructor by saving time spent generating solutions for assigned problem sets and aids the student by offering a convenient independent source to check their understanding of the material. Most solutions have been worked in the same detail as the in-chapter sample exercises to help guide students in their studies.

When using this manual, keep in mind that the numerical result of any calculation is influenced by the precision of the numbers used in the calculation. In this manual, for example, atomic masses and physical constants are typically expressed to four significant figures, or at least as precisely as the data given in the problem. If students use slightly different values to solve problems, their answers will differ slightly from those listed in the appendix of the text or this manual. This is a normal and a common occurrence when comparing results from different calculations or experiments.

Rounding methods are another source of differences between calculated values. In this manual, when a solution is given in steps, intermediate results will be rounded to the correct number of significant figures; however, unrounded numbers will be used in subsequent calculations. By following this scheme, calculators need not be cleared to re-enter rounded intermediate results in the middle of a calculation sequence. The final answer will appear with the correct number of significant figures. This may result in a small discrepancy in the last significant digit between student-calculated answers and those given in this manual. Variations due to rounding can occur in any analysis of numerical data.

The first step in checking your solution and resolving differences between your answer and the listed value is to look for similarities and differences in problem-solving methods. Ultimately,

Introduction

resolving the small numerical differences described above is less important than understanding the general method for solving a problem. The goal of this manual is to provide a reference for sound and consistent problem-solving methods in addition to accurate answers to text exercises.

Extraordinary efforts have been made to keep this manual as error-free as possible. All exercises were worked and proof-read by at least three chemists to ensure clarity in methods and accuracy in mathematics. The work and advice of Ms. Renee Rice, Ms. Kate Vigour, Dr. Christopher Musto and Dr. Timothy Kucharski have been invaluable to this project. However, in a written work as technically challenging as this manual, typos and errors inevitably creep in. Please help us find and eliminate them. We hope that both instructors and students will find this manual accurate, helpful and instructive.

Roxy B. Wilson, Ph.D.
1829 Maynard Dr.
Champaign, IL 61822
rbwilson@uiuc.edu

1 Introduction: Matter and Measurement

Visualizing Concepts

1.2   After a *physical change*, the identities of the substances involved are the same as their identities before the change. That is, molecules retain their original composition. During a *chemical change*, at least one new substance is produced; rearrangement of atoms into new molecules occurs.

The diagram represents a chemical change because the molecules after the change are different than the molecules before the change.

1.4   (a)   time   (b)   mass   (c)   temperature   (d)   area   (e)   length

(f)   area   (g)   temperature   (h)   density   (i)   volume

1.6   Results (bullet holes) that are close to each other are *precise*. Results that are close to the "true value" (the bull's-eye) are *accurate*.

(a)   The results on target A are precise but not accurate. The individual shots landed close to each other, but not close to the bull's-eye. The results on target B are both precise and accurate. The individual shots landed close together and in the center of the target. The results on target C are neither precise nor accurate. The individual shots are scattered around the target.

(b)   The precise grouping on A is high and to the right of center. To improve accuracy, the sighting mechanism on the gun should be adjusted down and slightly left from its current position.

To improve the results on C, someone who can produce a precise grouping must shoot the student's target rifle. Then, the position of the sighting mechanism can be adjusted to produce an accurate shot. After adjusting the sight, the student needs more practice to improve precision.

1.8   (a)   Volume = length × width × height. Because the operation is multiplication, the dimension with fewest significant figures (sig figs) determines the number of sig figs in the result. The dimension "2.5 cm" has 2 sig figs, so the volume is reported with 2 sig figs.

(b)   Density = mass/volume. Because the operation is division, again the datum with fewer significant figures determines the number of sig figs in the result. While mass, 104.72 g, has 5 sig figs, volume [from (a)] has 2 sig figs, so density is also reported to 2 sig figs.

1

1.10   Given: m/s Find: mi/hr. Both the given and desired units have distance in the numerator and time in the denominator. Use appropriate conversion factors to change 'm' to 'mi' in the numerator and 's' to 'hr' in the denominator.

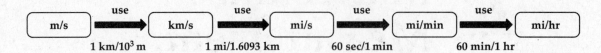

1.12   Compounds are pure substances, they have constant composition and properties throughout. The agate stone cannot be a compound, because materials with different properties appear as irregular rings in the stone. Ellen is correct.

Classification and Properties of Matter (Sections 1.2 and 1.3)

1.14   (a)   homogeneous mixture

            (b)   heterogeneous mixture (particles in liquid)

            (c)   pure substance

            (d)   heterogeneous mixture

1.16   (a)   C   (b) N   (c) Ti   (d) Zn   (e) Fe   (f) phosphorus

            (g)   calcium (h) helium   (i) lead   (j) silver

1.18   Gold, Au, is an element and "fool's gold," FeS_2, is a compound; both are solids and pure substances. Take advantage of differences in physical and or chemical properties between the two substances. Density and melting point measurements are often used to identify solids. For these two substances, melting points are very high, but densities are easy to measure. Gold is much denser than "fool's gold." Gold is much less chemically reactive than FeS_2, so relative reactivity with acids and bases can be observed. Of these experiments, density measurement is the most definitive and does not destroy the sample. (Note that neither substance is attracted to a magnet, so this test will not identify the gold.)

1.20   (a)   *Physical properties*: melting point = 420°C; hardness = 2.5 Mohs; density = 7.13 g/cm^3 at 25 °C. *Chemical properties*: granules react with dilute hydrochloric acid to produce hydrogen gas; at elevated temperatures, reacts slowly with oxygen gas to produce ZnO.

            (b)   From the photo, we see that zinc is a shiny dark gray solid at atmospheric conditions, (physical state, color). These are physical properties.

1.22   (a)   chemical

            (b)   physical

            (c)   physical (The production of H_2O is a chemical change, but its *condensation* is a physical change.)

            (d)   physical (The production of soot is a chemical change, but its *deposition* is a physical change.)

1.24   First heat the liquid in each beaker to 100°C to evaporate the water. The beaker with no residue contained pure water. The other two beakers have a solid, white residue. Measure the melting point of each solid. Sugar has a much lower melting point than salt, so the beaker with the lower-melting residue contained sugar water and that with the higher-melting residue contained salt water. (If confirmation is required, measure the densities of the two white residues.)

Units and Measurement (Section 1.4)

1.26   (a) $2.3 \times 10^{-10}\,\text{L} \times \dfrac{1\,\text{nL}}{1 \times 10^{-9}\,\text{L}} = 0.23\,\text{nL}$

        (b) $4.7 \times 10^{-6}\,\text{g} \times \dfrac{1\,\mu\text{g}}{1 \times 10^{-6}\,\text{g}} = 4.7\,\mu\text{g}$

        (c) $1.85 \times 10^{-12}\,\text{m} \times \dfrac{1\,\text{pm}}{1 \times 10^{-12}\,\text{m}} = 1.85\,\text{pm}$

        (d) $16.7 \times 10^{6}\,\text{s} \times \dfrac{1\,\text{Ms}}{1 \times 10^{6}\,\text{s}} = 16.7\,\text{Ms}$

        (e) $15.7 \times 10^{3}\,\text{g} \times \dfrac{1\,\text{kg}}{1 \times 10^{3}\,\text{g}} = 15.7\,\text{kg}$

        (f) $1.34 \times 10^{-3}\,\text{m} \times \dfrac{1\,\text{mm}}{1 \times 10^{-3}\,\text{m}} = 1.34\,\text{mm}$

        (g) $1.84 \times 10^{2}\,\text{cm} \times \dfrac{1\,\text{m}}{1 \times 10^{2}\,\text{cm}} = 1.84\,\text{m}$

1.28   (a)   °C = 5/9 (87°F – 32°) = 31 °C

        (b)   K = 25 °C + 273.15 = 298 K; °F = 9/5 (25 °C) + 32 = 77 °F

        (c)   °C = 5/9 (400 °F – 32°) = 204.444 = 204 °C

                K = °C + 273.15 = 204.444 °C + 273.15 = 478 K

        (d)   °C = 77 K – 273.15 = –196.15 = –196 °C; °F = 9/5 (–196.15 °C) + 32 = –321 °F

1.30   (a)   volume = length^{3} (cm^{3}); density = mass/volume (g/cm^{3})

                volume = (1.500)^{3} cm^{3} = 3.375 cm^{3}

                $\text{density} = \dfrac{76.31\,\text{g}}{3.375\,\text{cm}^{3}} = 22.61\,\text{g/cm}^{3}\ \text{osmium}$

        (b) $125.0\,\text{mL} \times \dfrac{1\,\text{cm}^{3}}{1\,\text{mL}} \times \dfrac{4.51\,\text{g}}{1\,\text{cm}^{3}} = 563.75 = 564\,\text{g titanium}$

        (c) $0.1500\,\text{L} \times \dfrac{1\,\text{mL}}{1 \times 10^{-3}\,\text{L}} \times \dfrac{0.8787\,\text{g}}{1\,\text{mL}} = 131.8\,\text{g benzene}$

1.32   (a) $\dfrac{21.95\,\text{g}}{25.0\,\text{mL}} = 0.878\,\text{g/mL}$

The tabulated value has four significant figures, while the experimental value has three. The tabulated value rounded to three figures is 0.879. The values agree within one in the last significant figure of the experimental value; the two results agree. The liquid could be benzene.

     (b) $15.0\,\text{g} \times \dfrac{1\,\text{mL}}{0.7781\,\text{g}} = 19.3\,\text{mL cyclohexane}$

     (c) $r = d/2 = 5.0\,\text{cm}/2 = 2.5\,\text{cm}$

$V = 4/3\,\pi\,r^3 = 4/3 \times \pi \times (2.5)^3\,\text{cm}^3 = 65.4498 = 65\,\text{cm}^3$

$65.4498\,\text{cm}^3 \times \dfrac{11.34\,\text{g}}{\text{cm}^3} = 7.4 \times 10^2\,\text{g}$

(The answer has two significant figures because the diameter had only two significant figures.)

Note: This is the first exercise where "intermediate rounding" occurs. In this manual, when a solution is given in steps, the intermediate result will be rounded to the correct number of significant figures. However, the **unrounded** number will be used in subsequent calculations. The final answer will appear with the correct number of significant figures. That is, calculators need not be cleared and new numbers entered in the middle of a calculation sequence. This may result in a small discrepancy in the last significant digit between student-calculated answers and those given in the manual. These variations occur in any analysis of numerical data.

For example, in this exercise the volume of the sphere, 65.4498 cm^3, is rounded to 65 cm^3, but 65.4498 is retained in the subsequent calculation of mass, 7.4×10^2 g. In this case, 65 cm^3 × 11.34 g/cm^3 also yields 7.4×10^2 g. In other exercises, the correctly rounded results of the two methods may not be identical.

1.34   (a)   The wafers have the same diameter as the boule, so the question becomes 'how many 0.75 mm wafers can be cut from the 2 m boule?'

$\dfrac{2.0\,\text{m}}{\text{boule}} \times \dfrac{1\,\text{mm}}{1 \times 10^{-3}\,\text{m}} \times \dfrac{1\,\text{wafer}}{0.75\,\text{mm}} = 2667 = 2.7 \times 10^3\,\text{wafers}$

     (b)   Calculate the volume of the wafer in cm^3. $V = \pi r^2 h$

$r = \dfrac{d}{2} = \dfrac{300\,\text{mm}}{2} \times \dfrac{1\,\text{cm}}{10\,\text{mm}} = 15\,\text{cm}; \quad h = 0.75\,\text{mm} \times \dfrac{1\,\text{cm}}{10\,\text{mm}} = 7.5 \times 10^{-2}\,\text{cm}$

$V = \pi r^2 h = \pi (15\,\text{cm})^2 (7.5 \times 10^{-2}\,\text{cm}) = 53.0144 = 53\,\text{cm}^3$

Density = mass / V;   mass = density × V

$\dfrac{2.33\,\text{g}}{\text{cm}^3} \times 53.0144\,\text{cm}^3 = 123.52 = 1.2 \times 10^2\,\text{g}$

Uncertainty in Measurement (Section 1.5)

1.36   Exact: (b), (e) (The number of students is exact on any given day.)

1.38   (a)   4   (b) 3   (c) 4   (d) 5   (e) 6   (f) 2

1.40   (a) 7.93×10^3 mi   (b) 4.001×10^4 km

1.42   (a) $[320.5 - 6104.5/2.3] = -2.3 \times 10^3$ (The intermediate result has two significant figures, so only the thousand and hundred places in the answer are significant.)

        (b) $[285.3 \times 10^5 - 0.01200 \times 10^5] \times 2.8954 = 8.260 \times 10^7$ (Because subtraction depends on decimal places, both numbers must have the same exponent to determine decimal places/sig figs. The intermediate result has 1 decimal place and 4 sig figs, so the answer has 4 sig figs.)

        (c) $(0.0045 \times 20,000.0)$ + (2813×12) $= 3.4 \times 10^4$
              2 sig figs /0 dec pl   2 sig figs /first 2 digits

        (d)   863 $\times$ [1255   − (3.45×108)] $= 7.62 \times 10^5$
                                            (3 sig figs /0 dec pl)

              3 sig figs $\times$ [0 dec pl/3 sig figs] $= 3$ sig figs

1.44   The volume in the graduated cylinder is 19.5 mL. Liquid volumes are read at the bottom of the meniscus, so the volume is slightly less than 20 mL. Volumes in this cylinder can be read with certainty to 1 mL, and with some uncertainty to 0.1 mL, so this measurement has 3 sig figs.

Dimensional Analysis (Section 1.6)

1.46   In each conversion factor, the old unit appears in the denominator, so it cancels, and the new unit appears in the numerator.

        (a) $\mu m \rightarrow mm:$ $\dfrac{1 \times 10^{-6}\ m}{1\,\mu m} \times \dfrac{1\,mm}{1 \times 10^{-3}\ m} = 1 \times 10^{-3}\ mm/\mu m$

        (b) $ms \rightarrow ns:$ $\dfrac{1 \times 10^{-3}\ s}{1\,ms} \times \dfrac{1\,ns}{1 \times 10^{-9}\ s} = 1 \times 10^6\ ns/ms$

        (c)   mi $\rightarrow$ km: 1.6093 km/mi

        (d) $ft^3 \rightarrow L:$ $\dfrac{(12)^3\ in^3}{1\,ft^3} \times \dfrac{(2.54)^3\ cm^3}{1\,in^3} \times \dfrac{1\,L}{1000\ cm^3} = 28.3\ L/ft^3$

1.48   (a) $\dfrac{2.998 \times 10^8\,m}{s} \times \dfrac{1\,km}{1000\ m} \times \dfrac{1\,mi}{1.6093\ km} \times \dfrac{60\ s}{1\,min} \times \dfrac{60\ min}{1\ hr} = 6.707 \times 10^8$ mi/hr

        (b)   1454 ft $\times$ $\dfrac{1\,yd}{3\,ft} \times \dfrac{1\,m}{1.0936\ yd} = 443.18 = 443.2\ m$

        (c)   3,666,500 m^3 $\times$ $\dfrac{1^3\,dm^3}{(1 \times 10^{-1})^3\ m^3} \times \dfrac{1\,L}{1\,dm^3} = 3.6665 \times 10^9\,L$

(d) $\dfrac{242\text{ mg cholesterol}}{100\text{ mL blood}} \times \dfrac{1\text{ mL}}{1\times10^{-3}\text{ L}} \times 5.2\text{ L} \times \dfrac{1\times10^{-3}\text{ g}}{1\text{ mg}} = 12.58 = 13\text{ g cholesterol}$

1.50   One uncertainty is the definition of "Grand Rapids" and "Detroit." These are cities with a finite area, so the distance uncertainty can be as large as the sum of the radii of the two towns. Another uncertainty is the route itself. And, as with any measured quantity, there is the uncertainty of the measuring device itself.

         To make the number more precise, one would need to specify the route (shortest, fastest, largest highways, etc.), the endpoint in each city, and use the most precise measuring device available.

1.52   (a) $0.105\text{ in} \times \dfrac{2.54\text{ cm}}{\text{in}} \times \dfrac{1\times10^{-2}\text{ m}}{\text{cm}} \times \dfrac{1\text{ mm}}{1\times10^{-3}\text{ m}} = 2.667 = 2.67\text{ mm}$

        (b) $0.650\text{ qt} \times \dfrac{1\text{ L}}{1.057\text{ qt}} \times \dfrac{1\text{ mL}}{1\times10^{-3}\text{ L}} = 614.94 = 615\text{ mL}$

        (c) $\dfrac{8.75\,\mu\text{m}}{\text{s}} \times \dfrac{1\times10^{-6}\text{ m}}{1\,\mu\text{m}} \times \dfrac{1\text{ km}}{1\times10^{3}\text{ m}} \times \dfrac{60\text{ s}}{1\text{ min}} \times \dfrac{60\text{ min}}{1\text{ hr}} = 3.15\times10^{-5}\text{ km/hr}$

        (d) $1.955\text{ m}^3 \times \dfrac{(1.0936)^3\text{ yd}^3}{1\text{ m}^3} = 2.55695 = 2.557\text{ yd}^3$

        (e) $\dfrac{\$3.99}{\text{lb}} \times \dfrac{2.205\text{ lb}}{1\text{ kg}} = 8.798 = \$8.80/\text{kg}$

        (f) $\dfrac{8.75\text{ lb}}{\text{ft}^3} \times \dfrac{453.59\text{ g}}{1\text{ lb}} \times \dfrac{1\text{ ft}^3}{12^3\text{ in}^3} \times \dfrac{1\text{ in}^3}{2.54^3\text{ cm}^3} \times \dfrac{1\text{ cm}^3}{1\text{ mL}} = 0.140\text{ g/mL}$

1.54   (a) $1257\text{ mi} \times \dfrac{1\text{ km}}{0.62137\text{ mi}} \times \dfrac{\text{charge}}{225\text{ km}} = 8.99\text{ charges}$

        Because charges are integral events, 9 total charges are required. The trip begins with a full charge, so 8 additional charges during the trip are needed.

        (b) $\dfrac{14\text{ m}}{\text{s}} \times \dfrac{1\text{ km}}{1\times10^{3}\text{ m}} \times \dfrac{1\text{ mi}}{1.6093\text{ km}} \times \dfrac{60\text{ s}}{1\text{ min}} \times \dfrac{60\text{ min}}{1\text{ hr}} = 31\text{ mi/hr}$

        (c) $450\text{ in}^3 \times \dfrac{(2.54)^3\text{ cm}^3}{1\text{ in}^3} \times \dfrac{1\text{ mL}}{1\text{ cm}^3} \times \dfrac{1\times10^{-3}\text{ L}}{1\text{ mL}} = 7.37\text{ L}$

        (d) $2.4\times10^5\text{ barrels} \times \dfrac{42\text{ gal}}{1\text{ barrel}} \times \dfrac{4\text{ qt}}{1\text{ gal}} \times \dfrac{1\text{ L}}{1.057\text{ qt}} = 3.8\times10^7\text{ L}$

1.56 $10.6\text{ ft} \times 14.8\text{ ft} \times 20.5\text{ ft} = 3216.04 = 3.22\times10^3\text{ ft}^3$

        $3216.04\text{ ft}^3 \times \dfrac{(1\text{ yd})^3}{(3\text{ ft})^3} \times \dfrac{(1\text{ m})^3}{(1.0936\text{ yd})^3} \times \dfrac{48\,\mu\text{g CO}}{1\text{ m}^3} \times \dfrac{1\times10^{-6}\text{ g}}{1\,\mu\text{g}} = 4.4\times10^{-3}\text{ g CO}$

1.58   A wire is a very long, thin cylinder of volume, $V = \pi r^2 h$, where h is the length of the wire and πr^2 is the cross-sectional area of the wire.

        Strategy: 1) Calculate total volume of copper in cm^3 from mass and density

    2) $h \text{ (length in cm)} = \dfrac{V}{\pi \, r^2}$

    3) Change cm → ft

$$150 \text{ lb Cu} \times \frac{453.6 \text{ g}}{1 \text{ lb Cu}} \times \frac{1 \text{ cm}^3}{8.94 \text{ g}} = 7610.7 = 7.61 \times 10^3 \text{ cm}^3$$

$$r = d/2 = 7.50 \text{ mm} \times \frac{1 \text{ cm}}{10 \text{ mm}} \times \frac{1}{2} = 0.375 \text{ cm}$$

$$h = \frac{V}{\pi r^2} = \frac{7610.7 \text{ cm}^3}{\pi (0.375)^2 \text{ cm}^2} = 1.7227 \times 10^4 = 1.72 \times 10^4 \text{ cm}$$

$$1.7227 \times 10^4 \text{ cm} \times \frac{1 \text{ in}}{2.54 \text{ cm}} \times \frac{1 \text{ ft}}{12 \text{ in}} = 565 \text{ ft}$$

(too difficult to estimate)

Additional Exercises

1.59   (a)   A gold coin is probably a *solid solution*. Pure gold (element 79) is too soft and too valuable to be used for coinage, so other metals are added. However, the simple term "gold coin" does not give a specific indication of the other metals in the mixture.

                A cup of coffee is a *solution* if there are no suspended solids (coffee grounds). It is a heterogeneous mixture if there are grounds. If cream or sugar is added, the homogeneity of the mixture depends on how thoroughly the components are mixed.

                A wood plank is a *heterogeneous mixture* of various cellulose components. The different domains in the mixture are visible as wood grain or knots.

    (b)   The ambiguity in each of these examples is that the name of the substance does not provide a complete description of the material. We must rely on mental images, and these vary from person to person.

1.60   (a)   A *hypothesis* is a possible explanation for certain phenomena based on preliminary experimental data. A *theory* may be more general, and has a significant body of experimental evidence to support it; a theory has withstood the test of experimentation.

    (b)   A scientific *law* is a summary or statement of natural behavior; it tells how matter behaves. A *theory* is an explanation of natural behavior; it attempts to explain why matter behaves the way it does.

1.62   (a)   (l) → (g)

    (b)   °F = 9/5 (°C) + 32 °; 9/5 (12) + 32 = 53.6 = 54 °F

    (c)   0.765 g/mL × 103.5 mL = 79.178 = 79.2 g ethyl chloride

1.64   (a)   Appropriate. The number 22,727,000 implies a precision of one part per thousand, or 0.1%. This is an appropriate level of precision for the accounting records of a company like Apple Computer.

(b) Appropriate. Rainfall data can be measured to a precision of at least one decimal place. Calculating annual rainfall and average annual rainfall involves addition which dictates that significant figures are determined by the least number of decimal places in the data being summed.

(c) Appropriate. The percentage has three significant figures. In a population as large as that of the United States, the number of people named Brown can surely be counted by census data or otherwise to a precision of three significant figures.

(d) Inappropriate. Letter grades are posted at most to two decimal places and three significant figures (if plus and minus modifiers are quantified). The grade-point-average, obtained by addition and division, cannot have more decimal places or significant figures than the numbers being averaged.

1.66 (a) $\dfrac{m}{s^2}$ (b) $\dfrac{kg-m}{s^2}$ (c) $\dfrac{kg-m}{s^2} \times m = \dfrac{kg-m^2}{s^2}$

(d) $\dfrac{kg-m}{s^2} \times \dfrac{1}{m^2} = \dfrac{kg}{m-s^2}$ (e) $\dfrac{kg-m^2}{s^2} \times \dfrac{1}{s} = \dfrac{kg-m^2}{s^3}$

(f) $\dfrac{m}{s}$ (g) $kg \times \left(\dfrac{m}{s}\right)^2 = \dfrac{kg-m^2}{s^2}$

1.67 (a) $2.4 \times 10^5 \text{ mi} \times \dfrac{1.609 \text{ km}}{1 \text{ mi}} \times \dfrac{1000 \text{ m}}{1 \text{ km}} = 3.862 \times 10^8 = 3.9 \times 10^8 \text{ m}$

(b) $2.4 \times 10^5 \text{ mi} \times \dfrac{1.609 \text{ km}}{1 \text{ mi}} \times \dfrac{1 \text{ hr}}{350 \text{ km}} \times \dfrac{60 \text{ min}}{1 \text{ hr}} \times \dfrac{60 \text{ s}}{1 \text{ min}} = 4.0 \times 10^6 \text{ s}$

(c) $3.862 \times 10^8 \text{ m} \times 2 \times \dfrac{1 \text{ s}}{3.00 \times 10^8 \text{ m}} = 2.574 = 2.6 \text{ s}$

(d) $\dfrac{29.783 \text{ km}}{s} \times \dfrac{1 \text{ mi}}{1.6093 \text{ km}} \times \dfrac{60 \text{ s}}{1 \text{ min}} \times \dfrac{60 \text{ min}}{1 \text{ hr}} = 6.6624 \times 10^4 \text{ mi/hr}$

1.70 (a) $\dfrac{\$1950}{acre-ft} \times \dfrac{1 \text{ acre}}{4840 \text{ yd}^2} \times \dfrac{3 \text{ ft}}{1 \text{ yd}} \times \dfrac{(1.094 \text{ yd})^3}{(1 \text{ m})^3} \times \dfrac{(1 \text{ m})^3}{(10 \text{ dm})^3} \times \dfrac{(1 \text{ dm})^3}{1 \text{ L}} =$

$\$1.583 \times 10^{-3}/\text{L or } 0.1583 \text{ ¢/L}$ (0.158 ¢/L to 3 sig figs)

(b) $\dfrac{\$1950}{acre-ft} \times \dfrac{1 \text{ acre}-ft}{2 \text{ households}-year} \times \dfrac{1 \text{ year}}{365 \text{ days}} \times 1 \text{ household} = \dfrac{\$2.671}{day} = \dfrac{\$2.67}{day}$

1.71 Select a common unit for comparison, in this case the kg.

1 kg > 2 lb, 1 L ≈ 1 qt

5 lb potatoes < 2.5 kg

5 kg sugar = 5 kg

1 gal = 4 qt ≈ 4 L; 1 mL H_2O = 1 g H_2O; 1 L = 1000 g, 4 L = 4000 g = 4 kg

The order of mass from lightest to heaviest is 5 lb potatoes < 1 gal water < 5 kg sugar.

1.72   There are 347 degrees between the freezing and boiling points on the oleic acid (O) scale and 100 degrees on the celsius (C) scale. Also, 13°C = 0°O. By analogy with °F and °C,

$$°O = \frac{100}{347}(°C - 13) \quad \text{or} \quad °C = \frac{347}{100}(°O) + 13$$

These equations correctly relate the freezing point (and boiling point) of oleic acid on the two scales.

$$\text{f.p. of } H_2O: \quad °O = \frac{100}{347}(0°C - 13) = -3.746 = -4°O$$

1.74   Density is the ratio of mass and volume. For samples with the same volume, in this case spheres with the same diameter, the denser ball will have a greater mass. The heavier ball, the red one on the right in the diagram is more dense.

1.75   The mass of water in the bottle does not change with temperature, but the density (ratio of mass to volume) does. That is, the amount of volume occupied by a certain mass of water changes with temperature. Calculate the mass of water in the bottle at 25°C, and then the volume occupied by this mass at −10°C.

     (a) $25°C: \ 1.50\,L\,H_2O \times \dfrac{1000\ cm^3}{1\,L} \times \dfrac{0.997\ g\ H_2O}{1\ cm^3} = 1.4955 \times 10^3 = 1.50 \times 10^3\ g\ H_2O$

            $-10°C: \ 1.4955 \times 10^3\ g\ H_2O \times \dfrac{1\ cm^3}{0.917\ g\ H_2O} \times \dfrac{1\,L}{1000\ cm^3} = 1.6309 = 1.63\,L$

     (b)   No. If the soft-drink bottle is completely filled with 1.50 L of water, the 1.63 L of ice **cannot** be contained in the bottle. The extra volume of ice will push through any opening in the bottle, or crack the bottle to create an opening.

1.77   (a) $V = 4/3\,\pi\,r^3 = 4/3\,\pi\,(28.9\ cm)^3 = 1.0111 \times 10^5 = 1.01 \times 10^5 = 1.01 \times 10^5\ cm^3$.

            $1.0111 \times 10^5\ cm^3 \times \dfrac{19.3\ g}{cm^3} \times \dfrac{1\ lb}{453.59\ g} = 4302 = 4.30 \times 10^3\ lb$

     (b)   No. The sphere weighs 4300 pounds, more than two tons. The student is unlikely to be able to carry it without assistance.

1.78 $1.00\ gal\ battery\ acid \times \dfrac{4\ qt}{1\ gal} \times \dfrac{1000\ mL}{1.0567\ qt} \times \dfrac{1.28\ g}{mL} = 4845.3 = 4.85 \times 10^3\ g\ battery\ acid$

         $4.8453 \times 10^3\ g\ battery\ acid \times \dfrac{38.1\ g\ sulfuric\ acid}{100\ g\ battery\ acid} = 1846 = 1.85 \times 10^3\ g\ sulfuric\ acid$

1.80 $8.0\ oz \times \dfrac{1\ lb}{16\ oz} \times \dfrac{453.6\ g}{lb} \times \dfrac{1\ cm^3}{2.70\ g} = 84.00 = 84\ cm^3$

     $\dfrac{84\ cm^3}{50\ ft^2} \times \dfrac{1^2\ ft^2}{12^2\ in^2} \times \dfrac{1^2\ in^2}{2.54^2\ cm^2} \times \dfrac{10\ mm}{1\ cm} = 0.018\ mm$

1.81   The total solar flux, the power provided by the sun, is the average solar flux times the area of the disc in sunlight at any instant. However, because the earth rotates, each disc is in sunlight for only half of the day, so only half of this flux can be collected.

$$\frac{680 \text{ W}}{\text{m}^2} \times 1.28 \times 10^{14} \text{ m}^2 \times 1/2 = 4.352 \times 10^{16} = 4.4 \times 10^{16} \text{ W}$$

Collection is 10% efficient, so only 10% of this power is available, 4.4×10^{15} W

$$15 \text{TW} \times \frac{1 \times 10^{12} \text{ W}}{1 \text{ TW}} = 15 \times 10^{12} \text{ W needed}$$

$$\frac{15 \times 10^{12} \text{ W needed}}{4.352 \times 10^{15} \text{ W available}} \times 100 = 0.3447 = 0.34\% \text{ Earth's surface needed}$$

1.82   Warren's first hypothesis was that the bacterium he observed was involved in causing ulcers. This was a radical hypothesis in 1979, when it was commonly accepted that the acidic environment of the stomach would not support microorganisms. Strong evidence was needed to change the opinion and practice of the medical community.

To test his hypothesis, Warren first collected many more tissue samples, from patients with and without ulcers. He established that healthy tissue did not contain the bacterium, while many of the ulcerated samples did. From there, with Marshall and other collaborators, he established two different ways to diagnose the infection (determine the presence of the bacterium in patients with potential ulcers), and learned to culture (grow in the laboratory) the new bacterium so that its effects could be studied. Finally, in patients diagnosed and treated for bacterial ulcers, he demonstrated that the recurrence of ulcers was rare.

This information was obtained from the autobiography of J. Robin Warren, http://www.nobelprize.org/nobel_prizes/medicine/laureates/2005/warren -autobio.html

1.84   (a)   Let x = mass of Au in jewelry

9.85 - x = mass of Ag in jewelry

The total volume of jewelry = volume of Au + volume of Ag

$$0.675 \text{ cm}^3 = x \text{ g} \times \frac{1 \text{ cm}^3}{19.3 \text{ g}} + (9.85 - x) \text{g} \times \frac{1 \text{ cm}^3}{10.5 \text{ g}}$$

$$0.675 = \frac{x}{19.3} + \frac{9.85 - x}{10.5} \quad \text{(To solve, multiply both sides by (19.3)(10.5))}$$

$$0.675 (19.3)(10.5) = 10.5 \, x + (9.85 - x)(19.3)$$

$$136.79 = 10.5 \, x + 190.105 - 19.3 \, x$$

$$-53.315 = -8.8 \, x$$

x = 6.06 g Au; 9.85 g total – 6.06 g Au = 3.79 g Ag

$$\text{mass \% Au} = \frac{6.06\,\text{g Au}}{9.85\,\text{g jewelry}} \times 100 = 61.5\%\,\text{Au}$$

   (b) $24\,\text{carats} \times 0.615 = 15\,\text{carat gold}$

1.86   (a)   False. Air and water are not elements. Air is a homogeneous mixture of gases and pure water is a compound.

       (b)   False. Mixtures can contain any number of pure substances, either elements or compounds or both.

       (c)   True.

       (d)   True.

       (e)   False. When yellow stains in a kitchen sink are treated with bleach water, a chemical change occurs.

       (f)   True.

       (g)   False. The number 0.0033 has the same number of significant figures as 0.033.

       (h)   True. (In a conversion factor, the quantity in the numerator is equal to the quantity in the denominator, so the overall numerical value is one.)

       (i)   True.

1.87   The densities are:

carbon tetrachloride (methane, tetrachloro) – $1.5940\,\text{g/cm}^3$

hexane – $0.6603\,\text{g/cm}^3$

benzene – $0.87654\,\text{g/cm}^3$

methylene iodide (methane, diiodo) – $3.3254\,\text{g/cm}^3$

Only methylene iodide will separate the two granular solids. The undesirable solid $(2.04\,\text{g/cm}^3)$ is less dense than methylene iodide and will float; the desired material is more dense than methylene iodide and will sink. The other three liquids are less dense than both solids and will not produce separation.

2 Atoms, Molecules, and Ions

Visualizing Concepts

2.2   (a) $\% \text{ abundance} = \dfrac{\# \text{ of mass number} \times \text{particles}}{\text{total number of particles}} \times 100$

12 red ^{293}Nv particles

8 blue ^{295}Nv particles

20 total particles

$\% \text{ abundance } ^{293}\text{Nv} = \dfrac{12}{20} \times 100 = 60\%$

$\% \text{ abundance } ^{295}\text{Nv} = \dfrac{8}{20} \times 100 = 40\%$

(b)   Atomic weight (AW) is the same as average atomic mass.

Atomic weight (average atomic mass) = $\sum$ fractional abundance $\times$ mass of isotope

AW of Nv = 0.60(293.15) + 0.40(295.15) = 293.95 amu

(Because % abundance was calculated by counting exact numbers of particles, assume % abundance is an exact number. Then, the number of significant figures in the AW is determined by the number of sig figs in the masses of the isotopes.)

2.3   In general, metals occupy the left side of the chart, and nonmetals the right side.

metals: red and green   *nonmetals*: blue and yellow

alkaline earth metal: red   *noble gas*: yellow

2.5   In a solid, particles are close together and their relative positions are fixed. In a liquid, particles are close but moving relative to each other. In a gas, particles are far apart and moving. All ionic compounds are solids because of the strong forces among charged particles. Molecular compounds can exist in any state: solid, liquid, or gas.

Because the molecules in *ii* are far apart, *ii* must be a molecular compound. The particles in *i* are near each other and exist in a regular, ordered arrangement, so *i* is likely to be an ionic compound.

2.7   See Figure 2.18. yellow box: 1+ (group 1A); blue box: 2+ (group 2A)

black box: 3+ (a metal in Group 3A); red box: 2– (a nonmetal in group 6A);

green box: 1– (a nonmetal in group 7A)

2.9　　These two compounds are isomers. They have the same chemical formula, C_4H_9Cl, but different arrangements of atoms. That is, they have different chemical structures. In the first isomer, the Cl atom is bound to the second C atom from the left. In the second isomer, the Cl atom is bound to the right-most C atom.

Atomic Theory and the Discovery of Atomic Structure (Sections 2.1–2.2)

2.12　　(a)　　6.500 g compound – 0.384 g hydrogen = 6.116 g sulfur

　　　　(b)　　*Conservation of mass*

　　　　(c)　　According to postulate 3 of the atomic theory, atoms are neither created nor destroyed during a chemical reaction. If 0.384 g of H are recovered from a compound that contains only H and S, the remaining mass must be sulfur.

2.14　　(a)　　1:　$\dfrac{3.56 \text{ g fluorine}}{4.75 \text{ g iodine}} = 0.749$ g fluorine/1 g iodine

　　　　　　　2:　$\dfrac{3.43 \text{ g fluorine}}{7.64 \text{ g iodine}} = 0.449$ g fluorine/1 g iodine

　　　　　　　3:　$\dfrac{9.86 \text{ g fluorine}}{9.41 \text{ g iodine}} = 1.05$ g fluorine/1 g iodine

　　　　(b)　　To look for integer relationships among these values, divide each one by the smallest.

　　　　　　　If the quotients aren't all integers, multiply by a common factor to obtain all integers.

　　　　　　　1: $0.749/0.449 = 1.67$; $1.67 \times 3 = 5$

　　　　　　　2: $0.449/0.449 = 1.00$; $1.00 \times 3 = 3$

　　　　　　　3: $1.05/0.449 = 2.34$; $2.34 \times 3 = 7$

　　　　　　　The ratio of g fluorine to g iodine in the three compounds is 5:3:7. These are in the ratio of small whole numbers and, therefore, obey the *law of multiple proportions*. This integer ratio indicates that the combining fluorine "units" (atoms) are indivisible entities.

2.16　　Because the unknown particle is deflected in the opposite direction from that of a negatively charged beta (β) particle, it is attracted to the (–) plate and repelled by the (+) plate. The unknown particle is positively charged. The magnitude of the deflection is less than that of the β particle, or electron, so the unknown particle has greater mass than the electron. The unknown is a positively charged particle of greater mass than the electron.

2.18　　(a)　　The droplets carry different total charges because there may be 1, 2, 3, or more electrons on the droplet.

　　　　(b)　　The electronic charge is likely to be the lowest common factor in all the observed charges.

　　　　(c)　　Assuming this is so, we calculate the apparent electronic charge from each drop as follows:

A: $1.60 \times 10^{-19} / 1 = 1.60 \times 10^{-19}$ C

B: $3.15 \times 10^{-19} / 2 = 1.58 \times 10^{-19}$ C

C: $4.81 \times 10^{-19} / 3 = 1.60 \times 10^{-19}$ C

D: $6.31 \times 10^{-19} / 4 = 1.58 \times 10^{-19}$ C

The reported value is the average of these four values. Because each calculated charge has three significant figures, the average will also have three significant figures.

$(1.60 \times 10^{-19} \text{ C} + 1.58 \times 10^{-19} \text{ C} + 1.60 \times 10^{-19} \text{ C} + 1.58 \times 10^{-19} \text{ C}) / 4 = 1.59 \times 10^{-19}$ C

Modern View of Atomic Structure; Atomic Weights (Sections 2.3–2.4)

2.20   (a) $r = d/2; \; r = \dfrac{2.7 \times 10^{-8} \text{ cm}}{2} \times \dfrac{1 \text{ Å}}{1 \times 10^{-8} \text{ cm}} = 1.35 = 1.4 \text{ Å}$

$r = \dfrac{2.7 \times 10^{-8} \text{ cm}}{2} \times \dfrac{1 \text{ m}}{100 \text{ cm}} = 1.35 \times 10^{-10} = 1.4 \times 10^{-10} \text{ m}$

(b)   Aligned Rh atoms have **diameters** touching. $d = 2.7 \times 10^{-8}$ cm $= 2.7 \times 10^{-10}$ m

$6.0 \, \mu\text{m} \times \dfrac{1 \times 10^{-6} \text{ m}}{1 \, \mu\text{m}} \times \dfrac{1 \text{ Rh atom}}{2.7 \times 10^{-10} \text{ m}} = 2.2 \times 10^{4}$ Rh atoms

(c) $V = 4/3 \, \pi \, r^3; \; r = 1.35 \times 10^{-10} = 1.4 \times 10^{-10}$ m

$V = (4/3)[(\pi(1.35 \times 10^{-10})^3] \text{ m}^3 = 1.031 \times 10^{-29} = 1.0 \times 10^{-29} \text{ m}^3$

2.22   (a)   The nucleus has most of the mass **but occupies very little** of the volume of an atom.

(b)   True

(c)   The number of electrons in a neutral atom is equal to the number of **protons** in the atom.

(d)   True

2.24   (a) $^{210}_{83}\text{Bi}$ has one more proton and one fewer neutron than $^{210}_{82}\text{Pb}$.

(b) $^{15}_{7}\text{N}$ has the same number of protons and one more neutron than $^{14}_{7}\text{N}$.

(c) $^{40}_{18}\text{Ar}$ has eight more protons and twelve more neutrons than $^{20}_{10}\text{Ne}$.

2.26   (a) $^{31}_{16}\text{X}$ and $^{32}_{16}\text{X}$ are isotopes of the same element, because they have identical atomic numbers.

(b)   These are isotopes of the element sulfur, S, atomic number = 16.

2.28   (a) ^{32}P has 15 p, 17 n   (b) ^{51}Cr has 24 p, 27 n

(c) ^{60}Co has 27 p, 33 n   (d) ^{99}Tc has 43 p, 56 n

(e) ^{131}I has 53 p, 78 n   (f) ^{201}Tl has 81 p, 120 n

2.30

Symbol	^{112}Cd	^{96}Sr	^{87}Sr	^{81}Kr	^{235}U
Protons	48	38	38	36	92
Neutrons	64	58	49	45	143
Electrons	48	38	38	36	92
Mass No.	112	96	87	81	235

2.32   Because the two nuclides are atoms of the same element, by definition they have the same number of protons, 54. They differ in mass number (and mass) because they have different numbers of neutrons. ^{129}Xe has 75 neutrons and ^{130}Xe has 76 neutrons.

2.34   (a)   12 amu

        (b)   The atomic weight of carbon reported on the front-inside cover of the text is the abundance-weighted average of the atomic masses of the two naturally occurring isotopes of carbon, ^{12}C and ^{13}C. The mass of a ^{12}C atom is exactly 12 amu, but the atomic weight of 12.011 takes into account the presence of some ^{13}C atoms in every natural sample of the element.

2.36   Atomic weight (average atomic mass) = Σ fractional abundance × mass of isotope

        Atomic weight = 0.7215(84.9118) + 0.2785(86.9092) = 85.4681 = 85.47 amu

        (The result has 2 decimal places and 4 sig figs because each term in the sum has 4 sig figs and 2 decimal places.)

2.38   (a)   The purpose of the magnet in the mass spectrometer is to change the path of the moving ions. The magnitude of the deflection is inversely related to mass, which is the basis of the discrimination by mass.

        (b)   The atomic weight of Cl, 35.5, is an average atomic mass. It is the average of the masses of two naturally occurring isotopes, weighted by their abundances.

        (c)   The single peak at mass 31 in the mass spectrum of phosphorus indicates that the sample contains a single isotope of P, and the mass of this isotope is 31 amu.

2.40   (a)   Three peaks: $^1H - {}^1H$, $^1H - {}^2H$, $^2H - {}^2H$

        (b) $^1H - {}^1H = 2(1.00783) = 2.01566$ amu

             $^1H - {}^2H = 1.00783 + 2.01410 = 3.02193$ amu

             $^2H - {}^2H = 2(2.01410) = 4.02820$ amu

             The mass ratios are 1 : 1.49923 : 1.99845 or 1 : 1.5 : 2.

        (c) $^1H - {}^1H$ is largest, because there is the greatest chance that two atoms of the more abundant isotope will combine.

             $^2H - {}^2H$ is the smallest, because there is the least chance that two atoms of the less abundant isotope will combine.

The Periodic Table; Molecules and Ions (Sections 2.5–2.7)

2.42   (a)   lithium, 3 (metal)   (b)   scandium, 21 (metal)

         (c)   germanium, 32 (metalloid)   (d)   ytterbium, 70 (metal)

         (e)   manganese, 25 (metal)   (f)   antimony, 51 (metalloid)

         (g)   xenon, 54 (nonmetal)

2.44   C, carbon, nonmetal; Si, silicon, metalloid; Ge, germanium, metalloid; Sn, tin, metal;

        Pb, lead, metal

2.46   Compounds with the same empirical but different molecular formulas differ by the integer number of empirical formula units in the respective molecules. Thus, they can have very different molecular structure, size, and mass, resulting in very different physical properties.

2.48   No. Two substances with the same molecular and empirical formulas can be isomers. They are not necessarily the same compound.

2.50   A molecular formula contains all atoms in a molecule. An empirical formula shows the simplest ratio of atoms in a molecule or elements in a compound.

         (a)   molecular formula: C_6H_6; empirical formula: CH

         (b)   molecular formula: $SiCl_4$; empirical formula: $SiCl_4$ (1:4 is the simplest ratio)

         (c)   molecular: B_2H_6; empirical: BH_3

         (d)   molecular: $C_6H_{12}O_6$; empirical: CH_2O

2.52   (a)   6   (b) 6   (c)   9

2.54   (a) C_2H_5Br

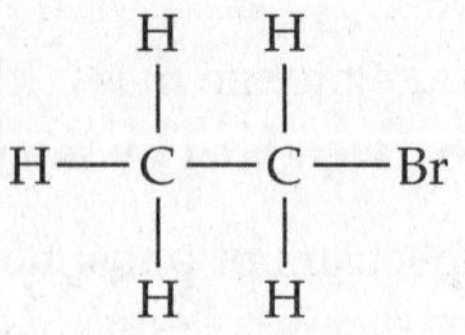

$$H\!-\!\underset{\underset{H}{|}}{\overset{\overset{H}{|}}{C}}\!-\!\underset{\underset{H}{|}}{\overset{\overset{H}{|}}{C}}\!-\!Br$$

        (b) C_2H_7N

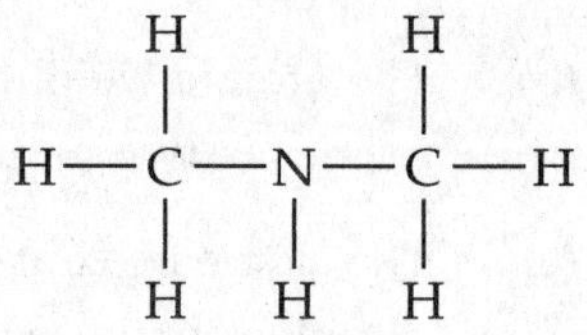

$$H\!-\!\underset{\underset{H}{|}}{\overset{\overset{H}{|}}{C}}\!-\!\underset{\underset{H}{|}}{N}\!-\!\underset{\underset{H}{|}}{\overset{\overset{H}{|}}{C}}\!-\!H$$

        (c) CH_2Cl_2

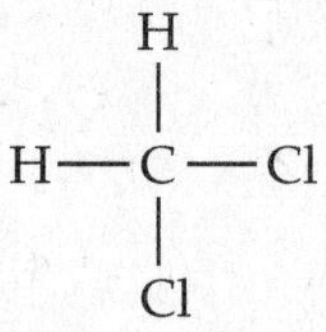

$$H\!-\!\underset{\underset{Cl}{|}}{\overset{\overset{H}{|}}{C}}\!-\!Cl$$

        (d) NH_2OH

$$H\!-\!\underset{\underset{H}{|}}{N}\!-\!\underset{\underset{H}{|}}{O}$$

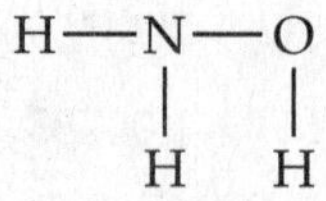

2.56

Symbol	$^{31}P^{3-}$	$^{79}Se^{2-}$	$^{119}Sn^{4+}$	$^{197}Au^{3+}$
Protons	15	34	50	79
Neutrons	16	45	69	118
Electrons	18	36	46	76
Net Charge	3–	2–	4+	3+

2.58   (a) Ga^{3+} (b) Sr^{2+} (c) As^{3-} (d) Br^{-} (e) Se^{2-}

2.60   (a) ScI_3 (b) Sc_2S_3 (c) ScN

2.62   (a) $CrBr_3$ (b) Fe_2O_3 (c) Hg_2CO_3 (d) $Ca(ClO_3)_2$ (e) $(NH_4)_3PO_4$

2.64

Ion	Na^+	Ca^{2+}	Fe^{2+}	Al^{3+}
O^{2-}	Na_2O	CaO	FeO	Al_2O_3
NO_3^-	$NaNO_3$	$Ca(NO_3)_2$	$Fe(NO_3)_2$	$Al(NO_3)_3$
SO_4^{2-}	Na_2SO_4	$CaSO_4$	$FeSO_4$	$Al_2(SO_4)_3$
AsO_4^{3-}	Na_3AsO_4	$Ca_3(AsO_4)_2$	$Fe_3(AsO_4)_2$	$AlAsO_4$

2.66   Molecular (all elements are nonmetals):

     (a) PF_5 (c) SCl_2 (h) N_2O_4

     Ionic (formed from ions, usually contains a metal cation):

     (b) NaI (d) $Ca(NO_3)_2$ (e) $FeCl_3$ (f) LaP (g) $CoCO_3$

Naming Inorganic Compounds; Organic Molecules (Sections 2.8–2.9)

2.68   (a)   selenate   (b) selenide   (c) hydrogen selenide (biselenide)

     (d)   hydrogen selenite (biselenite)

2.70   (a)   copper, 2+; sulfide, 2–   (b) silver, 1+; sulfate, 2–

     (c)   aluminum, 3+; chlorate, 1–   (d) cobalt, 2+; hydroxide, 1–

     (e)   lead, 2+; carbonate, 2–

2.72   (a)   potassium cyanide   (b) sodium bromite

     (c)   strontium hydroxide   (d) cobalt(II) telluride (cobaltous telluride)

     (e)   iron(III) carbonate (ferric carbonate)

     (f)   chromium(III) nitrate (chromic nitrate)

     (g)   ammonium sulfite   (h) sodium dihydrogen phosphate

     (i)   potassium permanganate   (j) silver dichromate

2.74   (a) Na_3PO_4 (b) $Zn(NO_3)_2$ (c) $Ba(BrO_3)_2$ (d) $Fe(ClO_4)_2$

     (e) $Co(HCO_3)_2$ (f) $Cr(CH_3COO)_3$ (g) $K_2Cr_2O_7$

2.76   (a) HI (b) $HClO_3$ (c) HNO_2

     (d)   carbonic acid   (e) perchloric acid   (f) acetic acid

2.78   (a)   dinitrogen monoxide   (b) nitrogen monoxide   (c) nitrogen dioxide

     (d)   dinitrogen pentoxide   (e) dinitrogen tetroxide

2.80   (a)   NaHCO$_3$ (b)   Ca(ClO)$_2$ (c)   HCN

        (d)   Mg(OH)$_2$ (e)   SnF$_2$ (f)   CdS, H$_2$SO$_4$, H$_2$S

2.82   (a)   Isomers are molecules with the same molecular formula, but different structural formulas. Isomers have the same number and kinds of atoms, but these atoms are arranged in different ways.

        (b)   Butane and pentane are both capable of existing in isomeric forms. There is more than one way to arrange the four C atoms and ten H atoms of butane, and more than one way to arrange the five C atoms and twelve H atoms of pentane. There is only one way to arrange the two C atoms and six H atoms of ethane and only one way to arrange the three C atoms and eight H atoms of propane.

2.84   (a)   They both have two carbon atoms in their molecular backbone, or chain.

        (b)   In 1-propanol one of the H atoms on an outer (terminal) C atom has been replaced by an —OH group.

2.86

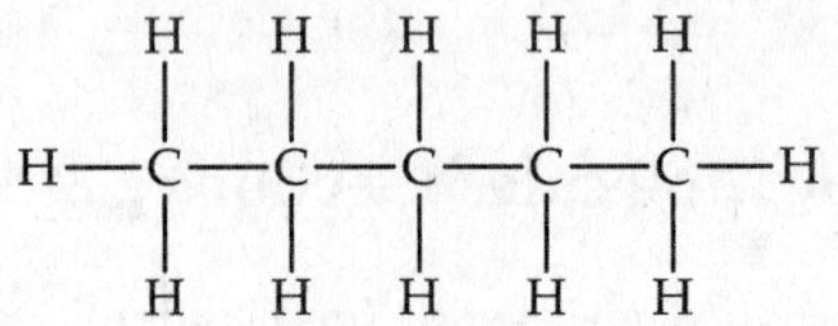

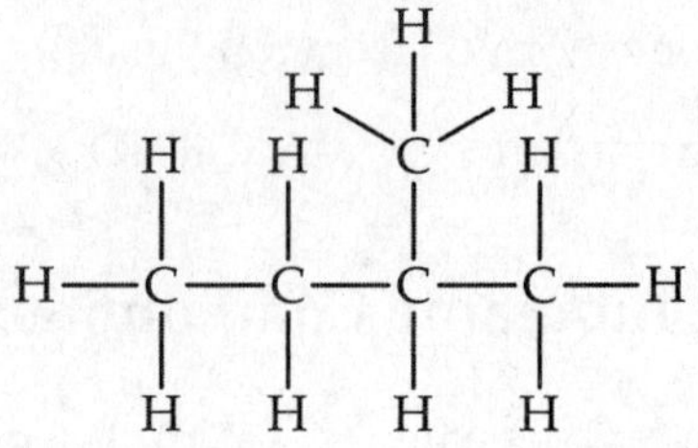

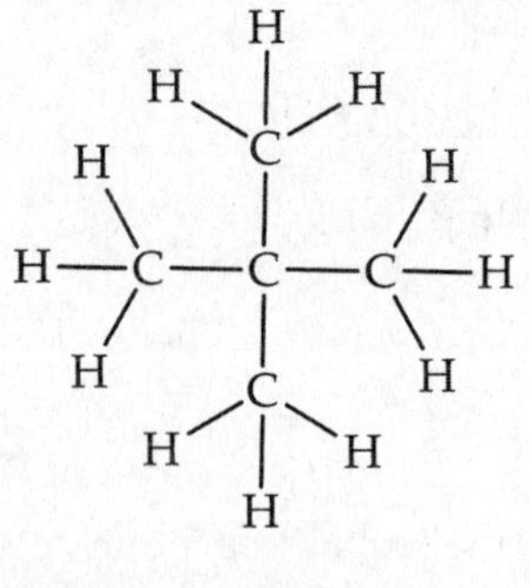

Additional Exercises

2.87   (a)   Droplet D would fall most slowly. It carries the most negative charge, so it would be most strongly attracted to the upper (+) plate and most strongly repelled by the lower (–) plate. These electrostatic forces would provide the greatest opposition to gravity.

        (b)   Calculate the lowest common factor.

              A: $3.84 \times 10^{-8} / 2.88 \times 10^{-8} = 1.33$; $1.33 \times 3 = 4$

              B: $4.80 \times 10^{-8} / 2.88 \times 10^{-8} = 1.67$; $1.67 \times 3 = 5$

              C: $2.88 \times 10^{-8} / 2.88 \times 10^{-8} = 1.00$; $1.00 \times 3 = 3$

              D: $8.64 \times 10^{-8} / 2.88 \times 10^{-8} = 3.00$; $3.00 \times 3 = 9$

The total charge on the drops is in the ratio of 4:5:3:9. Divide the total charge on each drop by the appropriate integer and average the four values to get the charge of an electron in warmombs.

A: $3.84 \times 10^{-8} / 4 = 9.60 \times 10^{-9}$ wa

B: $4.80 \times 10^{-8} / 5 = 9.60 \times 10^{-9}$ wa

C: $2.88 \times 10^{-8} / 3 = 9.60 \times 10^{-9}$ wa

D: $8.64 \times 10^{-8} / 9 = 9.60 \times 10^{-9}$ wa

The charge on an electron is 9.60×10^{-9} wa

(c)   The number of electrons on each drop are the integers calculated in part (b). A has $4\,e^-$, B has $5\,e^-$, C has $3\,e^-$ and D has $9\,e^-$.

(d) $\dfrac{9.60 \times 10^{-9}\ \text{wa}}{1\,e^-} \times \dfrac{1\,e^-}{1.60 \times 10^{-16}\ \text{C}} = 6.00 \times 10^{7}$ wa/C

2.89   (a)   Calculate the mass of a single gold atom, then divide the mass of the cube by the mass of the gold atom.

$\dfrac{197.0\ \text{amu}}{\text{gold atom}} \times \dfrac{1\,\text{g}}{6.022 \times 10^{23}\ \text{amu}} = 3.2713 \times 10^{-22} = 3.271 \times 10^{-22}$ g/gold atom

$\dfrac{19.3\ \text{g}}{\text{cube}} \times \dfrac{1\,\text{gold atom}}{3.271 \times 10^{-22}\ \text{g}} = 5.90 \times 10^{22}$ Au atoms in the cube

(b)   The shape of atoms is spherical; spheres cannot be arranged into a cube so that there is no empty space. The question is, how much empty space is there? We can calculate the two limiting cases, no empty space and maximum empty space. The true diameter will be somewhere in this range.

No empty space: volume cube/number of atoms = volume of one atom

$V = 4/3\pi\,r^3$; $r = (3\pi\,V/4)^{1/3}$; $d = 2r$

volume of cube $= (1.0 \times 1.0 \times 1.0) = \dfrac{1.0\ \text{cm}^3}{5.90 \times 10^{22}\ \text{Au atoms}} = 1.695 \times 10^{-23}$

$= 1.7 \times 10^{-23}\ \text{cm}^3$

$r = [\pi\,(1.695 \times 10^{-23}\ \text{cm}^3)/4]^{1/3} = 3.4 \times 10^{-8}$ cm; $d = 2r = 6.8 \times 10^{-8}$ cm

Maximum empty space: Assume atoms are arranged in rows in all three directions so they are touching across their diameters. That is, each atom occupies the volume of a cube, with the atomic diameter as the length of the side of the cube. The number of atoms along one edge of the gold cube is then

$(5.90 \times 10^{22})^{1/3} = 3.893 \times 10^{7} = 3.89 \times 10^{7}$ atoms/1.0 cm.

The diameter of a single atom is $1.0\ \text{cm}/3.89 \times 10^{7}$ atoms $= 2.569 \times 10^{-8}$

$= 2.6 \times 10^{-8}$ cm.

The diameter of a gold atom is between 2.6×10^{-8} cm and 6.8×10^{-8} cm ($2.6 – 6.8$ Å).

(c)   Some atomic arrangement must be assumed, because none is specified. The solid state is characterized by an orderly arrangement of particles, so it isn't surprising that atomic arrangement is required to calculate the density of a solid. A more detailed discussion of solid-state structure and density appears in Chapter 11.

2.91   (a)   diameter of nucleus $= 1 \times 10^{-4}$ Å; diameter of atom $= 1$ Å

$V = 4/3\, \pi\, r^3; r = d/2; r_n = 0.5 \times 10^{-4}$ Å; $r_a = 0.5$ Å

volume of nucleus $= 4/3\, \pi\, (0.5 \times 10^{-4})^3$ Å^3

volume of atom $= 4/3\, \pi\, (0.5)^3$ Å^3

volume fraction of nucleus $= \dfrac{\text{volume of nucleus}}{\text{volume of atom}} = \dfrac{4/3\, \pi\, (0.5 \times 10^{-4})^3 \text{ Å}^3}{4/3\, \pi\, (0.5)^3 \text{ Å}^3} = 1 \times 10^{-12}$

diameter of atom $= 5$ Å, $r_a = 2.5$ Å

volume fraction of nucleus $= \dfrac{4/3\, \pi\, (0.5 \times 10^{-4})^3 \text{ Å}^3}{4/3\, \pi\, (2.5)^3 \text{ Å}^3} = 8 \times 10^{-15}$

Depending on the radius of the atom, the volume fraction of the nucleus is between 1×10^{-12} and 8×10^{-15}, that is, between 1 part in 10^{12} and 8 parts in 10^{15}.

(b)   mass of proton $= 1.0073$ amu

$1.0073 \text{ amu} \times 1.66054 \times 10^{-24} \text{ g/amu} = 1.6727 \times 10^{-24}$ g

diameter $= 1.0 \times 10^{-15}$ m, radius $= 0.50 \times 10^{-15} \text{ m} \times \dfrac{100 \text{ cm}}{1 \text{ m}} = 5.0 \times 10^{-14}$ cm

Assuming a proton is a sphere, $V = 4/3\, \pi\, r^3$.

density $= \dfrac{g}{cm^3} = \dfrac{1.6727 \times 10^{-24} \text{ g}}{4/3\, \pi\, (5.0 \times 10^{-14})^3 \text{ cm}^3} = 3.2 \times 10^{15}$ g/cm^3

2.92   The integer on the lower left of a nuclide is the atomic number; it is the number of protons in any atom of the element and gives the element's identity. The number of neutrons is the mass number (upper left) minus atomic number.

(a)   As, 33 protons, 41 neutrons

(b)   I, 53 protons, 74 neutrons

(c)   Eu, 63 protons, 89 neutrons

(d)   Bi, 83 protons, 126 neutrons

2.93   (a) ^{6}Li, 3 protons, 3 neutrons; ^{7}Li, 3 protons, 4 neutrons

(b)   Average atomic weight of sample $= \Sigma$ fractional abundance $\times$ mass of isotope

Av. atomic weight $= 0.01442(6.015122) + 0.98558(7.016004) = 7.001571 = 7.002$ amu

2.95   Atomic weight (average atomic mass) $= \Sigma$ fractional abundance $\times$ mass of isotope

Atomic weight $= 0.014(203.97302) + 0.241(205.97444) + 0.221(206.97587) +$

$0.524(207.97663) = 207.22 = 207$ amu

(The result has 0 decimal places and 3 sig figs because the fourth term in the sum has 3 sig figs and 0 decimal places.)

2.97   (a)   There are 24 known isotopes of Ni, from ^{51}Ni to ^{74}Ni.

         (b,c)   The five most abundant isotopes (b) and their natural abundances (c) are

             ^{58}Ni, 57.935346 amu, 68.077%

             ^{60}Ni, 59.930788 amu, 26.223%

             ^{62}Ni, 61.928346 amu, 3.634%

             ^{61}Ni, 60.931058 amu, 1.140%

             ^{64}Ni, 63.927968 amu, 0.926%

             Data from *Handbook of Chemistry and Physics*, 74th edition [Data may differ slightly in other editions.]

2.98   (a)   A Br_2 molecule could consist of two atoms of the same isotope or one atom of each of the two different isotopes. This second possibility is twice as likely as the first. Therefore, the second peak (twice as large as peaks 1 and 3) represents a Br_2 molecule containing different isotopes. The mass numbers of the two isotopes are determined from the masses of the two smaller peaks. Because $157.836 \approx 158$, the first peak represents a ^{79}Br—^{79}Br molecule. Peak 3, $161.832 \approx 162$, represents a ^{81}Br—^{81}Br molecule. Peak 2 then contains one atom of each isotope, ^{79}Br—^{81}Br, with an approximate mass of 160 amu.

         (b)   The mass of the lighter isotope is 157.836 amu/2 atoms, or 78.918 amu/atom. For the heavier one, 161.832 amu/2 atoms = 80.916 amu/atom.

         (c)   The relative size of the three peaks in the mass spectrum of Br_2 indicates their relative abundance. The average mass of a Br_2 molecule is

             $0.2569(157.836) + 0.4999(159.834) + 0.2431(161.832) = 159.79$ amu.

             (Each product has four significant figures and two decimal places, so the answer has two decimal places.)

         (d) $\dfrac{159.79 \text{ amu}}{\text{avg. } Br_2 \text{ molecule}} \times \dfrac{1\,Br_2 \text{ molecule}}{2\,Br \text{ atoms}} = 79.895$ amu

         (e)   Let x = the abundance of ^{79}Br, 1 – x = abundance of ^{81}Br. From (b), the masses of the two isotopes are 78.918 amu and 80.916 amu, respectively. From (d), the mass of an average Br atom is 79.895 amu.

                 $x(78.918) + (1 - x)(80.916) = 79.895,\ x = 0.5110$

                 ^{79}Br = 51.10%, ^{81}Br = 48.90%

2.100   (a)   an alkali metal: K   (b) an alkaline earth metal: Ca   (c) a noble gas: Ar

         (d)   a halogen: Br   (e) a metalloid: Ge   (f) a nonmetal in 1A: H

         (g)   a metal that forms a 3+ ion: Al   (h)   a nonmetal that forms a 2– ion: O

         (i)   an element that resembles Al: Ga

2.101   (a) $^{266}_{106}$Sg has 106 protons, 160 neutrons and 106 electrons

         (b)   Sg is in Group 6B (or 6) and immediately below tungsten, W. We expect the chemical properties of Sg to most closely resemble those of W.

2.102   Strontium is an alkaline earth metal, similar in chemical properties to calcium and magnesium. Calcium is ubiquitous in biological organisms, humans included. It is a vital nutrient required for formation and maintenance of healthy bones and teeth. As such, there are efficient pathways for calcium uptake and distribution in the body, pathways that are also available to chemically similar strontium. Harmful strontium imitates calcium and then behaves badly when the body tries to use it as it uses calcium.

2.103   (a)   chlorine gas, Cl_2: ii   (b)   propane, C_3H_8: v   (c)   nitrate ion, NO_3^- : i

          (d)   sulfur trioxide, SO_3: iii   (e)   methylchloride, CH_3Cl: iv

2.105

Cation	Anion	Formula	Name
Li^+	O^{2-}	Li_2O	Lithium oxide
Fe^{2+}	PO_4^{3-}	$Fe_3(PO_4)_2$	Iron(II) phosphate
Al^{3+}	SO_4^{2-}	$Al_2(SO_4)_3$	Aluminum sulfate
Cu^{2+}	NO_3^-	$Cu(NO_3)_2$	Copper(II) nitrate
Cr^{3+}	I^-	CrI_3	Chromium(III) iodide
Mn^{1+}	ClO_2^-	$MnClO_2$	Manganese(I) chlorite
NH_4^+	CO_3^{2-}	$(NH_4)_2CO_3$	Ammonium carbonate
Zn^{2+}	ClO_4^-	$Zn(ClO_4)_2$	Zinc perchlorate

2.106   (a)   Empirical formula, CH_3

          The empirical and molecular formulas of propane are C_3H_8. Propane has two more H atoms than cyclopropane, so the empirical and molecular formulas are different.

          (b)   The solid wedges indicate bonds from C atoms to H atoms that are above the plane of the page; the dashed wedges show bonds from C atoms to H atoms that are behind the plane of the page.

          (c)   To illustrate chlorocyclopropane, replace any one of the H atoms on cyclopropane with a Cl atom. There are no isomers of chlorocyclopropane, because a structure with a Cl atom at any one of the six positions can be rotated into the original structure.

2.108   Carbonic acid: H_2CO_3; the cation is H^+ because it is an acid; the anion is carbonate because the acid reacts with lithium hydroxide to form lithium carbonate.
          Lithium hydroxide: LiOH; lithium carbonate: Li_2CO_3

2.109 (a) sodium chloride (b) sodium bicarbonate (or sodium hydrogen carbonate)

(c) sodium hypochlorite (d) sodium hydroxide

(e) ammonium carbonate (f) calcium sulfate

2.111 (a) CaS, $Ca(HS)_2$ (b) HBr, $HBrO_3$ (c) AlN, $Al(NO_2)_3$ (d) FeO, Fe_2O_3

(e) NH_3, NH_4^+ (f) K_2SO_3, $KHSO_3$ (g) Hg_2Cl_2, $HgCl_2$ (h) $HClO_3$, $HClO_4$

2.112 In the nucleus. The strong force holds the protons together against the repulsive electrostatic force.

3 Chemical Reactions and Reaction Stoichiometry

Visualizing Concepts

3.2 (a) There are four CH_3OH molecules in the products box. CO is the only source of C atoms for the reaction, so there must be four CO molecules in the reactants box.

(b) $CO + 2H_2 \rightarrow CH_3OH$

3.4 The box contains 4 C atoms and 10 H atoms, so the empirical formula of the hydrocarbon is C_2H_5.

3.6 *Analyze.* Given: 4.0 mol CH_4. Find: mol CO and mol H_2

Plan. Examine the boxes to determine the CH_4:CO mol ratio and CH_4:H_2O mole ratio.

Solve. There are 2 CH_4 molecules in the reactant box and 2 CO molecules in the product box. The mole ratio is 2:2 or 1:1. Therefore, 4.0 mol CH_4 can produce 4.0 mol CO. There are 2 CH_4 molecules in the reactant box and 6 H_2 molecules in the product box. The mole ratio is 2:6 or 1:3. So, 4.0 mol CH_4 can produce 12:0 mol H_2.

Check. Use proportions. 2 mol CH_4/2 mol CO = 4 mol CH_4/4 mol CO;

2 mol CH_4/6 mol H_2 = 4 mol CH_4/12 mol H_2.

3.8 (a) $2NO + O_2 \rightarrow 2NO_2$, $O_2 = $ ⃝⃝, $NO_2 = $ ⃝●⃝

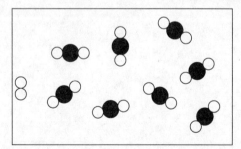

Each NO molecule reacts with 1 O atom (1/2 of an O_2 molecule) to produce 1 NO_2 molecule. Eight NO molecules react with 8 O atoms (4 O_2 molecules) to produce 8 NO_2 molecules. One O_2 molecule doesn't react (is in excess). NO is the limiting reactant.

(b) $\%\ \text{yield} = \dfrac{\text{actual yield}}{\text{theoretical yield}} \times 100$; $\quad \text{actual yield} = \dfrac{\%\ \text{yield}}{100} \times \text{theoretical yield}$

The theoretical yield from part (a) is 8 NO_2 molecules. If the percent yield is 75%, then 0.75(8) = 6 NO_2 would appear in the products box.

Balancing Chemical Equations (Section 3.1)

3.10   (a) $CaO(s) + H_2O(l) \rightarrow Ca(OH)_2(aq)$

        (b)   The only way to write a balanced equation with CaOH(aq) as a product is to include OH^-(aq) as a second product. Otherwise, the ratio of elements in the product is never the same as the ratio of elements in the reactants.

             $CaO(s) + H_2O(l) \rightarrow CaOH(aq) + OH^-(aq)$

3.12   (a) $6\,Li(s) + N_2(g) \rightarrow 2\,Li_3N(s)$

        (b) $TiCl_4(l) + 2\,H_2O(l) \rightarrow TiO_2(s) + 4\,HCl(aq)$

        (c) $2\,NH_4NO_3(s) \rightarrow 2\,N_2(g) + O_2(g) + 4\,H_2O(g)$

        (d) $2\,AlCl_3(s) + Ca_3N_2(s) \rightarrow 2\,AlN(s) + 3\,CaCl_2(s)$

3.14   (a) $Ca_3P_2(s) + 6\,H_2O(l) \rightarrow 3\,Ca(OH)_2(aq) + 2\,PH_3(g)$

        (b) $2\,Al(OH)_3(s) + 3\,H_2SO_4(aq) \rightarrow Al_2(SO_4)_3(aq) + 6\,H_2O(l)$

        (c) $2\,AgNO_3(aq) + Na_2CO_3(aq) \rightarrow Ag_2CO_3(s) + 2\,NaNO_3(aq)$

        (d) $4\,C_2H_5NH_2(g) + 15\,O_2(g) \rightarrow 8\,CO_2(g) + 14\,H_2O(g) + 2\,N_2(g)$

3.16   (a) $SO_3(g) + H_2O(l) \rightarrow H_2SO_4(aq)$

        (b) $B_2S_3(s) + 6\,H_2O(l) \rightarrow 2\,H_3BO_3(aq) + 3\,H_2S(g)$

        (c) $4\,PH_3(g) + 8\,O_2(g) \rightarrow P_4O_{10}(s) + 6\,H_2O(g)$

        (d) $2\,Hg(NO_3)_2(s) \xrightarrow{\Delta} 2\,HgO(s) + 4\,NO_2(g) + O_2(g)$

        (e) $Cu(s) + 2\,H_2SO_4(aq) \rightarrow CuSO_4(aq) + SO_2(g) + 2\,H_2O(l)$

Patterns of Chemical Reactivity (Section 3.2)

3.18   (a) $O_2(g)$. Combustion is reaction with oxygen.

        (b)   The products are $CO_2(g)$ and $H_2O(l)$.

        (c)   The sum of the balanced coefficients is 11. (Remember that the coefficient in front of $C_3H_6O(l)$ is one.)

            $C_3H_6O(l) + 4\,O_2(g) \rightarrow 3\,CO_2(g) + 3\,H_2O(l)$

3.20   (a) $2\,Ti(s) + O_2(g) \rightarrow 2\,TiO(s)$ [or $Ti(s) + O_2(g) \rightarrow TiO_2(s)$]

        (b) $2\,Ag_2O(s) \xrightarrow{\Delta} 4\,Ag(s) + O_2(g)$

        (c) $2\,C_3H_7OH(l) + 9\,O_2(g) \rightarrow 6\,CO_2(g) + 8\,H_2O(l)$

        (d) $2\,C_5H_{12}O(l) + 15\,O_2(g) \rightarrow 10\,CO_2(g) + 12\,H_2O(l)$

3.22   (a) $PbCO_3(s) \rightarrow PbO(s) + CO_2(g)$ decomposition

        (b) $C_2H_4(g) + 3\,O_2(g) \rightarrow 2\,CO_2(g) + 2\,H_2O(g)$ combustion

        (c) $3\,Mg(s) + N_2(g) \rightarrow Mg_3N_2(s)$ combination

    (d) $C_7H_8O_2(l) + 8\,O_2(g) \rightarrow 7\,CO_2(g) + 4\,H_2O(g)$ combustion

    (e) $2\,Al(s) + 3\,Cl_2(g) \rightarrow 2\,AlCl_3(s)$ combination

Formula Weights (Section 3.3)

3.24   Formula weight in amu to 1 decimal place.

    (a) N_2O: FW = 2(14.0) + 1(16.0) = 44.0 amu

    (b) $HC_7H_5O_2$: 7(12.0) + 6(1.0) + 2(16.0) = 122.0 amu

    (c) $Mg(OH)_2$: 1(24.3) + 2(16.0) + 2(1.0) = 58.3 amu

    (d) $(NH_2)_2CO$: 2(14.0) + 4(1.0) + 1(12.0) + 1(16.0) = 60.0 amu

    (e) $CH_3CO_2C_5H_{11}$: 7(12.0) + 14(1.0) + 2(16.0) = 130.0 amu

3.26   (a) C_2H_2: FW = 2(12.0) + 2(1.0) = 26.0 amu

$$\% C = \frac{2(12.0)\,amu}{26.0\,amu} \times 100 = 92.3\%$$

    (b) $HC_6H_7O_6$: FW = 6(12.0) + 8(1.0) + 6(16.0) = 176.0 amu

$$\% H = \frac{8(1.0)\,amu}{176.0\,amu} \times 100 = 4.5\%$$

    (c) $(NH_4)_2SO_4$: FW = 2(14.0) + 8(1.0) + 1(32.1) + 4(16.0) = 132.1 amu

$$\% H = \frac{8(1.0)\,amu}{132.1\,amu} \times 100 = 6.1\%$$

    (d) $PtCl_2(NH_3)_2$: FW = 1(195.1) + 2(35.5) + 2(14.0) + 6(1.0) = 300.1 amu

$$\% Pt = \frac{1(195.1)\,amu}{300.1\,amu} \times 100 = 65.01\%$$

    (e) $C_{18}H_{24}O_2$: FW = 18(12.0) + 24(1.0) + 2(16.0) = 272.0 amu

$$\% O = \frac{2(16.0)\,amu}{272.0\,amu} \times 100 = 11.8\%$$

    (f) $C_{18}H_{27}NO_3$: FW = 18(12.0) + 27(1.0) + 1(14.0) + 3(16.0) = 305.0 amu

$$\% C = \frac{18(12.0)\,amu}{305.0\,amu} \times 100 = 70.8\%$$

3.28   (a) CO_2: FW = 1(12.0) + 2(16.0) = 44.0 amu

$$\% C = \frac{12.0\,amu}{44.0\,amu} \times 100 = 27.3\%$$

    (b) CH_3OH: FW = 1(12.0) + 4(1.0) + 1(16.0) = 32.0 amu

$$\% C = \frac{12.0\,amu}{32.0\,amu} \times 100 = 37.5\%$$

    (c) C_2H_6: FW = 2(12.0) + 6(1.0) = 30.0 amu

$$\% C = \frac{2(12.0)\,amu}{30.0\,amu} \times 100 = 80.0\%$$

(d) $CS(NH_2)_2$: FW $= 1(12.0) + 1(32.1) + 2(14.0) + 4(1.0) = 76.1$ amu

$$\% C = \frac{12.0 \, \text{amu}}{76.1 \, \text{amu}} \times 100 = 15.8\%$$

Avogadro's Number and the Mole (Section 3.4)

3.30   (a)   <u>exactly</u> 12 g   (b) 6.0221421×10^{23}, Avogadro's number

3.32   42 g $NaHCO_3$ (molar mass = 84 g) contains (6 atoms $\times$ 0.5 mol) = 3 mol atoms

1.5 mol CO_2 contains (3 atoms $\times$ 1.5 mol) = 4.5 mol atoms

6.0×10^{24} Ne atoms contains (1 atoms $\times$ 10 mol) = 10 mol atoms

3.34   314 million $= 314 \times 10^6 = 3.14 \times 10^8$ people

$$\frac{6.022 \times 10^{23} \cancel{\text{¢}}}{3.14 \times 10^8 \text{ people}} \times \frac{\$1}{100 \cancel{\text{¢}}} = \frac{\$6.022 \times 10^{21}}{3.14 \times 10^8 \text{ people}} = 1.918 \times 10^{13} = \$1.92 \times 10^{13}/\text{person}$$

$$\$15.1 \text{ trillion} = \$1.51 \times 10^{13} \qquad \frac{\$1.918 \times 10^{13}}{\$1.51 \times 10^{13}} = 1.270 = 1.27$$

Each person would receive an amount that is 1.27 times the dollar amount of the national debt.

3.36   (a)   molar mass $= 1(112.41) + 1(32.07) = 144.48$ g

$$1.50 \times 10^{-2} \text{ mol CdS} \times \frac{144.48 \, \text{g}}{1 \, \text{mol}} = 2.17 \text{ g CdS}$$

(b)   molar mass $= 1(14.01) + 4(1.008) + 1(35.45) = 53.49$ g/mol

$$86.6 \text{ g } NH_4Cl \times \frac{1 \, \text{mol}}{53.49 \, \text{g}} = 1.6190 = 1.62 \text{ mol } NH_4Cl$$

(c) $8.447 \times 10^{-2} \text{ mol } C_6H_6 \times \dfrac{6.02214 \times 10^{23} \text{ molecules}}{1 \, \text{mol}} = 5.087 \times 10^{22} \, C_6H_6 \text{ molecules}$

(d) $6.25 \times 10^{-3} \text{ mol } Al(NO_3)_3 \times \dfrac{9 \, \text{mol O}}{1 \, \text{mol } Al(NO_3)_3} \times \dfrac{6.022 \times 10^{23} \text{ O atoms}}{1 \, \text{mol}}$

$$= 3.39 \times 10^{22} \text{ O atoms}$$

3.38   (a) $Fe_2(SO_4)_3$ molar mass $= 2(55.845) + 3(32.07) + 12(16.00) = 399.900 = 399.9$ g/mol

$$1.223 \text{ mol } Fe_2(SO_4)_3 \times \frac{399.9 \text{ g } Fe_2(SO_4)_3}{1 \, \text{mol}} = 489.077 = 489.1 \text{ g } Fe_2(SO_4)_3$$

(b) $(NH_4)_2CO_3$ molar mass $= 2(14.007) + 8(1.008) + 12.011 + 3(15.9994) = 96.0872$

$$= 96.087 \text{ g/mol}$$

$$6.955 \text{ g } (NH_4)_2CO_3 \times \frac{1 \, \text{mol}}{96.087 \text{ g } (NH_4)_2CO_3} \times \frac{2 \, \text{mol } NH_4{}^+}{1 \, \text{mol } (NH_4)_2CO_3} = 0.1448 \text{ mol } NH_4{}^+$$

(c) $C_9H_8O_4$ molar mass = 9(12.01) + 8(1.008) + 4(16.00) = 180.154 = 180.2 g/mol

$$1.50 \times 10^{21} \text{ molecules} \times \frac{1 \text{ mol}}{6.022 \times 10^{23} \text{ molecules}} \times \frac{180.2 \text{ g } C_9H_8O_4}{1 \text{ mol aspirin}} = 0.449 \text{ g } C_9H_8O_4$$

(d) $\dfrac{15.86 \text{ g diazepam}}{0.05570 \text{ mol}} = 284.7$ g diazepam/ mol

3.40   (a) $C_{14}H_{18}N_2O_5$ molar mass = 14(12.01) + 18(1.008) + 2(14.01) + 5(16.00)

$$= 294.30 \text{ g/mol}$$

(b) $1.00 \text{ mg aspartame} \times \dfrac{1 \times 10^{-3} \text{ g}}{1 \text{ mg}} \times \dfrac{1 \text{ mol}}{294.3 \text{ g}} = 3.398 \times 10^{-6} = 3.40 \times 10^{-6}$ mol aspartame

(c) 3.398×10^{-6} mol aspartame $\times \dfrac{6.022 \times 10^{23} \text{ molecules}}{1 \text{ mol}} = 2.046 \times 10^{18}$

$$= 2.05 \times 10^{18} \text{ aspartame molecules}$$

(d) 2.046×10^{18} aspartame molecules $\times \dfrac{18 \text{ H atoms}}{1 \text{ aspartame molecule}} = 3.68 \times 10^{19}$ H atoms

3.42   (a) 3.88×10^{21} H atoms $\times \dfrac{19 \text{ C atoms}}{28 \text{ H atoms}} = 2.63 \times 10^{21}$ C atoms

(b) 3.88×10^{21} H atoms $\times \dfrac{1 \, C_{19}H_{28}O_2 \text{ molecule}}{28 \text{ H atoms}} = 1.3857 \times 10^{20}$

$$= 1.39 \times 10^{20} \ C_{19}H_{28}O_2 \text{ molecules}$$

(c) $1.3857 \times 10^{20} \ C_{19}H_{28}O_2$ molecules $\times \dfrac{1 \text{ mol}}{6.022 \times 10^{23} \text{ molecules}} = 2.301 \times 10^{-4}$

$$= 2.30 \times 10^{-4} \ \text{mol } C_{19}H_{28}O_2$$

(d) $C_{19}H_{28}O_2$ molar mass = 19(12.01) + 28(1.008) + 2(16.00) = 288.41 = 288.4 g/mol

$$2.301 \times 10^{-4} \text{ mol } C_{19}H_{28}O_2 \times \frac{288.4 \text{ g } C_{19}H_{28}O_2}{1 \text{ mol}} = 0.0664 \text{ g } C_{19}H_{28}O_2$$

3.44 $25 \times 10^{-6} \text{ g } C_{21}H_{30}O_2 \times \dfrac{1 \text{ mol } C_{21}H_{30}O_2}{314.5 \text{ g } C_{21}H_{30}O_2} = 7.95 \times 10^{-8} = 8.0 \times 10^{-8}$ mol $C_{21}H_{30}O_2$

$$7.95 \times 10^{-8} \text{ mol } C_{21}H_{30}O_2 \times \frac{6.022 \times 10^{23} \text{ molecules}}{1 \text{ mol}} = 4.8 \times 10^{16} \ C_{21}H_{30}O_2 \text{ molecules}$$

Empirical Formulas (Section 3.5)

3.46   (a)   Calculate the simplest ratio of moles.

0.104 mol K / 0.052 = 2

0.052 mol C / 0.052 = 1

0.156 mol O / 0.052 = 3

The empirical formula is K_2CO_3.

(b)   Calculate moles of each element present, then the simplest ratio of moles.

$$5.28 \text{ g Sn} \times \frac{1 \text{ mol Sn}}{118.7 \text{ g Sn}} = 0.04448 \text{ mol Sn};\ 0.04448/0.04448 = 1$$

$$3.37 \text{ g F} \times \frac{1 \text{ mol F}}{19.00 \text{ g F}} = 0.1774 \text{ mol F};\ 0.1774/0.04448 \approx 4$$

The integer ratio is 1 Sn : 4 F; the empirical formula is SnF_4.

(c)   Assume 100 g sample, calculate moles of each element, find the simplest ratio of moles.

$$87.5\% \text{ N} = 87.5 \text{ g N} \times \frac{1 \text{ mol N}}{14.01 \text{ g}} = 6.25 \text{ mol N};\ 6.25/6.25 = 1$$

$$12.5\% \text{ H} = 12.5 \text{ g H} \times \frac{1 \text{ mol}}{1.008 \text{ g}} = 12.4 \text{ mol H};\ 12.4/6.25 \approx 2$$

The empirical formula is NH_2.

3.48   See Solution 3.47 for stepwise problem-solving approach.

(a) $$55.3 \text{ g K} \times \frac{1 \text{ mol K}}{39.10 \text{ g K}} = 1.414 \text{ mol K}; 1.414/0.4714 \approx 3$$

$$14.6 \text{ g P} \times \frac{1 \text{ mol P}}{30.97 \text{ g P}} = 0.4714 \text{ mol P};\ 0.4714/0.4714 = 1$$

$$30.1 \text{ g O} \times \frac{1 \text{ mol O}}{16.00 \text{ g O}} = 1.881 \text{ mol O};\ 1.881/0.4714 \approx 4$$

The empirical formula is K_3PO_4.

(b) $$24.5 \text{ g Na} \times \frac{1 \text{ mol Na}}{22.99 \text{ g Na}} = 1.066 \text{ mol Na};\ 1.066/0.5304 \approx 2$$

$$14.9 \text{ g Si} \times \frac{1 \text{ mol Si}}{28.09 \text{ Si}} = 0.5304 \text{ mol si};\ 0.5304/0.5304 = 1$$

$$60.6 \text{ g F} \times \frac{1 \text{ mol F}}{19.00 \text{ g F}} = 3.189 \text{ mol F}; 3.189/0.5304 \approx 6$$

The empirical formula is Na_2SiF_6.

(c)   The mass of O is [100 g total – (62.1 g C + 5.21 g H + 12.1 g N)] = 20.59 = 20.6 g O

$$62.1 \text{ g C} \times \frac{1 \text{ mol C}}{12.01 \text{ g C}} = 5.17 \text{ mol C};\ 5.17/0.864 \approx 6$$

$$5.21 \text{ g H} \times \frac{1 \text{ mol H}}{1.008 \text{ g H}} = 5.17 \text{ mol O};\ 5.17/0.864 \approx 6$$

$$12.1 \text{ g N} \times \frac{1 \text{ mol N}}{14.01 \text{ g N}} = 0.864 \text{ mol N};\ 0.864/0.864 = 1$$

$$20.6 \text{ g O} \times \frac{1 \text{ mol O}}{16.00 \text{ g O}} = 1.29 \text{ mol O};\ 1.29/0.864 \approx 1.5$$

Multiplying by 2, the empirical formula is $C_{12}H_{12}N_2O_3$.

3.50   Follow the logic in Solution 3.49. Match the calculated atomic mass to that of an element.

mol Cl = 75.0/35.453 = 2.1155 = 2.12;   mol X = 2.1155/4 = 0.52887 = 0.529

0.52887 mol X = 25.0 g X/AM X; AM X = 25.0 g X/0.52887 mol X = 47.271 = 47.3 g/mol

The element with atomic mass closest to 47.3 is **Ti**, atomic mass = 47.867 g/mol.

3.52   (a)   FW HCO_2 = 12.01 + 1.008 + 2(16.00) = 45.0 $\dfrac{MM}{FW} = \dfrac{90.0}{45.0} = 2$

The molecular formula is $C_2H_2O_4$.

(b)   FW C_2H_4O = 2(12) + 4(1) + 16 = 44. $\dfrac{MM}{FW} = \dfrac{88}{44} = 2$

The molecular formula is $C_4H_8O_2$.

3.54   Assume 100 g in the following problems.

(a) $75.69 \text{ g C} \times \dfrac{1 \text{ mol C}}{12.01 \text{ g C}} = 6.30 \text{ mol C}$; 6.30/0.969 = 6.5

$8.80 \text{ g H} \times \dfrac{1 \text{ mol H}}{1.008 \text{ g H}} = 8.73 \text{ mol H}$; 8.73/0.969 = 9.0

$15.51 \text{ g O} \times \dfrac{1 \text{ mol O}}{16.00 \text{ g O}} = 0.969 \text{ mol O}$; 0.969/0.969 = 1

Multiply by 2 to obtain the integer ratio 13:18:2. The empirical formula is $C_{13}H_{18}O_2$, FW = 206 g. Because the empirical formula weight and the molar mass are equal (206 g), the empirical and molecular formulas are $C_{13}H_{18}O_2$.

(b) $58.55 \text{ g C} \times \dfrac{1 \text{ mol C}}{12.01 \text{ g C}} = 4.875 \text{ mol C}$;   4.875/1.956 ≈ 2.5

$13.81 \text{ g H} \times \dfrac{1 \text{ mol H}}{1.008 \text{ g H}} = 13.700 \text{ mol H}$;   13.700/1.956 ≈ 7.0

$27.40 \text{ g N} \times \dfrac{1 \text{ mol N}}{14.01 \text{ g N}} = 1.956 \text{ mol N}$;   1.956/1.956 = 1.0

Multiply by 2 to obtain the integer ratio 5:14:2. The empirical formula is $C_5H_{14}N_2$; FW = 102. Because the empirical formula weight and the molar mass are equal (102 g), the empirical and molecular formulas are $C_5H_{14}N_2$.

(c) $59.0 \text{ g C} \times \dfrac{1 \text{ mol C}}{12.01 \text{ g C}} = 4.91 \text{ mol C}$;   4.91 / 0.550 ≈ 9

$7.1 \text{ g H} \times \dfrac{1 \text{ mol H}}{1.008 \text{ g H}} = 7.04 \text{ mol H}$;   7.04 / 0.550 ≈ 13

$26.2 \text{ g O} \times \dfrac{1 \text{ mol O}}{16.00 \text{ g O}} = 1.64 \text{ mol O}$;   1.64 / 0.550 ≈ 3

$7.7 \text{ g N} \times \dfrac{1 \text{ mol N}}{14.01 \text{ g N}} = 0.550 \text{ mol N}$;   0.550 / 0.550 = 1

The empirical formula is $C_9H_{13}O_3N$, FW = 183 amu (or g). Because the molecular weight is approximately 180 amu, the empirical formula and molecular formula are the same, $C_9H_{13}O_3N$.

3.56   (a)   *Plan.* Calculate mol C and mol H, then g C and g H; get g O by subtraction.

Solve.

$$6.32\times10^{-3}\text{ g CO}_2\times\frac{1\text{ mol CO}_2}{44.01\text{ g CO}_2}\times\frac{1\text{ mol C}}{1\text{ mol CO}_2}=1.436\times10^{-4}=1.44\times10^{-4}\text{ mol C}$$

$$2.58\times10^{-3}\text{ g H}_2\text{O}\times\frac{1\text{ mol H}_2\text{O}}{18.02\text{ g H}_2\text{O}}\times\frac{2\text{ mol H}}{1\text{ mol H}_2\text{O}}=2.863\times10^{-4}=2.86\times10^{-4}\text{ mol H}$$

$$1.436\times10^{-4}\text{ mol C}\times\frac{12.01\text{ g C}}{1\text{ mol C}}=1.725\times10^{-3}\text{ g C}=1.73\text{ mg C}$$

$$2.863\times10^{-4}\text{ mol H}\times\frac{1.008\text{ g H}}{1\text{ mol H}}=2.886\times10^{-4}\text{ g H}=0.289\text{ mg H}$$

mass of O = 2.78 mg sample – (1.725 mg C + 0.289 mg H) = 0.77 mg O

$$0.77\times10^{-3}\text{ g O}\times\frac{1\text{ mol O}}{16.00\text{ g O}}=4.81\times10^{-5}\text{ mol O}.\text{ Divide moles by }4.81\times10^{-5}.$$

$$\text{C}:\frac{1.44\times10^{-4}}{4.81\times10^{-5}}\approx3;\quad\text{H}:\frac{2.86\times10^{-4}}{4.81\times10^{-5}}\approx6;\quad\text{O}:\frac{4.81\times10^{-5}}{4.81\times10^{-5}}=1$$

The empirical formula is C_3H_6O.

(b)   *Plan.* Calculate mol C and mol H, then g C and g H. In this case, get N by subtraction.   *Solve.*

$$14.242\times10^{-3}\text{ g CO}_2\times\frac{1\text{ mol CO}_2}{44.01\text{ g CO}_2}\times\frac{1\text{ mol C}}{1\text{ mol CO}_2}=3.2361\times10^{-4}\text{ mol C}$$

$$4.083\times10^{-3}\text{ g H}_2\text{O}\times\frac{1\text{ mol H}_2\text{O}}{18.02\text{ g H}_2\text{O}}\times\frac{2\text{ mol H}}{1\text{ mol H}_2\text{O}}=4.5316\times10^{-4}=4.532\times10^{-4}\text{ mol H}$$

$$3.2361\times10^{-4}\text{ mol C}\times\frac{12.01\text{ g C}}{1\text{ mol H}}=3.8866\times10^{-3}\text{ g C}=3.8866\text{ mg C}$$

$$4.532\times10^{-4}\text{ mol H}\times\frac{1.008\text{ g H}}{1\text{ mol H}}=0.45683\times10^{-3}\text{ g H}=0.4568\text{ mg H}$$

mass of N = 5.250 mg sample – (3.8866 mg C + 0.4568 mg H) = 0.9066

$$= 0.907\text{ mg N}$$

$$0.9066\times10^{-3}\text{ g N}\times\frac{1\text{ mol N}}{14.01\text{ g N}}=6.47\times10^{-5}\text{ mol N}.\text{ Divide moles by }6.47\times10^{-5}.$$

$$\text{C}:\frac{3.24\times10^{-4}}{6.47\times10^{-5}}\approx5;\quad\text{H}:\frac{4.53\times10^{-4}}{6.47\times10^{-5}}\approx7;\quad\text{N}:\frac{6.47\times10^{-5}}{6.47\times10^{-5}}=1$$

The empirical formula is C_5H_7N, FW = 81. A molar mass of 160 ± 5 indicates a factor of 2 and a molecular formula of $C_{10}H_{14}N_2$.

3.58   Follow the logic in Solution 3.57.

$$0.374 \text{ g CO}_2 \times \frac{1 \text{ mol CO}_2}{44.01 \text{ g CO}_2} \times \frac{1 \text{ mol C}}{1 \text{ mol CO}_2} = 8.498 \times 10^{-3} = 8.50 \times 10^{-3} \text{ mol C}$$

$$0.102 \text{ g H}_2\text{O} \times \frac{1 \text{ mol H}_2\text{O}}{18.02 \text{ g H}_2\text{O}} \times \frac{2 \text{ mol H}}{1 \text{ mol H}_2\text{O}} = 0.01132 = 0.0113 \text{ mol H}$$

$$8.498 \times 10^{-3} \text{ mol C} \times \frac{12.01 \text{ g C}}{1 \text{ mol C}} = 0.10206 \text{ g C} = 0.102 \text{ g C}$$

$$0.01132 \text{ mol H} \times \frac{1.008 \text{ g H}}{1 \text{ mol H}} = 0.01141 \text{ g H} = 0.0114 \text{ g H}$$

mass of O = 0.2033 g sample – (0.10206 g C + 0.01141 g H) = 0.08983 = 0.090 g O

$$0.08983 \text{ g O} \times \frac{1 \text{ mol O}}{16.00 \text{ g O}} = 5.614 \times 10^{-3} = 5.6 \times 10^{-3} \text{ mol O.} \text{ Divide moles by } 5.614 \times 10^{-3}.$$

$$\text{C: } \frac{8.498 \times 10^{-3}}{5.614 \times 10^{-3}} \approx 1.5; \quad \text{H: } \frac{0.01132}{5.614 \times 10^{-3}} \approx 2; \quad \text{O: } \frac{5.614 \times 10^{-3}}{5.614 \times 10^{-3}} = 1$$

Multiply these ratios by 2 to obtain whole number subscripts. The empirical formula is $C_3H_4O_2$.

3.60   The reaction involved is $MgSO_4 \cdot xH_2O(s) \rightarrow MgSO_4(s) + xH_2O(g)$. First, calculate the number of moles of product $MgSO_4$; this is the same as the number of moles of starting hydrate.

$$2.472 \text{ g MgSO}_4 \times \frac{1 \text{ mol MgSO}_4}{120.4 \text{ g MgSO}_4} \times \frac{1 \text{ mol MgSO}_4 \cdot xH_2O}{1 \text{ mol MgSO}_4} = 0.02053 \text{ mol MgSO}_4 \cdot xH_2O$$

$$\text{Thus, } \frac{5.061 \text{ g MgSO}_4 \cdot xH_2O}{0.02053} = 246.5 \text{ g/mol} = \text{FW of MgSO}_4 \cdot xH_2O$$

FW of $MgSO_4 \cdot xH_2O$ = FW of $MgSO_4$ + x(FW of H_2O).

246.5 = 120.4 + x(18.02). x = 6.998. The hydrate formula is $MgSO_4 \cdot \underline{7}H_2O$.

Alternatively, we could calculate the number of moles of water represented by weight loss: (5.061 – 2.472) = 2.589 g H_2O lost.

$$2.589 \text{ g H}_2\text{O} \times \frac{1 \text{ mol H}_2\text{O}}{18.02 \text{ g H}_2\text{O}} = 0.1437 \text{ mol H}_2\text{O}; \quad \frac{\text{mol H}_2\text{O}}{\text{mol MgSO}_4} = \frac{0.1437}{0.02053} = 7.000$$

Again the correct formula is $MgSO_4 \cdot \underline{7}H_2O$.

Calculations Based on Chemical Equations (Section 3.6)

3.62 $4 KO_2 + 2 CO_2 \rightarrow 2 K_2CO_3 + 3 O_2$

(a) $0.400 \text{ mol KO}_2 \times \dfrac{3 \text{ mol O}_2}{4 \text{ mol KO}_2} = 0.300 \text{ mol O}_2$

(b) $7.50 \text{ g O}_2 \times \dfrac{1 \text{ mol O}_2}{32.00 \text{ g O}_2} \times \dfrac{4 \text{ mol KO}_2}{3 \text{ mol O}_2} \times \dfrac{71.10 \text{ g KO}_2}{1 \text{ mol KO}_2} = 22.2 \text{ g KO}_2$

(c) $7.50 \text{ g O}_2 \times \dfrac{1 \text{ mol O}_2}{32.00 \text{ g O}_2} \times \dfrac{2 \text{ mol CO}_2}{3 \text{ mol O}_2} \times \dfrac{44.01 \text{ g CO}_2}{1 \text{ mol CO}_2} = 6.88 \text{ g CO}_2$

3.64   (a) $Fe_2O_3(s) + 3\,CO(g) \rightarrow 2\,Fe(s) + 3\,CO_2(g)$

       (b) $0.350\text{ kg Fe}_2O_3 \times \dfrac{1000\text{ g}}{1\text{ kg}} \times \dfrac{1\text{ mol Fe}_2O_3}{159.688\text{ g Fe}_2O_3} = 2.1918 = 2.19\text{ mol Fe}_2O_3$

             $2.1918\text{ mol Fe}_2O_3 \times \dfrac{3\text{ mol CO}}{1\text{ mol Fe}_2O_3} \times \dfrac{28.01\text{ g CO}}{1\text{ mol CO}} = 184.17 = 184\text{ g CO}$

       (c) $2.1918\text{ mol Fe}_2O_3 \times \dfrac{2\text{ mol Fe}}{1\text{ mol Fe}_2O_3} \times \dfrac{55.845\text{ g Fe}}{1\text{ mol Fe}} = 244.80 = 245\text{ g Fe}$

             $2.1918\text{ mol Fe}_2O_3 \times \dfrac{3\text{ mol CO}_2}{1\text{ mol Fe}_2O_3} \times \dfrac{44.01\text{ g CO}_2}{1\text{ mol CO}_2} = 289.38 = 289\text{ g CO}_2$

       (d)   reactants: $350\text{ g Fe}_2O_3 + 184.17\text{ g CO} = 534.17 = 534\text{ g}$

             products: $244.80\text{ g Fe} + 289.38\text{ g CO}_2 = 534.18 = 534\text{ g}$

             Mass is conserved.

3.66   (a) $CaH_2(s) + 2\,H_2O(l) \rightarrow Ca(OH)_2(aq) + 2\,H_2(g)$

       (b) $4.500\text{ g H}_2 \times \dfrac{1\text{ mol H}_2}{2.016\text{ g H}_2} \times \dfrac{1\text{ mol CaH}_2}{2\text{ mol H}_2} \times \dfrac{42.10\text{ g CaH}_2}{1\text{ mol CaH}_2} = 46.99\text{ g CaH}_2$

3.68 $2\,C_8H_{18}(l) + 25\,O_2(g) \rightarrow 16\,CO_2(g) + 18\,H_2O(l)$

       (a) $1.50\text{ mol C}_8H_{18} \times \dfrac{25\text{ mol O}_2}{2\text{ mol C}_8H_{18}} = 18.75 = 18.8\text{ mol O}_2$

       (b) $10.0\text{ g C}_8H_{18} \times \dfrac{1\text{ mol C}_8H_{18}}{114.2\text{ g C}_8H_{18}} \times \dfrac{25\text{ mol O}_2}{2\text{ mol C}_8H_{18}} \times \dfrac{32.00\text{ g O}_2}{1\text{ mol O}_2} = 35.0\text{ g O}_2$

       (c) $15.0\text{ gal C}_8H_{18} \times \dfrac{3.7854\text{ L}}{1\text{ gal}} \times \dfrac{1000\text{ mL}}{1\text{ L}} \times \dfrac{0.692\text{ g}}{1\text{ mL}} = 39{,}292 = 3.93 \times 10^4\text{ g C}_8H_{18}$

             $3.9292 \times 10^4\text{ g C}_8H_{18} \times \dfrac{1\text{ mol C}_8H_{18}}{114.2\text{ g C}_8H_{18}} \times \dfrac{25\text{ mol O}_2}{2\text{ mol C}_8H_{18}} \times \dfrac{32.00\text{ g O}_2}{1\text{ mol O}_2} = 137{,}627\text{ g}$

                                                    $= 1.38 \times 10^5\text{ g O}_2$

       (d) $3.9292 \times 10^4\text{ g C}_8H_{18} \times \dfrac{1\text{ mol C}_8H_{18}}{114.2\text{ g C}_8H_{18}} \times \dfrac{16\text{ mol CO}_2}{2\text{ mol C}_8H_{18}} \times \dfrac{44.01\text{ g CO}_2}{1\text{ mol O}_2} = 121{,}139\text{ g}$

                                                    $= 1.21 \times 10^5\text{ g CO}_2$

3.70   (a)   *Plan.* Calculate a "mole ratio" between nitroglycerine and total moles of gas produced. $(12 + 6 + 1 + 10) = 29$ mol gas; 4 mol nitro: 29 total mol gas.   *Solve.*

             $2.00\text{ mL nitro} \times \dfrac{1.592\text{ g}}{\text{mL}} \times \dfrac{1\text{ mol nitro}}{227.1\text{ g nitro}} \times \dfrac{29\text{ mol gas}}{4\text{ mol nitro}} = 0.10165 = 0.102\text{ mol gas}$

       (b) $0.10165\text{ mol gas} \times \dfrac{55\text{ L}}{\text{mol}} = 5.5906 = 5.6\text{ L}$

       (c) $2.00\text{ mL nitro} \times \dfrac{1.592\text{ g}}{\text{mL}} \times \dfrac{1\text{ mol nitro}}{227.1\text{ g nitro}} \times \dfrac{6\text{ mol N}_2}{4\text{ mol nitro}} \times \dfrac{28.01\text{ g N}_2}{1\text{ mol N}_2} = 0.589\text{ g N}_2$

Limiting Reactants; Theoretical Yields (Section 3.7)

3.72   (a)   *Theoretical yield* is the maximum amount of product possible, as predicted by stoichiometry, assuming that the limiting reactant is converted entirely to product.

                 Actual yield is the amount of product actually obtained, less than or equal to the theoretical yield. *Percent yield* is the ratio of (actual yield to theoretical yield) $\times$ 100.

       (b)   No reaction is perfect. Not all reactant molecules come together effectively to form products; alternative reaction pathways may produce secondary products and reduce the amount of desired product actually obtained, or it might not be possible to completely isolate the desired product from the reaction mixture. In any case, these factors reduce the actual yield of a reaction.

       (c)   No, 110% actual yield is not possible. Theoretical yield is the maximum possible amount of pure product, assuming all available limiting reactant is converted to product, and that all product is isolated. If an actual yield of 110% is obtained, the product must contain impurities which increase the experimental mass.

3.74   (a) $48{,}775 \text{ L beverage} \times \dfrac{1 \text{ bottle}}{0.355 \text{ L}} = 137{,}394.37 = 1.37 \times 10^5$ portions of beverage

                 (The uncertainty in 355 mL limits the precision of the number of portions we can reasonably expect to deliver to three significant figures.)

                 126,515 bottles; 108,500 caps; 1.09×10^5 bottles can be filled and capped.

       (b)   126,515 empty bottles – 108,500 caps = 18,015 bottles remain

                 137,394 portions – 108,500 caps = 28,894 = 2.89×10^4 portions remain

                 (Uncertainty in the number of portions available limits the results to 3 sig fig.)

       (c)   The caps limit production.

3.76   (a) $C_3H_8 + 5\,O_2 \rightarrow 3\,CO_2(g) + 4\,H_2O$

       (b) O_2 is the limiting reactant. According to the balanced equation, one molecule of C_3H_8 requires five molecules of O_2 for reaction. The three C_3H_8 molecules in the box would require fifteen molecules of O_2 for complete reaction, but only ten O_2 molecules are present. C_3H_8 is present in excess and O_2 limits.

       (c)   If the reaction goes to completion, there will be six molecules of CO_2, eight molecules of H_2O, one molecule of C_3H_8 (excess reactant), and zero molecules of O_2 (the limiting reactant is completely consumed).

3.78 $0.500 \text{ mol Al(OH)}_3 \times \dfrac{3 \text{ mol H}_2\text{SO}_4}{2 \text{ mol Al(OH)}_3} = 0.750 \text{ mol H}_2\text{SO}_4$ needed for complete reaction

       Only 0.500 mol H_2SO_4 available, so H_2SO_4 limits.

       $0.500 \text{ mol H}_2\text{SO}_4 \times \dfrac{1 \text{ mol Al}_2(\text{SO}_4)_3}{3 \text{ mol H}_2\text{SO}_4} = 0.1667 = 0.167 \text{ mol Al}_2(\text{SO}_4)_3$ can form

$$0.500 \text{ mol } H_2SO_4 \times \frac{2 \text{ mol } Al(OH)_3}{3 \text{ mol } H_2SO_4} = 0.3333 = 0.333 \text{ mol } Al(OH)_3 \text{ react}$$

$$0.500 \text{ mol } Al(OH)_3 \text{ initial} - 0.333 \text{ mol react} = 0.167 \text{ mol } Al(OH)_3 \text{ remain}$$

3.80 $4 \, NH_3(g) + 5 \, O_2(g) \rightarrow 4 \, NO(g) + 6 \, H_2O(g)$

     (a)   Follow the approach in Sample Exercise 3.19.

$$2.00 \text{ g } NH_3 \times \frac{1 \text{ mol } NH_3}{17.03 \text{ g } NH_3} = 0.11744 = 0.117 \text{ mol } NH_3$$

$$2.50 \text{ g } O_2 \times \frac{1 \text{ mol } O_2}{32.00 \text{ g } O_2} = 0.07813 = 0.0781 \text{ mol } O_2$$

$$0.07813 \text{ mol } O_2 \times \frac{4 \text{ mol } NH_3}{5 \text{ mol } O_2} = 0.06250 = 0.0625 \text{ mol } NH_3 \text{ required}$$

            More than 0.0625 mol NH_3 is available, so O_2 is the limiting reactant.

     (b) $$0.07813 \text{ mol } O_2 \times \frac{4 \text{ mol } NO}{5 \text{ mol } O_2} \times \frac{30.01 \text{ g } NO}{1 \text{ mol } NO} = 1.8756 = 1.88 \text{ g } NO \text{ produced}$$

$$0.07813 \text{ mol } O_2 \times \frac{6 \text{ mol } H_2O}{5 \text{ mol } O_2} \times \frac{18.02 \text{ g } H_2O}{1 \text{ mol } H_2O} = 1.6894 = 1.69 \text{ g } H_2O \text{ produced}$$

     (c) 0.11744 mol $NH_3 - 0.0625$ mol NH_3 reacted $= 0.05494 = 0.0549$ mol NH_3 remain

$$0.05494 \text{ mol } NH_3 \times \frac{17.03 \text{ g } NH_3}{1 \text{ mol } NH_3} = 0.93563 = 0.936 \text{ g } NH_3 \text{ remain}$$

     (d)   mass products = 1.8756 g NO + 1.6894 g H_2O + 0.9356 g NH_3 remaining = 4.50g products

            mass reactants = 2.00 g NH_3 + 2.50 g O_2 = 4.50 g reactants

            (For comparison purposes, the mass of excess reactant can be either added to the products, as above, or subtracted from reactants.)

3.82   *Plan.* Write balanced equation; determine limiting reactant; calculate amounts of excess reactant remaining and products, based on limiting reactant.

      Solve. $H_2SO_4(aq) + Pb(C_2H_3O_2)_2(aq) \rightarrow PbSO_4(s) + 2 \, HC_2H_3O_2(aq)$

$$5.00 \text{ g } H_2SO_4 \times \frac{1 \text{ mol } H_2SO_4}{98.09 \text{ g } H_2SO_4} = 0.05097 = 0.0510 \text{ mol } H_2SO_4$$

$$5.00 \text{ g } Pb(C_2H_3O_2)_2 \times \frac{1 \text{ mol } Pb(C_2H_3O_2)_2}{325.3 \text{ g } Pb(C_2H_3O_2)_2} = 0.015370 = 0.0154 \text{ mol } Pb(C_2H_3O_2)_2$$

      1 mol H_2SO_4:1 mol $Pb(C_2H_3O_2)_2$, so $Pb(C_2H_3O_2)_2$ is the limiting reactant.

      0 mol $Pb(C_2H_3O_2)_2$, (0.05097 − 0.01537) = 0.0356 mol H_2SO_4, 0.0154 mol $PbSO_4$,

      (0.01537 × 2) = 0.0307 mol $HC_2H_3O_2$ are present after reaction

      0.03560 mol H_2SO_4 × 98.09 g/mol = 3.4920 = 3.49 g H_2SO_4

$0.01537 \text{ mol PbSO}_4 \times 303.3 \text{ g/mol} = 4.6619 = 4.66 \text{ g PbSO}_4$

$0.03074 \text{ mol HC}_2\text{H}_3\text{O}_2 \times 60.05 \text{ g/mol} = 1.8460 = 1.85 \text{ g HC}_2\text{H}_3\text{O}_2$

Check. The initial mass of reactants was 10.00 g; and the final mass of excess reactant and products is 10.00 g; mass is conserved.

3.84   (a) $C_2H_6 + Cl_2 \rightarrow C_2H_5Cl + HCl$

$$125 \text{ g C}_2\text{H}_6 \times \frac{1 \text{ mol C}_2\text{H}_6}{30.07 \text{ g C}_2\text{H}_6} = 4.157 = 4.16 \text{ mol C}_2\text{H}_6$$

$$255 \text{ g Cl}_2 \times \frac{1 \text{ mol Cl}_2}{70.91 \text{ g Cl}_2} = 3.596 = 3.60 \text{ mol Cl}_2$$

Because the reactants combine in a 1:1 mole ratio, Cl_2 is the limiting reactant. The theoretical yield is:

$$3.596 \text{ mol Cl}_2 \times \frac{1 \text{ mol C}_2\text{H}_5\text{Cl}}{1 \text{ mol Cl}_2} \times \frac{64.51 \text{ g C}_2\text{H}_5\text{Cl}}{1 \text{ mol C}_2\text{H}_5\text{Cl}} = 231.98 = 232 \text{ g C}_2\text{H}_5\text{Cl}$$

  (b) $\% \text{ yield} = \dfrac{206 \text{ g C}_2\text{H}_5\text{Cl actual}}{232 \text{ g C}_2\text{H}_5\text{Cl theoretical}} \times 100 = 88.8\%$

3.86 $H_2S(g) + 2 \text{ NaOH}(aq) \rightarrow Na_2S(aq) + 2 H_2O(l)$

$$1.25 \text{ g H}_2\text{S} \times \frac{1 \text{ mol H}_2\text{S}}{34.08 \text{ g H}_2\text{S}} = 0.03668 = 0.0367 \text{ mol H}_2\text{S}$$

$$2.00 \text{ g NaOH} \times \frac{1 \text{ mol NaOH}}{40.00 \text{ g NaOH}} = 0.0500 \text{ mol NaOH}$$

By inspection, twice as many mol NaOH as H_2S are needed for exact reaction, but mol NaOH given is less than twice mol H_2S, so NaOH limits.

$$0.0500 \text{ mol NaOH} \times \frac{1 \text{ mol Na}_2\text{S}}{2 \text{ mol NaOH}} \times \frac{78.05 \text{ g Na}_2\text{S}}{1 \text{ mol Na}_2\text{S}} = 1.95125 = 1.95 \text{ g Na}_2\text{S theoretical}$$

$$\frac{92.0\%}{100} \times 1.95125 \text{ g Na}_2\text{S theoretical} = 1.7951 = 1.80 \text{ g Na}_2\text{S actual}$$

Additional Exercises

3.88 $C_2H_5OH(l) + 3 O_2(g) \rightarrow 2 CO_2(g) + 3 H_2O(g)$

    $C_3H_8(g) + 5 O_2(g) \rightarrow 3 CO_2(g) + 4 H_2O(g)$

    $CH_3CH_2COCH_3(l) + 11/2 O_2(g) \rightarrow 4 CO_2(g) + 4 H_2O(l)$

In a combustion reaction, all H in the fuel is transformed to H_2O in the products. The reactant with most mol H/mol fuel will produce the most H_2O. C_3H_8 and $CH_3CH_2COCH_3$ (C_4H_8O) both have 8 mol H/mol fuel, so 1.5 mol of either fuel will produce the same amount of H_2O. 1.5 mol C_2H_5OH will produce less H_2O.

3.89   The formulas of the fertilizers are NH_3, NH_4NO_3, $(NH_4)_2SO_4$ and $(NH_2)_2CO$. Qualitatively, the more heavy, non-nitrogen atoms in a molecule, the smaller the mass % of N. By inspection, the mass of NH_3 is dominated by N, so it will have the greatest % N, $(NH_4)_2SO_4$ will have the least. In order of increasing % N:

$(NH_4)_2SO_4 < NH_4NO_3 < (NH_2)_2CO < NH_3$.

Check by calculation:

$(NH_4)_2SO_4$: FW = 2(14.0) + 8(1.0) + 1(32.1) + 4(16.0) = 132.1 amu

% N = [2(14.0)/132.1] × 100 = 21.2%

NH_4NO_3: FW = 2(14.0) + 4(1.0) + 3(16.0) = 80.0 amu

% N = [2(14.0)/80.0] × 100 = 35.0%

$(NH_2)_2CO$: FW = 2(14.0) + 4(1.0) = 1(12.0) + 1(16.0) = 60.0 amu

% N = [2(14.0)/60.0] × 100 = 46.7% N

NH_3: FW = 1(14.0) + 3(1.0) = 17.0 amu

% N = [14.0/17.0] × 100 = 82.4 % N

3.90 (a) $0.500 \text{ g } C_9H_8O_4 \times \dfrac{1 \text{ mol } C_9H_8O_4}{180.2 \text{ g } C_9H_8O_4} = 2.7747 \times 10^{-3} = 2.77 \times 10^{-3} \text{ mol } C_9H_8O_4$

(b) $0.0027747 \text{ mol } C_9H_8O_4 \times \dfrac{6.022 \times 10^{23} \text{ molecules}}{1 \text{ mol}} = 1.67 \times 10^{21} \, C_9H_8O_4 \text{ molecules}$

(c) $1.67 \times 10^{21} \, C_9H_8O_4 \text{ molecules} \times \dfrac{9 \text{ C atoms}}{1 \, C_9H_8O_4 \text{ molecule}} = 1.50 \times 10^{22} \text{ C atoms}$

3.92 (a) $\dfrac{5.342 \times 10^{-21} \text{ g}}{1 \text{ molecule penicillin G}} \times \dfrac{6.0221 \times 10^{23} \text{ molecules}}{1 \text{ mol}} = 3217 \text{ g/mol penicillin G}$

(b) 1.00 g hemoglobin (hem) contains 3.40×10^{-3} g Fe.

$\dfrac{1.00 \text{ g hem}}{3.40 \times 10^{-3} \text{ g Fe}} \times \dfrac{55.85 \text{ g Fe}}{1 \text{ mol Fe}} \times \dfrac{4 \text{ mol Fe}}{1 \text{ mol hem}} = 6.57 \times 10^4 \text{ g/mol hemoglobin}$

3.93 *Plan.* Assume 100 g, calculate mole ratios, empirical formula, then molecular formula from molar mass. *Solve.*

$68.2 \text{ g C} \times \dfrac{1 \text{ mol C}}{12.01 \text{ g C}} = 5.68 \text{ mol C; } 5.68/0.568 \approx 10$

$6.86 \text{ g H} \times \dfrac{1 \text{ mol H}}{1.008 \text{ g H}} = 6.81 \text{ mol H; } 6.81/0.568 \approx 12$

$15.9 \text{ g N} \times \dfrac{1 \text{ mol N}}{14.01 \text{ g N}} = 1.13 \text{ mol N; } 1.13/0.568 \approx 2$

$9.08 \text{ g O} \times \dfrac{1 \text{ mol O}}{16.00 \text{ g O}} = 0.568 \text{ mol O; } 0.568/0.568 = 1$

The empirical formula is $C_{10}H_{12}N_2O$, FW = 176 amu (or g). Because the molar mass is 176, the empirical and molecular formula are the same, $C_{10}H_{12}N_2O$.

3.94 *Plan.* Assume 1.000 g and get mass O by subtraction. *Solve.*

(a) $0.7787 \text{ g C} \times \dfrac{1 \text{ mol C}}{12.01 \text{ g C}} = 0.06484 \text{ mol C}$

$$0.1176 \text{ g H} \times \frac{1 \text{ mol H}}{1.008 \text{ g H}} = 0.1167 \text{ mol H}$$

$$0.1037 \text{ g O} \times \frac{1 \text{ mol C}}{16.00 \text{ g O}} = 0.006481 \text{ mol O}$$

Dividing through by the smallest of these values we obtain $C_{10}H_{18}O$.

(b)   The formula weight of $C_{10}H_{18}O$ is 154. Thus, the empirical formula is also the molecular formula.

3.96   *Plan.* Because different sample sizes were used to analyze the different elements, calculate mass % of each element in the sample.

i.   Calculate mass % C from g CO_2.

ii.   Calculate mass % Cl from AgCl.

iii.   Get mass % H by subtraction.

iv.   Calculate mole ratios and the empirical formulas.

Solve.

i. $$3.52 \text{ g CO}_2 \times \frac{1 \text{ mol CO}_2}{44.01 \text{ g CO}_2} \times \frac{1 \text{ mol C}}{1 \text{ mol CO}_2} \times \frac{12.01 \text{ g C}}{1 \text{ mol C}} = 0.9606 = 0.961 \text{ g C}$$

$$\frac{0.9606 \text{ g C}}{1.50 \text{ g sample}} \times 100 = 64.04 = 64.0\% \text{ C}$$

ii. $$1.27 \text{ g AgCl} \times \frac{1 \text{ mol AgCl}}{143.3 \text{ g AgCl}} \times \frac{1 \text{ mol Cl}}{1 \text{ mol AgCl}} \times \frac{35.45 \text{ g Cl}}{1 \text{ mol Cl}} = 0.3142 = 0.314 \text{ g Cl}$$

$$\frac{0.3142 \text{ g Cl}}{1.00 \text{ g sample}} \times 100 = 31.42 = 31.4\% \text{ Cl}$$

iii.   % H = 100.0 − (64.04% C + 31.42% Cl) = 4.54 = 4.5% H

iv.   Assume 100 g sample.

$$64.04 \text{ g C} \times \frac{1 \text{ mol C}}{12.01 \text{ g C}} = 5.33 \text{ mol C}; \quad 5.33 / 0.886 = 6.02$$

$$31.42 \text{ g Cl} \times \frac{1 \text{ mol Cl}}{35.45 \text{ g Cl}} = 0.886 \text{ mol Cl}; \quad 0.886 / 0.886 = 1.00$$

$$4.54 \text{ g H} \times \frac{1 \text{ mol H}}{1.008 \text{ g H}} = 4.50 \text{ mol H}; \quad 4.50 / 0.886 = 5.08$$

The empirical formula is probably C_6H_5Cl.

The subscript for H, 5.08, is relatively far from 5.00, but C_6H_5Cl makes chemical sense. More significant figures in the mass data are required for a more accurate mole ratio.

3.97   The mass percentage is determined by the relative number of atoms of the element times the atomic weight, divided by the total formula mass. Thus, the mass percent of bromine in $KBrO_x$ is given by $0.5292 = \dfrac{79.91}{39.10 + 79.91 + x(16.00)}$.

Solving for x, we obtain x = 2.00. Thus, the formula is $KBrO_2$.

3.98   (a)   Let AW = the atomic weight of X.

According to the chemical reaction, moles XI_3 reacted = moles XCl_3 produced

0.5000 g $XI_3 \times 1$ mol $XI_3 / (AW + 380.71)$ g XI_3

$$= 0.2360 \text{ g } XCl_3 \times \frac{1 \text{ mol } XCl_3}{(AW + 106.36) \text{ g } XCl_3}$$

$0.5000 \,(AW + 106.36) = 0.2360 \,(AW + 380.71)$

$0.5000 \,AW + 53.180 = 0.2360 \,AW + 89.848$

$0.2640 \,AW = 36.67; \; AW = 138.9$ g

(b)   X is lanthanum, La, atomic number 57.

3.100 $2 \text{ NaCl}(aq) + 2 \text{ H}_2\text{O}(l) \rightarrow 2 \text{ NaOH}(aq) + \text{H}_2(g) + \text{Cl}_2(g)$

Calculate mol Cl_2 and relate to mol H_2, mol NaOH.

$$1.5 \times 10^6 \text{ kg} \times \frac{1000 \text{ g}}{1 \text{ kg}} \times \frac{1 \text{ mol } Cl_2}{70.91 \text{ g } Cl_2} = 2.115 \times 10^7 = 2.1 \times 10^7 \text{ mol } Cl_2$$

$$2.115 \times 10^7 \text{ mol } Cl_2 \times \frac{1 \text{ mol } H_2}{1 \text{ mol } Cl_2} \times \frac{2.016 \text{ g } H_2}{1 \text{ mol } H_2} = 4.26 \times 10^7 \text{ g } H_2 = 4.3 \times 10^4 \text{ kg } H_2$$

$$4.3 \times 10^7 \text{ g} \times \frac{1 \text{ metric ton}}{1 \times 10^6 \text{ g } (1 \text{ Mg})} = 43 \text{ metric tons } H_2$$

$$2.115 \times 10^7 \text{ mol } Cl_2 \times \frac{2 \text{ mol NaOH}}{1 \text{ mol } Cl_2} \times \frac{40.0 \text{ g NaOH}}{1 \text{ mol NaOH}} = 1.69 \times 10^9 = 1.7 \times 10^9 \text{ g NaOH}$$

1.7×10^9 g NaOH $= 1.7 \times 10^6$ kg NaOH $= 1.7 \times 10^3$ metric tons NaOH

3.101 $2 \text{ C}_{57}\text{H}_{110}\text{O}_6 + 163 \text{ O}_2 \rightarrow 114 \text{ CO}_2 + 110 \text{ H}_2\text{O}$

molar mass of fat = $57(12.01) + 110(1.008) + 6(16.00) = 891.5$

$$1.0 \text{ kg fat} \times \frac{1000 \text{ g}}{1 \text{ kg}} \times \frac{1 \text{ mol fat}}{891.5 \text{ g fat}} \times \frac{110 \text{ mol } H_2O}{2 \text{ mol fat}} \times \frac{18.02 \text{ g } H_2O}{1 \text{ mol } H_2O} \times \frac{1 \text{ kg}}{1000 \text{ g}} = 1.1 \text{ kg } H_2O$$

3.102   (a)   *Plan.* Calculate the total mass of C from g CO and g CO_2. Calculate the mass of H from g H_2O. Calculate mole ratios and the empirical formula.   *Solve.*

$$0.467 \text{ g CO} \times \frac{1 \text{ mol CO}}{28.01 \text{ g CO}} \times \frac{1 \text{ mol C}}{1 \text{ mol CO}} \times 12.01 \text{ g C} = 0.200 \text{ g C}$$

$$0.733 \text{ g CO}_2 \times \frac{1 \text{ mol CO}_2}{44.01 \text{ g CO}_2} \times \frac{1 \text{ mol C}}{1 \text{ mol CO}_2} \times 12.01 \text{ g C} = 0.200 \text{ g C}$$

Total mass C is 0.200 g + 0.200 g = 0.400 g C.

$$0.450 \text{ g H}_2O \times \frac{1 \text{ mol } H_2O}{18.02 \text{ g } H_2O} \times \frac{2 \text{ mol H}}{1 \text{ mol } H_2O} \times \frac{1.008 \text{ g H}}{1 \text{ mol H}} = 0.0503 \text{ g H}$$

(Because hydrocarbons contain only the elements C and H, g H can also be obtained by subtraction: 0.450 g sample − 0.400 g C = 0.050 g H.)

$$0.400 \, g \, C \times \frac{1 \, mol \, C}{12.01 \, g \, C} = 0.0333 \, mol \, C; \; 0.0333 / 0.0333 = 1.0$$

$$0.0503 \, g \, H \times \frac{1 \, mol \, H}{1.008 \, g \, H} = 0.0499 \, mol \, H; \; 0.0499 / 0.0333 = 1.5$$

Multiplying by a factor of 2, the empirical formula is C_2H_3.

(b)   Mass is conserved. Total mass products − mass sample = mass O_2 consumed.

0.467 g CO + 0.733 g CO_2 + 0.450 g H_2O − 0.450 g sample = 1.200 g O_2 consumed

(c)   For complete combustion, 0.467 g CO must be converted to CO_2.

$$2 \, CO(g) + O_2(g) \rightarrow 2 \, CO_2(g)$$

$$0.467 \, g \, CO \times \frac{1 \, mol \, CO}{28.01 \, g \, CO} \times \frac{1 \, mol \, O_2}{2 \, mol \, CO} \times \frac{32.00 \, g \, O_2}{1 \, mol \, O_2} = 0.267 \, g \, O_2$$

The total mass of O_2 required for complete combustion is
1.200 g + 0.267 g = 1.467 g O_2.

3.104   All of the O_2 is produced from $KClO_3$; get g $KClO_3$ from g O_2. All of the H_2O is produced from $KHCO_3$; get g $KHCO_3$ from g H_2O. The g H_2O produced also reveals the g CO_2 from the decomposition of $KHCO_3$. The remaining CO_2 (13.2 g CO_2 − g CO_2 from $KHCO_3$) is due to K_2CO_3 and g K_2CO_3 can be derived from it.

$$4.00 \, g \, O_2 \times \frac{1 \, mol \, O_2}{32.00 \, g \, O_2} \times \frac{2 \, mol \, KClO_3}{3 \, mol \, O_2} \times \frac{122.6 \, g \, KClO_3}{1 \, mol \, KClO_3} = 10.22 = 10.2 \, g \, KClO_3$$

$$1.80 \, g \, H_2O \times \frac{1 \, mol \, H_2O}{18.02 \, g \, H_2O} \times \frac{2 \, mol \, KHCO_3}{1 \, mol \, H_2O} \times \frac{100.1 \, g \, KHCO_3}{1 \, mol \, KHCO_3} = 20.00 = 20.0 \, g \, KHCO_3$$

$$1.80 \, g \, H_2O \times \frac{1 \, mol \, H_2O}{18.02 \, g \, H_2O} \times \frac{2 \, mol \, CO_2}{1 \, mol \, H_2O} \times \frac{44.01 \, g \, CO_2}{1 \, mol \, CO_2} = 8.792 = 8.79 \, g \, CO_2 \; from \; KHCO_3$$

13.20 g CO_2 total − 8.792 CO_2 from $KHCO_3$ = 4.408 = 4.41 g CO_2 from K_2CO_3

$$4.408 \, g \, CO_2 \times \frac{1 \, mol \, CO_2}{44.01 \, g \, CO_2} \times \frac{1 \, mol \, K_2CO_3}{1 \, mol \, CO_2} \times \frac{138.2 \, g \, K_2CO_3}{1 \, mol \, K_2CO_3} = 13.84 = 13.8 \, g \, K_2CO_3$$

100.0 g mixture − 10.22 g $KClO_3$ − 20.00 g $KHCO_3$ − 13.84 g K_2CO_3 = 55.9 g KCl

3.105   (a) $2 \, C_2H_2(g) + 5 \, O_2(g) \rightarrow 4 \, CO_2(g) + 2 \, H_2O(g)$

(b)   Following the approach in Sample Exercise 3.18,

$$10.0 \, g \, C_2H_2 \times \frac{1 \, mol \, C_2H_2}{26.04 \, g \, C_2H_2} \times \frac{5 \, mol \, O_2}{2 \, mol \, C_2H_2} \times \frac{32.00 \, g \, O_2}{1 \, mol \, O_2} = 30.7 \, g \, O_2 \; required$$

Only 10.0 g O_2 are available, so O_2 limits.

(c)   Because O_2 limits, 0.0 g O_2 remain.

Next, calculate the g C_2H_2 consumed and the amounts of CO_2 and H_2O produced by reaction of 10.0 g O_2.

$$10.0 \text{ g O}_2 \times \frac{1 \text{ mol O}_2}{32.00 \text{ g O}_2} \times \frac{2 \text{ mol C}_2\text{H}_2}{5 \text{ mol O}_2} \times \frac{26.04 \text{ g C}_2\text{H}_2}{1 \text{ mol C}_2\text{H}_2} = 3.26 \text{ g C}_2\text{H}_2 \text{ consumed}$$

$$10.0 \text{ g C}_2\text{H}_2 \text{ initial} - 3.26 \text{ g consumed} = 6.74 = 6.7 \text{ g C}_2\text{H}_2 \text{ remain}$$

$$10.0 \text{ g O}_2 \times \frac{1 \text{ mol O}_2}{32.00 \text{ g O}_2} \times \frac{4 \text{ mol CO}_2}{5 \text{ mol O}_2} \times \frac{44.01 \text{ g CO}_2}{1 \text{ mol CO}_2} = 11.0 \text{ g CO}_2 \text{ produced}$$

$$10.0 \text{ g O}_2 \times \frac{1 \text{ mol O}_2}{32.00 \text{ g O}_2} \times \frac{2 \text{ mol H}_2\text{O}}{5 \text{ mol O}_2} \times \frac{18.02 \text{ g H}_2\text{O}}{1 \text{ mol H}_2\text{O}} = 2.25 \text{ g H}_2\text{O produced}$$

Integrative Exercises

3.107   (a)   *Plan.* volume of Ag cube $\xrightarrow{\text{density}}$ mass of Ag → mol Ag → Ag atoms

               Solve. $(1.000)^3 \text{ cm}^3 \text{ Ag} \times \dfrac{10.5 \text{ g Ag}}{1 \text{ cm}^3 \text{ Ag}} \times \dfrac{1 \text{ mol Ag}}{107.87 \text{ g Ag}} \times \dfrac{6.022 \times 10^{23} \text{ atoms}}{1 \text{ mol}}$

$$= 5.8618 \times 10^{22} = 5.86 \times 10^{22} \text{ Ag atoms}$$

       (b) 1.000 cm^3 cube volume, 74% is occupied by Ag atoms

            $0.74 \text{ cm}^3 = $ volume of 5.86×10^{22} Ag atoms

$$\frac{0.7400 \text{ cm}^3}{5.8618 \times 10^{22} \text{ Ag atoms}} = 1.2624 \times 10^{-23} = 1.3 \times 10^{-23} \text{ cm}^3 / \text{Ag atom}$$

            Because atomic dimensions are usually given in Å, we will show this conversion.

$$1.2624 \times 10^{-23} \text{ cm}^3 \times \frac{(1 \times 10^{-2})^3 \text{ m}^3}{1 \text{ cm}^3} \times \frac{1 \text{ Å}^3}{(1 \times 10^{-10})^3 \text{ m}^3} = 12.62 = 13 \text{ Å}^3 / \text{Ag atom}$$

       (c) $V = 4/3 \pi r^3; \; r^3 = 3V/4\pi; \; r = (3V/4\pi)^{1/3}$

            $r_A = (3 \times 12.62 \text{ Å}^3 / 4\pi)^{1/3} = 1.444 = 1.4 \text{ Å}$

3.108   (a)   *Analyze.* Given: gasoline = C_8H_{18}, density = 0.69 g/mL, 20.5 mi/gal, 225 mi.

            Find: kg CO_2.

            Plan. Write and balance the equation for the combustion of octane. Change mi →

            gal octane → mL → g octane. Use stoichiometry to calculate g and kg CO_2 from

            g octane.

            Solve. $2 \text{ C}_8\text{H}_{18}(l) + 25 \text{ O}_2(g) \rightarrow 16 \text{ CO}_2(g) + 18 \text{ H}_2\text{O}(l)$

$$225 \text{ mi} \times \frac{1 \text{ gal}}{20.5 \text{ mi}} \times \frac{3.7854 \text{ L}}{1 \text{ gal}} \times \frac{1 \text{ mL}}{1 \times 10^{-3} \text{ L}} \times \frac{0.69 \text{ g octane}}{1 \text{ mL}} = 2.8667 \times 10^4 \text{ g}$$

$$= 29 \text{ kg octane}$$

$$2.8667 \times 10^4 \text{ g C}_8\text{H}_{18} \times \frac{1 \text{ mol C}_8\text{H}_{18}}{114.2 \text{ g C}_8\text{H}_{18}} \times \frac{16 \text{ mol CO}_2}{2 \text{ mol C}_8\text{H}_{18}} \times \frac{44.01 \text{ g CO}_2}{1 \text{ mol CO}_2} = 8.8382 \times 10^4 \text{ g}$$

$$= 88 \text{ kg CO}_2$$

Check. $\left(\dfrac{225\times4\times0.7}{20}\right)\times10^3=(45\times0.7)\times10^3=30\times10^3\,g=30\,kg\,octane$

$\dfrac{44}{114}\approx\dfrac{1}{3};\ \dfrac{30\,kg\times8}{3}\approx80\,kg\,CO_2$

(b) *Plan.* Use the same strategy as part (a). *Solve.*

$225\,mi\times\dfrac{1\,gal}{5\ mi}\times\dfrac{3.7854\,L}{1\,gal}\times\dfrac{1\,mL}{1\times10^{-3}\,L}\times\dfrac{0.69\,g\,octane}{1\,mL}=1.1754\times10^5$

$=1\times10^2\,kg\,octane$

$1.1754\times10^5\,g\,C_8H_{18}\times\dfrac{1\,mol\,C_8H_{18}}{114.2\,g\,C_8H_{18}}\times\dfrac{16\,mol\,CO_2}{2\,mol\,C_8H_{18}}\times\dfrac{44.01\,g\,CO_2}{1\,mol\,CO_2}=$

$3.624\times10^5\,g=4\times10^2\,kg\,CO_2$

Check. Mileage of 5 mi/gal requires ~4 times as much gasoline as mileage of 20.5 mi/gal, so it should produce ~4 times as much CO_2. 90 kg CO_2 [from (a)] × 4 = 360 = 4×10^2 kg CO_2 [from (b)].

3.109 Structural isomers, like 1-propanol and 2-propanol, have the same number and kinds of atoms, but different arrangements of these atoms. Because molecular weight is the sum of atomic weights, and number and kinds of atoms are the same, the molecular weights of structural isomers are the same. Again, because number and kinds of atoms are the same, percent composition and therefore combustion analysis results will be the same. Physical properties, like boiling point and density, are influenced by structure as well as molecular weight, and are different for structural isomers.

The properties (a) boiling point and (d) density will distinguish between 1-propanol and 2-propanol. This is confirmed by comparing these properties from either Wolfram Alpha (WA) or the CRC Handbook of Chemistry and Physics (CRC).

Compound	Boiling Point (WA)	Boiling Point (CRC)	Density (WA)	Density (CRC)
1-propanol	97°C	97.4°C	0.804 g/cm³	0.8035 g/cm³
2-propanol	82°C	82.4°C	0.785 g/cm³	0.7855 g/cm³

3.111 (a) *Plan.* Calculate the kg of air in the room and then the mass of HCN required to produce a dose of 300 mg HCN/kg air. *Solve.*

12 ft × 15 ft × 8.0 ft = 1440 = 1.4×10^3 ft³ of air in the room

$1440\,ft^3\,air\times\dfrac{(12\,in)^3}{1\,ft^3}\times\dfrac{(2.54\,cm)^3}{1\,in^3}\times\dfrac{0.00118\,g\,air}{1\,cm^3\,air}\times\dfrac{1\,kg}{1000\,g}=48.12=48\,kg\,air$

$48.12\,kg\,air\times\dfrac{300\,mg\,HCN}{1\,kg\,air}\times\dfrac{1\,g}{1000\,mg}=14.43=14\,g\,HCN$

(b) $2\,NaCN(s)+H_2SO_4(aq)\rightarrow Na_2SO_4(aq)+2\,HCN(g)$

The question can be restated as: What mass of NaCN is required to produce 14 g of HCN according to the above reaction?

$$14.43 \text{ g HCN} \times \frac{1 \text{ mol HCN}}{27.03 \text{ g HCN}} \times \frac{2 \text{ mol NaCN}}{2 \text{ mol HCN}} \times \frac{49.01 \text{ g NaCN}}{1 \text{ mol NaCN}} = 26.2 = 26 \text{ g NaCN}$$

(c) $12 \text{ ft} \times 15 \text{ ft} \times \dfrac{1 \text{ yd}^2}{9 \text{ ft}^2} \times \dfrac{30 \text{ oz}}{1 \text{ yd}^2} \times \dfrac{1 \text{ lb}}{16 \text{ oz}} \times \dfrac{454 \text{ g}}{1 \text{ lb}} = 17,025$

$$= 1.7 \times 10^4 \text{ g CH}_2\text{CHCN in the room}$$

50% of the carpet burns, so the starting amount of CH_2CHCN is
$0.50(17,025) = 8,513 = 8.5 \times 10^3 \text{ g}$

$$8513 \text{ g CH}_2\text{CHCN} \times \frac{50.9 \text{ g HCN}}{100 \text{ g CH}_2\text{CHCN}} = 4333 = 4.3 \times 10^3 \text{ g HCN possible}$$

If the actual yield of combustion is 20%, actual g HCN = 4,333(0.20) = 866.6
$= 8.7 \times 10^2$ g HCN produced. From part (a), 14 g of HCN is a lethal dose. The fire produces much more than a lethal dose of HCN.

3.112   (a) $N_2(g) + O_2(g) \rightarrow 2 \text{ NO}(g); \; 2 \text{ NO}(g) + O_2(g) \rightarrow 2 \text{ NO}_2(g)$

       (b) $1 \text{ million} = 1 \times 10^6$

$$22 \times 10^6 \text{ tons NO}_2 \times \frac{2000 \text{ lb}}{1 \text{ ton}} \times \frac{453.6 \text{ g}}{1 \text{ lb}} = 1.996 \times 10^{13} = 2.0 \times 10^{13} \text{ g NO}_2$$

       (c)   *Plan.* Calculate g O_2 needed to burn 500 g octane. This is 85% of total O_2 in the engine. 15% of total O_2 is used to produce NO_2, according to the second equation in part (a).

Solve. $2 \text{ C}_8\text{H}_{18}(l) + 25 \text{ O}_2(g) \rightarrow 16 \text{ CO}_2(g) + 18 \text{ H}_2\text{O}(l)$

$$500 \text{ g C}_8\text{H}_{18} \times \frac{1 \text{ mol C}_8\text{H}_{18}}{114.2 \text{ g C}_8\text{H}_{18}} \times \frac{25 \text{ mol O}_2}{2 \text{ mol C}_8\text{H}_{18}} \times \frac{32.00 \text{ g O}_2}{\text{mol O}_2} = 1751 = 1.75 \times 10^3 \text{ g O}_2$$

$$\frac{1751 \text{ g O}_2}{\text{total g O}_2} = 0.85; \; 2060 = 2.1 \times 10^3 \text{ g O}_2 \text{ total in engine}$$

2060 g O_2 total × 0.15 = 309.1 = 3.1×10^2 g O_2 used to produce NO_2. One mol O_2 produces 2 mol NO. Then 2 mol NO react with a second mol O_2 to produce 2 mol NO_2. Two mol O_2 are required to produce 2 mol NO_2; one mol O_2 per mol NO_2.

$$309.1 \text{ g O}_2 \times \frac{1 \text{ mol O}_2}{32.00 \text{ g}} \times \frac{1 \text{ mol NO}_2}{1 \text{ mol O}_2} \times \frac{46.01 \text{ g NO}_2}{1 \text{ mol O}_2} = 444.4 = 4.4 \times 10^2 \text{ g NO}_2$$

4 Reactions in Aqueous Solution

Visualizing Concepts

4.2 *Analyze/Plan.* Correlate the neutral molecules, cations, and anions in the diagrams with the definitions of strong, weak, and nonelectrolytes. *Solve.*

(a) AX is a nonelectrolyte because no ions form when the molecules dissolve.

(b) AY is a weak electrolyte because a few molecules ionize when they dissolve, but most do not.

(c) AZ is a strong electrolyte because all molecules break up into ions when they dissolve.

4.3 *Analyze/Plan.* From the molecular representations, write molecular formulas for the compounds. Using Table 4.2 and molecular formulas (there are no ionic compounds in this exercise), classify the compounds as strong acid, strong base, weak acid, weak base (NH_3), or nonelectrolyte. Strong acids and bases are strong electrolytes, weak acids and bases are weak electrolytes. *Solve.*

(a) HCOOH. The molecule has a –COOH group; it is a weak acid and weak electrolyte (it is not one of the strong acids listed in Table 4.2).

(b) HNO_3. The molecule is a strong acid (Table 4.2) and a strong electrolyte.

(c) CH_3CH_2OH. The molecule is neither an acid nor a base; it is a nonelectrolyte.

4.4 Statement (b) is most correct. Statement (a) is incorrect because, at equilibrium, the chemical reactions are ongoing. Statement (c) is incorrect because the concentration of the product (or reactant) is changing until equilibrium is reached, but not at equilibrium. Equilibrium is a state of dynamic constancy.

4.6 Certain pairs of ions form precipitates because their attraction is so strong that they cannot be surrounded and separated by solvent molecules. That is, the attraction between solute particles is greater than the stabilization offered by interaction of individual ions with solvent molecules. These precipitates are insoluble ionic solids.

4.8 *Analyze/Plan.* Given three metal powders and three 1 *M* solutions, use Table 4.5, the activity series of metals, to find a scheme to distinguish the metals.

Solve. In the activity series, any metal on the list can be oxidized by the ions of elements below it. The nitric acid solution contains $H^+(aq)$. This solution will oxidize and thus dissolve Zn(s) and Pb(s), which appear above $H_2(g)$ on the list. Platinum, Pt(s), is distinguished by its lack of reaction with nitric acid.

To distinguish between Zn and Pb, use a metal ion that occurs between them on the list. We have such an ion, $Ni^{2+}(aq)$ in the nickel nitrate solution. $Ni^{2+}(aq)$ will oxidize and thus dissolve Zn(s), which is above it on the list. $Ni^{2+}(aq)$ will not oxidize or dissolve Pb(s), which is below it on the list.

To summarize, Pt(s) will neither be oxidized by nor dissolve in any of the three available solutions. Pb(s) is oxidized by and will dissolve in the nitric acid solution, but not the nickel nitrate solution. Zn(s) is oxidized by and will dissolve in both nitric acid and nickel nitrate solutions.

4.9   In a redox reaction, one reactant loses electrons and a different reactant gains electrons; electrons are transferred. Acids ionize in aqueous solution to produce (donate) hydrogen ions (H^+, protons). Bases are substances that react with or accept protons (H^+). In an neutralization reaction, protons are transferred from an acid to a base. We characterize redox reactions by tracking electron transfer using oxidation numbers. We characterize neutralization reactions by tracking H^+ (proton) transfer via molecular formulas of reactants and products.

4.10   Concentration is a ratio of amount of solute to amount of solution or solvent. Thus there are two ways to double the concentration of a solution: double the amount of solute, keeping volume constant or reduce the volume of solution by half, keeping the amount of solute the same.

4.11   *Analyze/Plan.* The plot shows indicator color versus volume of standard solution added. Consider how the indicator changes color in a titration like the one in Figure 4.18, and relate this behavior to the shapes shown in the graph.

         Solve. The "green" data set is expected from a titration like the one in Figure 4.18. The indicator color remains constant until the reaction is very near the equivalence point. Within a very small volume of standard solution added, the indicator color changes rapidly. This behavior is shown in green. The red graph shows a constantly changing indicator color.

4.12   *Analyze/Plan.* The purpose of every titration is to determine the equivalence point. Based on the indicator behavior described in the exercise, decide how the amount of reactants and products in the titration beaker relate to the equivalence point. Given this reaction mixture, design an experiment to reach the equivalence point.

         Solve. This indicator is colorless in acid and blue in base. When it is added to the beaker, the solution is dark blue, so the solution is quite basic. This means that the amount of base added is already greater than the amount of acid initially present. To reach the equivalence point, more acid must be added to the titration beaker.

         First, record the volume of base added. Then, find a standard acid solution (an acid of very well-known concentration) or standardize an acid solution (probably HCl). Rinse and fill a clean buret with the standard acid. Carefully titrate the mixture in the beaker until the blue color fades and finally disappears. Record the volume of standard acid added. Subtract the amount of added acid from the total amount of base added. The remaining amount of base is the volume required to reach equivalence with the original acid sample. This procedure is called "back titration."

General Properties of Aqueous Solutions (Section 4.1)

4.14   (a)   False. Methanol is an organic alcohol, and the –OH group is not ionizable. The neutral methanol molecules in solution do not support the movement of charge. The solution does not conduct electricity.

(b) True. CH_3COOH is a weak electrolyte. When it dissolves in water, a small percentage of the molecules ionize to form H^+ and CH_3COO^- ions. The presence of a small concentration of ions produces a weakly conducting solution, and the specific presence of H^+ makes the solution acidic.

4.16   Anions are negatively charged. They will be attracted to and thus physically closer to the partially positive portion of the water molecule, which is the hydrogens.

4.18   (a) $MgI_2(aq) \rightarrow Mg^{2+}(aq) + 2\,I^-(aq)$

        (b) $K_2CO_3(aq) \rightarrow 2\,K^+(aq) + CO_3^{2-}(aq)$

        (c) $HClO_4(aq) \rightarrow H^+(aq) + ClO_4^-(aq)$

        (d) $NaCH_3COO(aq) \rightarrow Na^+(aq) + CH_3COO^-(aq)$

4.20   (a)   acetone (nonelectrolyte): $CH_3COCH_3(aq)$ molecules only; hypochlorous acid (weak electrolyte): $HClO(aq)$ molecules, $H^+(aq)$, $ClO^-(aq)$; ammonium chloride (strong electrolyte): $NH_4^+(aq)$, $Cl^-(aq)$

        (b) NH_4Cl, 0.2 mol solute particles; $HClO$, between 0.1 and 0.2 mol particles; CH_3COCH_3, 0.1 mol of solute particles

Precipitation Reactions (Section 4.2)

4.22   According to Table 4.1:

        (a)   A**gI**: insoluble (an exception to the generally soluble iodides)

        (b) $Na_2\mathbf{CO_3}$: soluble, Na^+ is an exception to insoluble carbonates

        (c) $Ba\mathbf{Cl_2}$: soluble

        (d) $Al\mathbf{(OH)_3}$: insoluble

        (e) $Zn\mathbf{(CH_3COO)_2}$: soluble

4.24   In each reaction, the precipitate is in bold type.

        (a)   No precipitate. [CH_3COO^- gains H^+ to form $CH_3COOH(aq)$, which is soluble.]

        (b) $Cu(NO_3)_2(aq) + 2\,KOH(aq) \rightarrow \mathbf{Cu(OH)_2(s)} + 2\,KNO_3(aq)$

        (c) $Na_2S(aq) + CdSO_4(aq) \rightarrow \mathbf{CdS(s)} + Na_2SO_4(aq)$

4.26   Spectator ions are those that do not change during reaction.

        (a) $2\,Cr^{3+}(aq) + 3\,CO_3^{2-}(aq) \rightarrow Cr_2(CO_3)_3(s)$; spectators: NH_4^+, SO_4^{2-}

        (b) $Ba^{2+}(aq) + SO_4^{2-}(aq) \rightarrow BaSO_4(s)$; spectators: K^+, NO_3^-

        (c) $Fe^{2+}(aq) + 2\,OH^-(aq) \rightarrow Fe(OH)_2(s)$; spectators: K^+, NO_3^-

4.28 Br^- and NO_3^- can be ruled out because the $BaBr_2$ is soluble and all NO_3^- salts are soluble. CO_3^{2-} forms insoluble salts with the three cations given; it must be the anion in question.

4.30   Consider all the possible combinations of Pb^{2+}(aq), Na^+(aq), Ca^{2+}(aq), CH_3COO^-(aq), S^{2-}(aq), and Cl^-(aq).   Which of these compounds are insoluble ionic compounds that will precipitate when the solutions are combined?

PbS(s) and $PbCl_2$(s) will precipitate.

Acids, Bases, and Neutralization Reactions (Section 4.3)

4.32 NH_3(aq) is a weak base, whereas KOH and $Ba(OH)_2$ are strong bases. NH_3(aq) is only slightly ionized, so even (a) 0.6 M NH_3 is less basic than (b) 0.150 M KOH. $Ba(OH)_2$ has twice as many OH^- per mole as KOH, so (c) 0.100 M $Ba(OH)_2$ is more basic than (b) 0.150 M KOH. The most basic solution is (c) 0.100 M $Ba(OH)_2$.

4.34   (a)   True. NH_3 produces OH^- in aqueous solution by reacting with H_2O (hydrolysis): NH_3(aq) + H_2O(l) $\rightleftharpoons$ NH_4^+(aq) + OH^- (aq). The OH^- causes the solution to be basic. NH_3(aq) attracts an H^+ from water, leaving OH^-(aq) in the solution.

          (b)   False. According to Table 4.2, HF is not one of the strong acids.

          (c)   True. H_2SO_4 is a **diprotic** acid; it has two ionizable hydrogens. The first hydrogen completely ionizes to form H^+ and HSO_4^-, but HSO_4^- only **partially** ionizes into H^+ and SO_4^{2-} (HSO_4^- is a weak electrolyte). Thus, an aqueous solution of H_2SO_4 contains a mixture of H^+, HSO_4^-, and SO_4^{2-}, with the concentration of HSO_4^- greater than the concentration of SO_4^{2-}.

4.36   Because the solution does conduct some electricity, but less than an equimolar NaCl solution (a strong electrolyte), the unknown solute must be a weak electrolyte. The weak electrolytes in the list of choices are NH_3 and H_3PO_3; because the solution is acidic, the unknown must be **H_3PO_3**.

4.38   (a) $LiClO_4$: strong   (b) HClO: weak   (c) $CH_3CH_2CH_2OH$: non

          (d) $HClO_3$: strong   (e) $CuSO_4$: strong   (f) $C_{12}H_{22}O_{11}$: non

4.40   (a)   2 CH_3COOH(aq) + $Ba(OH)_2$(aq) $\rightarrow$ $Ba(CH_3COO)_2$(aq) + 2 H_2O(l)

                 CH_3COOH(aq) + OH^-(aq) $\rightarrow$ CH_3COO^-(aq) + H_2O(l)

          (b) $Cr(OH)_3$(s) + 3 HNO_2(aq) $\rightarrow$ $Cr(NO_2)_3$(aq) + 3 H_2O(l)

                 $Cr(OH)_3$(s) + 3 HNO_2(aq) $\rightarrow$ 3 H_2O(l) + Cr^{3+}(aq) + 3 NO_2^-(aq)

          (c) HNO_3(aq) + NH_3(aq) $\rightarrow NH_4NO_3$(aq)

                 H^+(aq) + NH_3(aq) $\rightarrow NH_4^+$(aq)

4.42   (a)   FeO(s) + 2 H^+(aq) $\rightarrow H_2O$(l) + Fe^{2+}(aq)

          (b)   NiO(s) + 2 H^+(aq) $\rightarrow H_2O$(l) + Ni^{2+}(aq)

4.44   (a) K_2O(aq) + H_2O(l) $\rightarrow$ 2 KOH(aq), molecular; O^{2-}(aq) + H_2O(l) $\rightarrow$ 2 OH^-(aq), net ionic

          (b)   base (H^+ ion acceptor): O^{2-}(aq);

          (c)   acid (H^+ ion donor): H_2O(aq);

          (d)   spectator: K^+

Oxidation–Reduction Reactions (Section 4.4)

4.46   (a)   True. Oxidation is loss of electrons; it can occur in the presence of any electron acceptor, not just oxygen.

(b)   False. Oxidation and reduction can only occur together, not separately. When a substance is oxidized, it loses electrons, but free electrons do not exist under normal conditions. If electrons are lost by one substance they must be gained by another, and vice versa.

4.48   (a) $BaSO_4$; +6   (b) H_2SO_3; +4   (c) SrS; –2   (d) H_2S; –2

(e)   Sulfur is the third row of group 6A, the third column from the right on the periodic table. That is in region D on the designated chart.

(f)   Based on these compounds, the range of oxidation numbers for sulfur is +6 to –2. Sulfur and other nonmetals in region D can adopt both positive and negative oxidation numbers. This is also true for the metalloids in region C. These elements have properties of both metals and nonmetals, and can thus adopt both positive and negative oxidation numbers.

4.50   (a) +3   (b) +3   (c) –2   (d) –3   (e) +3   (f) +6

4.52   (a)   oxidation–reduction reaction; P is oxidized, Cl is reduced

(b)   oxidation–reduction reaction; K is oxidized, Br is reduced

(c)   oxidation–reduction reaction; C is oxidized, O is reduced

(d)   precipitation reaction

4.54   (a) $2 HCl(aq) + Ni(s) \rightarrow NiCl_2(aq) + H_2(g)$; $Ni(s) + 2 H^+(aq) \rightarrow Ni^{2+}(aq) + H_2(g)$

(b) $H_2SO_4(aq) + Fe(s) \rightarrow FeSO_4(aq) + H_2(g)$; $Fe(s) + 2 H^+(aq) \rightarrow Fe^{2+}(aq) + H_2(g)$

Products with the metal in a higher oxidation state are possible, depending on reaction conditions and acid concentration.

(c) $2 HBr(aq) + Mg(s) \rightarrow MgBr_2(aq) + H_2(g)$; $Mg(s) + 2 H^+(aq) \rightarrow Mg^{2+}(aq) + H_2(g)$

(d) $2 CH_3COOH(aq) + Zn(s) \rightarrow Zn(CH_3COO)_2(aq) + H_2(g)$;

$Zn(s) + 2 CH_3COOH(aq) \rightarrow Zn^{2+}(aq) + 2 CH_3COO^-(aq) + H_2(g)$

4.56   (a) $Ni(s) + Cu(NO_3)_2(aq) \rightarrow Ni(NO_3)_2(aq) + Cu(s)$

(b) $Zn(NO_3)_2(aq) + MgSO_4(aq) \rightarrow NR$

(c) $Au(s) + HCl(aq) \rightarrow NR$

(d) $2 Cr(s) + 3 CoCl_2(aq) \rightarrow 2 CrCl_3(aq) + 3 Co(s)$

(e) $H_2(g) + 2 AgNO_3(aq) \rightarrow 2 Ag(s) + 2 HNO_3(aq)$

4.58 $Br_2 + 2 NaI \rightarrow 2 NaBr + I_2$ indicates that Br_2 is more easily reduced than I_2.

$Cl_2 + 2 NaBr \rightarrow 2 NaCl + Br_2$ shows that Cl_2 is more easily reduced than Br_2.

The order for ease of reduction is $Cl_2 > Br_2 > I_2$. Conversely, the order for ease of oxidation is $I^- > Br^- > Cl^-$.

(a) From the information above, the halogen I_2 is most stable (less likely to react) when mixed with other halides, X^-.

(b) $Cl_2 + 2\,KI \rightarrow 2\,KCl + I_2$

(c) $Br_2 + LiCl \rightarrow$ no reaction

Concentrations of Solutions (Section 4.5)

4.60 (a) No, he is not correct. The concentration of the solution is not 1.5 M.

(b) Molarity is the *ratio* of moles solute to liters solution. The correct molarity is

1.5 mol NaOH/1.5 L solution = 1.0 M NaOH

4.62 (a) $M = \dfrac{\text{mol solute}}{\text{L solution}}$; $\dfrac{12.5\text{ g Na}_2\text{CrO}_4}{0.750\text{ L}} \times \dfrac{1\text{ mol Na}_2\text{CrO}_4}{161.97\text{ g Na}_2\text{CrO}_4} = 0.103\ M\text{ Na}_2\text{CrO}_4$

(b) mol = M × L; $\dfrac{0.112\text{ mol KBr}}{1\text{ L}} \times 0.150\text{ L} = 1.68 \times 10^{-2}\text{ mol KBr}$

(c) L = $\dfrac{\text{mol}}{M}$; $\dfrac{0.150\text{ mol HCl}}{6.1\text{ mol HCl/L}} = 2.5 \times 10^{-2}\text{ L or 25 mL}$

4.64 Calculate the mol of Na^+ at the two concentrations; the difference is the mol NaCl required to increase the Na^+ concentration to the desired level.

$\dfrac{0.118\text{ mol}}{\text{L}} \times 4.6\text{ L} = 0.5428 = 0.54\text{ mol Na}^+$

$\dfrac{0.138\text{ mol}}{\text{L}} \times 4.6\text{ L} = 0.6348 = 0.63\text{ mol Na}^+$

$(0.6348 - 0.5428) = 0.092 = 0.09$ mol NaCl (2 decimal places and 1 sig fig)

$0.092\text{ mol NaCl} \times \dfrac{58.5\text{ g NaCl}}{\text{mol}} = 5.38 = 5\text{ g NaCl}$

4.66 *Analyze.* Given: BAC (definition from Exercise 4.65), vol of blood. Find: mass alcohol in bloodstream.

Plan. Change BAC (g/100 mL) to (g/L), then times vol of blood in L.

Solve. BAC = 0.10 g/100 mL

$\dfrac{0.10\text{ g alcohol}}{100\text{ mL blood}} \times \dfrac{1000\text{ mL}}{1\text{ L}} \times 5.0\text{ L blood} = 5.0\text{ g alcohol}$

4.68 $M = \dfrac{\text{mol}}{\text{L}}$; mol $= \dfrac{\text{g}}{\text{MM}}$ (MM is the symbol for molar mass in this manual.)

$M = \dfrac{\text{mol solute}}{\text{L solution}}$; $\dfrac{124\text{ mg C}_6\text{H}_8\text{O}_6}{0.2366\text{ L}} \times \dfrac{1\text{ g}}{1000\text{ mg}} \times \dfrac{1\text{ mol C}_6\text{H}_8\text{O}_6}{176.12\text{ g C}_6\text{H}_8\text{O}_6} = 2.98 \times 10^{-3}\ M\text{ C}_6\text{H}_8\text{O}_6$

4.70 (a) 0.10 M BaI$_2$ = 0.2 M I$^-$; 0.25 M KI = 0.25 M I$^-$

0.25 M KI has the higher I$^-$ concentration.

(b) 0.10 M KI = 0.1 M I$^-$; 0.040 M ZnI$_2$ = 0.080 M I$^-$; 0.10 M KI has a higher I$^-$ concentration than 0.040 M ZnI$_2$. Total volume does not affect concentration.

(c) $3.2\ M$ HI $= 3.2\ M$ I⁻

$$145\ \text{g NaI} \times \frac{1\ \text{mol NaI}}{149.9\ \text{g NaI}} \times \frac{1}{0.150\ \text{L}} = 6.45\ M\ \text{NaI} = 6.45\ M\ \text{I}^-$$

The NaI solution has the higher I⁻ concentration.

4.72   (a)   *Plan.* These two solutions have common ions. Find the ion concentration resulting from each solution, then add.

Solve. total volume $= 42.0\ \text{mL} + 37.6\ \text{mL} = 79.6\ \text{mL}$

$$\frac{0.170\ M\ \text{NaOH} \times 42.0\ \text{mL}}{79.6\ \text{mL}} = 0.08970 = 0.0897\ M\ \text{NaOH};$$

$0.0897\ M$ Na⁺, $0.0897\ M$ OH⁻

$$\frac{0.400\ M\ \text{NaOH} \times 37.6\ \text{mL}}{79.6\ \text{mL}} = 0.18894 = 0.189\ M\ \text{NaOH};$$

$0.189\ M$ Na⁺, $0.189\ M$ OH⁻

M Na⁺ $= 0.08970\ M + 0.18894\ M = 0.27864 = 0.2786\ M$ Na⁺

M OH⁻ $= M$ Na⁺ $= 0.2786\ M$ OH⁻

(b)   *Plan.* No common ions; just dilution.

Solve. $44.0\ \text{mL} + 25.0\ \text{mL} = 69.0\ \text{mL}$

$$\frac{0.100\ M\ \text{Na}_2\text{SO}_4 \times 44.0\ \text{mL}}{69.0\ \text{mL}} = 0.06377 = 0.0638\ M\ \text{Na}_2\text{SO}_4$$

$2 \times (0.06377\ M) = 0.1275 = 0.128\ M$ Na⁺; $0.0638\ M\ \text{SO}_4^{2-}$

$$\frac{0.150\ M\ \text{KCl} \times 25.0\ \text{mL}}{69.0\ \text{mL}} = 0.054348 = 0.0543\ M\ \text{KCl}$$

$0.0543\ M$ K⁺, $0.0543\ M$ Cl⁻

(c)   *Plan.* Calculate concentration of K⁺ and Cl⁻ due to the added solid. Then sum to get total concentration of Cl⁻.

Solve. $\dfrac{3.60\ \text{g KCl}}{75.0\ \text{mL so ln}} \times \dfrac{1\ \text{mol KCl}}{74.55\ \text{g KCl}} \times \dfrac{1000\ \text{mL}}{1\ \text{L}} = 0.6439 = 0.644\ M\ \text{KCl}$

$0.250\ M$ CaCl₂; $2(0.250\ M) = 0.500\ M$ Cl⁻

total Cl⁻ $= 0.644\ M + 0.500\ M = 1.144\ M$ Cl⁻, $0.644\ M$ K⁺, $0.250\ M$ Ca²⁺

4.74   (a) $V_1 = M_2 V_2/M_1;\quad \dfrac{0.500\ M\ \text{HNO}_3 \times 0.110\ \text{L}}{6.0\ M\ \text{HNO}_3} = 0.00917\ \text{L} = 9.2\ \text{mL}\ 6.0\ M\ \text{HNO}_3$

(b) $M_2 = M_1 V_1/V_2;\quad \dfrac{6.0\ M\ \text{HNO}_3 \times 10.0\ \text{mL}}{250\ \text{mL}} = 0.240\ M\ \text{HNO}_3$

4.76   (a)   The amount of AgNO₃ needed is:

$0.150\ M \times 0.2000\ \text{L} = 0.03000 = 0.0300\ \text{mol AgNO}_3$

$$0.03000\ \text{mol AgNO}_3 \times \frac{169.88\ \text{g AgNO}_3}{1\ \text{mol AgNO}_3} = 5.0964 = 5.10\ \text{g AgNO}_3$$

(b)   Dilute the 3.6 M HNO_3 to prepare 100 mL of 0.50 M HNO_3. To determine the volume of 3.6 M HNO_3 needed, calculate the moles HNO_3 present in 100 mL of 0.50 M HNO_3 and then the volume of 3.6 M solution that contains this number of moles.

0.100 L × 0.50 M = 0.050 mol HNO_3 needed;

$$L = \frac{mol}{M}; \ L \text{ of } 3.6 \ M \ HNO_3 = \frac{0.050 \text{ mol needed}}{3.6 \ M} = 0.01389 \ L = 14 \ mL$$

The dilution requires 14 mL of the 3.6 M HNO_3 and enough water to produce 100 mL of the 0.50 M solution. Note that this may not be exactly 86 mL of water, because solution volumes are not necessarily additive.

4.78 $50.000 \text{ mL glycerol} \times \dfrac{1.2656 \text{ g glycerol}}{1 \text{ mL glycerol}} = 63.280 \text{ g glycerol}$

$63.280 \text{ g } C_3H_8O_3 \times \dfrac{1 \text{ mol } C_3H_8O_3}{92.094 \text{ g } C_3H_8O_3} = 0.687124 = 0.68712 \text{ mol } C_3H_8O_3$

$M = \dfrac{0.687124 \text{ mol } C_3H_8O_3}{0.25000 \text{ L solution}} = 2.7485 \ M \ C_3H_8O_3$

Solution Stoichiometry and Chemcial Analysis (Section 4.6)

4.80   *Plan.* $M \times L$ = mol $Cd(NO_3)_2$; balanced equation → mol ratio → mol NaOH → g NaOH

Solve. $\dfrac{0.500 \text{ mol } Cd(NO_3)_2}{1 \ L} \times 0.0350 \ L = 0.0175 \text{ mol } Cd(NO_3)_2$

$Cd(NO_3)_2(aq) + 2 \ NaOH(aq) \rightarrow Cd(OH)_2(s) + 2 \ NaNO_3(aq)$

$0.0175 \text{ mol } Cd(NO_3)_2 \times \dfrac{2 \text{ mol NaOH}}{1 \text{ mol } Cd(NO_3)_2} \times \dfrac{40.00 \text{ g NaOH}}{1 \text{ mol NaOH}} = 1.40 \text{ g NaOH}$

4.82   (a) $2 \ HCl(aq) + Ba(OH)_2(aq) \rightarrow BaCl_2(aq) + 2 \ H_2O(l)$

$\dfrac{0.101 \text{ mol } Ba(OH)_2}{1 \ L \ Ba(OH)_2} \times 0.0500 \ L \ Ba(OH)_2 \times \dfrac{2 \text{ mol HCl}}{1 \text{ mol } Ba(OH)_2} \times \dfrac{1 \ L \ HCl}{0.120 \text{ mol HCl}}$

$= 0.0842 \ L \text{ or } 84.2 \ mL \ HCl \text{ soln}$

(b) $H_2SO_4(aq) + 2 \ NaOH(aq) \rightarrow Na_2SO_4(aq) + 2 \ H_2O(l)$

$0.200 \text{ g NaOH} \times \dfrac{1 \text{ mol NaOH}}{40.00 \text{ g NaOH}} \times \dfrac{1 \text{ mol } H_2SO_4}{2 \text{ mol NaOH}} \times \dfrac{1 \ L \ H_2SO_4}{0.125 \text{ mol } H_2SO_4}$

$= 0.0200 \ L \text{ or } 20.0 \ mL \ H_2SO_4 \text{ soln}$

(c) $BaCl_2(aq) + Na_2SO_4(aq) \rightarrow BaSO_4(s) + 2 \ NaCl(aq)$

$752 \text{ mg} = 0.752 \text{ g } Na_2SO_4 \times \dfrac{1 \text{ mol } Na_2SO_4}{142.1 \text{ g } Na_2SO_4} \times \dfrac{1 \text{ mol } BaCl_2}{1 \text{ mol } Na_2SO_4} \times \dfrac{1}{0.0558 \ L}$

$= 0.0948 \ M \ BaCl_2$

(d) $2\,HCl(aq) + Ca(OH)_2(aq) \rightarrow CaCl_2(aq) + 2\,H_2O(l)$

$$0.0427\,L\,HCl \times \frac{0.208\,mol\,HCl}{1\,L\,HCl} \times \frac{1\,mol\,Ca(OH)_2}{2\,mol\,HCl} \times \frac{74.10\,g\,Ca(OH)_2}{1\,mol\,Ca(OH)_2} = 0.329\,g\,Ca(OH)_2$$

4.84   See Exercise 4.81(a) for a more detailed approach.

$$\frac{0.115\,mol\,NaOH}{1\,L} \times 0.0425\,L \times \frac{1\,mol\,CH_3COOH}{1\,mol\,NaOH} \times \frac{60.05\,g\,CH_3COOH}{1\,mol\,CH_3COOH}$$
$$= 0.29349 = 0.293\,g\,CH_3COOH \text{ in } 3.45\,mL$$

$$1.00\,qt\,vinegar \times \frac{1\,L}{1.057\,qt} \times \frac{1000\,mL}{1\,L} \times \frac{0.29349\,g\,CH_3COOH}{3.45\,mL\,vinegar} = 80.5\,g\,CH_3COOH/qt$$

4.86   *Analyze/Plan.* Follow the logic in Exercise 4.85. The unknown is a group 2A metal hydroxide, general formula $M(OH)_2$. Two mol HCL are required to neutralize 1 mol $M(OH)_2$. *Solve.*

(a) $$\frac{2.50\,mol\,HCl}{1\,L} \times 0.0569\,L \times \frac{1\,mol\,M(OH)_2}{2\,mol\,HCl} = 0.071125 = 0.0711\,mol\,M(OH)_2$$

$$MM\text{ of }M(OH)_2 = \frac{8.65\,g\,M(OH)_2}{0.071125\,mol\,M(OH)_2} = 121.62 = 122\,g/mol$$

(b)   MM of group 2A metal = MM of $M(OH)_2 - 2(17.01\,g)$ *Solve.*

MM of group 2A metal = $(121.62\,g/mol - 34.02\,g/mol) = 87.60 = 88\,g/mol$

The experimental molar mass most closely fits that of Sr^{2+}, $87.62\,g/mol$

Check. The experimental molar mass matches one of the group 2A metals.

4.88   (a) $2\,HNO_3(aq) + Sr(OH)_2(s) \rightarrow Sr(NO_3)_2(aq) + 2\,H_2O(l)$

(b)   Determine the limiting reactant, then the identity and concentration of ions remaining in solution. Assume that the $H_2O(l)$ produced by the reaction does **not** increase the total solution volume.

$$15.0\,g\,Sr(OH)_2 \times \frac{1\,mol\,Sr(OH)_2}{121.64\,g\,Sr(OH)_2} = 0.1233 = 0.123\,mol\,Sr(OH)_2$$

$mol\,OH^- = 2(0.1233)\,mol\,Sr(OH)_2 = 0.2466 = 0.247\,mol\,OH^-$

$0.200\,M\,HNO_3 \times 0.0550\,L\,HNO_3 = 0.0110\,mol\,HNO_3$.

Two mol HNO_3 react with one mol $Sr(OH)_2$, so HNO_3 is the limiting reactant. No excess H^+ remains in solution. The remaining ions are OH^- (excess reactant), Sr^{2+}, and NO_3^- (spectators).

OH^-: $0.2466\,mol\,OH^-$ initial $- 0.0110\,mol\,OH^-$ react $= 0.2356 = 0.236\,mol\,OH^-$ remain

$0.2356\,mol\,OH^-/0.0550\,L\,soln = 4.28\,M\,OH^-(aq)$

Sr^{2+}: $0.123\,mol\,Sr^{2+}/0.0550\,L\,soln = 2.24\,M\,Sr^{2+}(aq)$

NO_3^-: $0.0110\,mol\,NO_3^-/0.0550\,L = 0.200\,M\,NO_3^-(aq)$

(c)   The resulting solution is **basic** because of the large excess of $OH^-(aq)$.

4.90   *Plan.* $CaCO_3(s) + 2\ HCl(aq) \rightarrow CaCl_2(aq) + H_2O(l) + CO_2(g)$

$HCl(aq) + NaOH(aq) \rightarrow NaCl(aq) + H_2O(l)$

total mol HCl – excess mol HCl = mol HCl reacted; mol $CaCO_3$ = (mol HCl)/2;

g $CaCO_3$ = mol $CaCO_3$ × molar mass; mass % = (g $CaCO_3$/g sample) × 100

Solve:

$$\frac{1.035\ \text{mol HCl}}{1\ \text{L soln}} \times 0.03000\ \text{L} = 0.031050 = 0.03105\ \text{mol HCl total}$$

$$\frac{1.010\ \text{mol NaOH}}{1\ \text{L soln}} \times 0.01156\ \text{L} = 0.011676 = 0.01168\ \text{mol HCl excess}$$

0.031050 total – 0.011676 excess = 0.019374 = 0.01937 mol HCl reacted

$$0.019374\ \text{mol HCl} \times \frac{1\ \text{mol CaCO}_3}{2\ \text{mol HCl}} \times \frac{100.09\ \text{g CaCO}_3}{1\ \text{mol CaCO}_3} = 0.96959 = 0.9696\ \text{g CaCO}_3$$

$$\text{mass \% CaCO}_3 = \frac{\text{g CaCO}_3}{\text{g rock}} \times 100 = \frac{0.96959}{1.248} \times 100 = 77.69\%\ \text{CaCO}_3$$

Additional Exercises

4.92   (a)   The precipitate is CdS(s).

(b) $Na^+(aq)$ and $NO_3^-(aq)$ are spectator ions and remain in solution. Any excess reactant ions also remain in solution.

(c) $Cd^{2+}(aq) + S^{2-}(aq) \rightarrow CdS(s)$.

(d)   This is not a redox reaction. (It is a metathesis reaction.)

4.93   The two precipitates formed are AgCl(s) and $SrSO_4(s)$. Because no precipitate forms on addition of hydroxide ion to the remaining solution, the other two possibilities, Ni^{2+} and Mn^{2+}, are absent.

4.94   (a,b) Expt. 1 No reaction

Expt. 2 $2\ Ag^+(aq) + CrO_4^{2-}(aq) \rightarrow Ag_2CrO_4(s)$ red precipitate

Expt. 3 $Ca^{2+}(aq) + CrO_4^{2-}(aq) \rightarrow CaCrO_4(s)$ yellow precipitate

Expt. 4 $2\ Ag^+(aq) + C_2O_4^{2-}(aq) \rightarrow Ag_2C_2O_4(s)$ white precipitate

Expt. 5 $Ca^{2+}(aq) + C_2O_4^{2-}(aq) \rightarrow CaC_2O_4(s)$ white precipitate

Expt. 6 $Ag^+(aq) + Cl^-(aq) \rightarrow AgCl(s)$ white precipitate

4.96 $4\ NH_3(g) + 5\ O_2(g) \rightarrow 4\ NO(g) + 6\ H_2O(g)$.
         N = –3   O = 0   N = +2   O = –2

(a)   redox reaction (b)   N is oxidized, O is reduced

$2\ NO(g) + O_2(g) \rightarrow 2\ NO_2(g)$.
         N = +2   O = 0   N = +4, O = –2

(a)   redox reaction (b)   N is oxidized, O is reduced

$3 NO_2(g) + H_2O(l) \rightarrow 2 HNO_3(aq) + NO(g).$

$N = +4 \qquad\qquad N = +5 \qquad N = +2$

(a)   redox reaction

(b)   N is oxidized ($NO_2 \rightarrow HNO_3$), N is reduced ($NO_2 \rightarrow NO$). A reaction where the same element is both oxidized and reduced is called disproportionation.

(c) 0.150 M $HNO_3 \times 1000.0$ L $= 150.00 = 150$ mol HNO_3 required

$$150 \text{ mol } HNO_3 \times \frac{3 \text{ mol } NO_2}{2 \text{ mol } HNO_3} \times \frac{2 \text{ mol } NO}{2 \text{ mol } NO_2} \times \frac{4 \text{ mol } NH_3}{4 \text{ mol } NO} = 225 \text{ mol } NH_3$$

$$225 \text{ mol } HNO_3 \times \frac{17.0305 \text{ g } NH_3}{1 \text{ mol } NH_3} = 3831.86 = 3.83 \times 10^3 \text{ g } NH_3$$

4.97   A metal on Table 4.5 can be oxidized by ions of the elements below it. Or, a metal on the table is able to displace the (metal) cations below it from their compounds.

(a) $Zn(s) + 2 HNO_3(aq) \rightarrow Zn(NO_3)_2(aq) + H_2(g)$

The substance that inflates the balloon is $H_2(g)$. Of Zn, Cu, and Hg, only Zn is above H on Table 4.5, so only Zn can displace H from HCl.

(b) $35.0 \text{ g } Zn \times \dfrac{1 \text{ mol } Zn}{65.39 \text{ g } Zn} = 0.53525 = 0.535 \text{ mol } Zn$

$\dfrac{3.00 \text{ mol } HNO_3}{L} \times 0.150 \text{ L} = 0.450 \text{ mol } HNO_3$

One mol Zn reacts with 2 mol HNO_3, so HNO_3 is the limiting reactant.

$0.450 \text{ mol } HNO_3 \times \dfrac{1 \text{ mol } H_2}{2 \text{ mol } HNO_3} \times \dfrac{2.016 \text{ g } H_2}{1 \text{ mol } H_2} = 0.4536 = 0.454 \text{ g } H_2$

(c)   Both Zn and Cu are above Ag on Table 4.5, so both Zn and Cu can displace Ag from $AgNO_3$. Note that $H_2(g)$ would also displace Ag, but $H^+(aq)$ will not.

$Zn(s) + 2 AgNO_3(aq) \rightarrow 2 Ag(s) + Zn(NO_3)_2(aq)$; $Zn^{2+}(aq)$ and $NO_3^-(aq)$ remain

$Cu(s) + 2 AgNO_3(aq) \rightarrow 2 Ag(s) + Cu(NO_3)_2(aq)$; $Cu^{2+}(aq)$ and $NO_3^-(aq)$ remain

(d) $0.535 \text{ mol } Zn$ [from part (b)]; $42.0 \text{ g } Zn \times \dfrac{1 \text{ mol } Zn}{63.546 \text{ g } Cu} = 0.6609 = 0.661 \text{ mol } Cu$

$\dfrac{0.750 \text{ mol } AgNO_3}{L} \times 0.150 \text{ L} = 0.1125 = 0.113 \text{ mol } AgNO_3$

One mol metal reacts with 2 mol $AgNO_3$, so $AgNO_3$ is the limiting reactant for both Zn and Cu.

$0.1125 \text{ mol } AgNO_3 \times \dfrac{107.87 \text{ g } Ag}{1 \text{ mol } Ag} = 12.135 = 12.1 \text{ g } Ag$ in both reactions

4.98   (a)   A : La_2O_3 Metals often react with the oxygen in air to produce metal oxides.

B : $La(OH)_3$ When metals react with water (HOH) to form H_2, OH^- remains.

C : $LaCl_3$ Most chlorides are soluble.

D : $La_2(SO_4)_3$ Sulfuric acid provides SO_4^{2-} ions.

(b) $4\,La(s) + 3\,O_2(g) \rightarrow 2\,La_2O_3(s)$

$2\,La(s) + 6\,H_2O(l) \rightarrow 2\,La(OH)_3(s) + 3\,H_2(g)$

(There are no spectator ions in either of these reactions.)

molecular: $La_2O_3(s) + 6\,HCl(aq) \rightarrow 2\,LaCl_3(aq) + 3\,H_2O(l)$

net ionic: $La_2O_3(s) + 6\,H^+(aq) \rightarrow 2\,La^{3+}(aq) + 3\,H_2O(l)$

molecular: $La(OH)_3(s) + 3\,HCl(aq) \rightarrow LaCl_3(aq) + 3\,H_2O(l)$

net ionic: $La(OH)_3(s) + 3\,H^+(aq) \rightarrow La^{3+}(aq) + 3\,H_2O(l)$

molecular: $2\,LaCl_3(aq) + 3\,H_2SO_4(aq) \rightarrow La_2(SO_4)_3(s) + 6\,HCl(aq)$

net ionic: $2\,La^{3+}(aq) + 3\,SO_4^{2-}(aq) \rightarrow La_2(SO_4)_3(s)$

(c)   La metal is oxidized by water to produce $H_2(g)$, so La is definitely above H_2 on the activity series. In fact, because an acid is not required to oxidize La, it is probably one of the more active metals.

4.100   (a) $\dfrac{50\ pg}{1\ mL} \times \dfrac{1\times10^{-12}\,g}{1\ pg} \times \dfrac{1\times10^3\ mL}{L} \times \dfrac{1\ mol\ Na}{23.0\ g\ Na} = 2.17\times10^{-9} = 2.2\times10^{-9}\ M\ Na^+$

(b) $\dfrac{2.17\times10^{-9}\ mol\ Na^+}{1\ L\ soln} \times \dfrac{1\ L}{1\times10^3\ cm^3} \times \dfrac{6.022\times10^{23}\ Na^+}{1\ mol\ Na^+} = 1.3\times10^{12}\ Na^+\ ions/\ cm^3$

(c) $\dfrac{50\ pg}{1\ mL} \times \dfrac{1\times10^{-12}\,g}{1\ pg} \times \dfrac{1\times10^3\ mL}{L} \times 1\times10^3\,L = 5.0\times10^{-5}\,g\ Na^+\ (50\ \mu g\ Na^+)$

4.101   (a) Na^+ must replace the total positive (+) charge due to Ca^{2+} and Mg^{2+}. Think of this as moles of charge rather than moles of particles.

$\dfrac{0.020\ mol\ Ca^{2+}}{1\ L\ water} \times 1.5\times10^3\ L \times \dfrac{2\ mol\ (+)\ charge}{1\ mol\ Ca^{2+}} = 60\ mol\ of\ (+)\ charge$

$\dfrac{0.0040\ mol\ Mg^{2+}}{1\ L\ water} \times 1.5\times10^3\ L \times \dfrac{2\ mol\ (+)\ charge}{1\ mol\ Mg^{2+}} = 12\ mol\ of\ (+)\ charge$

72 moles of (+) charge must be replaced; 72 mol Na^+ are needed.

(b) $72\ mol\ Na^+ \times \dfrac{1\ mol\ Na^+}{1\ mol\ NaCl} \times \dfrac{58.44\ g\ NaCl}{1\ mol\ NaCl} = 4208\ g = 4.2\times10^3\ g\ NaCl$

4.102 $H_2C_4H_4O_6 + 2\,OH^-(aq) \rightarrow C_4H_4O_6^{2-}(aq) + 2\,H_2O(l)$

$0.02465\ L\ NaOH\ soln \times \dfrac{0.2500\ mol\ NaOH}{1\ L} \times \dfrac{1\ mol\ H_2C_4H_4O_6}{2\ mol\ NaOH} \times \dfrac{1}{0.0500\ L\ H_2C_4H_4O_6}$

$= 0.06163\ M\ H_2C_4H_4O_6\ soln$

4.104 mol OH^- from NaOH(aq) + mol OH^- from $Zn(OH)_2$(s) = mol H^+ from HBr

mol H^+ = M HBr × L HBr = 0.500 M HBr × 0.350 L HBr = 0.175 mol H^+

mol OH^- from NaOH = M NaOH × L NaOH = 0.500 M NaOH × 0.0885 L NaOH

$$= 0.04425 = 0.0443 \text{ mol } OH^-$$

mol OH^- from $Zn(OH)_2$(s) = 0.175 mol H^+ − 0.04425 mol OH^- from NaOH = 0.13075

$$= 0.131 \text{ mol } OH^- \text{ from } Zn(OH)_2$$

$$0.13075 \text{ mol } OH^- \times \frac{1 \text{ mol } Zn(OH)_2}{2 \text{ mol } OH^-} \times \frac{99.41 \text{ g } Zn(OH)_2}{1 \text{ mol } Zn(OH)_2} = 6.50 \text{ g } Zn(OH)_2$$

Integrative Exercises

4.105 (a) A metal can be oxidized by ions of the elements below it on Table 4.5. Of the three substances given, K^+(aq) is above Mg(s), but Ag^+(aq) is below it, so $AgNO_3$(aq) will react with Mg(s).

(b) $Mg(s) + 2\,Ag^+(aq) \rightarrow Mg^{2+}(aq) + 2\,Ag(s)$

(c) g Mg → mol Mg → [via mole ratio] mol Ag^+ → [via (mol/M)] vol $AgNO_3$(aq)

$$5.00 \text{ g Mg} \times \frac{1 \text{ mol Mg}}{24.305 \text{ g Mg}} = 0.2057 = 0.206 \text{ mol Mg}$$

$$0.2057 \text{ mol Mg} \times \frac{2 \text{ mol } AgNO_3}{1 \text{ mol Mg}} \times \frac{1.00 \text{ L}}{2.00 \text{ mol } AgNO_3} = 0.2057 = 0.206 \text{ L } AgNO_3(aq)$$

(d) $\dfrac{0.2057 \text{ mol } Mg^{2+}}{0.2057 \text{ L soln}} = 1.00\ M\ Mg^{2+}(aq)$

4.106 (a) At the equivalence point of a titration, mol NaOH added = mol H^+ present

$$M_{NaOH} \times L_{NaOH} = \frac{\text{g acid}}{\text{MM acid}} \text{ (for an acid with 1 acidic hydrogen)}$$

$$\text{MM acid} = \frac{\text{g acid}}{M_{NaOH} \times L_{NaOH}} = \frac{0.2053 \text{ g}}{0.1008\ M \times 0.0150 \text{ L}} = 136 \text{ g/mol}$$

(b) Assume 100 g of acid.

$$70.6 \text{ g C} \times \frac{1 \text{ mol C}}{12.01 \text{ g C}} = 5.88 \text{ mol C}; \; 5.88 / 1.47 \approx 4$$

$$5.89 \text{ g H} \times \frac{1 \text{ mol H}}{1.008 \text{ g H}} = 5.84 \text{ mol H}; \; 5.84 / 1.47 \approx 4$$

$$23.5 \text{ g O} \times \frac{1 \text{ mol O}}{16.00 \text{ g O}} = 1.47 \text{ mol O}; \; 1.47 / 1.47 = 1$$

The empirical formula is C_4H_4O.

$$\frac{\text{MM}}{\text{FW}} = \frac{136}{68.1} = 2; \text{ the molecular formula is } 2 \times \text{the empirical formula.}$$

The molecular formula is $C_8H_8O_2$.

4.107 $Ba^{2+}(aq) + SO_4^{2-}(aq) \rightarrow BaSO_4(s)$

$$0.2815 \text{ g } BaSO_4 \times \frac{137.3 \text{ g Ba}}{233.4 \text{ g } BaSO_4} = 0.16560 = 0.1656 \text{ g Ba}$$

$$\text{mass \%} = \frac{\text{g Ba}}{\text{g sample}} \times 100 = \frac{0.16560 \text{ g Ba}}{3.455 \text{ g sample}} \times 100 = 4.793 \text{ \% Ba}$$

4.108   *Plan.* Abbreviate the commercial aqueous ammonia solution as NH_3 soln. Abbreviate citric acid as H_3Cit. Write the balanced equation.

vol NH_3 soln $\xrightarrow{\text{density}}$ mass NH_3 soln $\xrightarrow{\text{mass \%}}$ mass NH_3

mass $NH_3 \rightarrow$ mol $NH_3 \rightarrow$ mol $H_3Cit \rightarrow$ mass H_3Cit

Solve. $H_3Cit(aq) + 3 NH_3(aq) \rightarrow 3 NH_4^+(aq) + Cit^{3-}(aq)$

$$3.43 \times 10^4 \text{ gal } NH_3 \text{ soln} \times \frac{3.785 \text{ L}}{\text{gal}} \times \frac{1000 \text{ mL}}{\text{L}} \times \frac{0.88 \text{ g soln}}{\text{mL soln}} = 1.14246 \times 10^8 = 1.1 \times 10^8 \text{ g } NH_3 \text{ soln}$$

$$1.14246 \times 10^8 \text{ g } NH_3 \text{ soln} \times \frac{30 \text{ g } NH_3}{100 \text{ g } NH_3 \text{ soln}} = 3.4274 \times 10^7 = 3.4 \times 10^7 \text{ g } NH_3$$

$$3.4274 \times 10^7 \text{ g } NH_3 \times \frac{1 \text{ mol } NH_3}{17.0305 \text{ g } NH_3} \times \frac{1 \text{ mol } H_3Cit}{3 \text{ mol } NH_3} \times \frac{192.124 \text{ g } H_3Cit}{1 \text{ mol } H_3Cit} = 1.3 \times 10^8 \text{ g } H_3Cit$$

4.110   Write a balanced equation for the reaction between PbO and Na_2S. Calculate the mass of PbO in the goblet using the mass% PbO. Solve the stoichiometry problem to calculate the mass of Na_2S required to react with this mass of PbO.

$$PbO + Na_2S \rightarrow PbS + Na_2O$$

$$286 \text{ g goblet} \times \frac{27 \text{ g PbO}}{100 \text{ g goblet}} = 77.22 = 77 \text{ g PbO}$$

$$77.22 \text{ g PbO} \times \frac{1 \text{ mol PbO}}{223.3 \text{ g PbO}} \times \frac{1 \text{ mol } Na_2S}{1 \text{ mol PbO}} \times \frac{78.045 \text{ g } Na_2S}{1 \text{ mol } Na_2S} = 27 \text{ g } Na_2S$$

4.111   *Plan.* Calculate the mass of gold that has a value of $5000. Find the volume of seawater that contains this mass of gold. One liter of seawater contains 100 fmol of gold and $1 \text{ fmol} = 1 \times 10^{-15}$. If the process is 50% efficient, we need twice that volume of seawater. Assume 100 fmol has 3 sig figs and the volume of seawater also has 3 sig figs.

Solve. $\$5000 \times \dfrac{1 \text{ tr oz}}{\$1764.20} \times \dfrac{31.103 \text{ g}}{1 \text{ tr oz}} = 88.15044 = 88.15 \text{ g Au}$

$$88.15044 \text{ g Au} \times \frac{1 \text{ mol Au}}{196.967 \text{ g Au}} \times \frac{1 \text{ fmol}}{1 \times 10^{-15} \text{ mol}} \times \frac{1 \text{ L seawater}}{100 \text{ fmol}} = 4.4754 \times 10^{12} = 4.48 \times 10^{12} \text{ L}$$

If the process is 50% efficient, we need twice that volume of seawater, or 8.95×10^{12} L seawater. (for 1 sig fig, 9×10^{12} L seawater)

4.112 $Ag^+(aq) + Cl^-(aq) \rightarrow AgCl(s)$

$$\frac{0.2997 \text{ mol Ag}^+}{1 \text{ L}} \times 0.04258 \text{ L} \times \frac{1 \text{ mol Cl}^-}{1 \text{ mol Ag}^+} \times \frac{35.453 \text{ g Cl}^-}{1 \text{ mol Cl}^-} = 0.45242 = 0.4524 \text{ g Cl}^-$$

$$25.00 \text{ mL seawater} \times \frac{1.025 \text{ g}}{\text{mL}} = 25.625 = 25.63 \text{ g seawater}$$

$$\text{mass \% Cl}^- = \frac{0.45242 \text{ g Cl}^-}{25.625 \text{ g seawater}} \times 100 = 1.766\% \text{ Cl}^-$$

4.114   *Analyze.* Given 10 ppb AsO_4^{3-}, find mass Na_3AsO_4 in 1.00 L of drinking water.

Plan. Use the definition of ppb to calculate g AsO_4^{3-} in 1.0 L of water. Convert g $AsO_4^{3-} \rightarrow$ g Na_3AsO_4 using molar masses. Assume the density of H_2O is 1.00 g/mL.

Solve. 1 billion $= 1 \times 10^9$; 1 ppb $= \dfrac{1 \text{ g solute}}{1 \times 10^9 \text{ g solution}}$

$$\frac{1 \text{ g solute}}{1 \times 10^9 \text{ g solution}} \times \frac{1 \text{ g solution}}{1 \text{ mL solution}} \times \frac{1 \times 10^3 \text{ mL}}{1 \text{ L solution}} = \frac{\text{g AsO}_4^{3-}}{1 \times 10^6 \text{ L H}_2\text{O}}$$

$$10 \text{ ppb AsO}_4^{3-} = \frac{10 \text{ g AsO}_4}{1 \times 10^6 \text{ L H}_2\text{O}} \times 1 \text{ L H}_2\text{O} = 1.0 \times 10^{-5} \text{ g As/L.}$$

$$1.0 \times 10^{-5} \text{ g AsO}_4^{3-} \times \frac{1 \text{ mol AsO}_4^{3-}}{138.92 \text{ g AsO}_4^{3-}} \times \frac{1 \text{ mol Na}_3\text{AsO}_4}{1 \text{ mol As}} \times \frac{207.89 \text{ g Na}_3\text{AsO}_4^{3-}}{1 \text{ mol Na}_3\text{AsO}_4}$$

$$= 1.5 \times 10^{-5} \text{ g Na}_3\text{AsO}_4 \text{ in } 1.00 \text{ L H}_2\text{O}$$

4.115   (a)   mol HCl initial – mol NH_3 from air = mol HCl remaining

                       = mol NaOH required for titration

mol NaOH $= 0.0588 \, M \times 0.0131 \text{ L} = 7.703 \times 10^{-4} = 7.70 \times 10^{-4}$ mol NaOH

                       $= 7.70 \times 10^{-4}$ mol HCl remain

mol HCl initial – mol HCl remaining = mol NH_3 from air

$(0.0105 \, M \text{ HCl} \times 0.100 \text{ L}) - 7.703 \times 10^{-4}$ mol HCl = mol NH_3

10.5×10^{-4} mol HCl $- 7.703 \times 10^{-4}$ mol HCl $= 2.80 \times 10^{-4} = 2.8 \times 10^{-4}$ mol NH_3

$$2.8 \times 10^{-4} \text{ mol NH}_3 \times \frac{17.03 \text{ g NH}_3}{1 \text{ mol NH}_3} = 4.77 \times 10^{-3} = 4.8 \times 10^{-3} \text{ g NH}_3$$

     (b)   ppm is defined as molecules of $NH_3 / 1 \times 10^6$ molecules in air.

Calculate molecules NH_3 from mol NH_3.

$$2.80 \times 10^{-4} \text{ mol NH}_3 \times \frac{6.022 \times 10^{23} \text{ molecules}}{1 \text{ mol}} = 1.686 \times 10^{20}$$

$$= 1.7 \times 10^{20} \text{ NH}_3 \text{ molecules}$$

Calculate total volume of air processed, then g air using density, then molecules air using molar mass.

$$\frac{10.0\ \text{L}}{1\ \text{min}} \times 10.0\ \text{min} \times \frac{1.20\ \text{g air}}{1\ \text{L air}} \times \frac{1\ \text{mol air}}{29.0\ \text{g air}} \times \frac{6.022 \times 10^{23}\ \text{molecules}}{1\ \text{mol}}$$

$$= 2.492 \times 10^{24} = 2.5 \times 10^{24}\ \text{air molecules}$$

$$\text{ppm NH}_3 = \frac{1.686 \times 10^{20}\ \text{NH}_3\ \text{molecules}}{2.492 \times 10^{24}\ \text{air molecules}} \times 1 \times 10^6 = 68\ \text{ppm NH}_3$$

(c)   68 ppm > 50 ppm. The manufacturer is **not** in compliance.

5 Thermochemistry

Visualizing Concepts

5.2   (a)   The caterpillar uses energy produced by its metabolism of food to climb the twig and increase its potential energy.

        (b)   Heat, q, is the energy transferred from a hotter to a cooler object. Without knowing the temperature of the caterpillar and its surroundings, we cannot predict the sign of q. It is likely that q is approximately zero, because a small creature like a caterpillar is unlikely to support a body temperature much different from its environmental temperature.

        (c)   Work, w, is the energy transferred when a force moves an object. When the caterpillar climbs the twig, it does work as its body moves against the force of gravity.

        (d)   No. The amount of work is independent of time and therefore independent of speed (assuming constant caterpillar speed).

        (e)   No. Potential energy depends only on the caterpillar's position, so the change in potential energy depends only on the distance climbed, not on the speed of the climb.

5.3   (a)   The internal energy, E, of the products is greater than that of the reactants, so the diagram represents an increase in the internal energy of the system.

        (b) ΔE for this process is positive, (+).

        (c)   If no work is associated with the process, it is endothermic.

5.5   (a).   No. This distance traveled to the top of a mountain depends on the path taken by the hiker. Distance is a path function, not a state function.

        (b)   Yes. Change in elevation depends only on the location of the base camp and the height of the mountain, not on the path to the top. Change in elevation is a state function, not a path function.

5.6   (a)   State B

        (b) ΔE_{AB} = energy difference between State A and State B.

            $\Delta E_{AB} = \Delta E_1 + \Delta E_2$ or $\Delta E_{AB} = \Delta E_3 + \Delta E_4$

        (c) ΔE_{CD} = energy difference between State C and State D.

            $\Delta E_{CD} = \Delta E_2 - \Delta E_4$ or $\Delta E_{CD} = \Delta E_3 - \Delta E_1$

        (Note that the sign of ΔE depends on the definition of initial and final state, but the magnitude is the absolute value of the difference in energy.)

(d)   The energy of State E is $\Delta E_1 + \Delta E_4$, whereas the energy of State B is $\Delta E_1 + \Delta E_2$. Because $\Delta E_4 > \Delta E_2$, State E is above State B on the diagram; State E would be the highest energy on the diagram.

5.8   (a)   The temperature of the system and surroundings will equalize, so the temperature of the hotter system will decrease, and the temperature of the colder surroundings will increase. The system loses heat by decreasing its temperature, so the sign of q_{sys} is (–). The surrounding gains heat by increasing its temperature, so the sign of q_{surr} is (+). From the system's perspective, the process is exothermic because it loses heat.

        (b)   If neither volume nor pressure of the system changes, $w = 0$ and $\Delta E = q = \Delta H$. The change in internal energy is equal to the change in enthalpy.

5.9   (a) $w = -P\Delta V$. Because ΔV for the process is (–), the sign of w is (+).

        (b) $\Delta E = q + w$. At constant pressure, $\Delta H = q$. If the reaction is endothermic, the signs of ΔH and q are (+). From (a), the sign of w is (+), so the sign of ΔE is (+). The internal energy of the system increases during the change. (This situation is described by the diagram (ii) in Exercise 5.4.)

5.11   (a) $\Delta H_A = \Delta H_B + \Delta H_C$. The net enthalpy change associated with going from the initial state to the final state does not depend on path. The diagram shows that the change can be accomplished via reaction A or via two successive reactions, B then C, with the same net enthalpy change. $\Delta H_A = \Delta H_B + \Delta H_C$ because ΔH is a state function, independent of path.

        (b) $\Delta H_Z = \Delta H_X + \Delta H_Y$. The diagram indicates that Reaction Z can be written as the sum of reactions X and Y.

        (c)   Hess's law states that the enthalpy change for a net reaction is the sum of the enthalpy changes of the component steps, regardless of whether the reaction actually occurs via this path. The diagrams are a visual statement of Hess's law.

5.12   Because mass must be conserved in the reaction A → B, the component elements of A and B must be the same. Further, if $\Delta H_f^\circ > 0$ for both A and B, the energies of both A and B are above the energies of their component elements on the energy diagram.

        (a)   The bold arrow shows the reaction as written; combination of the two thin arrows shows an alternate route from A to B.

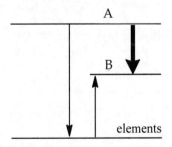

        (b) $\Delta H_{rxn}^\circ = \Delta H_f^\circ\, B - \Delta H_f^\circ\, A$. If the overall reaction is exothermic, the sign of ΔH is (–) and $\Delta H_f^\circ\, A > \Delta H_f^\circ\, B$. This means that the enthalpy of A is the highest energy level on the diagram. This is the situation pictured in the diagram above, but nothing in the given information requires this arrangement. If the reaction is endothermic, $\Delta H_f^\circ\, B > \Delta H_f^\circ\, A$ and the enthalpy of B would be the highest energy level on the diagram.

5 Thermochemistry

Solutions to Exercises

The Nature of Energy (Section 5.1)

5.14   (a)   The kinetic energy of the ball **decreases** as it moves higher. As the ball moves higher and opposes gravity, kinetic energy is changed into potential energy.

      (b)   The potential energy of the ball **increases** as it moves higher.

      (c)   The heavier ball would go **half as high** as the tennis ball. At the apex of the trajectory, all initial kinetic energy has been changed into potential energy. The magnitude of the change in potential energy is m × g × d, which is equal to the energy initially imparted to the ball. If the same amount of energy is imparted to a ball with twice the mass, m doubles so d is half as large.

5.16   (a)   *Analyze.* Given: mass and speed of ball.   Find: kinetic energy.

           Plan. Because $1\,J = 1\,kg \cdot m^2/s^2$, convert oz to kg and mph to m/s to obtain E_k in J.

           Solve. $5.13\,oz \times \dfrac{1\,lb}{16\,oz} \times \dfrac{1\,kg}{2.205\,lb} = 0.14541 = 0.145\,kg$

$$\frac{95.0\,mi}{1\,h} \times \frac{1.6093\,km}{1\,mi} \times \frac{1000\,m}{1\,km} \times \frac{1\,h}{60\,min} \times \frac{1\,min}{60\,sec} = 42.468 = 42.5\,m/s$$

$$E_k = 1/2\,mv^2 = 1/2 \times 0.14541\,kg \times \left(\frac{42.468\,m}{1\,s}\right)^2 = \frac{131\,kg \cdot m^2}{1\,s^2} = 131\,J$$

           Check. $1/2(0.15 \times 1600) \approx 1/2(160+80) \approx 120\,J$

      (b)   Kinetic energy is related to velocity squared (v^2); if the speed of the ball decreases to 55.0 mph, the kinetic energy of the ball will decrease by a factor of $(55.0/95.0)^2$. (The conversion factors to m/s apply to both speeds and will cancel in the ratio.) The numerical multiplier is $(55/95)^2 = 0.335$. The kinetic energy decreases by approximately a factor of 3.

      (c)   As the ball hits the catcher's glove, its speed (and hence its kinetic energy) drops to 0. Some of the kinetic energy is transferred to the glove, which deforms when the ball lands. Some is transferred to the catcher's body (mostly the arm), which recoils while catching the ball. As usual, some energy is released as heat through friction between the ball and the glove.

5.18   (a)   *Analyze.* Given: 1 kwh; 1 watt = 1 J/s; 1 watt · s = 1 J. Find: conversion factor for joules and kwh.

           Plan. kwh → wh → ws → J

           Solve. $1\,kwh \times \dfrac{1000\,w}{1\,kw} \times \dfrac{60\,min}{h} \times \dfrac{60\,s}{min} \times \dfrac{1\,J}{1\,w \cdot s} = 3.6 \times 10^6\,J$

           $1\,kwh = 3.6 \times 10^6\,J$

      (b)   *Analyze.* Given: 100 watt bulb. Find: heat in kcal radiated by bulb or person in 24 hr.

           Plan. 1 watt = 1 J/s; 1 kcal = 4.184×10^3 J; watt → J/s → J → kcal.   *Solve.*

$$100\,watt = \frac{100\,J}{1\,s} \times \frac{60\,sec}{min} \times \frac{60\,min}{h} \times 24\,h \times \frac{1\,kcal}{4.184 \times 10^3\,J} = 2065 = 2.1 \times 10^3\,kcal$$

           24 hr has 2 sig figs, but 100 watt is ambiguous. The answer to 1 sig fig would be 2×10^3 kcal.

62

Copyright © 2015 Pearson Education, Inc.

5.20   (a)   The system is *open* because it exchanges both mass and energy with the surroundings. Mass exchange occurs when solution flows into and out of the apparatus. The apparatus is not insulated, so energy exchange also occurs. Closed systems exchange energy but not mass, whereas isolated systems exchange neither.

        (b)   If the system is defined as shown, it can be closed by blocking the flow in and out but leaving the flask full of solution.

5.22   (a)   Electrostatic attraction; no work is done because the particles are held apart at a constant distance.

        (b)   Magnetic attraction; work is done because the nail is moved a distance in opposition to the force of magnetic attraction.

The First Law of Thermodynamics (Section 5.2)

5.24   (a) $\Delta E = q + w$

        (b)   The quantities q and w are negative when the system loses heat to the surroundings (it cools) or does work on the surroundings.

5.26   In each case, evaluate q and w in the expression $\Delta E = q + w$. For an exothermic process, q is negative; for an endothermic process, q is positive.

        (a)   q is negative and w is positive. $\Delta E = -0.655 \text{ kJ} + 0.382 \text{ kJ} = -0.273 \text{ kJ}$. The process is exothermic.

        (b)   q is positive and w is essentially zero. $\Delta E = 322 \text{ J}$. The process is endothermic.

5.28 $E_{el} = \dfrac{\kappa Q_1 Q_2}{r^2}$ For two oppositely charged particles, the sign of E_{el} is negative; the closer the particles, the greater the magnitude of E_{el}.

        (a)   The potential energy becomes less negative as the particles are separated (r increases).

        (b) ΔE for the process is positive; the internal energy of the system increases as the oppositely charged particles are separated.

        (c)   Work is done on the system to separate the particles so w is positive. We have no direct knowledge of the change in q, except that it cannot be large and negative, because overall $\Delta E = q + w$ is positive.

5.30   (a)   Independent. Potential energy is a state function.

        (b)   Dependent. Some of the energy released could be employed in performing work, as is done in the body when sugar is metabolized; heat is not a state function.

        (c)   Dependent. The work accomplished depends on whether the gasoline is used in an engine, burned in an open flame, or in some other manner. Work is not a state function.

Enthalpy (Sections 5.3 and 5.4)

5.32 $P = 0.857 \text{ atm}$. $\Delta V = 1.26 \text{ L} - 5.00 \text{ L} = -3.74 \text{ L}$

        $w = -0.857 \text{ atm}(-3.74 \text{ L}) = 3.2052 = 3.21 \text{ L-atm}$;

        $3.2052 \text{ L-atm} \times 101.3 \text{ J/L-atm} = 324.69 = 325 \text{ J}$

5.34   (a)   When a process occurs under constant external pressure and only P-V work occurs, the enthalpy change (ΔH) equals the amount of heat transferred. $\Delta H = q_p$.

        (b) $\Delta H = q_p$. If the system releases heat, q and ΔH are negative, and the enthalpy of the system decreases.

        (c)   If $\Delta H = 0$, $q_p = 0$ and $\Delta E = w$.

5.36   (a)   At constant volume ($\Delta V = 0$), $\Delta E = q_v$.

        (b) ΔE will be larger than ΔH.

        (c)   According to the definition of enthalpy, $H = E + PV$, so $\Delta H = \Delta E + \Delta(PV)$. For an ideal gas at constant temperature and volume, $\Delta PV = V\Delta P = RT\Delta n$. For this reaction, there are 2 mol of gaseous product and 3 mol of gaseous reactants, so $\Delta n = -1$. Thus $V\Delta P$ or $\Delta(PV)$ is negative. Because $\Delta H = \Delta E + \Delta(PV)$, the negative $\Delta(PV)$ term means that ΔE is larger or less negative than ΔH.

5.38   The gas is the system. If 0.49 kJ of heat is added, q = +0.49 kJ. Work done by the system decreases the overall energy of the system, so w = –214 J = –0.214 kJ .

        $\Delta E = q + w = 0.49\ \text{kJ} - 0.214\ \text{kJ} = 0.276\ \text{kJ}$. $\Delta H = q = 0.49$ kJ (at constant pressure).

5.40   (a) $Ca(OH)_2(s) \rightarrow CaO(s) + H_2O(g)$ (b)

                $\Delta H = 109$ kJ

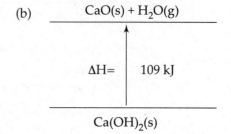

5.42   *Plan.* Consider the sign of an enthalpy change that would convert one of the substances into the other.   *Solve.*

        (a) $CO_2(s) \rightarrow CO_2(g)$. This change is sublimation, which is endothermic, $+\Delta H$. $CO_2(g)$ has the higher enthalpy.

        (b) $H_2 \rightarrow 2\ H$. Breaking the H–H bond requires energy, so the process is endothermic, $+\Delta H$. Two moles of H atoms have higher enthalpy.

        (c) $H_2O(g) \rightarrow H_2(g) + 1/2\ O_2(g)$. Decomposing H_2O into its elements requires energy and is endothermic, $+\Delta H$. One mole of $H_2(g)$ and 0.5 mol $O_2(g)$ at 25°C have the higher enthalpy.

        (d) $N_2(g)$ at 100 °C $\rightarrow N_2(g)$ at 300 °C. An increase in the temperature of the sample requires that heat is added to the system, +q and $+\Delta H$. $N_2(g)$ at 300 °C has the higher enthalpy.

5.44   (a)   The sign of ΔH is positive, so the reaction is endothermic.

        (b) $24.0\ \text{g CH}_3\text{OH} \times \dfrac{1\ \text{mol CH}_3\text{OH}}{32.04\ \text{g CH}_3\text{OH}} \times \dfrac{252.8\ \text{kJ}}{2\ \text{mol CH}_3\text{OH}} = 94.7\ \text{kJ heat absorbed}$

(c) $82.1\,\text{kJ} \times \dfrac{2\,\text{mol CH}_4}{252.8\,\text{kJ}} \times \dfrac{16.04\,\text{g CH}_4}{1\,\text{mol CH}_4} = 10.4\,\text{g CH}_4$ produced

(d)   The sign of ΔH is reversed for the reverse reaction: $\Delta H = -252.8$ kJ

$38.5\,\text{g CH}_4 \times \dfrac{1\,\text{mol CH}_4}{16.04\,\text{g CH}_4} \times \dfrac{-252.8\,\text{kJ}}{2\,\text{mol CH}_4} = -303$ kJ heat released

5.46   (a) $1.36\,\text{mol O}_2 \times \dfrac{-89.4\,\text{kJ}}{3\,\text{mol O}_2} = -40.53 = -40.5$ kJ

(b) $10.4\,\text{g KCl} \times \dfrac{1\,\text{mol KCl}}{74.55\,\text{g KCl}} \times \dfrac{-89.4\,\text{kJ}}{2\,\text{mol KCl}} = -6.2358 = -6.24$ kJ

(c)   Because the sign of ΔH is reversed for the reverse reaction, it seems reasonable that other characteristics would be reversed, as well. If the forward reaction proceeds spontaneously, the reverse reaction is probably not spontaneous. Also, we know from experience that KCl(s) does not spontaneously react with atmospheric $O_2(g)$, even at elevated temperature.

5.48   (a) $3\,C_2H_2(g) \rightarrow C_6H_6(l)$ $\Delta H = -630$ kJ

(b) $C_6H_6(l) \rightarrow 3\,C_2H_2(g)$ $\Delta H = +630$ kJ

ΔH for the formation of 3 mol of acetylene is 630 kJ. ΔH for the formation of 1 mol of C_2H_2 is then 630 kJ/3 = 210 kJ.

(c)   The exothermic reverse reaction is more likely to be thermodynamically favored.

(d)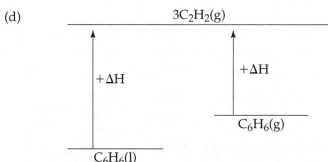

If the reactant is in the higher enthalpy gas phase, the overall ΔH for the reaction has a smaller positive value.

Calorimetry (Section 5.5)

The specific heat of water to four significant figures, **4.184 J/g‑K,** will be used in many of the following exercises; temperature units of K and °C will be used interchangeably.

5.50   *Analyze.* Both objects are heated to 100 °C. The two hot objects are placed in the same amount of cold water at the same temperature. Object A raises the water temperature more than object B. *Plan.* Apply the definition of heat capacity to heating the water and heating the objects to determine which object has the greater heat capacity.   *Solve.*

(a)   Both beakers of water contain the same mass of water, so they both have the same heat capacity. Object A raises the temperature of its water more than object

B, so more heat was transferred from object A than from object B. Because both objects were heated to the same temperature initially, object A must have absorbed more heat to reach the 100 °C temperature. The greater the heat capacity of an object, the greater the heat required to produce a given rise in temperature. Thus, object A has the greater heat capacity.

(b)　Because no information about the masses of the objects is given, we cannot compare or determine the specific heats of the objects.

5.52　(a)　In Table 5.2, Hg(l) has the smallest specific heat, so it will require the smallest amount of energy to heat 50.0 g of the substance 10 K.

(b)　$50.0 \text{ g Hg(l)} \times 10 \text{ K} \times \dfrac{0.14 \text{ J}}{\text{g-K}} = 70 \text{ J}$

5.54　(a)　$\text{specific heat} = \dfrac{\text{J}}{1 \text{ g-}^{\circ}\text{C}} = \dfrac{322 \text{ J}}{100.0 \text{ g} \times (50\,^{\circ}\text{C} - 25\,^{\circ}\text{C})} = 0.1288 = \dfrac{0.13 \text{ J}}{1 \text{ g-}^{\circ}\text{C}}$

(b)　In general, the greater the heat capacity, the more heat is required to raise the temperature of 1 gram of substance 1 °C. The specific heat of gold is 0.13 J/g-°C, whereas that of iron is 0.45 J/g-°C (Table 5.2). For gold and iron blocks with equal mass, same initial temperature and same amount of heat added, the one with the lower specific heat, gold, will require less heat per °C and have the higher final temperature.

(c)　$\dfrac{0.1288 \text{ J}}{1 \text{ g-}^{\circ}\text{C}} \times \dfrac{196.97 \text{ g Au}}{1 \text{ mol Au}} = 25.37 = \dfrac{25 \text{ J}}{\text{mol-}^{\circ}\text{C}}$

5.56　(a)　Follow the logic in Solution 5.55. The total mass of the solution is (60.0 g H_2O + 4.25 g NH_4NO_3) = 64.25 = 64.3 g. The temperature change of the solution is 22.0 – 16.9 = –5.1 °C. The heat lost by the surroundings is

$64.25 \text{ g solution} \times \dfrac{4.184 \text{ J}}{1 \text{ g-}^{\circ}\text{C}} \times -5.1\,^{\circ}\text{C} \times \dfrac{1 \text{ kJ}}{1000 \text{ J}} = -1.371 = -1.4 \text{ kJ}$

That is, 1.4 kJ is absorbed when 4.25 g NH_4NO_3(s) dissolves.

$\dfrac{+1.371 \text{ kJ}}{4.25 \text{ NH}_4\text{NO}_3} \times \dfrac{80.04 \text{ g NH}_4\text{NO}_3}{1 \text{ mol NH}_4\text{NO}_3} = +25.82 = +26 \text{ kJ/mol NH}_4\text{NO}_3$

(b)　This process is endothermic because the temperature of the surroundings decreases, indicating that heat is absorbed by the system.

5.58　(a)　$C_6H_5OH(s) + 7 O_2(g) \rightarrow 6 CO_2(g) + 3 H_2O(l)$

(b)　$q_{bomb} = -q_{rxn};\ \Delta T = 26.37\,^{\circ}\text{C} - 21.36\,^{\circ}\text{C} = 5.01\,^{\circ}\text{C}$

$q_{bomb} = \dfrac{11.66 \text{ kJ}}{1\,^{\circ}\text{C}} \times 5.01\,^{\circ}\text{C} = 58.417 = 58.4 \text{ kJ}$

At constant volume, $q_v = \Delta E$. ΔE and ΔH are very similar.

$\Delta H_{rxn} \approx \Delta E_{rxn} = q_{rxn} = -q_{bomb} = \dfrac{-58.417 \text{ kJ}}{1.800 \text{ g C}_6\text{H}_5\text{OH}} = -32.454 = -32.5 \text{ kJ/g C}_6\text{H}_5\text{OH}$

$$\Delta H_{rxn} = \frac{-32.454 \text{ kJ}}{1 \text{ g C}_6\text{H}_5\text{OH}} \times \frac{94.11 \text{ g C}_6\text{H}_5\text{OH}}{1 \text{ mol C}_6\text{H}_5\text{OH}} = \frac{-3.054 \times 10^3 \text{ kJ}}{\text{mol C}_6\text{H}_5\text{OH}}$$

$$= -3.05 \times 10^3 \text{ kJ/mol C}_6\text{H}_5\text{OH}$$

5.60 (a) $C = 2.760 \text{ g C}_6\text{H}_5\text{COOH} \times \dfrac{26.38 \text{ kJ}}{1 \text{ g C}_6\text{H}_5\text{COOH}} \times \dfrac{1}{8.33\,^{\circ}\text{C}} = 8.74055 = 8.74 \text{ kJ/}^{\circ}\text{C}$

    (b) $\dfrac{8.74055 \text{ kJ}}{^{\circ}\text{C}} \times 4.95\,^{\circ}\text{C} \times \dfrac{1}{1.440 \text{ g sample}} = 30.046 = 30.0 \text{ kJ/ g sample}$

    (c) If water is lost from the calorimeter, there is less water to heat, so the same amount of heat (kJ) from a reaction would cause a larger increase in the calorimeter temperature. The calorimeter constant, kJ/°C, would decrease, because °C is in the denominator of the expression.

Hess's Law (Section 5.6)

5.62 (a) *Analyze/Plan.* Arrange the reactions so that in the overall sum, B appears in both reactants and products and can be canceled. This is a general technique for using Hess's Law. *Solve.*

$$
\begin{array}{ll}
A \rightarrow B & \Delta H = +30 \text{ kJ} \\
\underline{B \rightarrow C} & \underline{\Delta H = +60 \text{ kJ}} \\
A \rightarrow C & \Delta H = +90 \text{ kJ}
\end{array}
$$

    (b)

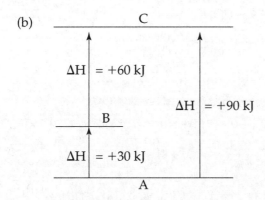

    Check. The process of A forming C can be described as A forming B and B forming C.

5.64

$$
\begin{array}{ll}
2\,\text{C}(s) + \text{O}_2(g) + 4\,\text{H}_2(g) \rightarrow 2\,\text{CH}_3\text{OH}(g) & \Delta H = -402.4 \text{ kJ} \\
\underline{2\,\text{CO}(g) \rightarrow \text{O}_2(g) + 2\,\text{C}(s)} & \underline{\Delta H = 221.0 \text{ kJ}} \\
2\,\text{CO}(g) + 4\,\text{H}_2(g) \rightarrow 2\,\text{CH}_3\text{OH}(g) & \Delta H = -181.4 \text{ kJ} \\
\text{CO}(g) + 2\,\text{H}_2(g) \rightarrow \text{CH}_3\text{OH}(g) & \Delta H = (-181.4)/2 = -90.7 \text{ kJ}
\end{array}
$$

5.66
$$N_2O(g) \rightarrow N_2(g) + 1/2\,O_2(g) \qquad \Delta H = 1/2\ (-163.2\ kJ)$$
$$NO_2(g) \rightarrow NO(g) + 1/2\,O_2(g) \qquad \Delta H = 1/2\,(113.1\ kJ)$$
$$N_2(g) + O_2(g) \rightarrow 2\,NO(g) \qquad \Delta H = 180.7\ kJ$$

$$N_2O(g) + NO_2(g) \rightarrow 3\,NO(g) \qquad \Delta H = 155.7\ kJ$$

Enthalpies of Formation (Section 5.7)

5.68   (a)   Tables of ΔH_f^o are useful because, according to Hess's law, the standard enthalpy of any reaction can be calculated from the standard enthalpies of formation for the reactants and products.

$$\Delta H_{rxn}^o = \Sigma \Delta H_f^o\ (products) - \Sigma \Delta H_f^o\ (reactants)$$

(b)   The standard enthalpy of formation for any element in its standard state is zero. Elements in their standard states are the reference point for the enthalpy of formation scale.

(c) $12\ C(s) + 11\ H_2(g) + 11/2\ O_2(g) \rightarrow C_{12}H_{22}O_{11}(s)$

5.70   (a) $H_2(g) + O_2(g) \rightarrow H_2O_2(g)$ $\qquad\qquad \Delta H_f^o = -136.10\ kJ$

(b) $Ca(s) + C(s) + 3/2\ O_2(g) \rightarrow CaCO_3(s)$ $\qquad \Delta H_f^o = -1207.1\ kJ$

(c) $1/4\ P_4(s) + 1/2\ O_2(g) + 3/2\ Cl_2(g) \rightarrow POCl_3(l)$ $\qquad \Delta H_f^o = -597.0\ kJ$

(d) $2\ C(s) + 3\ H_2(g) + 1/2\ O_2(g) \rightarrow C_2H_5OH(l)$ $\qquad \Delta H_f^o = -277.7\ kJ$

5.72   Use heats of formation to calculate ΔH° for the combustion of butane.

$$C_3H_8(g) + 5\ O_2(g) \rightarrow 3\ CO_2(g) + 4\ H_2O(l)$$

$$\Delta H_{rxn}^o = 3\ \Delta H_f^o\ CO_2(g) + 4\ \Delta H_f^o\ H_2O(l) - \Delta H_f^o\ C_3H_8(g) - 5\ \Delta H_f^o\ O_2(g)$$

$$\Delta H_{rxn}^o = 3(-393.5\ kJ) + 4(-285.83\ kJ) - (-103.85\ kJ) - 5(0) = -2219.97 = -2220.0\ kJ/\ mol\ C_3H_8$$

$$10.00\ g\ C_3H_8 \times \frac{1\ mol\ C_3H_8}{44.096\ g\ C_3H_8} \times \frac{-2219.97\ kJ}{1\ mol\ C_3H_8} = -503.4\ kJ$$

5.74   (a) $\Delta H_{rxn}^o = \Delta H_f^o\ CaCl_2(s) + \Delta H_f^o\ H_2O(g) - \Delta H_f^o\ CaO(s) - 2\ \Delta H_f^o\ HCl(g)$

$\qquad\qquad = -795.8\ kJ + (-241.82\ kJ) - (-635.5\ kJ) - 2(-92.30\ kJ) = -217.5\ kJ$

(b) $\Delta H_{rxn}^o = 2\ \Delta H_f^o\ Fe_2O_3(s) - 4\ \Delta H_f^o\ FeO(s) - \Delta H_f^o\ O_2(g)$

$\qquad\qquad = 2(-822.16\ kJ) - 4(-271.9\ kJ) - (0) = -556.7\ kJ$

(c) $\Delta H_{rxn}^o = \Delta H_f^o\ Cu_2O(s) + \Delta H_f^o\ NO_2(g) - 2\ \Delta H_f^o\ CuO(s) - \Delta H_f^o\ NO(g)$

$\qquad\qquad = -170.7\ kJ + (33.84\ kJ) - 2(-156.1\ kJ) - (90.37\ kJ) = 85.0\ kJ$

(d) $\Delta H_{rxn}^o = 2\ \Delta H_f^o\ N_2H_4(g) + 2\ \Delta H_f^o\ H_2O(l) - 4\ \Delta H_f^o\ NH_3(g) - \Delta H_f^o\ O_2(g)$

$\qquad\qquad = 2(95.40\ kJ) + 2(-285.83\ kJ) - 4(-46.19\ kJ) - (0) = -196.10\ kJ$

5.76 $\Delta H^\circ_{rxn} = \Delta H^\circ_f\ Ca(OH)_2(s) + \Delta H^\circ_f\ C_2H_2(g) - 2\ \Delta H^\circ_f\ H_2O(l) - \Delta H^\circ_f\ CaC_2(s)$

$-127.2\ kJ = -986.2\ kJ + 226.77\ kJ - 2(-285.83\ kJ) - \Delta H^\circ_f\ CaC_2(s)$

ΔH°_f for $CaC_2(s) = -60.6\ kJ$

5.78   (a) $C_4H_{10}O(l) + 6\ O_2(g) \rightarrow 4\ CO_2(g) + 5\ H_2O(l)$ $\Delta H^\circ = -2723.7\ kJ$

      (b) $\Delta H^\circ_{rxn} = 4\ \Delta H^\circ_f\ CO_2(g) + 5\ \Delta H^\circ_f\ H_2O(l) - \Delta H^\circ_f\ C_4H_{10}O(l) - 6\ \Delta H^\circ_f\ O_2(g)$

$-2723.7 = 4(-393.5\ kJ) + 5(-285.83\ kJ) - \Delta H^\circ_f\ C_4H_{10}O(l) - 6(0)$

$\Delta H^\circ_f\ C_4H_{10}O(l) = 4(-393.5\ kJ) + 5(-285.83\ kJ) + 2723.7\ kJ = -279.45 = -279.5\ kJ$

5.80   (a) $CH_3OH(l) + 3/2\ O_2(g) \rightarrow CO_2(g) + 2\ H_2O(g)$

      (b) $\Delta H^\circ_{rxn} = \Delta H^\circ_f\ CO_2(g) + 2\ \Delta H^\circ_f\ H_2O(g) - \Delta H^\circ_f\ CH_3OH(l) - 3/2\ \Delta H^\circ_f\ O_2(g)$

$= -393.5\ kJ + 2(-241.82\ kJ) - (-238.6\ kJ) - 3/2(0) = -638.54 = -638.5\ kJ$

      (c) $\dfrac{-638.54\ kJ}{mol\ CH_3OH} \times \dfrac{1\ mol\ CH_3OH}{32.04\ g} \times \dfrac{0.791\ g}{mL} \times \dfrac{1000\ mL}{L} = 1.58 \times 10^4\ kJ/L$ produced

      (d) $\dfrac{1\ mol\ CO_2}{-638.54\ kJ} \times \dfrac{44.0095\ g\ CO_2}{mol} = 0.06892\ g\ CO_2/kJ$ emitted

Foods and Fuels (Section 5.8)

5.82   (a)   Fats are appropriate for fuel storage because they are insoluble in water (and body fluids) and have a high fuel value.

      (b)   For convenience, assume 100 g of chips.

$12\ g\ protein \times \dfrac{17\ kJ}{1\ g\ protein} \times \dfrac{1\ Cal}{4.184\ kJ} = 48.76 = 49\ Cal$

$14\ g\ fat \times \dfrac{38\ kJ}{1\ g\ fat} \times \dfrac{1\ Cal}{4.184\ kJ} = 127.15 = 130\ Cal$

$74\ g\ carbohydrates \times \dfrac{17\ kJ}{1\ g\ carbohydrates} \times \dfrac{1\ Cal}{4.184\ kJ} = 300.67 = 301\ Cal$

total Cal = (48.76 + 127.15 + 300.67) = 476.58 = 480 Cal

$\%\ Cal\ from\ fat = \dfrac{127.15\ Cal\ fat}{476.58\ total\ Cal} \times 100 = 26.68 = 27\%$

(Because the conversion from kJ to Cal was common to all three components, we would have determined the same percentage by using kJ.)

      (c) $25\ g\ fat \times \dfrac{38\ kJ}{g\ fat} = x\ g\ protein \times \dfrac{17\ kJ}{g\ protein};$ $x = 56\ g\ protein$

5.84   Calculate the fuel value in a pound of M&M® candies.

$$96 \text{ fat} \times \frac{38 \text{ kJ}}{1 \text{ g fat}} = 3648 \text{ kJ} = 3.6 \times 10^3 \text{ kJ}$$

$$320 \text{ g carbohydrate} \times \frac{17 \text{ kJ}}{1 \text{ g carbohydrate}} = 5440 \text{ kJ} = 5.4 \times 10^3 \text{ kJ}$$

$$21 \text{ g protein} \times \frac{17 \text{ kJ}}{1 \text{ g protein}} = 357 \text{ kJ} = 3.6 \times 10^2 \text{ kJ}$$

total fuel value = 3648 kJ + 5440 kJ + 357 kJ = 9445 kJ = 9.4×10^3 kJ/lb

$$\frac{9445 \text{ kJ}}{\text{lb}} \times \frac{1 \text{ lb}}{453.6 \text{ g}} \times \frac{42 \text{ g}}{\text{serving}} = 874.5 \text{ kJ} = 8.7 \times 10^2 \text{ kJ/serving}$$

$$\frac{874.5 \text{ kJ}}{\text{serving}} \times \frac{1 \text{ kcal}}{4.184 \text{ kJ}} \times \frac{1 \text{ Cal}}{1 \text{ kcal}} = 209.0 \text{ Cal} = 2.1 \times 10^2 \text{ Cal/serving}$$

Check. 210 Cal is the approximate food value of a candy bar, so the result is reasonable.

5.86 $$177 \text{ mL} \times \frac{1.0 \text{ g wine}}{1 \text{ mL}} \times \frac{0.106 \text{ g ethanol}}{1 \text{ g wine}} \times \frac{1 \text{ mol ethanol}}{46.1 \text{ g ethanol}} \times \frac{1367 \text{ kJ}}{1 \text{ mol ethanol}} \times \frac{1 \text{ Cal}}{4.184 \text{ kJ}}$$

$$= 133 = 1.3 \times 10^2 \text{ Cal}$$

Check. A "typical" 6 oz. glass of wine has 150–250 Cal, so this is a reasonable result. Note that alcohol is responsible for most of the food value of wine.

5.88 $$\Delta H^{\circ}_{rxn} = \Delta H^{\circ}_f \text{ CO}_2(g) + 2 \Delta H^{\circ}_f \text{ H}_2\text{O}(g) - \Delta H^{\circ}_f \text{ CH}_4(g) - 2 \Delta H^{\circ}_f \text{ O}_2(g)$$

$$= -393.5 \text{ kJ} + 2(-241.82 \text{ kJ}) - (-74.8 \text{ kJ}) - 2(0) \text{ kJ} = -802.3 \text{ kJ}$$

$$\Delta H^{\circ}_{rxn} = \Delta H^{\circ}_f \text{ CF}_4(g) + 4 \Delta H^{\circ}_f \text{ HF}(g) - \Delta H^{\circ}_f \text{ CH}_4(g) - 4 \Delta H^{\circ}_f \text{ F}_2(g)$$

$$= -679.9 \text{ kJ} + 4(-268.61 \text{ kJ}) - (-74.8 \text{ kJ}) - 4(0) \text{ kJ} = -1679.5 \text{ kJ}$$

The second reaction is twice as exothermic as the first. The "fuel values" of hydrocarbons in a fluorine atmosphere are approximately twice those in an oxygen atmosphere. Note that the difference in ΔH° values for the two reactions is in the ΔH°_f for the products, because the ΔH°_f for the reactants is identical.

5.90   (a)   Use density to change L to g, molar mass to change g to mol, heat of combustion to change mol to kJ. Ethanol is C_2H_5OH, gasoline is C_8H_{18}. From Exercise 5.79 (c), heat of combustion of ethanol is –1234.8 kJ/mol.

$$1.0 \text{ L C}_2\text{H}_5\text{OH} \times \frac{1000 \text{ mL}}{1 \text{ L}} \times \frac{0.79 \text{ g}}{1 \text{ mL}} \times \frac{1 \text{ mol C}_2\text{H}_5\text{OH}}{46.07 \text{ g}} \times \frac{1234.8 \text{ kJ}}{1 \text{ mol C}_2\text{H}_5\text{OH}}$$

$$= 21,174 = 2.1 \times 10^4 \text{ kJ/L C}_2\text{H}_5\text{OH}$$

$$1.0 \text{ L C}_8\text{H}_{18} \times \frac{1000 \text{ mL}}{1 \text{ L}} \times \frac{0.70 \text{ g}}{1 \text{ mL}} \times \frac{1 \text{ mol C}_8\text{H}_{18}}{114.23 \text{ g C}_8\text{H}_{18}} \times \frac{5400 \text{ kJ}}{1 \text{ mol C}_8\text{H}_{18}}$$

$$= 33,091 = 3.3 \times 10^4 \text{ kJ/L C}_8\text{H}_{18}$$

(b) If density and heat of combustion of E85 are weighted averages of the values for the pure substances, than energy per liter E85 is also a weighted average of energy per liter for the two substances.

$$kJ/L \; E85 = 0.15(kJ/L \; C_8H_{18}) + 0.85(kJ/L \; C_2H_5OH)$$

$$kJ/L \; E85 = 0.15(33,091 \; kJ) + 0.85(21,174 \; kJ) = 22,962 = 2.3 \times 10^4 \; kJ/L \; E85$$

(c) Whether comparing gal or L, all conversion factors for the two fuels cancel, so we can apply the energy ratio directly to the volume under consideration.

The energy ratio for E85 to gasoline is $(22,962/33,091) = 0.6939 = 0.69$

$$10 \; gal \; gas \times \frac{kJ \; from \; E85}{0.6939 \; kJ \; from \; gas} = 14.41 = 14 \; gal \; E85$$

(d) If the E85/gasoline energy ratio is 0.69, the cost ratio must be 0.69 or less to "break-even" on price. $0.69(\$3.88) = \$2.68/gal \; E85$

Check. 10 gal gas(\$3.88/gal) = \$39; 14.4 gal E85(\$2.68/gal) = \$39.

Additional Exercises

5.92 (a) $E_p = mgd = 52.0 \; kg \times 9.81 \; m/s^2 \times 10.8 \; m = 5509.3 \; J = 5.51 \; kJ$

(b) $E_k = 1/2 \; mv^2; \; v = (2E_k/m)^{1/2} = \left(\dfrac{2 \times 5509.3 \; kg \cdot m^2/s^2}{52.0 \; kg} \right)^{1/2} = 14.6 \; m/s$

(c) Yes, the diver does work on entering (pushing back) the water in the pool.

5.94 Freezing is an exothermic process (the opposite of melting, which is clearly endothermic). When the system, the soft drink, freezes, it releases energy to the surroundings, the can. Some of this energy does the work of splitting the can.

5.95 (a) No work is done when the gas expands.

(b) No work is done because the evacuated flask is truly empty. There is no surrounding substance to be "pushed back."

(c) $\Delta E = q + w$. From part (b), no work is done when the gas expands. The flasks are perfectly insulated, so no heat flows. $\Delta E = 0 + 0 = 0$. The answer is a bit surprising, because a definite change occurred that required no work or heat transfer and consequently involved no energy change.

5.96 (a) $q = 0$, $w > 0$ (work done to system), $\Delta E > 0$

(b) Because the system (the gas) is losing heat, the sign of q is negative.

Two interpretations of the final state in (b) are possible. If the final state in (b) is identical to the final state in (a), $\Delta E(a) = \Delta E(b)$. If the final volumes are identical, case (b) requires either more (non-PV) work or heat input to compress the gas because some heat is lost to the surroundings. (The moral of this story is that the more energy lost by the system as heat, the greater the work on the system required to accomplish the desired change.)

Alternatively, if w is identical in the two cases and q is negative for case (b), then $\Delta E(b) < \Delta E(a)$. Assuming identical final volumes, the final temperature and pressure in (b) are slightly lower than those values in (a).

5.98   If a function sometimes depends on path, then it is simply not a state function. Enthalpy is a state function, so ΔH for the two pathways leading to the same change of state pictured in Figure 5.10 must be the same. However, q is not the same for the both. Our conclusion must be that $\Delta H \neq q$ for these pathways. The condition for $\Delta H = q_p$ (other than constant pressure) is that the only possible work on or by the system is pressure-volume work. Clearly, the work being done in this scenario is not pressure-volume work, so $\Delta H \neq q$, even though the two changes occur at constant pressure.

5.100   (a) $q_{Cu} = \dfrac{0.385 \text{ J}}{\text{g-K}} \times 121.0 \text{ g Cu} \times (30.1 \,^{\circ}C - 100.4 \,^{\circ}C) = -3274.9 = -3.27 \times 10^3 \text{ J}$

The negative sign indicates the 3.27×10^3 J are lost by the Cu block.

(b) $q_{H_2O} = \dfrac{4.184 \text{ J}}{\text{g-K}} \times 150.0 \text{ g } H_2O \times (30.1 \,^{\circ}C - 25.1 \,^{\circ}C) = 3138 = 3.1 \times 10^3 \text{ J}$

The positive sign indicates that 3.14×10^3 J are gained by the H_2O.

(c)   The difference in the heat lost by the Cu and the heat gained by the water is 3.275×10^3 J $- 3.138 \times 10^3$ J $= 0.137 \times 10^3$ J $= 1 \times 10^2$ J. The temperature change of the calorimeter is 5.0 °C. The heat capacity of the calorimeter in J/K is

$0.137 \times 10^3 \text{ J} \times \dfrac{1}{5.0 \,^{\circ}C} = 27.4 = 3 \times 10^1 \text{ J/ K}.$

Because q_{H_2O} is known to 1 decimal place, the difference has 1 decimal place and the result has 1 sig fig.

If the rounded results from (a) and (b) are used,

$C_{calorimeter} = \dfrac{0.2 \times 10^3 \text{ J}}{5.0 \,^{\circ}C} = 4 \times 10^1 \text{ J/ K}.$

(d) $q_{H_2O} = 3.275 \times 10^3 \text{ J} = \dfrac{4.184 \text{ J}}{\text{g-K}} \times 150.0 \text{ g} \times (\Delta T)$

$\Delta T = 5.22 \,^{\circ}C; \ T_f = 25.1 \,^{\circ}C + 5.22 \,^{\circ}C = 30.3 \,^{\circ}C$

5.101   (a)   From the mass of benzoic acid that produces a certain temperature change, we can calculate the heat capacity of the calorimeter.

$\dfrac{0.235 \text{ g benzoic acid}}{1.642 \,^{\circ}C \text{ change observed}} \times \dfrac{26.38 \text{ kJ}}{1 \text{ g benzoic acid}} = 3.7755 = 3.78 \text{ kJ/} \,^{\circ}C$

Now we can use this experimentally determined heat capacity with the data for caffeine.

$\dfrac{1.525 \,^{\circ}C \text{ rise}}{0.265 \text{ g caffeine}} \times \dfrac{3.7755 \text{ kJ}}{1 \,^{\circ}C} \times \dfrac{194.2 \text{ g caffeine}}{1 \text{ mol caffeine}} = 4.22 \times 10^3 \text{ kJ/ mol caffeine}$

(b)   The overall uncertainty is approximately equal to the sum of the uncertainties due to each effect. The uncertainty in the mass measurement is 0.001/0.235 or 0.001/0.265, about 1 part in 235 or 1 part in 265. The uncertainty in the temperature measurements is 0.002/1.642 or 0.002/1.525, about 1 part in 820 or 1 part in 760. Thus the uncertainty in heat of combustion from each measurement is

$\dfrac{4220}{235} = 18 \text{ kJ}; \quad \dfrac{4220}{265} = 16 \text{ kJ}; \quad \dfrac{4220}{820} = 5 \text{ kJ}; \quad \dfrac{4220}{760} = 6 \text{ kJ}$

The sum of these uncertainties is 45 kJ. In fact, the overall uncertainty is less than this because independent errors in measurement do tend to partially cancel.

5.103   (a)   For comparison, balance the equations so that 1 mole of CH_4 is burned in each.

$$CH_4(g) + O_2(g) \rightarrow C(s) + 2\,H_2O(l)$$

$$CH_4(g) + 3/2\,O_2(g) \rightarrow CO(g) + 2\,H_2O(l)$$

$$CH_4(g) + 2\,O_2(g) \rightarrow CO_2(g) + 2\,H_2O(l)$$

          (b) $\Delta H^{\circ}_{rxn} = \Delta H^{\circ}_f\ C(s) + 2\,\Delta H^{\circ}_f\ H_2O(l) - \Delta H^{\circ}_f\ CH_4(g) - \Delta H^{\circ}_f\ O_2(g)$

                      $= 0 + 2(-285.83\ kJ) - (-74.8) - 0 = -496.9\ kJ$

                $\Delta H^{\circ}_{rxn} = \Delta H^{\circ}_f\ CO(g) + 2\,\Delta H^{\circ}_f\ H_2O(l) - \Delta H^{\circ}_f\ CH_4(g) - 3/2\,\Delta H^{\circ}_f\ O_2(g)$

                      $= (-110.5\ kJ) + 2(-285.83\ kJ) - (-74.8\ kJ) - 3/2(0) = -607.4\ kJ$

                $\Delta H^{\circ}_{rxn} = \Delta H^{\circ}_f\ CO_2(g) + 2\,\Delta H^{\circ}_f\ H_2O(l) - \Delta H^{\circ}_f\ CH_4(g) - 2\,\Delta H^{\circ}_f\ O_2(g)$

                      $= -393.5\ kJ + 2(-285.83\ kJ) - (-74.8\ kJ) - 2(0) = -890.4\ kJ$

          (c)   Assuming that $O_2(g)$ is present in excess, the reaction that produces $CO_2(g)$ represents the most negative ΔH per mole of CH_4 burned. More of the potential energy of the reactants is released as heat during the reaction to give products of lower potential energy. The reaction that produces $CO_2(g)$ is the most "downhill" in enthalpy.

5.104

$$\begin{array}{ll}
2[CH_4(g) + 2\,O_2(g) \;\rightarrow\; CO_2(g) + 2\,H_2O(l)] & \Delta H^{\circ} = 2(-890.3\ kJ) \\
C_2H_6(g) \;\rightarrow\; C_2H_4(g) + H_2(g) & \Delta H^{\circ} = 136.3\ kJ \\
1/2[4\,CO_2(g) + 6\,H_2O(l) \;\rightarrow\; 2\,C_2H_6(g) + 7\,O_2(g)] & \Delta H^{\circ} = 1/2(3120.8\ kJ) \\
1/2[2\,H_2O(l) \;\rightarrow\; 2\,H_2(g) + O_2(g)] & \Delta H^{\circ} = 1/2(571.6\ kJ) \\
\hline
2\,CH_4(g) \;\rightarrow\; C_2H_4(g) + 2\,H_2(g) & \Delta H^{\circ} = 201.9\ kJ
\end{array}$$

5.105   For nitroethane:

$$\frac{1368\ kJ}{1\ mol\ C_2H_5NO_2} \times \frac{1\ mol\ C_2H_5NO_2}{75.072\ g\ C_2H_5NO_2} \times \frac{1.052\ g\ C_2H_5NO_2}{1\ cm^3} = 19.17\ kJ/cm^3$$

          For ethanol:

$$\frac{1367\ kJ}{1\ mol\ C_2H_5OH} \times \frac{1\ mol\ C_2H_5OH}{46.069\ g\ C_2H_5OH} \times \frac{0.789\ g\ C_2H_5OH}{1\ cm^3} = 23.4\ kJ/cm^3$$

          For methylhydrazine:

$$\frac{1307\ kJ}{1\ mol\ CH_6N_2} \times \frac{1\ mol\ CH_6N_2}{46.072\ g\ CH_6N_2} \times \frac{0.874\ g\ CH_6N_2}{1\ cm^3} = 24.8\ kJ/cm^3$$

          Thus, **methylhydrazine** would provide the most energy per unit volume, with ethanol a close second.

5.107   The reaction for which we want ΔH is:

$$4\,NH_3(l) + 3\,O_2(g) \rightarrow 2\,N_2(g) + 6\,H_2O(g)$$

          Before we can calculate ΔH for this reaction, we must calculate ΔH_f for $NH_3(l)$.

We know that ΔH_f for $NH_3(g)$ is –46.2 kJ/mol, and that for $NH_3(l) \rightarrow NH_3(g)$, $\Delta H = 23.2$ kJ/mol

Thus, $\Delta H_{vap} = \Delta H_f\, NH_3(g) - \Delta H_f\, NH_3(l)$.

23.2 kJ $= -46.2$ kJ $- \Delta H_f\, NH_3(l)$; $\Delta H_f\, NH_3(l) = -69.4$ kJ/mol

Then for the overall reaction, the enthalpy change is:

$\Delta H_{rxn} = 6\,\Delta H_f\, H_2O(g) + 2\,\Delta H_f\, N_2(g) - 4\,\Delta H_f\, NH_3(l) - 3\,\Delta H_f\, O_2$

$= 6(-241.82$ kJ$) + 2(0) - 4(-69.4$ kJ$) - 3(0) = -1173.3$ kJ

$$\frac{-1173.3\,\text{kJ}}{4\,\text{mol}\,NH_3} \times \frac{1\,\text{mol}\,NH_3}{17.0\,\text{g}\,NH_3} = \frac{0.81\,\text{g}\,NH_3}{1\,\text{cm}^3} \times \frac{1000\,\text{cm}^3}{1\,\text{L}} = \frac{-1.4\times10^4\,\text{kJ}}{\text{L}\,NH_3}$$

(This result has 2 significant figures because the density is expressed to 2 figures.)

$2\,CH_3OH(l) + 3\,O_2(g) \rightarrow 2\,CO_2(g) + 4\,H_2O(g)$

$\Delta H = 2(-393.5$ kJ$) + 4(-241.82$ kJ$) - 2(-239$ kJ$) - 3(0) = -1276$ kJ

$$\frac{-1276\,\text{kJ}}{2\,\text{mol}\,CH_3OH} \times \frac{1\,\text{mol}\,CH_3OH}{32.04\,\text{g}\,CH_3OH} \times \frac{0.792\,\text{g}\,CH_3OH}{1\,\text{cm}^3} \times \frac{1000\,\text{cm}^3}{1\,\text{L}} = \frac{-1.58\times10^4\,\text{kJ}}{\text{L}\,CH_3OH}$$

In terms of heat obtained per unit volume of fuel, methanol is a slightly better fuel than liquid ammonia.

5.108 **1,3-butadiene**, C_4H_6, MM = 54.092 g/mol

(a) $C_4H_6(g) + 11/2\,O_2(g) \rightarrow 4\,CO_2(g) + 3\,H_2O(l)$

$\Delta H^\circ_{rxn} = 4\,\Delta H^\circ_f\, CO_2(g) + 3\,\Delta H^\circ_f\, H_2O(l) - \Delta H^\circ_f\, C_4H_6(g) - 11/2\,\Delta H^\circ_f\, O_2(g)$

$= 4(-393.5$ kJ$) + 3(-285.83$ kJ$) - 111.9$ kJ $+ 11/2\,(0) = -2543.4$ kJ/mol C_4H_6

(b) $\dfrac{-2543.4\,\text{kJ}}{1\,\text{mol}\,C_4H_6} \times \dfrac{1\,\text{mol}\,C_4H_6}{54.092\,\text{g}} = 47.020 \rightarrow 47$ kJ/g

(c) % H $= \dfrac{6(1.008)}{54.092} \times 100 = 11.18\%$ H

1-butene, C_4H_8, MM = 56.108 g/mol

(a) $C_4H_8(g) + 6\,O_2(g) \rightarrow 4\,CO_2(g) + 4\,H_2O(l)$

$\Delta H^\circ_{rxn} = 4\,\Delta H^\circ_f\, CO_2(g) + 4\,\Delta H^\circ_f\, H_2O(l) - \Delta H^\circ_f\, C_4H_8(g) - 6\,\Delta H^\circ_f\, O_2(g)$

$= 4(-393.5$ kJ$) + 4(-285.83$ kJ$) - 1.2$ kJ $- 6(0) = -2718.5$ kJ/mol C_4H_8

(b) $\dfrac{-2718.5\,\text{kJ}}{1\,\text{mol}\,C_4H_8} \times \dfrac{1\,\text{mol}\,C_4H_8}{56.108\,\text{g}\,C_4H_8} = 48.451 \rightarrow 48$ kJ/g

(c) % H $= \dfrac{8(1.008)}{56.108} \times 100 = 14.37\%$ H

n-butane, $C_4H_{10}(g)$, MM = 58.124 g/mol

(a) $C_4H_{10}(g) + 13/2\ O_2(g) \rightarrow 4\ CO_2(g) + 5\ H_2O(l)$

$\Delta H^{\circ}_{rxn} = 4\ \Delta H^{\circ}_f\ CO_2(g) + 5\ \Delta H^{\circ}_f\ H_2O(l) - \Delta H^{\circ}_f\ C_4H_{10}(g) - 13/2\ \Delta H^{\circ}_f\ O_2(g)$

$= 4(-393.5\ kJ) + 5(-285.83\ kJ) - (-124.7\ kJ) - 13/2(0)$

$= -2878.5\ kJ/mol\ C_4H_{10}$

(b) $\dfrac{-2878.5\ kJ}{1\ mol\ C_4H_{10}} \times \dfrac{1\ mol\ C_4H_{10}}{58.124\ g\ C_4H_{10}} = 49.523 \rightarrow 50\ kJ/g$

(c) $\%\ H = \dfrac{10(1.008)}{58.124} \times 100 = 17.34\%\ H$

(d)   It is certainly true that as the mass % H increases, the fuel value (kJ/g) of the hydrocarbon increases, given the same number of C atoms. A graph of the data in parts (b) and (c) (see below) suggests that mass % H and fuel value are directly proportional when the number of C atoms is constant.

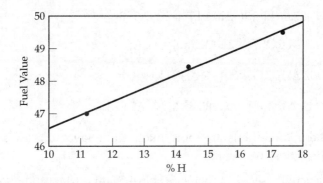

5.110   *Plan.* Use dimensional analysis to calculate the amount of solar energy supplied per m^2 in 1 hr. Use stoichiometry to calculate the amount of plant energy used to produce sucrose per m^2 in 1 hr. Calculate the ratio of energy for sucrose to total solar energy, per m^2 per hr.

Solve. 1 W = 1 J/s, 1 kW = 1 kJ/s

$$\frac{1.0\ kW}{m^2} = \frac{1.0\ kJ/s}{m^2} = \frac{1.0\ kJ}{m^2 \text{-}s} \times \frac{60\ s}{1\ min} \times \frac{60\ min}{1\ hr} = \frac{3.6 \times 10^3\ kJ}{m^2 \text{-} hr}$$

$$\frac{5645\ kJ}{mol\ sucrose} \times \frac{1\ mol\ sucrose}{342.3\ g\ sucrose} \times \frac{0.20\ g\ sucrose}{m^2 \text{-} hr} = 3.298 = 3.3\ kJ/m^2 \text{-} hr \text{ for sucrose production}$$

$$\frac{3.298\ kJ \text{ for sucrose}}{3.6 \times 10^3\ kJ \text{ total solar}} \times 100 = 0.092\% \text{ sunlight used to produce sucrose}$$

5.111   (a) $6\ CO_2(g) + 6\ H_2O(l) \rightarrow C_6H_{12}O_6(s) + 6\ O_2(g)$, $\Delta H^{\circ} = 2803\ kJ$

This is the reverse of the combustion of glucose (Section 5.8 and Exercise 5.89), so $\Delta H^{\circ} = -(-2803)\ kJ = +2803\ kJ$.

$$\frac{5.5 \times 10^{16}\ g\ CO_2}{yr} \times \frac{1\ mol\ CO_2}{44.01\ g\ CO_2} \times \frac{2803\ kJ}{6\ mol\ CO_2} = 5.838 \times 10^{17} = 5.8 \times 10^{17}\ kJ$$

(b) $1 W = 1 J/s; 1 W \cdot s = 1 J$

$$\frac{5.838 \times 10^{17} \text{ kJ}}{\text{yr}} \times \frac{1000 \text{ J}}{\text{kJ}} \times \frac{1 \text{ yr}}{365 \text{ d}} \times \frac{1 \text{ d}}{24 \text{ hr}} \times \frac{1 \text{ hr}}{60 \text{ min}} \times \frac{1 \text{ min}}{60 \text{ s}} \times \frac{1 W \cdot s}{J}$$

$$\times \frac{1 \text{ MW}}{1 \times 10^6 \text{ W}} = 1.851 \times 10^7 \text{ MW} = 1.9 \times 10^7 \text{ MW}$$

$$1.9 \times 10^7 \text{ MW} \times \frac{1 \text{ plant}}{10^3 \text{ MW}} = 1.9 \times 10^4 = 19{,}000 \text{ nuclear power plants}$$

Integrative Exercises

5.113   (a,b) $Ag^+(aq) + Li(s) \rightarrow Ag(s) + Li^+(aq)$

$$\Delta H° = \Delta H_f° \, Li^+(aq) - \Delta H_f° \, Ag^+(aq)$$

$$= -278.5 \text{ kJ} - 105.90 \text{ kJ} = -384.4 \text{ kJ}$$

$Fe(s) + 2 \, Na^+(aq) \rightarrow Fe^{2+}(aq) + 2 \, Na(s)$

$$\Delta H° = \Delta H_f° \, Fe^{2+}(aq) - 2 \, \Delta H_f° \, Na^+(aq)$$

$$= -87.86 \text{ kJ} - 2(-240.1 \text{ kJ}) = +392.3 \text{ kJ}$$

$2 \, K(s) + 2 \, H_2O(l) \rightarrow 2 \, KOH(aq) + H_2(g)$

$$\Delta H° = 2 \, \Delta H_f° \, KOH(aq) - 2 \, \Delta H_f° \, H_2O(l)$$

$$= 2(-482.4 \text{ kJ}) - 2(-285.83 \text{ kJ}) = -393.1 \text{ kJ}$$

(c)   Exothermic reactions are more likely to be favored, so we expect the first and third reactions be favored.

(d)   In the activity series of metals, Table 4.5, any metal can be oxidized by the cation of a metal below it on the table.

Ag$^+$ is below Li, so the first reaction will occur.

Na$^+$ is above Fe, so the second reaction will not occur.

H$^+$ (formally in H$_2$O) is below K, so the third reaction will occur.

These predictions agree with those in part (c).

5.115   (a) $\text{mol Cu} = M \times L = 1.00 \, M \times 0.0500 \text{ L} = 0.0500 \text{ mol}$

$g = \text{mol} \times MM = 0.0500 \times 63.546 = 3.1773 = 3.18 \text{ g Cu}$

(b)   The precipitate is copper(II) hydroxide, $Cu(OH)_2$.

(c) $CuSO_4(aq) + 2 \, KOH(aq) \rightarrow Cu(OH)_2(s) + K_2SO_4(aq)$, complete

$Cu^{2+}(aq) + 2 \, OH^-(aq) \rightarrow Cu(OH)_2(s)$, net ionic

(d)   The temperature of the calorimeter rises, so the reaction is exothermic and the sign of q is negative.

$$q = -6.2 \, °C \times 100 \text{ g} \times \frac{4.184 \text{ J}}{1 \text{ g-}°C} = -2.6 \times 10^3 \text{ J} = -2.6 \text{ kJ}$$

The reaction as carried out involves only 0.050 mol of $CuSO_4$ and the stoichiometrically equivalent amount of KOH. On a molar basis,

$$\Delta H = \frac{-2.6 \text{ kJ}}{0.050 \text{ mol}} = -52 \text{ kJ} \text{ for the reaction as written in part (c)}$$

5.117   (a) $21.83 \text{ g } CO_2 \times \dfrac{1 \text{ mol } CO_2}{44.01 \text{ g } CO_2} \times \dfrac{1 \text{ mol C}}{1 \text{ mol } CO_2} \times \dfrac{12.01 \text{ g C}}{1 \text{ mol C}} = 5.9572 = 5.957 \text{ g C}$

$4.47 \text{ g } H_2O \times \dfrac{1 \text{ mol } H_2O}{18.02 \text{ g } H_2O} \times \dfrac{2 \text{ mol H}}{1 \text{ mol } H_2O} \times \dfrac{1.008 \text{ g H}}{\text{mol H}} = 0.5001 = 0.500 \text{ g H}$

The sample mass is $(5.9572 + 0.5001) = 6.457 \text{ g}$

   (b) $5.957 \text{ g C} \times \dfrac{1 \text{ mol C}}{12.01 \text{ g C}} = 0.4960 \text{ mol C}; \ 0.4960/0.496 = 1$

$0.500 \text{ g H} \times \dfrac{1 \text{ mol H}}{1.008 \text{ g H}} = 0.496 \text{ mol H}; \ 0.496/0.496 = 1$

The empirical formula of the hydrocarbon is CH.

   (c)   Calculate "ΔH_f°" for 6.457 g of the sample.

$6.457 \text{ g sample} + O_2(g) \rightarrow 21.83 \text{ g } CO_2(g) + 4.47 \text{ g } H_2O(g), \ \Delta H_{comb}^{\circ} = -311 \text{ kJ}$

$\Delta H_{comb}^{\circ} = \Delta H_f^{\circ} \ CO_2(g) + \Delta H_f^{\circ} \ H_2O(g) - \Delta H_f^{\circ} \text{ sample} - \Delta H_f^{\circ} \ O_2(g)$

$\Delta H_f^{\circ} \text{ sample} = \Delta H_f^{\circ} \ CO_2(g) + \Delta H_f^{\circ} \ H_2O(g) - \Delta H_{comb}^{\circ}$

$\Delta H_f^{\circ} \ CO_2(g) = 21.83 \text{ g } CO_2 \times \dfrac{1 \text{ mol } CO_2}{44.01 \text{ g } CO_2} \times \dfrac{-393.5 \text{ kJ}}{\text{mol } CO_2} = -195.185 = -195.2 \text{ kJ}$

$\Delta H_f^{\circ} \ H_2O(g) = 4.47 \text{ g } H_2O \times \dfrac{1 \text{ mol } H_2O}{18.02 \text{ g } H_2O} \times \dfrac{-241.82 \text{ kJ}}{\text{mol } H_2O} = -59.985 = -60.0 \text{ kJ}$

$\Delta H_f^{\circ} \text{ sample} = -195.185 \text{ kJ} - 59.985 \text{ kJ} - (-311 \text{ kJ}) = 55.83 = 56 \text{ kJ}$

$H_f^{\circ} = \dfrac{55.83 \text{ kJ}}{6.457 \text{ g sample}} \times \dfrac{13.02 \text{ g}}{\text{CH unit}} = 112.6 = 1.1 \times 10^2 \text{ kJ/CH unit}$

   (d)   The hydrocarbons in Appendix C with empirical formula CH are C_2H_2 and C_6H_6.

substance	ΔH_f°/mol	ΔH_f°/CH unit
$C_2H_2(g)$	226.7 kJ	113.4 kJ
$C_6H_6(g)$	82.9 kJ	13.8 kJ
$C_6H_6(l)$	49.0 kJ	8.17 kJ
sample		1.1×10^2 kJ

The calculated value of ΔH_f°/CH unit for the sample is a good match with acetylene, $C_2H_2(g)$.

5.118   (a) $CH_4(g) \rightarrow C(g) + 4\,H(g)$ $\qquad\qquad\qquad$ (i) reaction given

$\qquad\qquad\qquad$ $CH_4(g) \rightarrow C(s) + 2\,H_2(g)$ $\qquad\qquad\qquad$ (ii) reverse of formation

$\qquad\qquad$ The differences are: the state of C in the products; the chemical form, atoms, or diatomic molecules, of H in the products.

$\qquad$ (b)   i. $\qquad$ $\Delta H° = \Delta H_f^° \; C(g) + 4\,\Delta H_f^° \; H(g) - \Delta H_f^° \; CH_4(g)$

$\qquad\qquad\qquad\qquad$ $= 718.4 \text{ kJ} + 4(217.94) \text{ kJ} - (-74.8) \text{ kJ} = 1665.0 \text{ kJ}$

$\qquad\qquad$ ii. $\qquad$ $\Delta H° = \Delta H_f^° \; CH_4 = -(-74.8) \text{ kJ} = 74.8 \text{ kJ}$

$\qquad\qquad$ The rather large difference in $\Delta H°$ values is due to the enthalpy difference between isolated gaseous C atoms and the orderly, bonded array of C atoms in graphite, C(s), as well as the enthalpy difference between isolated H atoms and H_2 molecules. In other words, it is due to the difference in the enthalpy stored in chemical bonds in C(s) and $H_2(g)$ versus the corresponding isolated atoms.

$\qquad$ (c) $CH_4(g) + 4\,F_2(g) \rightarrow CF_4(g) + 4\,HF(g)$ $\qquad$ $\Delta H° = -1679.5 \text{ kJ}$

$\qquad\qquad$ The $\Delta H°$ value for this reaction was calculated in Solution 5.88.

$$3.45 \text{ g } CH_4 \times \frac{1 \text{ mol } CH_4}{16.04 \text{ g } CH_4} \times 0.21509 = 0.215 \text{ mol } CH_4$$

$$1.22 \text{ g } F_2 \times \frac{1 \text{ mol } F_2}{38.00 \text{ g } F_2} = 0.03211 = 0.0321 \text{ mol } F_2$$

$\qquad\qquad$ There are fewer mol F_2 than CH_4, but 4 mol F_2 are required for every 1 mol of CH_4 reacted, so clearly F_2 is the limiting reactant.

$$0.03211 \text{ mol } F_2 \times \frac{-1679.5 \text{ kJ}}{4 \text{ mol } F_2} = -13.48 = -13.5 \text{ kJ heat evolved}$$

5.119   (a)   From Solution 5.17, 1 Btu = 1054 J = 1.054 kJ.

$$5.81 \times 10^{17} \text{ kJ} \times \frac{1 \text{ Btu}}{1.054 \text{ kJ}} \times \frac{1 \text{ Quad}}{1 \times 10^{12} \text{ Btu}} = 5.51 \times 10^5 \text{ Quads}$$

$\qquad$ (b)   From Solution 5.112(a), heat of combustion of methane, is –890.4 kJ/mol CH_4.

$$99.5 \text{ Quads} \times \frac{1 \times 10^{12} \text{ Btu}}{1 \text{ Quad}} \times \frac{1.054 \text{ kJ}}{1 \text{ Btu}} \times \frac{1 \text{ mol } CH_4}{890.4 \text{ kJ}} = 1.1778 \times 10^{11} = 1.18 \times 10^{11} \text{ mol } CH_4$$

$\qquad$ (c) $1.1778 \times 10^{11} \text{ mol } CH_4 \times \dfrac{1 \text{ mol } CO_2}{1 \text{ mol } CH_4} \times \dfrac{44.01 \text{ g } CO_2}{1 \text{ mol } CO_2} \times \dfrac{1 \text{ kg}}{1000 \text{ g}} = 5.18 \times 10^9 \text{ kg } CO_2$

$\qquad$ (d)   Exercise 5.111 states that Earth fixes 5.5×10^{16} g or 5.5×10^{13} kg CO_2 via photosynthesis. If only the United States produces CO_2, photosynthesis is an adequate means of maintain a stable level of CO_2 in the atmosphere. A more revealing comparison is to estimate the global production of CO_2 from the estimated energy consumption for the year 2015.

$$5.51 \times 10^5 \text{ Quads} \times \frac{5.18 \times 10^9 \text{ kg CO}_2}{99.5 \text{ Quads}} = 2.87 \times 10^{13} \text{ kg CO}_2$$

Earth fixes 5.5×10^{16} g or 5.5×10^{13} kg CO_2 per year. This is twice as much as the CO_2 projected to be produced globally in 2015. (Note that these projections are based on the assumptions that: burning fossil fuels is the only source of CO_2, the only fuel we burn is CH_4, and the burning is 100% efficient. All these assumptions underestimate the amount of CO_2 going into the atmosphere.) Photosynthesis may be adequate to maintain a stable level of CO_2 in the short term but it does not provide a large capacity for increased energy consumption and CO_2 production.

6 Electronic Structure of Atoms

Visualizing Concepts

6.1 (a) Speed is distance traveled per unit time. Measure the distance between the center point and a second reference point, possibly the edge of the container. Using a stop watch, measure the elapsed time between when a wave forms at the center and when it reaches the second reference point. Find the ratio of distance to time.

(b) Measure the distance between two wave crests (or troughs or any analogous points on two adjacent waves). Better yet, measure the distance between two crests (or analogous points) that are several waves apart and divide by the number of waves that separate them.

(c) Because speed is distance/time, and wavelength is distance, we can calculate frequency by dividing speed by wavelength, $v = c/\lambda$.

(d) We can measure frequency of the wave by dropping an object such as a cork in the water and counting the number of times per second it moves through a complete cycle of motion.

6.3 Wave (a) corresponds to higher energy radiation. The energy of electromagnetic radiation is directly proportional to frequency and inversely proportional to wavelength. Wave (a) has the shorter wavelength and thus the higher energy.

6.4 (a) The glowing stove burner is an example of black body radiation, the observational basis for Planck's quantum theory. The wavelengths emitted are related to temperature, with cooler temperatures emitting longer wavelengths and hotter temperatures emitting shorter wavelengths. At the hottest setting, the burner emits orange visible light. At the cooler low setting, the burner emits longer wavelengths out of the visible region, and the burner appears black.

(b) If the burner had a super high setting, the emitted wavelengths would be shorter than those of orange light and the glow color would be more yellowish, progressing to white. (See Figure 6.5.)

6.5 (a) (iii) *Betelgeuse* < (i) Sun < (ii) *Rigel*.

(b) Black body radiation

6.7 (a) $n = 1, n = 4$ (b) $n = 1, n = 2$

(c) Wavelength and energy are inversely proportional; the smaller the energy, the longer the wavelength. In order of increasing wavelength (and decreasing energy): (iii) $n = 2$ to $n = 4$ < (iv) $n = 3$ to $n = 1$ < (ii) $n = 3$ to $n = 2$ < (i) $n = 1$ to $n = 2$

6.8   (a) $\psi^2(x)$ will be positive or zero at all values of x and have two maxima with larger magnitudes than the maximum in $\psi(x)$.

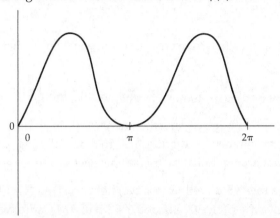

         (b)   The greatest probability of finding the electron is at the two maxima in $\psi^2(x)$ at $x = \pi/2$ and $3\pi/2$.

         (c)   There is zero probability of finding the electron at $x = \pi$. This value is called a node.

6.10   (a) $4 (d_{xy}, d_{yz}, d_{xz}, d_{x^2-y^2})$

         (b)   (iv) The d orbitals are named by the orientation of their lobes relative to the major axes.

6.11   (a)   In the left-most box, the two electrons cannot have the same spin. The *Pauli exclusion principle* states that no two electrons can have the same set of quantum numbers. Because the first three quantum numbers describe an orbital, the fourth quantum number, m_s, must have different values for two electrons in the same orbital; their "spins" must be opposite.

         (b)   Flip one of the arrows in the left-most box, so that one points up and the other down.

         (c)   Group 6A. The drawing shows three boxes or orbitals at the same energy, so it must represent p orbitals. Because some of these p orbitals are partially filled, they must be the valence orbitals of the element. Elements with four valence electrons in their p orbitals belong to group 6A.

6.12   (a)   Group 7A or 17, the halogens, the column second from the right

         (b)   Group 5A or 15

         (c)   Gallium, atomic number 31, at the intersection of row 4 and group 3A or 13

         (d)   All of the B groups, groups 3-12, in the middle of the major part of the table, not including the two rows of *f*-block elements

The Wave Nature of Light (Section 6.1)

6.14   (a)   Wavelength (λ) and frequency (ν) are inversely proportional; the proportionality constant is the speed of light (c). $\nu = c/\lambda$.

(b)   Light in the 210–230 nm range is in the ultraviolet region of the spectrum. These wavelengths are slightly shorter than the 400 nm short-wavelength boundary of the visible region.

6.16   (a)   False. The frequency of radiation decreases as the wavelength increases.

(b)   True.

(c)   False. Infrared light has lower frequencies than visible light.

(d)   False. The glow from a fireplace and the energy within a microwave oven are both forms of electromagnetic radiation. (A foghorn blast is a form of sound waves, which are not accompanied by oscillating electric and magnetic fields.)

6.18   Wavelength of (a) gamma rays < (d) yellow (visible) light < (e) red (visible) light < (b) 93.1 MHz FM (radio) waves < (c) 680 kHz or 0.680 MHz AM (radio) waves

6.20   (a) $\nu = c/\lambda$; $\dfrac{2.998 \times 10^8 \text{ m}}{\text{s}} \times \dfrac{1}{0.86 \text{ nm}} \times \dfrac{1}{1.0 \times 10^{-9} \text{ m}} = 3.5 \times 10^{17} \text{ s}^{-1}$

(b) $\lambda = c/\nu$; $\dfrac{2.998 \times 10^8 \text{ m}}{\text{s}} \times \dfrac{1 \text{ s}}{6.4 \times 10^{11}} = 4.7 \times 10^{-4} \text{ m}$

(c)   The radiation in (a) can be observed by an X-ray detector. The 8.6×10^{-10} m wavelength is within the range of 10^{-8} to 10^{-11} m for X-rays.

(d) $0.38 \text{ ps} \times \dfrac{1 \times 10^{-12} \text{ s}}{1 \text{ ps}} \times \dfrac{2.998 \times 10^8 \text{ m}}{\text{s}} = 1.1 \times 10^{-4} \text{ m} \ (0.11 \text{ mm})$

6.22   According to Figure 6.4, ultraviolet radiation has both higher frequency and shorter wavelength than infrared radiation. Looking forward to section 6.2, the energy of a photon is directly proportional to frequency (E = hν), so ultraviolet radiation yields more energy from a photovoltaic device.

Quantized Energy and Photons (Section 6.2)

6.24   Planck's original hypothesis was that energy could only be gained or lost in discrete amounts (quanta) with a certain minimum size. The size of the minimum energy change is related to the frequency of the radiation absorbed or emitted, ΔE = hν, and energy changes occur only in multiples of hν.

Einstein postulated that light itself is quantized, that the minimum energy of a photon (a quantum of light) is directly proportional to its frequency, E = hν. If a photon that strikes a metal surface has less than the threshold energy, no electron is emitted from the surface. If the photon has energy equal to or greater than the threshold energy, an electron is emitted and any excess energy becomes the kinetic energy of the electron.

6.26   *Analyze/Plan.* These questions deal with the relationships between energy, wavelength, and frequency. Use the relationships E = hν = hc/λ to calculate the desired quantities. Pay attention to units.   *Solve.*

(a) $\nu = c/\lambda$; $\dfrac{2.998 \times 10^8 \text{ m}}{\text{s}} \times \dfrac{1}{532 \text{ nm}} \times \dfrac{1 \text{ nm}}{1 \times 10^{-9} \text{ m}} = 5.64 \times 10^{14} \text{ s}^{-1}$

(b) $E = h\nu = 6.626 \times 10^{-34}$ J-s $\times$ 5.64×10^{14} s^{-1} = 3.73×10^{-19} J

(c) The energy gap between the ground and excited states is the energy of a single 532 nm photon emitted when one electron relaxes from the excited to the ground state. $\Delta E = 3.73 \times 10^{-19}$ J

6.28 $E = h\nu$

AM : 6.626×10^{-34} J-s $\times \dfrac{1010 \times 10^3}{1\,s} = 6.69 \times 10^{-28}$ J

FM : 6.626×10^{-34} J-s $\times \dfrac{98.3 \times 10^6}{1\,s} = 6.51 \times 10^{-26}$ J

The FM photon has about 100 times more energy than the AM photon.

6.30 $\dfrac{242 \times 10^3\,J}{mol\,Cl_2} \times \dfrac{1\,mol}{6.022 \times 10^{23}\,photons} = 4.0186 \times 10^{-19} = 4.02 \times 10^{-19}$ J/ photon

$\lambda = hc/E = \dfrac{6.626 \times 10^{-34}\,J\text{-}s}{4.0186 \times 10^{-19}\,J} \times \dfrac{2.998 \times 10^8\,m}{1\,s} = 4.94 \times 10^{-7}$ m = 494 nm

According to Figure 6.4, this is visible radiation.

6.32   (a) 3.55 mm = 3.55×10^{-3} m; the radiation is microwave.

(b) $E_{photon} = hc/\lambda = \dfrac{6.626 \times 10^{-34}\,J\text{-}s}{3.55 \times 10^{-3}\,m} \times \dfrac{2.998 \times 10^8\,m}{1\,s} = 5.5957 \times 10^{-23}$

$= 5.60 \times 10^{-23}$ J/photon

$\dfrac{5.5957 \times 10^{-23}\,J}{1\,photon} \times \dfrac{3.2 \times 10^8\,photons}{1\,s} \times \dfrac{60\,s}{1\,min} \times \dfrac{60\,min}{1\,hr} = 6.4463 \times 10^{-11}$

$= 6.4 \times 10^{-11}$ J/hr

6.34   (a) $\nu = E/h = \dfrac{6.94 \times 10^{-19}\,J}{6.626 \times 10^{-34}\,J\text{-}s} = 1.04739 \times 10^{15} = 1.05 \times 10^{15}$ s^{-1}

(b) $\lambda = hc/E = \dfrac{6.626 \times 10^{-34}\,J\text{-}s}{6.94 \times 10^{-19}\,J} \times \dfrac{2.998 \times 10^8\,m}{s} = 2.86 \times 10^{-7}$ m = 286 nm

(c) No. The maximum 286 nm wavelength required to eject an electron from titanium is in the ultraviolet (Figure 6.4).

(d) $E_{233} = hc/\lambda = \dfrac{6.626 \times 10^{-34}\,J\text{-}s}{233 \times 10^{-9}\,m} \times \dfrac{2.998 \times 10^8\,m}{s} = 8.5256 \times 10^{-19} = 8.53 \times 10^{-19}$ J

$E_K = E_{233} - E_{min} = 8.5256 \times 10^{-19}\,J - 6.94 \times 10^{-19}\,J = 1.5856 \times 10^{-19} = 1.59 \times 10^{-19}$ J

(e) One electron is emitted per photon. Calculate the number of 233 nm photons in 2.00 μJ. The excess energy in each photon will become the kinetic energy of the electron; it cannot be "pooled" to emit additional electrons.

$2.00\,\mu J \times \dfrac{1 \times 10^{-6}\,J}{\mu J} \times \dfrac{1\,photon}{8.5256 \times 10^{-19}\,J} \times \dfrac{1\,e^-}{1\,photon} = 2.35 \times 10^{12}$ electrons

Bohr's Model; Matter Waves (Sections 6.3 and 6.4)

6.36   (a)   Statements (i) and (iii) are true.

            (i) A hydrogen atom in the $n = 3$ state emits light by moving to a lower energy state. The only allowed destinations are $n = 2$ or $n = 1$, so only the two wavelengths corresponding to these transitions are allowed.

            (ii) As the value of n increases, energy becomes less negative.

            (iii) This is described mathematically in the Rydberg equation.

   (b)   When a hydrogen atom changes from the ground state to an excited state, the single electron moves farther away from the nucleus, so the atom "expands."

6.38   (a)   Absorbed.   (b)   Emitted.   (c)   Absorbed.

6.40   (a)   The value of n for the electron increases, so ΔE is positive.

   (b)   *Analyze/Plan.* Use Equation 6.6 to calculate ΔE, then $\lambda = hc/\Delta E$. *Solve.*

$$n_i = 4, n_f = 9; \quad \Delta E = -2.18 \times 10^{-18} \text{ J} \left[\frac{1}{n_f^2} - \frac{1}{n_i^2} \right] = -2.18 \times 10^{-18} \text{ J} (1/81 - 1/16)$$

$$\lambda = hc/E = \frac{6.626 \times 10^{-34} \text{ J-s} \times 2.998 \times 10^8 \text{ m/s}}{-2.18 \times 10^{-18} \text{ J} (1/81 - 1/16)} = 1.82 \times 10^{-6} \text{ m}$$

            This wavelength will be absorbed. Energy is required to move the electron to the higher energy $n = 9$ state. [Note that the denominator of the preceding equation has a positive sign, because $(1/81 - 1/16)$ is negative.]

   (c)   The light in (b) is in the infrared.

6.42   (a)   Transitions with $n_f = 1$ have larger ΔE values and shorter wavelengths than those with $n_f = 2$. These transitions will lie in the ultraviolet region.

   (b) $n_i = 2, n_f = 1;$ $\lambda = hc/E = \dfrac{6.626 \times 10^{-34} \text{ J-s} \times 2.998 \times 10^8 \text{ m/s}}{-2.18 \times 10^{-18} \text{ J} (1/1 - 1/4)} = 1.21 \times 10^{-7} \text{ m}$

            $n_i = 3, n_f = 1;$ $\lambda = hc/E = \dfrac{6.626 \times 10^{-34} \text{ J-s} \times 2.998 \times 10^8 \text{ m/s}}{-2.18 \times 10^{-18} \text{ J} (1/1 - 1/9)} = 1.03 \times 10^{-7} \text{ m}$

            $n_i = 4, n_f = 1;$ $\lambda = hc/E = \dfrac{6.626 \times 10^{-34} \text{ J-s} \times 2.998 \times 10^8 \text{ m/s}}{-2.18 \times 10^{-18} \text{ J} (1/1 - 1/16)} = 0.972 \times 10^{-7} \text{ m}$

6.44   (a) $1094 \text{ nm} \times \dfrac{1 \times 10^{-9} \text{ m}}{1 \text{ nm}} = 1.094 \times 10^{-6} \text{ m}$; this line is in the infrared.

   (b)   Absorption lines with $n_i = 1$ are in the ultraviolet and with $n_i = 2$ are in the visible. Thus, $n_i \geq 3$, but we do not know the exact value of n_i. Calculate the longest wavelength with $n_i = 3$ ($n_f = 4$). If this is less than 1094 nm, $n_i > 3$.

$$\lambda = hc/E = \frac{6.626 \times 10^{-34} \text{ J-s} \times 2.998 \times 10^8 \text{ m/s}}{-2.18 \times 10^{-18} \text{ J} (1/16 - 1/9)} = 1.875 \times 10^{-6} \text{ m}$$

This wavelength is longer than 1.094×10^{-6} m, but we are in the ballpark. Use $n_i = 3$ and solve for n_f as in Solution 6.43. Note that ΔE is positive because we are dealing with absorption.

$$n_f = \left(\frac{1}{n_i^2} - \frac{hc}{\lambda(2.18\times10^{-18} \text{ J})} \right)^{-1/2} = \left(1/9 - \frac{6.626\times10^{-34} \text{ J-s} \times 2.998\times10^8 \text{ m/s}}{1.094\times10^{-6} \text{ m} \times 2.18\times10^{-18} \text{ J}} \right)^{-1/2} = 6$$

$n_f = 6, n_i = 3$

6.46   According to Equation 6.6 and several preceding solutions, the greater the energy of light emitted, the lower the value of n_f. The greater the energy of light, the shorter its wavelength, $\Delta E = hc/\lambda$. The order of increasing wavelength (and decreasing energy) of light emitted is:

$n = 4$ to $n = 2$; $n = 3$ to $n = 2$; $n = 5$ to $n = 3$; $n = 7$ to $n = 4$

6.48   Find the mass of an electron on the inside back cover of the text. $\lambda = h/mv$; change velocity to m/s.

mass of muon $= 206.8 \times 9.1094 \times 10^{-31}$ g $= 1.8838 \times 10^{-28} = 1.88 \times 10^{-28}$ kg

$$\lambda = \frac{6.626\times10^{-34} \text{ kg-m}^2\text{-s}}{1 \text{ s}^2} \times \frac{1}{1.8838\times10^{-28} \text{ kg}} \times \frac{1 \text{ s}}{8.85\times10^3 \text{ m/s}} = 3.97\times10^{-10} \text{ m} = 3.97 \text{ Å}$$

6.50 $m_e = 9.1094 \times 10^{-31}$ kg (back inside cover of textbook)

$$\lambda = \frac{6.626 \times 10^{-34} \text{ kg-m}^2\text{-s}}{1 \text{ s}^2} \times \frac{1}{9.1094 \times 10^{-31} \text{ kg}} \times \frac{1 \text{ s}}{9.47 \times 10^6 \text{ m}} = 7.68 \times 10^{-11} \text{ m}$$

$$7.68 \times 10^{-11} \text{ m} \times \frac{1 \text{ Å}}{1 \times 10^{-10} \text{ m}} = 0.768 \text{ Å}$$

Because atomic radii and interatomic distances are on the order of 1–5 Å (Section 2.3), the wavelength of this electron is comparable to the size of atoms.

6.52 $\Delta x \geq = h/4\pi m\Delta v$; use masses in kg, Δv in m/s.

(a) $\dfrac{6.626 \times 10^{-34} \text{ J-s}}{4\pi(9.109 \times 10^{-31} \text{ kg})(0.01 \times 10^5 \text{ m/s})} = 6 \times 10^{-8}$ m

(b) $\dfrac{6.626 \times 10^{-34} \text{ J-s}}{4\pi(1.675 \times 10^{-27} \text{ kg})(0.01 \times 10^5 \text{ m/s})} = 3 \times 10^{-11}$ m

(c)   For particles moving with the same uncertainty in velocity, the more massive neutron has a much smaller uncertainty in position than the lighter electron. We know the position of the neutron with much greater precision.

Quantum Mechanics and Atomic Orbitals (Sections 6.5 and 6.6)

6.54   (a)   The Bohr model states with 100% certainty that the electron in H can be found 0.53 Å from the nucleus. The quantum mechanical model, taking the wave nature of the electron and the uncertainty principle into account, is a statistical model that states the probability of finding the electron in certain regions around the nucleus. Although 0.53 Å might be the radius with highest probability, that probability would always be less than 100%.

(b)   The equations of classical physics predict the instantaneous position, direction of motion, and speed of a macroscopic particle; they do not take quantum theory or the wave nature of matter into account. For macroscopic particles, these are not significant, but for microscopic particles like electrons, they are crucial. Schrödinger's equation takes these important theories into account to produce a statistical model of electron location given a specific energy.

(c)   The square of the wave function has the physical significance of an amplitude, or probability. The quantity ψ^2 at a given point in space is the probability of locating the electron within a small volume element around that point at any given instant. The total probability, that is, the sum of ψ^2 over all the space around the nucleus, must equal 1.

6.56   (a)   For $n = 3$, there are three values for l (2, 1, 0) and nine values for m_l ($l = 2$; $m_l = -2, -1, 0, 1, 2$; $l = 1$, $m_l = -1, 0, 1$; $l = 0$, $m_l = 0$).

        (b)   For $n = 5$, there are five values for l (4, 3, 2, 1, 0) and twenty-five values for m_l

                ($l = 4$, $m_l = -4$ to $+4$; $l = 3$, $m_l = -3$ to $+3$; $l = 2$, $m_l = -2$ to $+2$; $l = 1$, $m_l = -1$ to $+1$; $l = 0, = 0$).

                In general, for each principal quantum number n, there are n values for l and n^2 values for m_l. For each shell, there are n kinds of orbitals and n^2 total orbitals.

6.58   (a)   2, 1, 1; 2, 1, 0; 2, 1, –1   (b) 5, 2, 2; 5, 2, 1; 5, 2, 0; 5, 2, –1; 5, 2, –2

6.60   (a)   2, 3   (b) ½, –½

6.62

n	l	m_l	orbital
2	1	–1	2p (example
1	0	0	1s
3	–3	2	not allowed ($l < n$ and + only)
3	2	–2	3d
2	0	–1	not allowed ($m_l = -l$ to $+l$)
0	0	0	not allowed ($n \neq 0$)
4	2	1	4d
5	3	0	5f

6.64

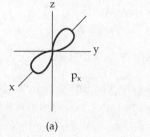

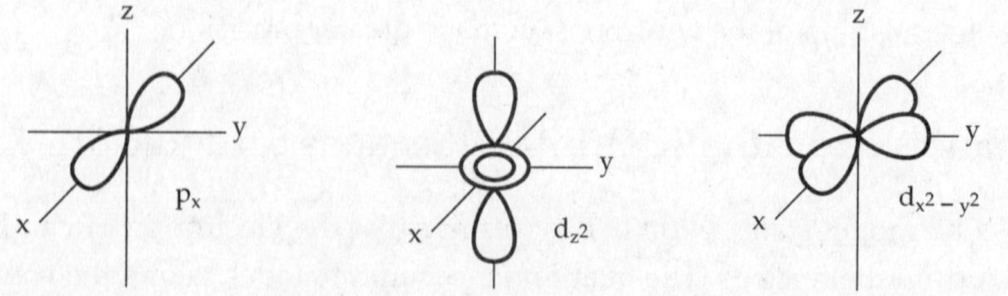

(a)

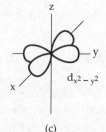

(b)

(c)

6.66   (a)   In an s orbital, there are $(n - 1)$ nodes.

(b)   The $2p_x$ orbital has one node (the yz plane passing through the nucleus of the atom). The 3s orbital has two nodes.

(c)   Probability density, $\psi^2(r)$, is the probability of finding an electron at a single point, r. The radial probability function, P(r), is the probability of finding an electron at any point that is distance r from the nucleus. Figure 6.19 contains plots of P(r) vs. r for 1s, 2s, and 3s orbitals. The most obvious features of these plots are the radii of maximum probability for the three orbitals, and the number and location of nodes for the three orbitals.

By comparing plots for the three orbitals, we see that as *n* increases, the number of nodes increases and the radius of maximum probability (orbital size) increases.

(d)   2s = 2p < 3s < 4d < 5s. In the hydrogen atom, orbitals with the same *n* value are degenerate and energy increases with increasing *n* value.

Many-Electron Atoms and Electron Configurations (Sections 6.7 – 6.9)

6.68   (a)   The electron with the greater average distance from the nucleus feels a smaller attraction for the nucleus and is higher in energy. Thus the 3p is higher in energy than 3s.

(b)   Because it has a larger *n* value, a 3s electron has a greater average distance from the chlorine nucleus than a 2p electron. The 3s electron experiences a smaller attraction for the nucleus and requires less energy to remove from the chlorine atom.

6.70   (a)   The Pauli exclusion principle states that no two electrons can have the same four quantum numbers.

(b,c)   An alternate statement of the Pauli exclusion principle is that a single orbital can hold a maximum of two electrons. Thus, the Pauli principle limits the maximum number of electrons in a main shell and its subshells, which determines when a new row of the periodic table begins.

6.72   (a)   2   (b)   14   (c)   2   (d)   2

6.74

Element	(a) N	(b) Si	(c) Cl
Electron Configuration	$[He]2s^2 2p^3$	$[Ne]3s^2 3p^2$	$[Ne]3s^2 3p^5$
Core electrons	2	10	10
Valence electrons	5	4	7
Unpaired electrons	3	2	1

6.76   (a)   Mg: $[Ne]3s^2$, 0 unpaired electrons

(b)   Ge: $[Ar]4s^2 3d^{10} 4p^2$, 2 unpaired electrons

(c)   Br: $[Ar]4s^2 3d^{10} 4p^5$, 1 unpaired electron

(d)   V: $[Ar]4s^2 3d^3$, 3 unpaired electrons

(e)   Y: $[Kr]5s^2 4d^1$, 1 unpaired electron

(f)   Lu: $[Xe]6s^2 4f^{14} 5d^1$, 1 unpaired electron

6.77   (a)   Be, 0 unpaired electrons   (b)   O, 2 unpaired electrons

        (c)   Cr, 6 unpaired electrons   (d)   Te , 2 unpaired electrons

6.78   (a)   7A (halogens), 1 unpaired electron

        (b)   4B, 2 unpaired electrons

        (c)   3A (row 4 and below), 1 unpaired electron

        (d)   the f-block elements Sm and Pu, 6 unpaired electrons

6.80   Count the total number of electrons to assign the element.

        (a)   F: $[He]2s^2 2p^5$ (b)   Ge: $[Ar]4s^2 3d^{10} 4p^2$ (c)   Nb: $[Kr]5s^2 4d^3$

Additional Exercises

6.82   (a) $v = c/\lambda = \dfrac{2.998 \times 10^8 \text{ m}}{\text{s}} \times \dfrac{1}{589 \text{ nm}} \times \dfrac{1 \text{ nm}}{1 \times 10^{-9} \text{ m}} = 5.0900 \times 10^{14} = 5.09 \times 10^{14} \text{ s}^{-1}$

        (b) $E = h v = 6.626 \times 10^{-34} \text{ J-s} \times 5.0900 \times 10^{14} \text{ s}^{-1} \times \dfrac{6.022 \times 10^{23} \text{ photons}}{\text{mol}}$

                                          $\times\, 0.1 \text{ mol} = 2.03 \times 10^4 \text{ J} = 20.3 \text{ kJ}$

        (c) $\Delta E = h v = \dfrac{hc}{\lambda} = \dfrac{6.626 \times 10^{-34} \text{ J-s} \times 2.998 \times 10^8 \text{ m/s}}{589 \text{ nm}} \times \dfrac{1 \text{ nm}}{1 \times 10^{-9} \text{ m}} = 3.37 \times 10^{-19} \text{ J}$

        (d)   The 589 nm light emission is characteristic of Na^+. If the pickle is soaked in a different salt long enough to remove all Na^+, the 589 nm light would not be observed. Emission at a different wavelength, characteristic of the new salt, would be observed.

6.83   (a)   Elements that emit in the visible: Ba (blue), Ca (violet-blue), K (violet), Na (yellow/orange). (The other wavelengths are in the ultraviolet.)

        (b)   Au: shortest wavelength, highest energy

            Na: longest wavelength, lowest energy

        (c) $\lambda = c/v = \dfrac{2.998 \times 10^8 \text{ m/s}}{9.23 \times 10^{14}/\text{s}} \times \dfrac{1 \text{ nm}}{1 \times 10^{-9} \text{ m}} = 325 \text{ nm}, \quad \text{Cu}$

6.85   (a) $v = c/\lambda = \dfrac{2.998 \times 10^8 \text{ m/s}}{320 \text{ nm}} \times \dfrac{1 \text{ nm}}{1 \times 10^{-9} \text{ m}} = 9.37 \times 10^{14} \text{ s}^{-1}$

        (b) $E = hc/\lambda = \dfrac{6.626 \times 10^{-34} \text{ J-s} \times 2.998 \times 10^8 \text{ m/s}}{3.20 \times 10^{-7} \text{ m}} \times \dfrac{1 \text{ kJ}}{1000 \text{ J}} \times \dfrac{6.022 \times 10^{23} \text{ photons}}{\text{mole}}$

                                          $= 374 \text{ kJ/mol}$

        (c)   UV-B photons have shorter wavelength and higher energy.

        (d)   Yes. The higher energy UV-B photons would be more likely to cause sunburn.

6 Electronic Structure of Atoms   **Solutions to Exercises**

6.87   (a)   If a plant appears orange, it absorbs the complementary (opposite) color on the color wheel. The plant most strongly absorbs blue light in the range 430-490 nm.

(b) $E = hc/\lambda = \dfrac{6.626 \times 10^{-34}\ \text{J-s}}{455\ \text{nm}} \times \dfrac{2.998 \times 10^{8}\ \text{m}}{1\ \text{s}} \times \dfrac{1\ \text{nm}}{1 \times 10^{-9}\ \text{m}} = 4.37 \times 10^{-19}\ \text{J}$

6.88   (a) $\nu = c/\lambda;\ \dfrac{2.998 \times 10^{8}\ \text{m}}{\text{s}} \times \dfrac{1}{542\ \text{nm}} \times \dfrac{1\ \text{nm}}{1 \times 10^{-9}\ \text{m}} = 5.5314 \times 10^{14} = 5.53 \times 10^{14}\ \text{s}^{-1}$

(b)   Calculate J/photon using $E = hc/\lambda$; change to kJ/mol.

$E_{photon} = \dfrac{6.626 \times 10^{-34}\ \text{J-s}}{542 \times 10^{-9}\ \text{m}} \times \dfrac{2.998 \times 10^{8}\ \text{m}}{\text{s}} = 3.6651 \times 10^{-19} = 3.67 \times 10^{-19}\ \text{J/ photon}$

$\dfrac{3.6651 \times 10^{-19}\ \text{J}}{\text{photon}} \times \dfrac{6.022 \times 10^{23}\ \text{photons}}{\text{mol}} \times \dfrac{1\ \text{kJ}}{1000\ \text{J}} = 220.71 = 221\ \text{kJ/ mol}$

(c)   Let E_{total} be the total energy of an incident photon, E_{min} be the minimum energy required to eject an electron, and E_k be the "extra" energy that becomes the kinetic energy of the ejected electron.

$E_{total} = E_{min} + E_k,\ E_k = E_{total} - E_{min} = h\nu - h\nu_o,\ E_k = h(\nu - \nu_o)$. The slope of the line is the value of h, Planck's constant.

6.89   (a)   When an electron is excited to $n = \infty$, it is completely removed from the atom. The end result of this process is ionization, the production of H^+.

(b) $n_i = 1, n_f = \infty;\ \Delta E = -2.18 \times 10^{-18}\ \text{J}\left[\dfrac{1}{n_f^2} - \dfrac{1}{n_i^2}\right] = -2.18 \times 10^{-18}\ \text{J}\ (1/\infty - 1/1) = 2.18 \times 10^{-18}\ \text{J}$

$\lambda = hc/E = \dfrac{6.626 \times 10^{-34}\ \text{J-s}}{2.18 \times 10^{-18}\ \text{J}} \times \dfrac{2.998 \times 10^{8}\ \text{m}}{1\ \text{s}} = 9.11 \times 10^{-8}\ \text{m} = 91.1\ \text{nm}$

(c)   If light with a wavelength shorter than 91.1 nm is used to excite the H atom, the excess energy will become the kinetic energy of the ejected electron. (The potential energy of the ejected electron is, by definition, zero. It no longer experiences electrostatic interactions with the H atom.)

(d)   The frequency associated with the wavelength calculated in part (b) is analogous to ν_0 on the plot in Exercise 6.88. Any excess kinetic energy imparted to the ejected electron corresponds to the sloped line to the right of ν_0.

6.90   (a)   "blue" cone, $\lambda_{max} = 450\ \text{nm} = 450 \times 10^{-9}\ \text{m}$

$E = hc/\lambda = \dfrac{6.626 \times 10^{-34}\ \text{J-s}}{450 \times 10^{-9}\ \text{m}} \times \dfrac{2.998 \times 10^{8}\ \text{m}}{1\ \text{s}} = 4.41 \times 10^{-19}\ \text{J}$

"green" cone, $\lambda_{max} = 545\ \text{nm} = 545 \times 10^{-9}\ \text{m}$

$E = hc/\lambda = \dfrac{6.626 \times 10^{-34}\ \text{J-s}}{545 \times 10^{-9}\ \text{m}} \times \dfrac{2.998 \times 10^{8}\ \text{m}}{1\ \text{s}} = 3.64 \times 10^{-19}\ \text{J}$

"red" cone, $\lambda_{max} = 585$ nm $= 585 \times 10^{-9}$ m

$$E = hc/\lambda = \frac{6.626 \times 10^{-34} \text{ J-s}}{585 \times 10^{-9} \text{ m}} \times \frac{2.998 \times 10^8 \text{ m}}{1 \text{ s}} = 3.40 \times 10^{-19} \text{ J}$$

(b)   "blue" scattering efficiency $= \left(\dfrac{1}{450}\right)^4$; "green" scattering efficiency $= \left(\dfrac{1}{545}\right)^4$

$$\text{ratio of "blue" to "green"} = \frac{\left(\dfrac{1}{450 \text{ nm}}\right)^4}{\left(\dfrac{1}{545 \text{ nm}}\right)^4} = \left(\frac{545}{450}\right)^4 = 2.15$$

(c)   Mainly, the shorter wavelengths perceived by the "blue" cone are scattered more efficiently, so there is more of the blue light to see. Also, the amplitude of the absorption curve for the "blue" cone is greater than the amplitudes of the other two curves. This indicates that our eyes are more sensitive to blue light than the other wavelengths. (It is also true that the intensities of the different wavelengths reaching Earth are not the same, but this information is not conveyed in the exercise.)

6.92   (a)   Gaseous atoms of various elements in the sun's atmosphere typically have ground state electron configurations. When these atoms are exposed to radiation from the sun, the electrons change from the ground state to one of several allowed excited states. Atoms absorb the wavelengths of light that correspond to these allowed energy changes. All other wavelengths of solar radiation pass through the atmosphere unchanged. Thus, the dark lines are the wavelengths that correspond to allowed energy changes in atoms of the solar atmosphere. The continuous background is all other wavelengths of solar radiation.

   (b)   The scientist should record the absorption spectrum of pure neon or other elements of interest. The Fraunhofer lines that belong to a particular element will appear at the same wavelength as the lines in the absorption spectrum of that element.

6.93   (a)   not valid, m_l cannot be greater than l.

   (b)   valid

   (c)   valid

   (d)   not valid, the only possible values for m_s are $+ \frac{1}{2}$ and $- \frac{1}{2}$.

   (e)   not valid, the maximum value for l is $(n - 1)$.

6.94   (a)   He$^+$ is hydrogen-like because it is a one-electron particle. An He atom has two electrons. The Bohr model is based on the interaction of a single electron with the nucleus but does not accurately account for additional interactions when two or more electrons are present.

   (b)   Divide each energy by the smallest value to find the integer relationship.

   H: $-2.18 \times 10^{-18} / -2.18 \times 10^{-18} = 1$; $Z = 1$

   He$^+$: $-8.72 \times 10^{-18} / -2.18 \times 10^{-18} = 4$; $Z = 2$

Li^{2+}: $-1.96 \times 10^{-17}/-2.18 \times 10^{-18} = 9$; $Z = 3$

The ground-state energies are in the ratio of 1:4:9, which is also the ratio Z^2, the square of the nuclear charge for each particle.

The ground state energy for hydrogen-like particles is:

$E = R_H Z^2$. (By definition, $n = 1$ for the ground state of a one-electron particle.)

(c) $Z = 6$ for C^{5+}. $E = -2.18 \times 10^{-18} J (6)^2 = -7.85 \times 10^{-17} J$

6.96   Heisenberg postulated that the dual nature of matter places a limitation on how precisely we can know both the position and momentum of an object. This limitation is significant at the subatomic particle level. The *Star Trek* transporter (presumably) disassembles humans into their protons, neutrons, and electrons, moves the particles at high speed (possibly the speed of light) to a new location, and reassembles the particles into the human. Heisenberg's uncertainty principle indicates that if we know the momentum (mv) of the moving particles, we can't precisely know their position (x). If a few of the subatomic particles don't arrive in exactly the correct location, the human would not be reassembled in their original form. So, the "Heisenberg compensator" is necessary to make sure that the transported human arrives at the new location intact.

6.97   (a)   A subatomic particle is so small that we must use light to measure its position. The interacting photons will impart some momentum to the subatomic particle, thus disturbing it. The shorter the wavelength of the photon, the more accurate the measurement but the greater the momentum imparted to the particle and the bigger the disturbance.

        (b)   An ongoing discussion in quantum theory is whether we can know the quantum states of a system without observing and thus disturbing the system. One interpretation of quantum theory indicated that a system could have multiple acceptable states before it was observed in a single state. That is, the act of observation defined the state. Schrodinger articulated this question on a macroscopic level with his "cat" paradox. Recently, physicists have devised clever ways to observe "cat states," where the act of observing does not destroy the simultaneous states.

6.98   (a)   Probability density, $[\psi(r)]^2$, is the probability of finding an electron at a single point at distance r from the nucleus. The radial probability function, $4\pi r^2$, is the probability of finding an electron at any point on the sphere defined by radius r. $P(r) = 4\pi r^2 [\psi(r)]^2$.

        (b)   The term $4\pi r^2$ explains the differences in plots of the two functions. Plots of the probability density, $[\psi(r)^2]$, for s orbitals shown in Figure 6.22 each have their maximum value at r = 0, with $(n-1)$ smaller maxima at greater values of r. The plots of radial probability, $P(r)$, for the same s orbitals shown in Figure 6.19 have values of zero at r = 0 and the size of the maxima increases. $P(r)$ is the product of $[\psi(r)]^2$ and $4\pi r^2$. At r = 0, the value of $[\psi(r)]^2$ is finite and large, but the value of $4\pi r^2$ is zero, so the value of $P(r)$ is zero. As r increases, the values of $[\psi(r)]^2$ vary as shown in Figure 6.22, but the values of $4\pi r^2$ increase continuously, leading to the increasing size of $P(r)$ maxima as r increases.

(c)

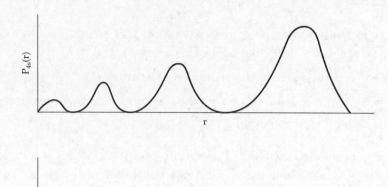

6.100   (a)   This is the frequency of radiowaves that excite the nuclei from one spin state to the other.

       (b) $\Delta E = h\nu = 6.626 \times 10^{-34} \text{ J} \cdot \text{s} \times \dfrac{450 \times 10^6}{\text{s}} = 2.98 \times 10^{-25} \text{ J}$

       (c)   Because $\Delta E = 0$ in the absence of a magnetic field, it is reasonable to assume that the stronger the external field, the greater ΔE. (In fact, ΔE is directly proportional to field strength). Because ΔE is relatively small [see part (b)], the two spin states are almost equally populated, with a very slight excess in the lower energy state. The stronger the magnetic field, the larger ΔE, the greater number of nuclei in the lower energy spin state. With more nuclei in the lower energy state, more are able to absorb the appropriate radio wave photons and reach the higher energy state. This increases the intensity of the NMR signal, which provides more information and more reliable information than a weak absorption signal.

6.101   If m_s had three allowed values instead of two, each orbital would hold three electrons instead of two. Assuming that the same orbitals are available (that there is no change in the n, l, and m_l values), the number of elements in each of the first four rows would be:

     1st row: 1 orbital $\times 3 =$ 3 elements

     2nd row: 4 orbitals $\times 3 = 12$ elements

     3rd row: 4 orbitals $\times 3 = 12$ elements

     4th row: 9 orbitals $\times 3 = 27$ elements

     The s-block would be 3 columns wide, the p-block 9 columns wide and the d-block 15 columns wide.

6.103   The core would be the electron configuration of element 118. If no new subshell begins to fill, the condensed electron configuration of element 126 would be similar to those of elements vertically above it on the periodic chart, Pu and Sm. The condensed configuration would be $[118]8s^2 6f^6$. On the other hand, the 5g subshell could begin to fill after 8s, resulting in the condensed configuration $[118]8s^2 5g^6$. Exceptions are also possible (likely).

6.104   (a)   Neutral H atoms have a single unpaired electron. For a beam of H atoms to be deflected by a magnetic field, the unpaired electrons must interact with the magnetic field. For a single beam of atoms to be split into two beams that are deflected in opposite directions, the unpaired electron on each atom must have some characteristic that can interact in two different ways with a magnetic field. We call that characteristic "electron spin." The significance of this observation is experimental evidence for electron spin.

(b)   If the strength of the magnetic field were increased, the magnitude of the deflection would increase.

(c)   If H atoms were replaced by He atoms, no deflection would occur. Helium has no unpaired electrons to interact with the magnetic field.

(d)   The electron configuration of Ag is $[Kr]5s^1 4d^{10}$. Neutral Ag atoms each have a single unpaired electron. Each unpaired electron has two possible m_s values and a beam of Ag atoms will be split in two by the magnetic field.

Integrative Exercises

6.106 $\Delta H^o_{rxn} = \Delta H^o_f\, O_2(g) + \Delta H^o_f\, O(g) - \Delta H^o_f\, O_3(g)$

$\Delta H^o_{rxn} = 0 + 247.5\,kJ - 142.3\,kJ = +105.2\,kJ$

$$\frac{105.2\,kJ}{mol\,O_3} \times \frac{1\,mol\,O_3}{6.022 \times 10^{23}\,molecules} \times \frac{1000\,J}{1\,kJ} = \frac{1.747 \times 10^{-19}\,J}{O_3\,molecule}$$

$$\Delta E = hc/\lambda;\ \lambda = \frac{hc}{\Delta E} = \frac{6.626 \times 10^{-34}\,J\text{-}s \times 2.998 \times 10^8\,m/s}{1.747 \times 10^{-19}\,J} = 1.137 \times 10^{-6}\,m$$

Radiation with this wavelength is in the infrared portion of the spectrum. (Clearly, processes other than simple photodissociation cause O_3 to absorb ultraviolet radiation.)

6.107   (a)   The electron configuration of Zr is $[Kr]5s^2 4d^2$ and that of Hf is $[Xe]6s^2 4f^{14} 5d^2$. Although Hf has electrons in f orbitals as the rare earth elements do, the 4f subshell in Hf is filled, and the 5d electrons primarily determine the chemical properties of the element. Thus, Hf should be chemically similar to Zr rather than the rare earth elements.

(b) $ZrCl_4(s) + 4\,Na(l) \rightarrow Zr(s) + 4\,NaCl(s)$

This is an oxidation–reduction reaction; Na is oxidized and Zr is reduced.

(c) $2\,ZrO_2(s) + 4\,Cl_2(g) + 3\,C(s) \rightarrow 2\,ZrCl_4(s) + CO_2(g) + 2\,CO(g)$

$$55.4\,g\,ZrO_2 \times \frac{1\,mol\,ZrO_2}{123.2\,g\,ZrO_2} \times \frac{2\,mol\,ZrCl_4}{2\,mol\,ZrO_2} \times \frac{233.0\,g\,ZrCl_4}{1\,mol\,ZrCl_4} = 105\,g\,ZrCl_4$$

(d)   In ionic compounds of the type MCl_4 and MO_2, the metal ions have a 4+ charge, indicating that the neutral atoms have lost 4 electrons. Zr, $[Kr]5s^2 4d^2$, loses the 4 electrons beyond its Kr core configuration. Hf, $[Xe]6s^2 4f^{14} 5d^2$, similarly loses its four 6s and 5d electrons, but not electrons from the "complete" 4f subshell.

6.108   (a)   Each oxide ion, O^{2-}, carries a 2- charge. Each metal oxide is a neutral compound, so the metal ion or ions must adopt a total positive charge equal to the total negative charge of the oxide ions in the compound. The table below lists the electron configuration of the neutral metal atom, the positive charge of each metal ion in the oxide, and the corresponding electron configuration of the metal ion.

     i.   K: $[Ar]\,4s^1$ 1+   [Ar]

     ii.   Ca: $[Ar]\,4s^2$ 2+   [Ar]

     iii.   Sc: $[Ar]\,4s^2 3d^1$ 3+   [Ar]

     iv.   Ti: $[Ar]\,4s^2 3d^2$ 4+   [Ar]

     v.   V: $[Ar]\,4s^2 3d^3$ 5+   [Ar]

     vi.   Cr: $[Ar]\,4s^1 3d^5$ 6+   [Ar]

Each metal atom loses all (valence) electrons beyond the Ar core configuration. In K_2O, Sc_2O_3, and V_2O_5, where the metal ions have odd charges, 2 metal ions are required to produce a neutral oxide.

   (b)   i.   potassium oxide

         ii.   calcium oxide

         iii.   scandium(III) oxide

         iv.   titanium (IV) oxide

         v.   vanadium (V) oxide

         vi.   chromium (VI) oxide

(Roman numerals are required to specify the charges on the transition metal ions, because more than one stable ion may exist.)

   (c)   Recall that $\Delta H_f^\circ = 0$ for elements in their standard states. In these reactions, M(s) and $H_2(g)$ are elements in their standard states.

     i. $K_2O(s) + H_2(g) \rightarrow 2\,K(s) + H_2O(g)$

         $\Delta H^\circ = \Delta H_f^\circ\,H_2O(g) + 2\,\Delta H_f^\circ\,K(s) - \Delta H\,K_2O(s) - \Delta H_f^\circ\,H_2(g)$

         $\Delta H^\circ = -241.82\ kJ + 2(0) - (-363.2\ kJ) - 0 = 121.4\ kJ$

     ii. $CaO(s) + H_2(g) \rightarrow Ca(s) + H_2O(g)$

         $\Delta H^\circ = \Delta H_f^\circ\,H_2O(g) + \Delta H_f^\circ\,Ca(s) - \Delta H_f^\circ\,CaO(s) - \Delta H_f^\circ\,H_2(g)$

         $\Delta H^\circ = -241.82\ kJ + 0 - (-635.1\ kJ) - 0 = 393.3\ kJ$

     iii. $TiO_2(s) + 2\,H_2(g) \rightarrow Ti(s) + 2\,H_2O(g)$

         $\Delta H^\circ = 2\,\Delta H_f^\circ\,H_2O(g) + \Delta H_f^\circ\,Ti(s) - \Delta H_f^\circ\,TiO_2(s) - 2\,\Delta H_f^\circ\,H_2(g)$

         $= 2(-241.82) + 0 - (-938.7) - 2(0) = 455.1\ kJ$

iv. $V_2O_5(s) + 5H_2(g) \rightarrow 2V(s) + 5H_2O(g)$

$$\Delta H^\circ = 5\,\Delta H_f^\circ\,H_2O(g) + 2\,\Delta H_f^\circ\,V(s) - \Delta H_f^\circ\,V_2O_5(s) - 5\,\Delta H_f^\circ\,H_2(g)$$

$$= 5(-241.82) + 2(0) - (-1550.6) - 5(0) = 341.5\text{ kJ}$$

(d) ΔH_f° becomes more negative moving from left to right across this row of the periodic chart. Because Sc lies between Ca and Ti, the median of the two ΔH_f° values is approximately –785 kJ/mol. However, the trend is clearly not linear. Dividing the ΔH_f° values by the positive charge on the pertinent metal ion produces the values –363, –318, –235, and –310. The value between Ca^{2+} (–318) and Ti^{4+} (–235) is Sc^{3+} (–277). Multiplying (–277) by 3, a value of approximately –830 kJ results. A reasonable range of values for ΔH_f° of $Sc_2O_3(s)$ is then –785 to –830 kJ/mol.

6.110   (a) ^{238}U: 92 p, 146 n, 92 e; ^{235}U: 92 p, 143 n, 92 e

In keeping with the definition isotopes, only the number of neutrons is different in the two nuclides. Because the two isotopes have the same number of electrons, they will have the same electron configuration.

(b)   U: $[Rn]7s^2 5f^4$

(c)   From Figure 6.31, the actual electron configuration is $[Rn]7s^2 5f^3 6d^1$. The energies of the 6d and 5f orbitals are very close, and electron configurations of many actinides include 6d electrons.

(d) $^{238}_{92}U \rightarrow\,^{234}_{90}Th +\,^4_2He$ ^{234}Th has 90 p, 144 n, 90 e. ^{238}U has lost 2 p, 2 n, 2 e.

These are organized into ^4_2He shown in the nuclear reaction above.

(e)   From Figure 6.31, the electron configuration of Th is $[Rn]7s^2 6d^2$. This is not really surprising because there are so many rare earth electron configurations that are exceptions to the expected orbital filling order. However, Th is the only rare earth that has two d valence electrons. Furthermore, the configuration of Th is different than that of Ce, the element above it on the periodic chart, so the electron configuration is at least interesting.

7 Periodic Properties of the Elements

Visualizing Concepts

7.1   (a)   The light bulb itself represents the nucleus of the atom. The brighter the bulb, the more nuclear charge the electron "sees." A frosted glass lampshade between the bulb and our eyes reduces the brightness of the bulb. The shade is analogous to core electrons in the atom shielding outer electrons (our eyes) from the full nuclear charge (the bare light bulb).

         (b)   Increasing the wattage of the light bulb mimics moving right along a row of the periodic table. The brighter bulb inside the same shade is analogous to having more protons in the nucleus with an unchanged core electron configuration.

         (c)   Moving down a family, both the nuclear charge and the core electron configuration change. To simulate the addition of core electrons farther from the nucleus, we would add larger frosted glass shades; to simulate the increase in Z, we would increase the wattage of the bulb. In our analogy, the brightness of the light should decrease, because the increased distance of the observer from the bulb and the additional frosted shades more than compensate for the increase in bulb wattage. This mimics the actual decrease in attraction of a valence electron for the nucleus as we move down a column. (Note that Z_{eff} does increase slightly going down a column, but this property is more than offset by the increased distance of valence electrons from the nucleus.)

7.3   (a) Mg^{2+} is isoelectronic with Ne, K^+, and Cl^- are isoelectronic with Ar, and Se^{2-} is isoelectronic with Kr. The atomic radii of the noble gases increase moving down the column, so this gives the rough order of size for the corresponding isoelectronic ions. The Cl^- ion is larger than K^+ because it has a smaller positive nuclear charge holding the same number and configuration of electrons. The order of ionic radii is then $Mg^{2+} < K^+ < Cl^- < Se^{2-}$; these ions match the spheres moving from left to right.

         (b) Ca^{2+} and S^{2-} are both isoelectronic with Ar, as are K^+ and Cl^-. For ions in an isoelectronic series, the larger the nuclear charge, the smaller the ionic radius. Ca^{2+} is smaller than K^+ and fits between the two leftmost spheres. S^{2-} is larger than Cl^- and fits between the two rightmost spheres.

7.4   The size of the red sphere decreases on reaction, so it loses one or more electrons and becomes a cation. Metals lose electrons when reacting with nonmetals, so the red sphere represents a metal. The size of the blue sphere increases on reaction, so it gains one or more electrons and becomes an anion. Nonmetals gain electrons when reacting with metals, so the blue sphere represents a nonmetal.

7.6   (a)   The 3*d* subshell is missing.

        (b)   Statement (ii) is the best description of why the 2*s* and 2*p* subshells have different energies in Na.

        (c)   In a Na atom, the highest energy electron is in the 3*s* subshell.

        (d)   In a Na vapor lamp, the highest energy 3*s* electron is excited into the empty 3*p* subshell.

7.7   (a)   Trends for bonding atomic radius (1) are shown in chart (iii).
                Trends in first ionization energy (2) are shown in chart (ii).
                Trends in metallic character (3) are shown in chart (iii).

        (b)   Generally (ignoring electron-electron repulsion), electron affinity for the main group elements becomes more negative moving up a column and right across a row. It increases (becomes more positive) moving down a column and left across a row. These are the trends shown in chart (iii).

Periodic Table; Effective Nuclear Charge (Sections 7.1 and 7.2)

7.10   Assuming *eka-* means one place below or under, *eka-manganese* on Figure 7.1 is technetium, Tc.

7.12   (a)   Moseley bombarded metal targets with high-energy electrons. These impinging electrons knock out core electrons from metal atoms in the target, creating an electron vacancy or hole. Outer, higher-energy electrons in the metal atoms then "drop down" into the lower-energy hole. Energy in the form of X-rays is emitted when atomic electrons move from a higher energy to a lower-energy state.

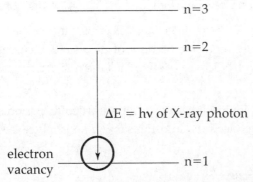

        (b)   The main determining factor of physical and especially chemical properties is electron configuration. For electrically neutral elements, the number of electrons equals the number of protons, which in turn is the atomic number of an element. Atomic weight is related to mass number, protons plus neutrons. The number of neutrons in its nucleus does influence the mass of an atom, but mass is a minor or non-factor in determining properties.

7.14   Statement (iii) is incorrect. Because of the nearly uniform spherical distribution of the core electrons, they screen much more effectively than valence electrons.

7.16   Follow the method in the preceding question to calculate Z_{eff} values.

(a)   Si: $Z = 14$; $[Ne]3s^2 3p^2$. 10 electrons in the Ne core. $Z_{eff} = 14 - 10 = 4$
Cl: $Z = 17$; $[Ne]3s^2 3p^5$. 10 electrons in the Ne core. $Z_{eff} = 17 - 10 = 7$

(b)   Si: $1s^2 2s^2 2p^6 3s^2 3p^2$. $S = 0.35(3) + 0.85(8) + 1(2) = 9.85$. $Z_{eff} = 14 - 9.85 = 4.15$
Cl: $1s^2 2s^2 2p^6 3s^2 3p^5$. $S = 0.35(6) + 0.85(8) + 1(2) = 10.90$. $Z_{eff} = 17 - 10.90 = 6.10$

(c)   The Slater values of 4.15 (Si) and 6.10 (Cl) are closer to the results of detailed calculations, 4.29 (Si) and 6.12 (Cl).

(d)   The Slater method of approximation more closely approximates the gradual increase in Z_{eff} moving across a row. The "core 100%-effective" approximation underestimates Z_{eff} for Si but overestimates it for Cl. Slater values are closer to detailed calculations and a better indication of the change in Z_{eff} moving from Si to Cl.

(e)   Relative to Si, P has one more proton $(Z + 1)$ and one more 3p electron $(S + 0.35)$, so it is reasonable to predict $Z_{eff} + 0.65$. That is, Z_{eff} for P will be $(4.15 + 0.65) = 4.80$.

7.18   Mg < P < K < Ti < Rh. The shielding of electrons in the $n = 3$ shell by 1s, 2s, and 2p core electrons in these elements is approximately equal, so the effective nuclear charge increases as Z increases.

Atomic and Ionic Radii (Section 7.3)

7.20   (a)   Because the quantum mechanical description of the atom does not specify the exact location of electrons, there is no specific distance from the nucleus where the last electron can be found. Rather, the electron density decreases gradually as the distance from the nucleus increases. There is no quantum mechanical "edge" of an atom.

(b)   When nonbonded atoms touch, it is their electron clouds that interact. These interactions are primarily repulsive because of the negative charges of electrons. Thus, the size of the electron clouds determines the nuclear approach distance of nonbonded atoms.

7.22   Statement (iv) is incorrect. Moving left to right in a particular period, the significant nuclear buildup while adding electrons into the same d subshell causes Z to increase and radii to decrease.

7.24   Bi–I = 2.81 Å = $r_{Bi} + r_I$. From Figure 7.7, $r_I = 1.39$ Å.

r_{Bi} = [Bi–I] – r_I = 2.81 Å – 1.39 Å = 1.42 Å.

7.26   (a)   Na < Ca < Ba   (b)   As < Sn < In

(c)   Be < Si < Al. This order assumes the increase in radius from the second to the third row is greater than the decrease moving right in the third row. Radii in Figure 7.7 confirm this assumption.

7.28   (a)   As Z stays constant and the number of electrons increases, the electron-electron repulsions increase, the electrons spread apart, and the anion becomes larger. The reverse is true for the cation, which becomes smaller than the neutral atom.

$I^- > I > I^+$

(b)   For cations with the same charge, ionic radii increase going down a column because there is an increase in the principle quantum number and the average distance from the nucleus of the outer electrons.

$Ca^{2+} > Mg^{2+} > Be^{2+}$

(c)   Fe: $[Ar]4s^2 3d^6$; Fe^{2+}: $[Ar]3d^6$; Fe^{3+}: $[Ar]3d^5$. The 4s valence electrons in Fe are on average farther from the nucleus than the 3d electrons, so Fe is larger than Fe^{2+}. Because there are five 3d orbitals, in Fe^{2+} at least one orbital must contain a pair of electrons. Removing one electron to form Fe^{3+} significantly reduces repulsion, increasing the nuclear charge experienced by each of the other d electrons and decreasing the size of the ion. $Fe > Fe^{2+} > Fe^{3+}$

7.30   (a) Cl^-: Ar   (b) Sc^{3+}: Ar

(c) Fe^{2+}: $[Ar]3d^6$. Fe^{2+} has 24 electrons. Neutral Cr has 24 electrons, $[Ar]4s^1 3d^5$. Because transition metals fill the s subshell first but also lose s electrons first when they form ions, many transition metal ions do not have neutral atoms with the same electron configuration.

(d) Zn^{2+}: $[Ar]3d^{10}$; no neutral atom with same configuration [same reason as (c)].

(e) Sn^{4+}: $[Kr]4d^{10}$; no neutral atom with same electron configuration [same reason as (c)], but Sn^{4+} does have the same configuration as Cd^{2+}.

7.32   (a) K^+ (larger Z) is smaller.

(b) Cl^- and K^+: $[Ne]3s^2 3p^6$. 10 core electrons
Cl^-, Z = 17. $Z_{eff} = 17 - 10 = 7$
K^+, Z = 19. $Z_{eff} = 19 - 10 = 9$

(c)   Valence electron, $n = 3$; seven other $n = 3$ electrons; eight $n = 2$ electrons; two $n = 1$ electrons. S = 0.35(7) + 0.85(8) + 1(2) = 11.25
Cl^-: $Z_{eff} = 17 - 11.25 = 5.75$. K^+: $Z_{eff} = 19 - 11.25 = 7.75$

(d)   For isoelectronic ions (without d electrons), the electron configurations and therefore shielding values (S) are the same. Only the nuclear charge changes. So, as nuclear charge (Z) increases, effective nuclear charge (Z_{eff}) increases and ionic radius decreases.

7.34   (a) $Se < Se^{2-} < Te^{2-}$ (b) $Co^{3+} < Fe^{3+} < Fe^{2+}$ (c) $Ti^{4+} < Sc^{3+} < Ca$ (d) $Be^{2+} < Na^+ < Ne$

7.36   Make a table of d(measured), d(ionic radii), d(covalent radii), as well as differences between measured and estimated values. The estimated distances are just the sum of the various ionic radii from Figure 7.8 and covalent radii from Figure 7.7. All distances and differences are given in Å. Use these values to judge accuracy in parts (b) and (c).

(a)

	d(meas)	(a) d(ion)	(b) Δ(ion – meas)	(c) d(cov)	(c) Δ(cov – meas)
Li–F	2.01	2.09	0.08	2.05	0.04
Na–Cl	2.82	2.83	0.01	2.53	–0.29
K–Br	3.30	3.34	0.04	3.10	–0.20
Rb–I	3.67	3.72	0.05	3.44	–0.23

(b)   The agreement between measured distances in specific ionic compounds and predicted distances based on ionic radii is not perfect. Differences for Na–Cl and K–Br are within the 0.04 Å experimental error, whereas those for Li–F and Rb–I are not. Ionic radii are averages compiled from distances in many ionic compounds containing the ion in question. The sum of these average radii may not give an exact match for the distance in any specific compound, but it will give good distance estimates for many ionic compounds. Although ionic radii provide a good estimate of ion-ion distances, the estimates are not the same as experimental values.

(c)   Distance estimates from bonding atomic radii are not as accurate as those from ionic radii, except for Li–F. This indicates that bonding in NaCl, KBr, and RbI is more accurately described as ionic, rather than covalent. The measured bond distance for LiF is midway between the two estimates, indicating that the bonding in LiF could have significant covalent character. The details of these two models will be discussed in Chapter 8.

Ionization Energies; Electron Affinities (Sections 7.4 and 7.5)

7.38   (a) $Pb(g) \rightarrow Pb^+(g) + 1e^-$; $Pb^+(g) \rightarrow Pb^{2+}(g) + 1e^-$

(b) $Zr^{3+}(g) \rightarrow Zr^{4+}(g) + 1e^-$

7.40   (a)   False. Ionization energies are always positive quantities.

(b)   False. F has a greater first ionization energy than O, because Z_{eff} for F is greater than Z_{eff} for O.

(c)   True.

(d)   False. Ionization energies are not cumulative. The third ionization energy is the energy needed to remove only the third electron from a neutral gas phase atom.

7.42   (a)   Moving from F to I in group 7A, first ionization energies decrease and atomic radii increase. The greater the atomic radius, the smaller the electrostatic attraction of an outer electron for the nucleus and the smaller the ionization energy of the element.

(b)   First ionization energies increase slightly going from K to Kr and atomic sizes decrease. As valence electrons are drawn closer to the nucleus (atom size decreases), it requires more energy to completely remove them from the atom (first ionization energy increases). Each trend has a discontinuity at Ga, owing to the increased shielding of the 4p electrons by the filled 3d subshell.

7.44   Greater distance of valence electrons from the nucleus predicts lower first ionization energy in all the pairs of elements below. Z_{eff} decreases moving left along a row but increases slightly moving down column. These trends are not (solely) predictive of first ionization energy for the pairs of elements in this exercise.

(a)   Ba. Recall that transition metals like Ti lose ns electrons first when forming ions. The 6s valence electrons in Ba are farther from the nucleus and have a smaller first ionization energy than the 4s valence electrons of Ti.

(b)　　Ag. Recall that transition elements lose ns electrons first when forming ions. The 5s valence electron of Ag is farther from the nucleus and has a lower first ionization energy than the 4s valence electron of Cu.

(c)　　Ge. The 4p valence electrons in Ge have a smaller first ionization energy than the 3p valence electrons in Cl. Going from Cl to Ge, the decrease in Z_{eff} moving four places to the left may more than compensate for the small increase moving one place down. If so, the trends in Z_{eff} and distance of valence electrons from the nucleus cooperate to produce the (significantly) lower first ionization energy for Ge.

(d)　　Pb. The 6p valence electrons in Pb are farther from the nucleus and have a smaller first ionization energy than the 5p valence electrons in Sb, despite the buildup in nuclear charge (Z) associated with filling the 4f subshell between Sb and Pb.

7.46　　(a)　　Ru^{3+}: [Kr]3d^5

　　　　(b)　　As^{3-}: [Ar]4s^{2}3d^{10}4p^6 = [Kr], noble-gas configuration

　　　　(c)　　Y^{3+}: [Kr], noble-gas configuration　　　　(d)　　Pd^{2+}: [Kr]3d^8

　　　　(e)　　Pb^{2+}: [Xe]6s^{2}4f^{14}5d^{10}　　　　　　(f)　　Au^{3+}: [Xe]4f^{14}5d^8

7.48　　The 2+ ions of group 8 metals and the 3+ ions of group 9 metals have the electrons configuration nd^6. (Other possibilities not listed on Figure 7.15 exist.)

Fe^{2+}: [Ar]3d^6;　　Ru^{2+}: [Kr]4d^6;　　Os^{2+}: [Xe]6s^{2}4f^{14}5d^6

Co^{3+}: [Ar]3d^6;　　Rh^{3+}: [Kr]4d^6;　　Ir^{3+}: [Xe]6s^{2}4f^{14}5d^6

7.50　　No. The process described by electron affinity can be written as:　$A + 1e^- \rightarrow A^-$
If ΔE for this process is negative, it means that the energy of A^- is lower than the total energy of A plus the energy of a free electron. If electron affinity is negative, the entity that is lower in energy, or more stable, is the added electron. An electron in an atom or ion is stabilized by its attraction for the atomic nucleus and is lower in energy than a free electron.

7.52　　Ionization energy of F^-:　　　$F^-(g) \rightarrow F(g) + 1e^-$

　　　　Electron affinity of F:　　　$F(g) + 1e^- \rightarrow F^-(g)$

　　　　The two processes are the reverse of each other. The energies are equal in magnitude but opposite in sign. $I_1 (F^-) = -E (F)$

7.54　　　$Ca^+(g) + 1e^- \rightarrow Ca(g)$
　　　　　[Ar] 4s^1　　　　　　　[Ar] 4s^2

　　　　Statements (i) and (ii) are true.

Properties of Metals and Nonmetals (Section 7.6)

7.56　　Element Y has the greater metallic character. Metallic character increases as ionization energy decreases.

7.58   Disagree. According to Figure 7.15, both Sb and Te are metalloids and commonly form ions. Sb forms cations and Te forms anions.

7.60   Follow the logic in Sample Exercise 7.8. Scandium is a metal, so we expect Sc_2O_3 to be ionic. Metal oxides are usually basic and react with acid to form a salt and water. We choose $HNO_3(aq)$ as the acid for our equation.

$Sc_2O_3(s) + 6\,HNO_3(aq) \rightarrow 2\,Sc(NO_3)_3(aq) + 3\,H_2O(l)$

The net ionic equation is:

$Sc_2O_3(s) + 6\,H^+(aq) \rightarrow 2\,Sc^{3+}(aq) + 3\,H_2O(l)$

7.62   The more nonmetallic the central atom, the more acidic the oxide. In order of increasing acidity: $CaO < Al_2O_3 < SiO_2 < CO_2 < P_2O_5 < SO_3$

7.64   (a) $XCl_4(l) + 2\,H_2O(l) \rightarrow XO_2(s) + 4\,HCl(g)$

The second product is $HCl(g)$.

(b)   If X were a metal, both the oxide and the chloride would be high melting solids. If X were a nonmetal, XO_2 would be a nonmetallic, molecular oxide and probably gaseous, like CO_2, NO_2, and SO_2. Neither of these statements describes the properties of XO_2 and XCl_4, so X is probably a metalloid.

(c)   Use the *Handbook of Chemistry* to find formulas and melting points of oxides, and formulas and boiling points of chlorides of selected metalloids.

metalloid	formula of oxide	m.p. of oxide	formula of chloride	b.p. of chloride
boron	B_2O_3	460 °C	BCl_3	12 °C
silicon	SiO_2	~1700 °C	$SiCl_4$	58 °C
germanium	GeO GeO_2	710 °C ~1100 °C	$GeCl_2$ $GeCl_4$	decomposes 84 °C
arsenic	As_2O_3 As_2O_5	315 °C 315 °C	$AsCl_3$	132 °C

Boron, arsenic, and, by analogy, antimony, do not fit the description of X because the formulas of their oxides and chlorides are wrong. Silicon and germanium, in the same family, have oxides and chlorides with appropriate formulas. Both SiO_2 and GeO_2 melt above 1000 °C, but the boiling point of $SiCl_4$ is much closer to that of XCl_4. Element X is silicon.

7.66   (a) $K_2O(s) + H_2O(l) \rightarrow 2\,KOH(aq)$

(b) $P_2O_3(l) + 3\,H_2O(l) \rightarrow 2\,H_3PO_3(aq)$

(c) $Cr_2O_3(s) + 6\,HCl(aq) \rightarrow 2\,CrCl_3(aq) + 3\,H_2O(l)$

(d) $SeO_2(s) + 2\,KOH(aq) \rightarrow K_2SeO_3(aq) + H_2O(l)$

Group Trends in Metals and Nonmetals (Sections 7.7 and 7.8)

7.68   Rb: [Kr]$5s^1$, r = 2.11 Å   Ag: [Kr]$5s^1 4d^{10}$, r = 1.53 Å

The electron configurations both have a [Kr] core and a single 5s electron; Ag has a completed 4d subshell as well. The smaller radius of Ag indicates that the 5s electron in Ag experiences a much greater effective nuclear charge than the 5s electron in Rb. Ag has a much larger Z (47 vs. 37), and although the 4d electrons in Ag shield the 5s electron somewhat, the increased shielding does not compensate for the large increase in Z. Ag is much less reactive (less likely to lose an electron) because its 5s electron experiences a much larger effective nuclear charge and is more difficult to remove.

7.70   (a) $2\,Cs(s) + 2\,H_2O(l) \rightarrow 2\,CsOH(aq) + H_2(g)$

       (b) $Sr(s) + 2\,H_2O(l) \rightarrow Sr(OH)_2(aq) + H_2(g)$

       (c) $2\,Na(s) + O_2(g) \rightarrow Na_2O_2(s)$ (See Equation 7.20.)

       (d) $Ca(s) + I_2(s) \rightarrow CaI_2(s)$

7.72   (a) $2\,K(s) + H_2(g) \rightarrow 2\,KH(s)$

       (b)

$K(g)$	$\rightarrow K^+(g)$	419 kJ	(I_1 of K)
$\underline{H(g) + e^- \rightarrow H^-(g)}$		$\underline{-73\ kJ}$	(E_1 of H)
$K(g) + H(g) \rightarrow K^+(g) + H^-(g)$		346 kJ	
$H(g)$	$\rightarrow H^+(g)$	1312 kJ	(I_1 of H)
$\underline{K(g) + e^- \rightarrow K^-(g)}$		$\underline{-48\ kJ}$	(E_1 of K)
$K(g) + H(g) \rightarrow K^-(g) + H^+(g)$		1264 kJ	

       (c)   Both reactions are endothermic; the first reaction is less unfavorable and therefore more favorable than the second.

       (d)   The more energetically favorable reaction in part (c) produces hydride ions (H^-) and potassium ions (K^+), so it is reasonable to describe potassium hydride as containing hydride ions.

7.74   *Plan.* Predict the physical and chemical properties of At based on the trends in properties in the halogen (7A) family. *Solve.*

       (a)   F, at the top of the column, is a diatomic gas; I, immediately above At, is a diatomic solid; the melting points of the halogens increase going down the column. At is likely to be a diatomic solid at room temperature.

       (b)   Like the other halogens, we expect it to be a nonmetal. According to Figure 7.13, there are no metalloids in row 6 of the periodic table, and At is a nonmetal. (Looking forward to Chapter 8, the most likely way for At to satisfy the octet rule is for it to gain an electron to form At^-, which makes it a nonmetal.)

       (c)   All halogens form ionic compounds with Na; they have the generic formula NaX. The compound formed by At will have the formula NaAt.

7.76   (a)   Xe has a lower ionization energy than Ne. The valence electrons in Xe are much farther from the nucleus than those of Ne ($n = 5$ vs $n = 2$) and much less tightly held by the nucleus; they are more "willing" to be shared than those in Ne. Also, Xe has empty 5d orbitals that can help to accommodate the bonding pairs of electrons, whereas Ne has all its valence orbitals filled.

    (b)   In the *CRC Handbook of Chemistry and Physics*, 79th edition, Xe – F bond distances in gas phase molecules are listed as: XeF_2, 1.977 Å; XeF_4, 1.94 Å; XeF_6, 1.89 Å. From Figure 7.7, the sum of atomic radii for Xe and F is (1.40 Å + 0.57 Å) = 1.97 Å. This number represents an "average" or "typical" distance and agrees well with the bond distance in XeF_2. Bond lengths in specific compounds are not exactly equal to the sum of covalent radii. Physical state, electronic, and steric factors affect bond lengths in specific compounds.

7.78   (a) $Cl_2(g) + H_2O(l) \rightarrow HCl(aq) + HOCl(aq)$

    (b) $Ba(s) + H_2(g) \rightarrow BaH_2(s)$

    (c) $2\,Li(s) + S(s) \rightarrow Li_2S(s)$

    (d) $Mg(s) + F_2(g) \rightarrow MgF_2(s)$

Additional Exercises

7.80   (a)   2s

    (b)   Slater's rules provide a method for calculating the shielding, S, and Z_{eff} experienced by a particular electron in an atom. Slater assigns a shielding value of 0.35 to electrons with the same n-value, assuming that s and p electrons shield each other to the same extent. However, because s electrons have a finite probability of being very close to the nucleus (Figure 7.4), they shield p electrons more than p electrons shield them. To account for this difference, assign a slightly larger shielding value to s electrons and a slightly smaller shielding value to the p electrons. This will produce a slightly greater S and smaller Z_{eff} for p electrons than for s electrons with the same n-value.

7.82   Atomic size (bonding atomic radius) is strongly correlated to Z_{eff}, which is determined by Z and S. Moving across the representative elements, electrons added to ns or np valence orbitals do not effectively screen each other. The increase in Z is not accompanied by a similar increase in S; Z_{eff} increases and atomic size decreases. Moving across the transition elements, electrons are added to $(n–1)$d orbitals and become part of the core electrons, which do significantly screen the ns valence electrons. The increase in Z is accompanied by a larger increase in S for the ns valence electrons; Z_{eff} increases more slowly and atomic size decreases more slowly.

7.83   (a)   The estimated distances in the table below are the sum of the radii of the group 5A elements and H from Figure 7.7.

bonded atoms	estimated distance	measured distance
P – H	1.38	1.419
As – H	1.50	1.519
Sb – H	1.70	1.707

In general, the estimated distances are very slightly shorter than the measured distances. (Recall that the radii in Figure 7.7 come from measuring many different molecules for each element, not just the bonds listed in this exercise.)

(b)   The principal quantum number of the outer electrons and thus the average distance of these electrons from the nucleus increases from P ($n = 3$) to As ($n = 4$) to Sb ($n = 5$). This causes the systematic increase in M – H distance.

7.84   She is correct. Xenon bonds with certain elements to form compounds, so its bonding atomic radius is an average of experimentally determined values. To date, no compound containing Ne has been observed, so its "atomic radius" is an estimate. Measured values are always more realistic than estimates.

7.86   The estimated A – B distance is $(r_A + r_B) = (A – A)/2 + (B – B)/2$. Because the AB_2 molecule is linear, the distance between the two terminal B atoms is twice the A – B distance, $2[(A – A)/2 + (B – B)/2] = (A – A) + (B – B)$. This is just the sum of the bond lengths of the two diatomic molecules. The separation between the two B nuclei in AB_2 is 2.36 Å + 1.94 Å = 4.30 Å.

7.88   Y: $[Kr]5s^2 4d^1$, Z = 39   Zr: $[Kr] 5s^2 4d^2$, Z = 40

La: $[Xe]6s^2 5d^1$, Z = 57   Hf: $[Xe] 6s^2 4f^{14} 5d^2$, Z = 72

The completed 4f subshell in Hf leads to a much larger change in Z going from Zr to Hf (72 – 40 = 32) than in going from Y to La (57 – 39 = 18). The 4f electrons in Hf do not completely shield the valence electrons, so there is also a larger increase in Z_{eff}. This significant increase in Z_{eff} going from Zr to Hf causes the two elements to have the same radii, even though the valence electrons of Hf have a larger n value than those of Zr. (This phenomenon is called the "lantanide contraction.")

7.89   (a) Co^{4+} is smaller.

(b) Co^{4+}, 0.67 Å < Co^{3+}, 0.75 Å < Li^+, 0.90 Å

Values from WebElements©, CN 6, high spin (for comparing equivalent ion environments)

(c)   As Li^+ ions are inserted, smaller Co^{4+} ions are reduced to larger Co^{3+} ions and the lithium colbalt electrode will expand.

(d)   "Sodium colbalt oxide" will probably not work as an electrode material, because Na^+ ions are much larger than Li^+ ions, which are larger than Co^{4+} and Co^{3+} ions. Na^+ ions would be too large to insert into the electrode without disrupting the structure of the material.

(e)   An alternative metal for a sodium version of the electrode would have redox-active ions with larger ionic radii than the Co^{4+} and Co^{3+} ions. Moving left along the fourth row of the periodic table, Fe^{3+}/Fe^{2+} and Mn^{3+}/Mn^{2+} ion couples are possibilities. Both have radii larger than Co^{4+}/Co^{3+} ions. Mn^{3+} is more redox-active than Fe^{3+} and may be a more effective electrode material.

7.90   (a) $2 Sr(s) + O_2(g) \rightarrow 2 SrO(s)$

(b)   Assume that the corners of the cube are at the centers of the outermost O^{2-} ions, and that the edges each pass through the center of one Sr^{2+} ion. The length of an edge is then $r(O^{2-}) + 2r(Sr^{2+}) + r(O^{2-}) = 2r(O^{2-}) + 2r(Sr^{2+}) = 2(1.32$ Å$) + 2(1.26$ Å$) = 5.16$ Å.

(c)   Density is the ratio of mass to volume.

$$d = \frac{\text{mass SrO in cube}}{\text{vol cube}} = \frac{\text{\# SrO units} \times \text{mass of SrO}}{\text{vol cube}}$$

Calculate the mass of 1 SrO unit in grams and the volume of the cube in cm^3; solve for number of SrO units.

$$\frac{103.62 \text{ g SrO}}{\text{mol}} \times \frac{1 \text{ mol SrO}}{6.022 \times 10^{23} \text{ SrO units}} = 1.7207 \times 10^{-22} = 1.721 \times 10^{-22} \text{ g/SrO unit}$$

$$V = (5.16)^3 \text{ Å}^3 \times \frac{(1 \times 10^{-8})^3 \text{ cm}^3}{\text{Å}^3} = 1.3739 \times 10^{-22} = 1.37 \times 10^{-22} \text{ cm}^3$$

$$d = \frac{\text{number of SrO units} \times 1.7207 \times 10^{-22} \text{ g/ SrO unit}}{1.3739 \times 10^{-22} \text{ cm}^3} = 5.10 \text{ g/ cm}^3$$

$$\text{number of SrO units} = 5.10 \text{ g/ cm}^3 \times \frac{1.3739 \times 10^{-22} \text{ cm}^3}{1.7207 \times 10^{-22} \text{ g/ SrO unit}} = 4.07 \text{ units}$$

Because the number of formula units must be an integer, there are four SrO formula units in the cube. Using average values for ionic radii to estimate the edge length probably leads to the small discrepancy.

7.92   Only statement (ii) is true.

We expect electron affinities for Group 4A to be more negative than Group 3A, based on increasing effective nuclear charge moving from left to right across a row. Electron-electron repulsion causes the electron affinities of Groups 5A and 6A to be less negative than expected from effective nuclear charge trends.

7.93

$$A(g) \rightarrow A^+(g) + e^- \qquad \text{ionization energy of A}$$
$$A(g) + e^- \rightarrow A^-(g) \qquad \text{electron affinity of A}$$
$$A(g) + A(g) \rightarrow A^+(g) + A^-(g) \qquad \text{ionization energy of A + electron affinity of A}$$

The energy change for the reaction is the ionization energy of A plus the electron affinity of A.

This process is endothermic for both nonmetals and metals. Considering data for Cl and Na from Figures 7.10 and 7.12, the endothermic ionization energy term dominates the exothermic electron affinity term, even for Cl, which has the most exothermic electron affinity listed.

7.94   (a)   O: $[He]2s^2 2p^4$

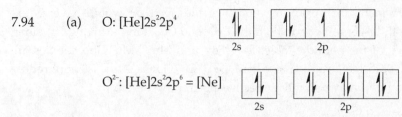

O^{2-}: $[He]2s^2 2p^6 = [Ne]$

(b) O^{3-}: $[Ne]3s^1$ The third electron would be added to the 3s orbital, which is farther from the nucleus and more strongly shielded by the [Ne] core. The overall attraction of this 3s electron for the O nucleus is not large enough for O^{3-} to be a stable particle.

7.95   (a)   P: [Ne] $3s^2 3p^3$; S: [Ne] $3s^2 3p^4$. In P, each 3p orbital contains a single electron, whereas in S one 3p orbital contains a pair of electrons. Removing an electron from S eliminates the need for electron pairing and reduces electrostatic repulsion, so the overall energy required to remove the electron is smaller than in P, even though Z is greater.

(b)   C: [He] $2s^2 2p^2$; N: [He] $2s^2 2p^3$; O: [He] $2s^2 2p^4$. An electron added to a N atom must be paired in a relatively small 2p orbital, so the additional electron-electron repulsion more than compensates for the increase in Z and the electron affinity is smaller (less exothermic) than that of C. In an O atom, one 2p orbital already contains a pair of electrons, so the additional repulsion from an extra electron is offset by the increase in Z and the electron affinity is greater (more exothermic). Note from Figure 7.12 that the electron affinity of O is only slightly more exothermic than that of C, although the value of Z has increased by 2.

(c) O^+: [He] $2s^2 2p^3$; O^{2+}: [He] $2s^2 2p^2$; F: [He] $2s^2 2p^5$; F^+: [He] $2s^2 2p^4$. Both 'core-only' [Z_{eff} (F) = 7; Z_{eff} (O^+) = 6] and Slater [Z_{eff} (F) = 5.2; Z_{eff} (O^+) = 4.9] predict that F has a greater Z_{eff} than O^+. Variation in Z_{eff} does not offer a satisfactory explanation. The decrease in electron-electron repulsion going from F to F^+ energetically favors ionization and causes it to be less endothermic than the second ionization of O, where there is no significant decrease in repulsion.

(d) Mn^{2+}: [Ar]$3d^5$; Mn^{3+}: [Ar] $3d^4$; Cr^{2+}: [Ar] $3d^4$; Cr^{3+}: [Ar] $3d^3$; Fe^{2+}: [Ar] $3d^6$; Fe^{3+}: [Ar] $3d^5$. The third ionization energy of Mn is expected to be larger than that of Cr because of the larger Z value of Mn. The third ionization energy of Fe is less than that of Mn because going from $3d^6$ to $3d^5$ reduces electron repulsions, making the process less endothermic than predicted by nuclear charge arguments.

7.97   (a)   (ii)   (b) (v)   (c) (i)

7.98   (a)   The group 2B metals have complete (n–1)d subshells. An additional electron would occupy an np subshell and be substantially shielded by both ns and (n–1)d electrons. Overall this is not a lower energy state than the neutral atom and a free electron.

(b)   Valence electrons in Group 1B elements experience a relatively large effective nuclear charge because of the buildup in Z with the filling of the (n–1)d subshell (and for Au, the 4f subshell.) Thus, the electron affinities are large and negative. Group 1B elements are exceptions to the usual electron filling order and have the generic electron configuration $ns^1(n-1)d^{10}$. The additional electron would complete the ns subshell and experience repulsion with the other ns electron. Going down the group, size of the ns subshell increases and repulsion effects decrease. That is, effective nuclear charge is greater going down the group because it is less diminished by repulsion, and electron affinities become more negative.

7.100   Because Xe reacts with F_2, and O_2 has approximately the same ionization energy as Xe, O_2 will probably react with F_2. Possible products would be O_2F_2, analogous to XeF_2, or OF_2.

$$O_2(g) + F_2(g) \rightarrow O_2F_2(g)$$
$$O_2(g) + 2\,F_2(g) \rightarrow 2\,OF_2(g)$$

7.101   The first ionization energies are: Ag, 731 kJ/mol; Mn, 717 kJ/mol. According to Figure 7.13, we define metallic character as showing the opposite trend as ionization energy. That is, the smaller the ionization energy, the greater the metallic character. Because Mn has the smaller ionization energy, it should have the greater metallic character. (It is difficult to predict the relative metallic character of these two elements from trends. Ag is one row lower but four columns further right than Mn; these are opposing trend directions.)

7.103   Statement (ii) is the best explanation.

Statements (ii) and (iii) are both true, assuming "smaller" in statement (iii) means more negative. In this reaction, Na or Cs loses an electron and acts like a metal. Statement (ii) about ionization energy is more directly applicable. Electron affinity is about an atom gaining an electron.

7.104   (a)   All alkali metals except Li form metal peroxides when they react with oxygen; the formation of a peroxide (or a superoxide) eliminates Li. The lilac-purple flame indicates that the metal is potassium (see Figure 7.22).

(b) $K_2O_2(s)$ + 2 $H_2O(l)$ → $H_2O_2(aq)$ + 2 KOH(aq)
potassium peroxide   hydrogen peroxide

2 $KO_2(s)$ + 2 $H_2O(l)$ → $H_2O_2(aq)$ + 2 KOH(aq) + $O_2(g)$
potassium superoxide   hydrogen peroxide

Both potassium peroxide and potassium superoxide react with water to form hydrogen peroxide. The white solid could be either potassium salt.

7.105   (a)   The pros are that Zn and Cd are in the same family, have the same electron configuration and thus similar chemical properties. The same can be said for Zn^{2+} and Cd^{2+} ions. Because of their chemical similarity, we expect Cd^{2+} to easily substitute for Zn^{2+} in flexible molecules. The main difference is that Zn^{2+}, with an ionic radius of 0.88 Å, is much smaller than Cd^{2+}, with an ionic radius of 1.09 Å. Although Zn^{2+} is beneficial in living systems, Cd^{2+} is toxic. This difference in biological function could be related to the size difference and is a definite con.

(b) Cu^+ is isoelectronic with Zn^{2+}. That is, the two ions have the same number of electrons and the same electron configurations. The ionic radius of Cu^+ is 0.91 Å, very similar to that of Zn^{2+}. We expect Cu^+ to be a reasonable substitute for Zn^{2+} in terms of chemical properties and size. Electrostatic interactions may vary, because of the difference in charges of the two ions. (All ionic radii are taken from WebElements©.)

7.106   (a)   *Plan.* Use qualitative physical (bulk) properties to narrow the range of choices, then match melting point and density to identify the specific element. *Solve.*

Hardness varies widely in metals and nonmetals, so this information is not too useful. The relatively high density, appearance, and ductility indicate that the element is probably less metallic than copper. Focus on the block of nine main group elements centered around Sn. Pb is not a possibility because it was used as a comparison standard. The melting point of the five elements closest to Pb are:

Tl, 303.5 °C; In, 156.1 °C; Sn, 232 °C; Sb, 630.5 °C; Bi, 271.3 °C

The best match is In. To confirm this identification, the density of In is 7.3 g/cm^3, also a good match to properties of the unknown element.

(b)   To write the correct balanced equation, determine the formula of the oxide product from the mass data, assuming the unknown is In.

5.08 g oxide – 4.20 g In = 0.88 g O

4.20 g In/114.82 g/mol = 0.0366 mol In; 0.0366/0.0366 = 1

0.88 g O/16.00 g/mol = 0.0550 mol O; 0.0550/0.0366 = 1.5

Multiplying by two produces an integer ratio of 2 In: 3 O and a formula of In_2O_3. The balanced equation is: $4\,In(s) + 3\,O_2(g) \rightarrow 2\,In_2O_3(s)$

(c)   According to Figure 7.1, the element In was discovered between 1843 and 1886. The investigator who first recorded this data in 1822 could have been the first to discover In.

7.108   (a)   Si and Ge are in group 4A and have 4 valence electrons. GaAs and GaP have their first element in group 3A with 3 valence electrons and their second element in group 5A with 5 valence electrons. Cd in CdS and CdSe is in group 2B and has 2 valence electrons, whereas S and Se are in group 6A with 6 valence electrons. In each case, the two elements in the compound semiconductor have an *average* of 4 valence electrons.

(b)   The roman numerals represent the number of valence electrons in the component elements of the compound semiconductor. CdS and CdSe are II-VI materials, whereas GaAs and GaP are III-V materials.

(c)   Replace Ga with In: InP, InAs, InSb; replace Se with Te: CdTe. It is problematic to replace Cd with Hg, because Hg is toxic. ZnS is ionic and an insulator, so Zn may not be a good substitute for Cd.

Integrative Exercises

7.109   (a) $\nu = c/\lambda$; $1\,Hz = 1\,s^{-1}$

Ne: $\nu = \dfrac{2.998 \times 10^8 \text{ m/s}}{14.610 \text{ Å}} \times \dfrac{1 \text{ Å}}{1 \times 10^{-10} \text{ m}} = 2.052 \times 10^{17} \text{ s}^{-1} = 2.052 \times 10^{17} \text{ Hz}$

Ca: $\nu = \dfrac{2.998 \times 10^8 \text{ m/s}}{3.358 \times 10^{-10} \text{ m}} = 8.928 \times 10^{17} \text{ Hz}$

Zn: $\nu = \dfrac{2.998 \times 10^8 \text{ m/s}}{1.435 \times 10^{-10} \text{ m}} = 20.89 \times 10^{17} \text{ Hz}$

Zr: $\nu = \dfrac{2.998 \times 10^8 \text{ m/s}}{0.786 \times 10^{-10} \text{ m}} = 38.14 \times 10^{17} = 38.1 \times 10^{17} \text{ Hz}$

Sn: $\nu = \dfrac{2.998 \times 10^8 \text{ m/s}}{0.491 \times 10^{-10} \text{ m}} = 61.06 \times 10^{17} = 61.1 \times 10^{17} \text{ Hz}$

(b)

Element	Z	v	$v^{1/2}$
Ne	10	2.052×10^{17}	4.530×10^{8}
Ca	20	8.928×10^{17}	9.449×10^{8}
Zn	30	20.89×10^{17}	14.45×10^{8}
Zr	40	38.14×10^{17}	19.5×10^{8}
Sn	50	61.06×10^{17}	24.7×10^{8}

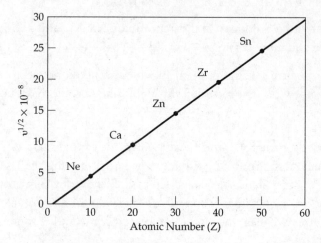

(c)   The plot in part (b) indicates that there is a linear relationship between atomic number and the square root of the frequency of the X-rays emitted by an element. Thus, elements with each integer atomic number should exist. This relationship allowed Moseley to predict the existence of elements that filled "holes" or gaps in the periodic table.

(d)   For Fe, Z = 26. From the graph, $v^{1/2} = 12.5 \times 10^{8}$, $v = 1.56 \times 10^{18}$ Hz.

$$\lambda = c/v = \frac{2.998 \times 10^{8} \text{ m/s}}{1.56 \times 10^{18} \text{ s}^{-1}} \times \frac{1 \text{ Å}}{1 \times 10^{-10} \text{ m}} = 1.92 \text{ Å}$$

(e) $\lambda = 0.980 \text{ Å} = 0.980 \times 10^{-10}$ m

$$v = c/\lambda = \frac{2.998 \times 10^{8} \text{ m/s}}{0.980 \times 10^{-10} \text{ m}} = 30.6 \times 10^{17} \text{ Hz}; \ v^{1/2} = 17.5 \times 10^{8}$$

From the graph, $v^{1/2} = 17.5 \times 10^{8}$, Z = 36. The element is krypton, Kr.

7.111   (a) $E = hc/\lambda$; 1 nm = 1×10^{-9} m; 58.4 nm = 58.4×10^{-9} m;

1 eV = 96.485 kJ/mol, 1 eV - mol = 96.485 kJ

$$E = \frac{6.626 \times 10^{-34} \text{ J-s} \times 2.998 \times 10^{8} \text{ m/s}}{58.4 \times 10^{-9} \text{ m}} = 3.4015 \times 10^{-18} = 3.40 \times 10^{-18} \text{ J/photon}$$

$$\frac{3.4015 \times 10^{-18} \text{ J}}{\text{photon}} \times \frac{1 \text{ kJ}}{1000 \text{ J}} \times \frac{6.022 \times 10^{23} \text{ photons}}{\text{mol}} \times \frac{1 \text{ eV - mol}}{96.485 \text{ kJ}} = 21.230 = 21.2 \text{ eV}$$

(b) $Hg(g) \rightarrow Hg^{+}(g) + 1e^{-}$

(c) $I_1 = E_{58.4} - E_K = 21.23 \text{ eV} - 10.75 \text{ eV} = 10.48 = 10.5 \text{ eV}$

$$10.48 \text{ eV} \times \frac{96.485 \text{ kJ}}{1 \text{ eV-mol}} = 1.01 \times 10^3 \text{ kJ/mol}$$

(d)   From Figure 7.10, iodine (I) appears to have the ionization energy closest to that of Hg, approximately 1000 kJ/mol.

7.112   (a)   The X-ray source had an energy of 1253.6 eV. Change eV to J/photon and use the relationship $\lambda = hc/E$ to find wavelength.

$$1253.6 \text{ eV} \times \frac{96.485 \text{ kJ}}{1 \text{ eV-mol}} \times \frac{1 \text{ mol}}{6.0221 \times 10^{23} \text{ photons}} \times \frac{1000 \text{ J}}{1 \text{ kJ}} = 2.0085 \times 10^{-16} \text{ J/photon}$$

$$\lambda = \frac{6.6261 \times 10^{-34} \text{ J-s} \times 2.9979 \times 10^8 \text{ m/s}}{2.0085 \times 10^{-16} \text{ J}} = 9.8902 \times 10^{-10} \text{ m} = 0.98902 \text{ nm} = 9.8902 \text{ Å}$$

(b)   Express energies of Hg 4f and O 1s electrons in terms of kJ/mol for comparison with data from Figure 7.10 of the text.

$$\text{Hg 4f: } 105 \text{ eV} \times \frac{96.485 \text{ kJ}}{1 \text{ eV-mol}} = 10{,}131 = 1.01 \times 10^4 \text{ kJ/mol}$$

$$\text{O 1s: } 531 \text{ eV} \times \frac{96.485 \text{ kJ}}{1 \text{ eV-mol}} = 51{,}234 = 5.12 \times 10^4 \text{ kJ/mol}$$

By definition, the first ionization energy is the minimum energy required to remove the first electron from an atom. This first electron is the highest energy valence electron in the neutral atom. We expect the energies of valence electrons to be higher than those of core electrons, and first ionization energies to be less than the energy required to remove a lower energy core electron.

For Hg, the first ionization energy is 1007 kJ/mol, whereas the XPS energy of the 4f electron is 10,100 kJ/mol. The energy required to remove a 4f core electron is 10 times the energy required to remove a 6s valence electron.

For O, the first ionization energy is 1314 kJ/mol, whereas the XPS energy of a 1s electron is 51,200 kJ/mol. The energy required to remove a 1s core electron is 50 times that required to remove a 2p valence electron.

(c) Hg^{2+}: $[Xe]4f^{14}5d^{10}$; valence electrons are 5d

O^{2-} : $[He]2s^2 2p^6$ or [Ne]; valence electrons are 2p

(d)   (Recall that Slater's rules are best applied to elements that do not contain d or f electrons.)

Hg^{2+} 5d valence: $n = 5$ (5s+5p+5d) has 18 e^- ; subtract one for the electron under consideration = 17 e^- with the same n value as the one under consideration.
$n = 4$ (4s+4p+4d+4f) has 32 e^- with $(n-1)$; $n = 3, 2, 1$ have (18+8+2) = 28 e^- with $(n-2)$ or less.

S (5d) = 0.35(17) + 0.85(32) + 1.0(28) = 61.15. $Z_{eff} = Z - S = 80 - 61.15 = 18.85$

Hg^{2+} 4f: The 4f electrons are not shielded by electrons with $n > 4$. $n = 4$ has $(32 - 1) = 31$ e⁻; $n = 3$ has 18 e⁻ with $(n - 1)$; $n = 2, 1$ have 10 e⁻ with $(n - 2)$ or less.

S (4f) = 0.35(31) + 0.85(18) + 1.0(10) = 36.15. $Z_{eff} = Z - S = 80 - 36.15 = 43.85$

O^{2-} 2p valence: $n = 2$ has $(8 - 1) = 7$ e⁻; $n = 1$ has 2 e⁻.

$S = 0.35(7) + 0.85(2) = 4.15$; $Z_{eff} = 8 - 4.15 = 3.85$

7.114   (a) $r_{Bi} = r_{BiBr_3} - r_{Br} = 2.63 \text{ Å} - 1.20 \text{ Å} = 1.43 \text{ Å}$

(b) $Bi_2O_3(s) + 6 \text{ HBr}(aq) \rightarrow 2 \text{ BiBr}_3(aq) + 3 H_2O(l)$

(c) Bi_2O_3 is soluble in acid solutions because it acts as a base and undergoes acid-base reactions like the one in part (b). It is insoluble in base because it cannot act as an acid. Thus, Bi_2O_3 is a basic oxide, the oxide of a metal. Based on the properties of its oxide, Bi is characterized as a metal.

(d)   Bi: $[Xe]6s^2 4f^{14} 5d^{10} 6p^3$. Bi has five outer electrons in the 6p and 6s subshells. If all five electrons participate in bonding, compounds such as BiF_5 are possible. Also, Bi has a large enough atomic radius (1.43 Å) and low-energy orbitals available to accommodate more than four pairs of bonding electrons.

(e)   The high ionization energy and relatively large negative electron affinity of F, coupled with its small atomic radius, make it the most electron withdrawing of the halogens. BiF_5 forms because F has the greatest tendency to attract electrons from Bi. Also, the small atomic radius of F reduces repulsions between neighboring bonded F atoms. The strong electron withdrawing properties of F are also the reason that only F compounds of Xe are known.

7.115   (a) $4 KO_2(s) + 2 CO_2(g) \rightarrow 2 K_2CO_3(s) + 3 O_2(g)$

(b)   K, +1; O, −1/2 (O_2^- is superoxide ion); C, +4; O, −2 → K, +1; C, +4; O, −2; O, 0

Oxygen (in the form of superoxide) is oxidized (to O_2) and reduced (to O^{2-}).

(c) $18.0 \text{ g } CO_2 \times \dfrac{1 \text{ mol } CO_2}{44.01 \text{ g } CO_2} \times \dfrac{4 \text{ mol } KO_2}{2 \text{ mol } CO_2} \times \dfrac{71.10 \text{ g } KO_2}{1 \text{ mol } KO_2} = 58.2 \text{ g } KO_2$

$18.0 \text{ g } CO_2 \times \dfrac{1 \text{ mol } CO_2}{44.01 \text{ g } CO_2} \times \dfrac{3 \text{ mol } O_2}{2 \text{ mol } CO_2} \times \dfrac{32.00 \text{ g } O_2}{1 \text{ mol } O_2} = 19.6 \text{ g } O_2$

8 Basic Concepts of Chemical Bonding

Visualizing Concepts

8.2 *Analyze.* Given the size and charge of four different ions, determine their ionic bonding characteristics.

Plan. The magnitude of lattice energy is directly proportional to the charges of the two ions and inversely proportional to their separation. $E_{el} = -Q_1Q_2/d$. Apply these concepts to A, B, X, and Y.

(a) AY and BX have a 1:1 ratio of cations and anions. In an ionic compound, the total positive and negative charges must be equal. To form a 1:1 compound, the magnitude of positive charge on the cation must equal the magnitude of negative charge on the anion. A^{2+} combines with Y^{2-} and B^+ combines with X^- to form 1:1 compounds.

(b) AY has the larger lattice energy. The A–Y and B–X separations are nearly equal. (A is smaller than B, but X is smaller than Y, so the differences in cation and anion radii approximately cancel.) In AY, $Q_1Q_2 = (2)(2) = 4$, whereas in BX, $Q_1Q_2 = (1)(1) = 1$.

8.3 *Analyze.* Given a schematic "slab" of NaCl(s), answer questions regarding the various ions and the electrostatic interactions among them. *Plan.* $E_{el} = -Q_1Q_2/d$. Use geometry to estimate or calculate distances when needed. *Solve.*

(a) The smaller purple balls represent Na^+ cations. Na^+ has a completed $n = 2$ shell, whereas Cl^- has a completed $n = 3$ shell.

(b) The larger green balls represent Cl^- anions.

(c) Four. Green-purple interactions are attractive; these are electrostatic attractions between two oppositely charged ions. The sign of E_{el} for these interactions is negative (–).

(d) Four. Green-green (and purple-purple) interactions are repulsive; these are electrostatic attractions between two ions with the same charge. The sign of E_{el} for these interactions is positive (+).

(e) Larger. Because the anions and cations have the same magnitude of charge (1– and 1+), the magnitude of their interactions depends on the distance between the ions; the shorter the distance, the larger the magnitude of the interaction. The distances between any green and any purple ion are the same, d. The magnitude between any two like-colored ions is the hypotenuse of a right triangle with distance $\sqrt{2}\, d$. The shorter attractive interactions have the greater magnitude. Because there are equal numbers of attractive and repulsive interactions, the sum of the attractive interactions is larger.

(f)   Positive. If this pattern of ions was extended indefinitely in two dimensions, the magnitude of the total attractive interactions would be greater than the magnitude of the total repulsive interactions. Lattice energy is the energy required to overcome attractive interactions and separate the particles into gas phase ions. The lattice energy would be positive.

8.5   *Analyze/Plan.* This question is a "reverse" Lewis structure. Count the valence electrons shown in the Lewis structure. For each atom, assume zero formal charge and determine the number of valence electrons an unbound atom has. Name the element.   *Solve.*

A: 1 shared e^- pair = 1 valence electron + 3 unshared pairs = 7 valence electrons, F

E: 2 shared pairs = 2 valence electrons + 2 unshared pairs = 6 valence electrons, O

D: 4 shared pairs = 4 valence electrons, C

Q: 3 shared pairs = 3 valence electrons + 1 unshared pair = 5 valence electrons, N

X: 1 shared pair = 1 valence electron, no unshared pairs, H

Z: same as X, H

Check. Count the valence electrons in the Lewis structure. Does the number correspond to the molecular formula CH_2ONF? 12 e^- pair in the Lewis structure. CH_2ONF = 4 + 2 + 6 + 5 + 7 = 24 e^-, 12 e^- pair. The molecular formula we derived matches the Lewis structure.

8.6   (a) HNO_2, 18 valence e^-, 9 e^- pairs NO_2^-, 18 valence e^-, 9 e^- pairs

$$H-\ddot{\underset{..}{O}}-\ddot{\underset{..}{N}}=\ddot{\underset{..}{O}} \qquad\qquad [:\ddot{\underset{..}{O}}-\underset{..}{N}=\ddot{\underset{..}{O}}]^-$$

(b)   The formal charge on N is zero, in both species.

(c) NO_2^- is expected to exhibit resonance; the double bond can be drawn to either oxygen atom. An alternate resonance structure for HNO_2 can be drawn, but it has nonzero formal charges on the oxygen atoms. This structure is less likely than the one shown above.

(d)   Assuming that the structure shown above is the main contributor to the structure of HNO_2, the N=O bond length in HNO_2 will be shorter than the N–O lengths in NO_2^-. Because there are two equivalent resonance structures for NO_2^-, the N–O lengths are approximately an average of N–O single and double bond lengths. These are longer than the full N=O double bond in HNO_2.

8.8   *Analyze/Plan.* Given an oxyanion of the type XO_4^{n-}, find the identity of X from elements in the third period. Use the generic Lewis structure to determine the identity of X, and to draw the ion-specific Lewis structures. Use the definition of formal charge, [# of valence electrons – # of nonbonding electrons – (# of bonding electrons/2)], to draw Lewis structures where X has a formal charge of zero. *Solve.*

(a)   According to the generic Lewis structure, each anion has 12 nonbonding and 4 bonding electron pairs for a total of 32 electrons. Of these 32 electrons, the 4 O atoms contribute (4 × 6) = 24, and the overall negative charges contribute 1, 2, or 3. # X electrons = 32 – 24 – n.

For n = 1–, X has (32 – 24 –1) = 7 valence electrons. X is Cl, and the ion is ClO_4^-.

For n = 2–, X has (32 – 24 – 2) = 6 valence electrons. X is S, and the ion is SO_4^{2-}.

For n = 3–, X has (32 – 24 – 3) = 5 valence electrons. X is P, and the ion is PO_4^{3-}.

Check. The identity of the ions is confirmed in Table 2.5.

(b) In the generic Lewis structure, X has 0 nonbonding electrons and (8/2) = 4 bonding electrons. Differences in formal charge are because of different numbers of valence electrons on X.

For PO_4^{3-}, formal charge of P is (5 – 4) = +1.

For SO_4^{2-}, formal charge of S is (6 – 4) = +2.

For ClO_4^-, formal charge of Cl is (7 – 4) = +3.

(c) To reduce the formal charge of X to zero, X must have more bonding electrons. This is accomplished by changing the appropriate number of lone pairs on O to multiple bonds between X and O.

Lewis Symbols (Section 8.1)

8.10   (a)   False. The valence shell of H is $n = 1$, which holds a maximum of 2 electrons.

           (b)   S: $[Ne]3s^2 3p^4$ A sulfur atom has 6 valence electrons, so it must gain 2 electrons to achieve an octet.

           (c) $1s^2 2s^2 2p^3 = [He]2s^2 2p^3$ The atom (N) has 5 valence electrons and must gain 3 electrons to achieve an octet.

8.12   (a)   Ti: $[Ar]4s^2 3d^2$. Ti has 4 valence electrons. These valence electrons are available for chemical bonding, whereas core electrons do not participate in chemical bonding.

           (b)   Hf: $[Xe]6s^2 4f^{14} 5d^2$

           (c)   If Hf and Ti both behave as if they have 4 valence electrons, the 6s and 5d orbitals in Hf behave as valence orbitals and the 4f behaves as a core orbital. This is reasonable because 4f is complete and 4f electrons are, on average, closer to the nucleus than 5d or 6s electrons. The core orbitals for Hf are then $[Xe]f^{14}$.

8.14   (a)   K·   (b)   ·Äs·   (c) $\left[:Sn\right]^{2+}$ (d) $\left[:\ddot{N}:\right]^{3-}$

Ionic Bonding (Section 8.2)

8.16   (a)

    (b) CaF_2

    (c)   2 electrons are transferred.

    (d)   Ca loses electrons.

8.18   (a) BaF_2 (b)   CsCl   (c) Li_3N (d) Al_2O_3

8.20   (a) Cd^{2+}: $[Kr]4d^{10}$

    (b) P^{3-}: $[Ne]3s^23p^6 = [Ar]$, noble-gas configuration

    (c) Zr^{4+}: $[Ar]4s^23d^{10}4p^6 = [Kr]$, noble-gas configuration

    (d) Ru^{3+}: $[Kr]4d^5$

    (e) As^{3-}: $[Ar]4s^23d^{10}4p^6 = [Kr]$, noble-gas configuration

    (f) Ag^+: $[Kr]4d^{10}$

8.22   (a)   NaCl, 788 kJ/mol; KF, 808 kJ/mol

          Given that crystal structure and ionic charges are the same for the two compounds, the difference in lattice energy is because of the difference in ion separation (d). Lattice energy is inversely proportional to ion separation (d), so we expect the compound with the smaller lattice energy, NaCl, to have the larger ion separation. That is, the Na–Cl distance should be longer than the K–F distance.

    (b)   Na–Cl, 1.16 Å + 1.67 Å = 2.83 Å

          K–F, 1.52 Å + 1.19 Å = 2.71 Å

          This estimate of the relative ion separations agrees with the estimate from lattice energies. Ionic radii indicate that the Na–Cl distance is longer than the K–F distance.

8.24   (a)   According to Equation 8.4, electrostatic attraction increases with increasing charges of the ions and decreases with increasing radius of the ions. Thus, lattice energy (i) increases as the charges of the ions increase and (ii) decreases as the sizes of the ions increase.

    (b)   KI < LiBr < MgS < GaN. Lattice energy increases as the charges on the ions increase. The ions in KI and LiBr all have 1+ and 1– charges. K^+ is larger than Li^+, and I^- is larger than Br^-. The ion separation is larger in KI, so it has the smaller lattice energy.

8.26   Trend (a) is because of differences in ionic radii. The ions have 1+ and 1– charges in all three compounds. In (b), the ions in the two compounds have different charges, which dominates the lattice energy trend. In (c), ion charge is the main influence, but differences in ionic radii predict the same trend.

8.28 $Ba(s) \rightarrow Ba(g)$; $Ba(g) \rightarrow Ba^+(g) + 1e^-$; $Ba^+(g) \rightarrow Ba^{2+}(g) + 1e^-$;

        $I_2(s) \rightarrow 2\,I(g)$; $2\,I(g) + 2e^- \rightarrow 2\,I^-(g)$, exothermic;

        $Ba^{2+}(g) + 2\,I^-(g) \rightarrow BaI_2(s)$, exothermic

8.30   (a) $MgCl_2$, 2326 kJ; $SrCl_2$, 2127 kJ. Because the ionic radius of Ca^{2+} is greater than that of Mg^{2+}, but less than that of Sr^{2+}, the ion separation (d) in $CaCl_2$ will be intermediate as well. We expect the lattice energy of $CaCl_2$ to be in the range 2200–2250 kJ.

       (b)   By analogy to Figure 8.5:

$$\Delta H_{latt} = -\Delta H_f^{\circ} CaCl_2 + \Delta H_f^{\circ} Ca(g) + 2\,\Delta H_f^{\circ} Cl(g) + I_1(Ca) + I_2(Ca) + 2\,E(Cl)$$
$$= -(-795.8\ kJ) + 179.3\ kJ + 2(121.7\ kJ) + 590\ kJ + 1145\ kJ + 2(-349\ kJ) = +2256\ kJ$$

       This value is near the range predicted in part (a).

Covalent Bonding, Electronegativity, and Bond Polarity
(Sections 8.3 and 8.4)

8.32   K and Ar. K is an active metal with 1 valence electron. It is most likely to achieve an octet by losing this single electron and to participate in ionic bonding. Ar has a stable octet of valence electrons; it is not likely to form chemical bonds of any type.

8.34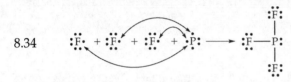

       (a)   5   (b) 7   (c) 8   (d) 8   (e) 3

8.36   (a)   The H atoms must be terminal because H can form only one bond.

              14 e⁻, 7 e⁻ pairs

              H—Ö—Ö—H

       (b)   There are two bonding electrons (one bonding electron pair) between the two O atoms.

       (c)   Longer. The oxygen atoms in H_2O_2 share one pair of electrons, whereas those in O_2 share two pairs (Solution 8.35). The fewer the number of shared electron pairs between two atoms, the longer the distance between them.

8.38   (a)   The electronegativity of the elements increases going from left to right across a row of the periodic table.

       (b)   Electronegativity generally decreases going down a family of the periodic table.

       (c)   False. Elements with the largest ionization energies are the most electronegative.

8.40   Electronegativity increases going up and to the right in the periodic table.

       (a)   O   (b) Al   (c) Cl   (d) F

8.42   The more different the electronegativity values of the two elements, the more polar the bond.

       (a)   O–F < C–F < Be–F. This order is clear from the periodic trend.

       (b)   S–Br < C–P < O–Cl. Refer to the electronegativity values in Figure 8.7 to confirm the order of bond polarity. The 3 pairs of elements all have the same positional relationship on the periodic table. The more electronegative element is one row above and one column to the left of the less electronegative element. This leads us to conclude that ΔEN is similar for the 3 bonds, which is confirmed by values in Figure 8.7. The most polar bond, O–Cl, involves the most electronegative element, O. Generally, the largest electronegativity differences tend to be between row 2 and row 3 elements. The 2 bonds in this exercise involving elements in row 2 and row 3 do have slightly greater ΔEN than the S–Br bond, between elements in rows 3 and 4.

(c)   C–S < N–O < B–F. You might predict that N–O is least polar because the elements are adjacent on the table. However, the big decrease going from the second row to the third means that the electronegativity of S is not only less than that of O, but essentially the same as that of C. C–S is the least polar.

8.44   (a)   The more electronegative element, Br, will have a stronger attraction for the shared electrons and adopt a partial negative charge.

       (b)   Q is the charge at either end of the dipole.

$$Q = \frac{\mu}{r} = \frac{1.21\,\text{D}}{2.49\,\text{Å}} \times \frac{1\,\text{Å}}{1 \times 10^{-10}\,\text{m}} \times \frac{3.34 \times 10^{-30}\,\text{C}\cdot\text{m}}{1\,\text{D}} \times \frac{1\,e}{1.60 \times 10^{-19}\,\text{C}} = 0.1014 = 0.101\,e$$

The charges on I and Br are 0.101 e.

8.46   Generally, compounds formed by a metal and a nonmetal are described as ionic, whereas compounds formed from two or more nonmetals are covalent. However, substances with metals in a high oxidation states often have properties of molecular compounds.

       (a) $TiCl_4$, metal and nonmetal, Ti(IV) is a relatively high oxidation state, molecular (by contrast with CaF_2, which is definitely ionic), titanium tetrachloride

            CaF_2, metal and nonmetal, ionic, calcium fluoride

       (b) ClF_3, two nonmetals, molecular, chlorine trifluoride

            VF_3, metal and nonmetal, ionic, vanadium(III) fluoride

       (c) $SbCl_5$, metalloid and nonmetal, molecular, antimony pentachloride

            AlF_3, metal and nonmetal, ionic, aluminum fluroide

Lewis Structures; Resonance Structures (Sections 8.5 and 8.6)

8.48   (a)   12 valence e^-, 6 e^- pairs   (b)   14 valence e^-, 7 e^- pairs

       (c)   50 valence e^-, 25 e^- pairs   (d)   26 valence e^-, 13 e^- pairs

(The Lewis structure that obeys the octet rule)

(e)   26 valence e^-, 13 e^- pairs   (f) NH_2Cl 14 e^-, 7 e^- pairs

H—Ö—S̈—Ö—H   H—N̈—C̈l:
        | |
      :Ö: H

(The Lewis structure that
obeys they octet rule)

8.50   (a)   26 e^-, 13 e^- pairs

:F̈—P̈—F̈:
        |
      :F̈:

The octet rule is satisfied for all atoms in the structure.

(b)   F is more electronegative than P. Assuming F atoms hold all shared electrons, the oxidation number of each F is –1. The oxidation number of P is +3.

(c)   Assuming perfect sharing, the formal charges on all F and P atoms are 0.

8.52   Formal charges are given near the atoms, oxidation numbers are listed below the structures.

(a)   18 e^-, 9 e^- pairs   (b)   24 e^-, 12 e^- pairs

                                                                        +2
:Ö—S̈═Ö   –1:Ö—S═Ö 0
 –1   +1   0 |
                                                        :Ö:
                                                         –1

ox. #: S, +4; O, –2   ox. #: S, +6; O, –2

(c)   26 e^-, 13 e^- pairs

$$\left[-1:\ddot{O}\!-\!\overset{+1}{\ddot{S}}\!-\!\ddot{O}:-1 \right]^{2-}$$
                      |
                    :Ö:
                     –1

ox. #: S, +4; O, –2

(d) $SO_2 < SO_3 < SO_3{}^{2-}$

Double bonds are shorter than single bonds. SO_2 has two resonance structures with alternating single and double bonds, for an approximate average "one-and-a-half" bond. SO_3 has three resonance structures with one double and two single bonds, for an approximately, "one-and-a-third" bond. $SO_3{}^{2-}$ has all single bonds. The order of increasing bond length is the order of decreasing bond type. SO_2 (1.5) $< SO_3$ (1.3) $< SO_3{}^{2-}$ (1.0).

8.54   (a)   18 e^-, 9 e^- pairs

$$\left[\begin{array}{c} :\ddot{O}: \\ | \\ H\!-\!C\!=\!\ddot{O} \end{array} \right]^{1-} \longleftrightarrow \left[\begin{array}{c} :O: \\ \| \\ H\!-\!C\!-\!\ddot{O}: \end{array} \right]^{1-}$$

(b)   Yes, resonance structures are required to describe the structure.

(c)   The Lewis structure of CO_2 (16 e$^-$, 8 e$^-$ pairs) is

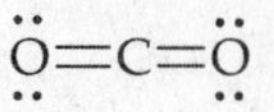

$$\ddot{O}=C=\ddot{O}$$

In CO_2, the C–O bonds are full double bonds with two shared pairs of electrons. In HCO_2^-, the two resonance structures indicate that the C–O bonds have partial, but not full, double bond character. The C–O bond lengths in formate will be longer than those in CO_2.

8.56   The Lewis structures are as follows:

5 e$^-$ pairs   9 e$^-$ pairs

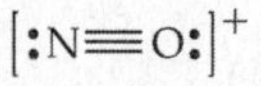

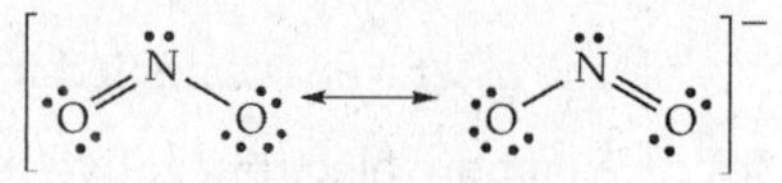

12 e$^-$ pairs

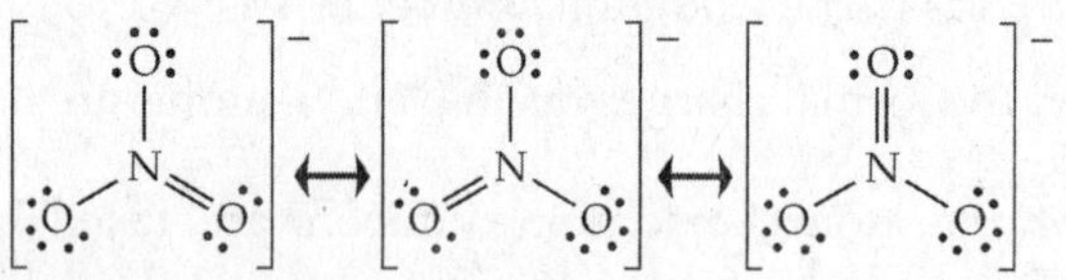

The average number of electron pairs in the N–O bond is 3.0 for NO^+, 1.5 for NO_2^-, and 1.33 for NO_3^-. The more electron pairs shared between two atoms, the shorter the bond. The order of N–O bond lengths from shortest to longest is: $NO^+ < NO_2^- < NO_3^-$.

8.58   (a)

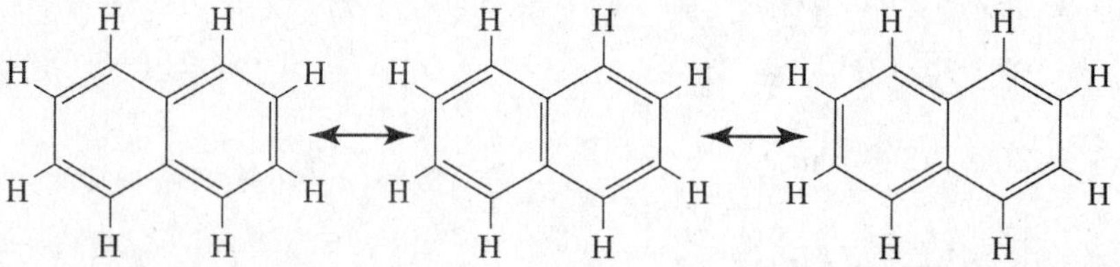

(b)   The resonance model of this molecule has bonds that are neither single nor double, but somewhere in between. This results in bond lengths that are intermediate between C–C single and C=C double bond lengths.

(c)   Four. Among the three resonance structures, there are four C–C bonds that appear twice as double bonds and once as a single bond. These are shorter than the others. (The other seven C–C bonds appear twice as single bonds and once as a double bond.)

Exceptions to the Octet Rule (Section 8.7)

8.60   (a)   7, 1   (b) 6, 2   (c) 5, 3   (d) 4, 4

8.62   The second friend is more correct. In the third row and beyond, atoms have the space and available orbitals to accommodate extra electrons. Because atomic radius increases going down a family, elements in the third period and beyond are less subject to destabilization from additional electron-electron repulsions. It is also true, but probably not as important, that elements in the third shell and beyond contain empty d orbitals that are relatively close in energy to valence orbitals (the ones that accommodate the octet).

8.64   (a)   11 e⁻, 5.5 e⁻ pairs

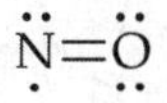

$\ddot{\text{N}}=\ddot{\text{O}}$

Does not obey the octet rule. N has only 7 electrons.

(b)   24 e⁻, 12 e⁻ pairs

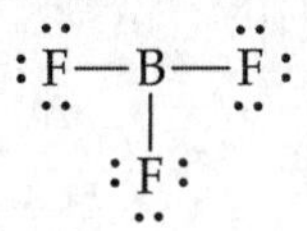

:F̈—B—F̈:
     |
    :F̈:

Does not obey the octet rule. B has only 6 electrons.

(c)   22 e⁻, 11 e⁻ pairs

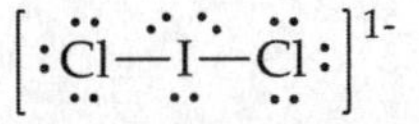

$\left[:\ddot{\text{Cl}}-\ddot{\text{I}}-\ddot{\text{Cl}}: \right]^{1-}$

Does not obey the octet rule. Central I has 10 electrons.

(d)   32 e⁻, 16 e⁻ pairs

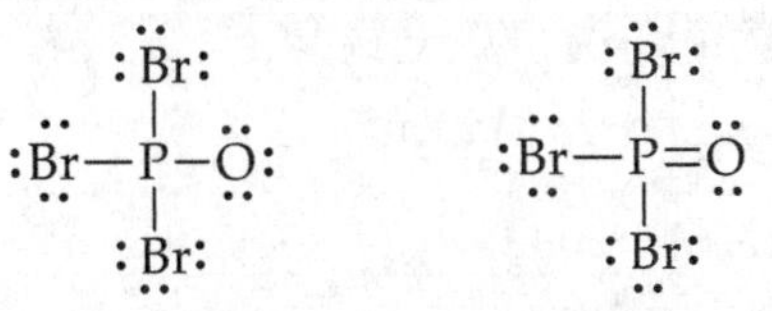

    :B̈r: :B̈r:
     | |
:Br—P—Ö: :Br—P=Ö
     | |
    :B̈r: :B̈r:

The structure on the left obeys the octet rule, whereas the one on the right minimizes formal charges but does not obey the octet rule. The P in the right structure has 10 electrons.

(e)   36 e⁻, 18 e⁻ pairs

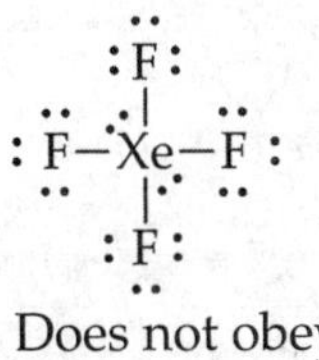

    :F̈:
     |
: F—Xe—F :
     |
    :F̈:

Does not obey the octet rule. Central Xe has 12 electrons.

8.66   (a)   26 e⁻, 13 e⁻ pairs

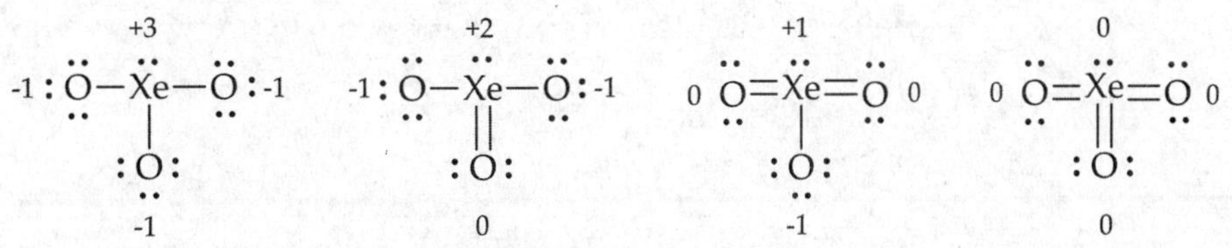

       +3   +2   +1   0

-1:Ö—Xe—Ö:-1   -1:Ö—Xe—Ö:-1   0 O=Xe=O 0   0 Ö=Xe=O 0
      | || | ||
     :O: :O: :O: :O:
     -1   0   -1   0

(b)   Yes, the structure with no double bonds obeys the octet rule for all atoms.

(c)   The structure with one double bond has 3 resonance structures (3 possible positions for the double bond), as does the structure with 2 double and 1 single bond (3 possible positions for the single bond). The total number of resonance structures is then 8.

(d)   The structure with 3 double bonds minimizes formal charges on all atoms.

8.68   (a)   32 e⁻, 16 e⁻ pairs   (b)

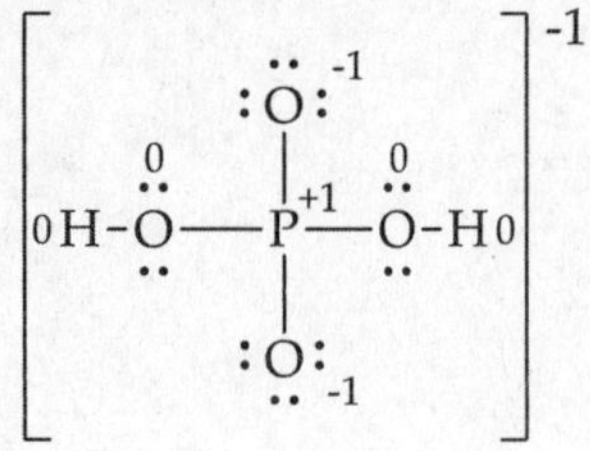

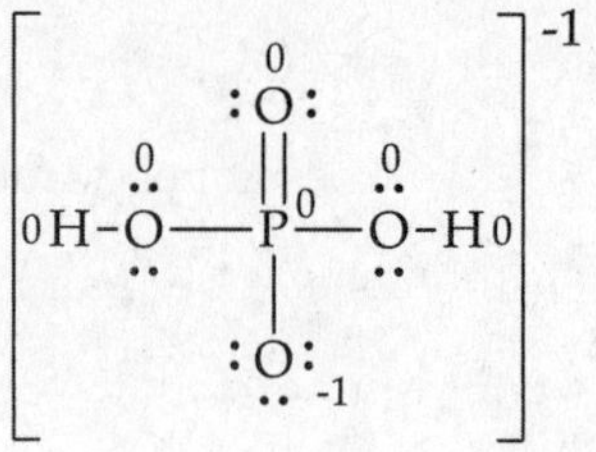

Bond Enthalpies (Section 8.8)

8.70   (a) $\Delta H = 3\,D(C-Br) + D(C-H) + D(Cl-Cl) - 3\,D(C-Br) - D(C-Cl) - D(H-Cl)$

$= D(C-H) + D(Cl-Cl) - D(C-Cl) - D(H-Cl)$

$\Delta H = 413 + 242 - 328 - 431 = -104\ kJ$

(b) $\Delta H = 4\,D(C-H) + 2\,D(C-S) + 2\,D(S-H) + D(C-C) + 2\,D(H-Br)$

$-4\,D(S-H) - D(C-C) - 2\,D(C-Br) - 4\,D(C-H)$

$= 2\,D(C-S) + 2\,D(H-Br) - 2\,D(S-H) - 2\,D(C-Br)$

$\Delta H = 2(259) + 2(366) - 2(339) - 2(276) = 20\ kJ$

(c) $\Delta H = 4\,D(N-H) + D(N-N) + D(Cl-Cl) - 4\,D(N-H) - 2\,D(N-Cl)$

$= D(N-N) + D(Cl-Cl) - 2D(N-Cl)$

$\Delta H = 163 + 242 - 2(200) = 5\ kJ$

8.72   *Plan.* Draw structural formulas when needed.   *Solve.*

(a)

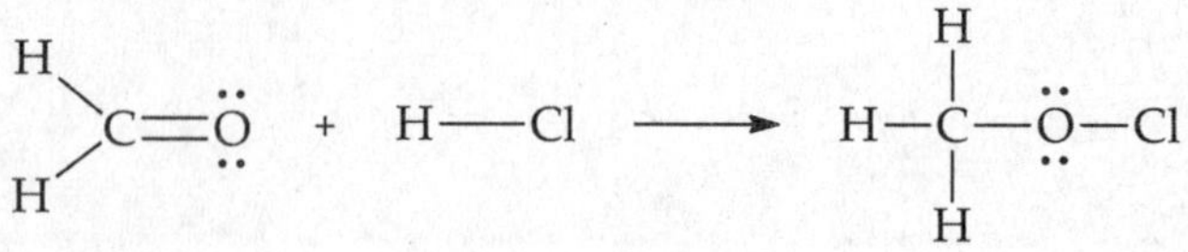

$\Delta H = D(C{=}O) + 2\,D(C-H) + D(H-Cl) - 3\,D(C-H) - D(C-O) - D(O-Cl)$

$\Delta H = D(C{=}O) + D(H-Cl) - D(C-H) - D(C-O) - D(O-Cl)$

$= (799) + (431) - (413) - (358) - (203) = 256\ kJ$

(b) $H-O-O-H + 2\,C{\equiv}O \rightarrow H-H + 2\,O{=}C{=}O$

$\Delta H = D(O-O) + 2\,D(O-H) + 2\,D(C{\equiv}O) - D(H-H) - 4\,D(C{=}O)$

$= 146 + 2(463) + 2(1072) - (436) - 4(799) = -416\ kJ$

(c)

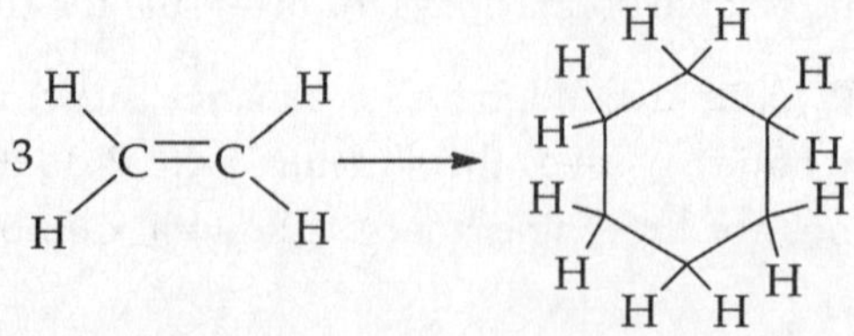

$\Delta H = 3\,D(C{=}C) + 12\,D(C-H) - 12\,D(C-H) - 6\,D(C-C)$

$\Delta H = 3\,D(C{=}C) - 6\,D(C-C)$

$\Delta H = 3(614) - 6(348) = -246\ kJ$

8.74   (a)

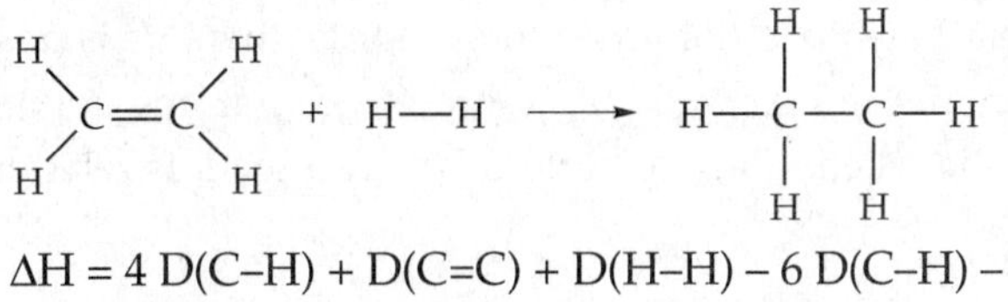

$$\Delta H = 4\,D(C–H) + D(C=C) + D(H–H) – 6\,D(C–H) – D(C–C)$$

$$= D(C=C) + D(H–H) – 2\,D(C–H) – D(C–C)$$

$$\Delta H = 614 + 436 – 2(413) – 348 = –124 \text{ kJ}$$

(b) $\Delta H° = \Delta H_f° \, C_2H_6(g) – \Delta H_f° \, C_2H_4(g) – \Delta H_f° \, H_2(g)$

$$= –84.68 – 52.30 – 0 = –136.98 \text{ kJ}$$

8.76   (a)   (i) 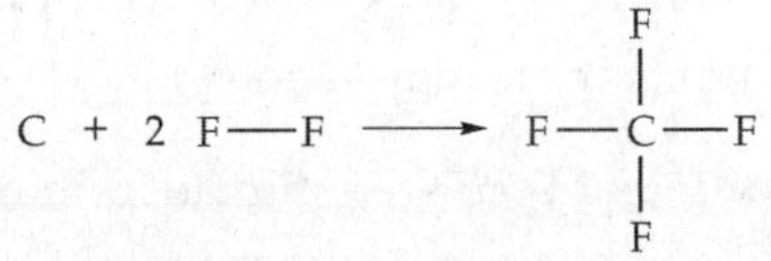 $C + 2\ F–F \longrightarrow F–\overset{\displaystyle F}{\underset{\displaystyle F}{C}}–F$

$$\Delta H = 2\,D(F–F) – 4\,D(C–F) = 2(155) – 4(485) = –1630 \text{ kJ}$$

(ii) 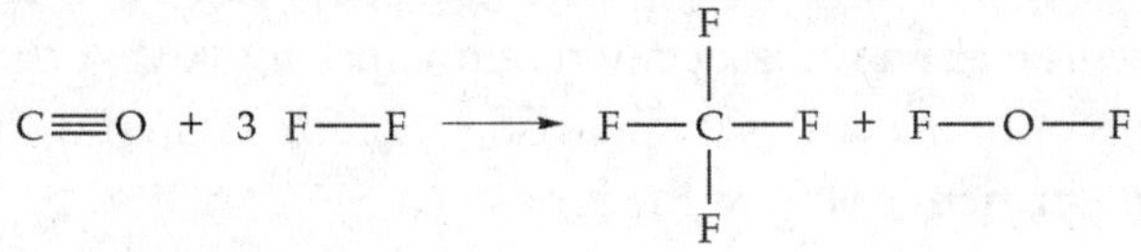 $C≡O + 3\ F–F \longrightarrow F–\overset{\displaystyle F}{\underset{\displaystyle F}{C}}–F + F–O–F$

$$\Delta H = D(C≡O) + 3\,D(F–F) – 4\,D(C–F) – 2\,D(O–F)$$

$$= 1072 + 3(155) – 4(485) – 2(190) = –783 \text{ kJ}$$

(iii) 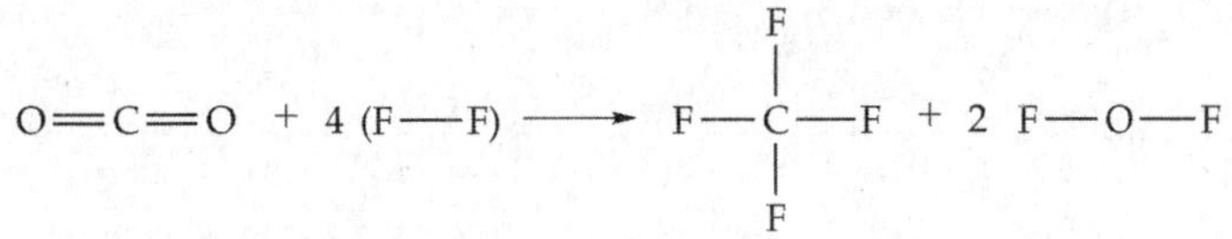 $O=C=O + 4\ (F–F) \longrightarrow F–\overset{\displaystyle F}{\underset{\displaystyle F}{C}}–F + 2\ F–O–F$

$$\Delta H = 2\,D(C=O) + 4\,D(F–F) – 4\,D(C–F) – 4\,D(O–F)$$

$$= 2(799) + 4(155) – 4(485) – 4(190) = –482 \text{ kJ}$$

Reaction (i) is most exothermic.

(b)

Substance	Oxidation State of C	Reaction Enthalpy, kJ
C(g)	0	-1630
CO(g)	2+	-783
CO₂(g)	4+	-482

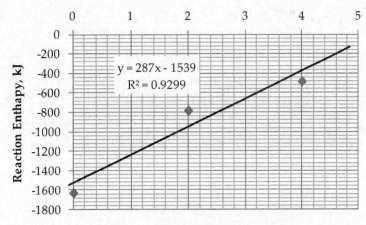

y = 287x - 1539
R² = 0.9299

Oxidation State of Carbon

Although the relationship is not linear, the slope of this line is positive. For these three carbon-containing reactants, the more oxygen atoms bound to carbon, the less exothermic the reaction in this series. [We can see this relationship without a graph.]

(c)   The oxidation number of C in CO_3^{2-} is the same as that of C in CO_2. From the graph, we expect the reaction enthalpies to be similar (approximately –500 kJ).

The reaction under consideration is: $CO_3^{2-}(g) + 5\ F_2(g) \rightarrow CF_4 + 3\ OF_2(g)$. Using average bond enthalpies, and ignoring the contribution of resonance to the bond enthalpies in $CO_3^{2-}(g)$,

$\Delta H = D(C=O) + 2\ D(C-O) + 5\ D(F-F) - 4\ D(C-F) - 6\ D(O-F)$

$\Delta H = 799 + 2(358) + 5(155) - 4(485) - 6(190) = -790$ kJ

This value is significantly different than the estimate from the graph. Using bond enthalpies for one single (C–O) bond and two double (C=O) bonds results in a value of –349 kJ, which is consistent with the series. This is an empirical result. A more accurate bond dissociation enthalpy for CO_3^{2-} (taking resonance into account) would make this comparison more meaningful. It is also possible that CO_3^{2-}, an ion, does not fit in the series.

Additional Exercises

8.78 $E = kQ_1Q_2/d;\quad k = 8.99 \times 10^9$ J-m/C^2

(a) Na^+, Br^-: $E = \dfrac{-8.99 \times 10^9\ \text{J-m}}{C^2} \times \dfrac{(1 \times 1.60 \times 10^{-19}\ C)^2}{(1.16 + 1.82) \times 10^{-10}\ \text{m}} = -7.72 \times 10^{-19}$ J

Na^+, Br^-: $E = \dfrac{-8.99 \times 10^9\ \text{J-m}}{C^2} \times \dfrac{(1 \times 1.60 \times 10^{-19}\ C)^2}{(1.16 + 1.82) \times 10^{-10}\ \text{m}} = -7.72 \times 10^{-19}$ J

The sign of E is negative because one of the interacting ions is an anion; this is an attractive interaction.

On a molar basis: $-7.723 \times 10^{-19} \times 6.022 \times 10^{23} = -4.65 \times 10^5$ J = –465 kJ

(b) Rb^+, Br^-: $E = \dfrac{-8.99 \times 10^9\ \text{J-m}}{C^2} \times \dfrac{(1 \times 1.60 \times 10^{-19}\ C)^2}{(1.66 + 1.82) \times 10^{-10}\ \text{m}} = -6.61 \times 10^{-19}$ J

On a molar basis: -3.98×10^5 J = –398 kJ

(c) Sr^{2+}, S^{2-}: $E = \dfrac{-8.99 \times 10^9\ \text{J-m}}{C^2} \times \dfrac{(2 \times 1.60 \times 10^{-19}\ C)^2}{(1.32 + 1.70) \times 10^{-10}\ \text{m}} = -3.05 \times 10^{-18}$ J

On a molar basis: -1.84×10^6 J = -1.84×10^3 kJ

8.79   (a)

Compound	Cation Radius, Å	Lattice Energy, kJ
BeH$_2$	0.59	3205
MgH$_2$	0.86	2791
CaH$_2$	1.14	2410
SrH$_2$	1.32	2250
BaH$_2$	1.49	2121
ZnH$_2$	?	2870

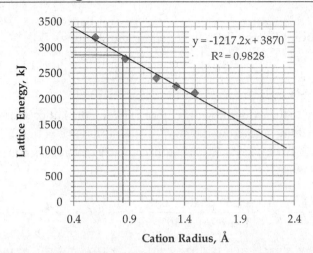

The slope of the line is negative. Lattice energy is proportional to Q_1Q_2/d. For each of these compounds, Q_1Q_2 is the same. The anion H^- is present in each compound, but the ionic radius of the cation increases going from Be to Ba. Thus, the value of d (the cation–anion separation) increases and the ratio Q_1Q_2/d decreases. This is reflected in the decrease in lattice energy going from BeH_2 to BaH_2.

(b) Again, Q_1Q_2 for ZnH_2 is the same as that for the other compounds in the series and the anion is H^-. The lattice energy of ZnH_2, 2870 kJ, is closest to that of MgH_2, 2791 kJ. The ionic radius of Zn^{2+} is similar to that of Mg^{2+}. We can calculate the cationic radius for Zn^{2+} using the equation for the best fit line.

$y = -1217.2 + 3870$, $y = 2870$; $x = (y-3871)/-1217.2 = (2870-3871)/-1217.2 = 0.822$ Å

8.80 (a)

Compound	Lattice Energy (kJ)		Compound	Lattice Energy (kJ)	
NaCl	788	⎤ 56 kJ	LiCl	834	⎤ 55 kJ
NaBr	732	⎦	**LiBr**	**779**	⎦
Na I	682		Li I	730	

106 kJ (for left group), 104 kJ (for right group)

The difference in lattice energy between LiCl and LiI is 104 kJ. The difference between NaCl and NaI is 106 kJ; the difference between NaCl and NaBr is 56 kJ, or 53% of the difference between NaCl and NaI. Applying this relationship to the Li salts, 0.53(104 kJ) = 55 kJ difference between LiCl and LiBr. The approximate lattice energy of LiBr is (834 – 55) kJ = 779 kJ.

(b)

Compound	Lattice Energy (kJ)		Compound	Lattice Energy (kJ)	
NaCl	788	⎤ 56 kJ	CsCl	657	⎤ 30 kJ
NaBr	732	⎦	**CsBr**	**627**	⎦
Na I	682		Cs I	600	

106 kJ (for left group), 57 kJ (for right group)

By analogy to the Na salts, the difference between lattice energies of CsCl and CsBr should be approximately 53% of the difference between CsCl and CsI. The lattice energy of CsBr is approximately 627 kJ.

(c)

		Lattice				Lattice	
	Compound	Energy (kJ)			Compound	Energy (kJ)	
	MgO	3795	381 kJ		$MgCl_2$	2326	131 kJ
578 kJ	CaO	3414		199 kJ	**$CaCl_2$**	**2195**	
	SrO	3217			$SrCl_2$	2127	

By analogy to the oxides, the difference between the lattice energies of $MgCl_2$ and $CaCl_2$ should be approximately 66% of the difference between $MgCl_2$ and $SrCl_2$. That is, 0.66(199 kJ) = 131 kJ. The lattice energy of $CaCl_2$ is approximately (2326 – 131) kJ = 2195 kJ.

8.82 $$E = \frac{-8.99 \times 10^9 \text{ J} \cdot \text{m}}{C^2} \times \frac{4(1.60 \times 10^{-19} \text{ C})^2}{(1.14 + 1.26) \times 10^{-10} \text{ m}} = -3.836 \times 10^{-18} = -3.84 \times 10^{-18} \text{ J}$$

On a molar basis: $(-3.836 \times 10^{-18} \text{ J})(6.022 \times 10^{23}) = -2.310 \times 10^6 \text{ J} = -2310 \text{ kJ}$

Note that the absolute value of this potential energy is less than the lattice energy of CaO, 3414 kJ/mol. The difference represents the added energy of putting all the $Ca^{2+}O^{2-}$ ion pairs together in a three-dimensional array, similar to the one in Figure 8.3.

8.83 By analogy to the Born-Haber cycle for NaCl(s), Figure 8.5, the enthalpy of formation for $NaCl_2$(s) is

$$\Delta H_f^o NaCl_2(s) = -\Delta H_{latt} NaCl_2 + \Delta H_f^o Na(g) + 2\Delta H_f^o Cl(g) + I_1(Na) + I_2(Na) + 2 E(Cl)$$

(a) $\Delta H_f^o NaCl_2(s) = -\Delta H_{latt} NaCl_2 + 107.7 \text{ kJ} + 2(121.7 \text{ kJ}) + 496 \text{ kJ} + 4562 \text{ kJ} + 2(-349 \text{ kJ})$

        $\Delta H_f^o NaCl_2(s) = -\Delta H_{latt} NaCl_2 + 4711 \text{ kJ}$

The collective energy of the "other" steps in the cycle (vaporization and ionization of Na^{2+}, dissociation of Cl_2 and electron affinity of Cl) is +4711 kJ. In order for the sign of $\Delta H_f^o NaCl_2$ to be negative, the lattice energy would have to be greater than 4711 kJ.

(b) $\Delta H_f^o NaCl_2(s) = -(2326 \text{ kJ}) + 4711 \text{ kJ} = 2385 \text{ kJ}$

This value is large and positive.

8.84 (a) Yes. If X and Y have different electronegativities, they have different attractions for the electrons in the molecule. The electron density around the more electronegative atom will be greater, producing a charge separation or dipole in the molecule.

(b) Yes. $\mu = Qr$. The dipole moment, μ, is the product of the magnitude of the separated charges, Q, and the distance between them, r. The longer the bond between X and Y, the larger the dipole moment.

8.86 (a) $$Q = \frac{\mu}{r} = \frac{1.24 \text{ D}}{1.60 \text{ Å}} \times \frac{1 \text{ Å}}{1 \times 10^{-10} \text{ m}} \times \frac{3.34 \times 10^{-30} \text{ C} \cdot \text{m}}{1 \text{ D}} \times \frac{1 e}{1.60 \times 10^{-19} \text{ C}} = 0.1618 = 0.162 e$$

(b) From Figure 8.7, the electronegativity of Cl is 3.0 and that of O is 3.5. Because O is the more electronegative element, we expect it to have a partial negative charge in the ClO molecule.

(c)   13 e⁻, 6.5 e⁻ pairs

$$+1 \; \cdot \ddot{Cl} - \ddot{\underset{..}{O}} : _{-1} \qquad\qquad 0 \; \cdot \ddot{Cl} = \ddot{\underset{..}{O}} \; 0$$

According to formal charge arguments, the Lewis structure on the right is dominant. In both structures, the less electronegative Cl atom is electron-deficient. However, the small electronegativity difference and calculated charges both point to a slightly polar covalent molecule. The true bonding situation is a blend of the two extreme Lewis structures, with the right-most structure making the larger contribution.

(d)   Because ClO⁻ has an overall charge of 1–, the sum of the formal charges in any correct Lewis structure is 1–. We expect the more electronegative O atom to carry the negative formal charge. The best Lewis structure for ClO⁻ is then

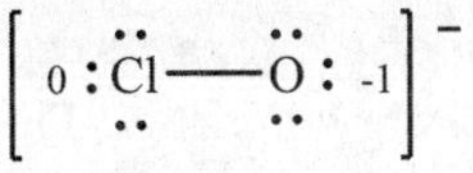

$$\left[_0 : \ddot{Cl} - \ddot{\underset{..}{O}} : _{-1} \right]^-$$

The formal charge on Cl in this structure is 0.

8.87   (a)   Estimate relative attraction for the bonding electron pair by calculating the relative electronegativity of the two atoms. From Figure 8.7, the electronegativity of Br is 2.8 and of Cl is 3.0.

             Br has 2.8/(3.0 + 2.8) = 0.48 of the charge of the bonding e⁻ pair.

             Cl has 3.0/(3.0 + 2.8) = 0.52 of the charge of the bonding e⁻ pair.

             This amounts to 0.52 × 2e = 1.04e on Cl or 0.04e more than a neutral Cl atom. This implies a –0.04 charge on Cl and +0.04 charge on Br.

   (b)   From Figure 7.7, the covalent radius of Br is 1.20 Å and of Cl is 1.02 Å. The Br–Cl separation is 2.22 Å.

$$\mu = Qr = 0.04e \times \frac{1.60 \times 10^{-19}\,C}{e} \times 2.22\,\text{Å} \times \frac{1 \times 10^{-10}\,m}{\text{Å}} \times \frac{1\,D}{3.34 \times 10^{-30}\,C\text{-}m} = 0.43\,D$$

   (c) $$Q = \frac{\mu}{r} = \frac{0.57\,D}{2.22\,\text{Å}} \times \frac{1\,\text{Å}}{1 \times 10^{-10}\,m} \times \frac{3.34 \times 10^{-30}\,C\text{-}m}{1\,D} \times \frac{1\,e}{1.60 \times 10^{-19}\,C} = 0.054\,e$$

             From this calculation, the partial charge on Br is +0.054 and on Cl is –0.054.

8.88   (a) $2\,NaAlH_4(s) \rightarrow 2\,NaH(s) + 2\,Al(s) + 3\,H_2(g)$

   (b)   Hydrogen is the only nonmetal in NaAlH₄, so we expect it to be most electronegative. (The position of H on the periodic table is problematic. Its electronegativity does not fit the typical trend for Gp 1A elements.) For the two metals, Na and Al, electronegativity increases moving up and to the right on the periodic table, so Al is more electronegative. The least electronegative element in the compound is Na.

   (c)   Covalent bonds hold polyatomic anions together; elements involved in covalent bonding have smaller electronegativity differences than those that are involved in ionic bonds. Possible covalent bonds in NaAlH₄ are Na–H and Al–H. Al and

H have a smaller electronegativity difference than Na and H and are more likely to form covalent bonds. The anion has an overall 1– charge, so it can be thought of as four hydride ions and one Al^{3+} ion. The formula is AlH_4^-. For the purpose of counting valence electrons, assume neutral atoms.

$8\,e^-\quad 4\,e^-$ pairs

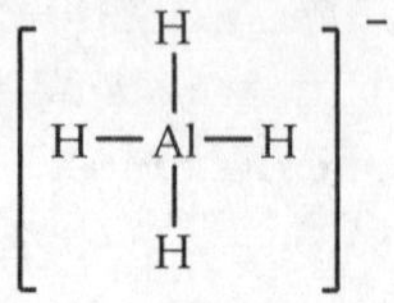

$$\left[\begin{array}{c} H \\ | \\ H-Al-H \\ | \\ H \end{array}\right]^-$$

(d)   The formal charge of H in AlH_4^- is 0. (The formal charge of Al is –1. This brings the sum of formal charges to –1, the overall charge of the polyatomic anion.)

8.89   Statement (b) is most correct. The Lewis structure of I_3^-, with three nonbonding and two bonding electron pairs about the central atom, is shown below. Fluorine is both too small and has no available d orbitals to accommodate an expanded octet.

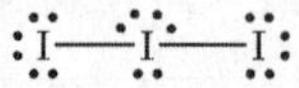

$$:\!\ddot{I}\!-\!\ddot{I}\!-\!\ddot{I}\!:$$

8.91   (a) $14e^-, 7\,e^-$ pairs $32\,e^-, 16\,e^-$ pairs

$$\left[:\!\ddot{C}l\!-\!\ddot{O}\!:\right]^-$$

$$\left[\begin{array}{c} :\ddot{O}: \\ | \\ :\ddot{O}-Cl-\ddot{O}: \\ | \\ :\ddot{O}: \end{array}\right]^-$$

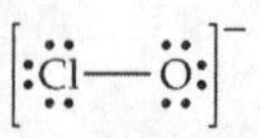

FC on Cl = 7 – [6 + 1/2(2)] = 0   FC on Cl = 7 – [0 + 1/2(8)] = +3

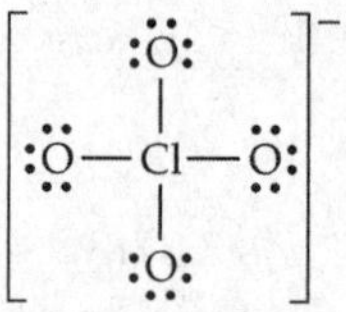

(b)   The oxidation number of Cl is +1 in ClO^- and +7 in ClO_4^-.

(c)   Oxidizing power is the tendency of a substance to be reduced, to gain electrons. Oxidation numbers show the maximum electron deficiency (or excess) of a substance. The higher the oxidation number of the central atom in an oxyanion, the greater its electron deficiency and oxidizing power. The greater oxidation number of Cl in perchlorate indicates that it is a stronger oxidizing agent than hypochlorite .

8.92   (a) $:N\!\equiv\!\!N\!-\!\ddot{O}: \longleftrightarrow :\ddot{N}\!-\!N\!\equiv\!O: \longleftrightarrow \ddot{N}\!=\!\!N\!=\!\ddot{O}$

                        0   +1   -1   -2   +1   +1   -1   +1   0

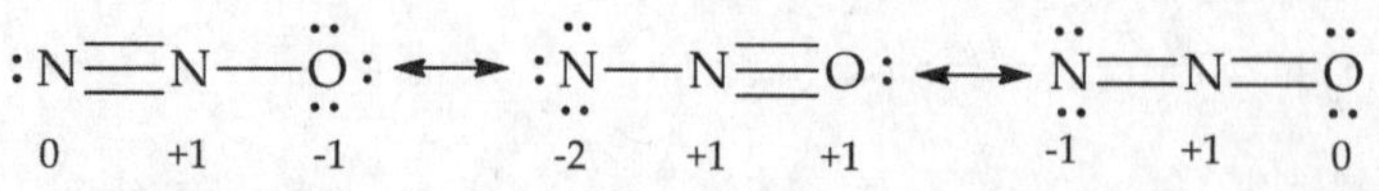

In the leftmost structure, the more electronegative O atom has the negative formal charge, so this structure is likely to be most important.

(b)   No single resonance structure rationalizes both observed bond lengths. In general, the more shared pairs of electrons between two atoms, the shorter the bond, and vice versa. That the N–N bond length in N_2O is slightly longer than the typical N≡N indicates that the middle and right resonance structures where the N atoms share less than three electron pairs are contributors to the true structure. That the N–O bond length is slightly shorter than a typical N=O indicates that the middle structure, where N and O share more than two electron pairs, does contribute to the true structure. This physical data indicates that although formal charge can be used to predict which resonance form will be more important to the observed structure, the influence of minor contributors on the true structure cannot be ignored.

8.93   (a) $12 + 3 + 15 = 30$ valence e^-, 15 e^- pairs.

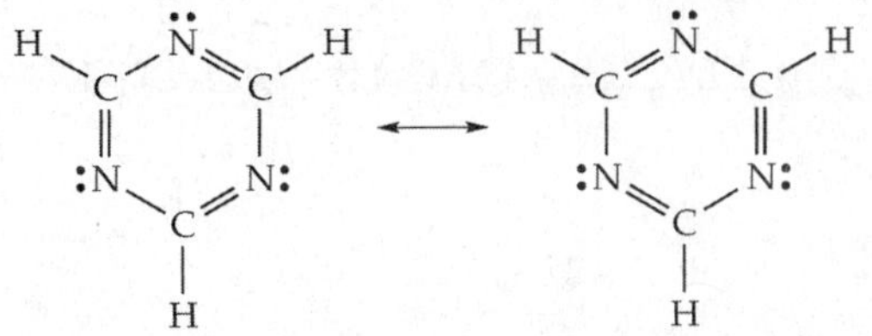

Structures with H bound to N and nonbonded electron pairs on C can be drawn, but the structures above minimize formal charges on the atoms.

    (b)   The resonance structures indicate that triazine will have six equal C–N bond lengths, intermediate between C–N single and C–N double bond lengths. (See Solutions 8.57 and 8.58.) From Table 8.5, an average C–N length is 1.43 Å, a C=N length is 1.38 Å. The average of these two lengths is 1.405 Å. The C–N bond length in triazine should be in the range 1.40–1.41 Å.

8.94   (a) $24 + 4 + 14 = 42$ valence e^-, 21 e^- pairs.   (b)

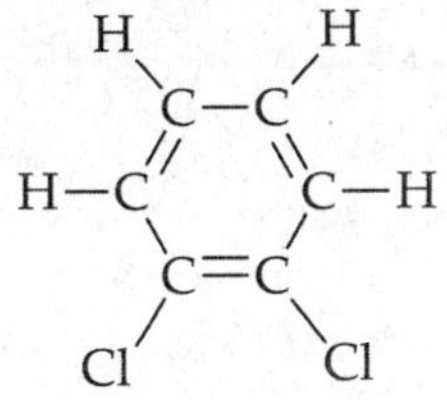

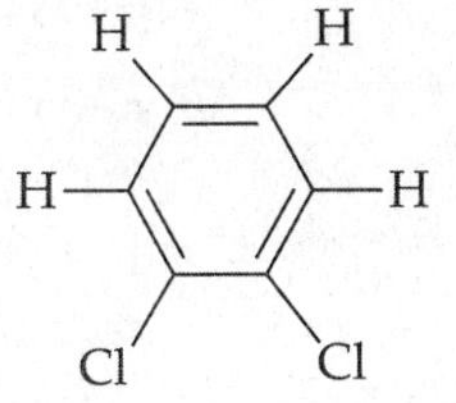

    (c)   No. In benzene, the six C atoms are equivalent. In ortho-dichlorobenzene, the two C atoms bound to Cl are not equivalent to the four C atoms bound to H. In the two resonance structures above, one has a double bond between the C atoms bound to Cl, and the other has a single bond in this position. The two ortho-dichlorobenzene resonance structures are not equivalent like the resonance structures of benzene.

8.96 $\Delta H = 8\ D(C–H) – D(C–C) – 6\ D(C–H) – D(H–H)$

        $= 2\ D(C–H) – D(C–C) – D(H–H)$

        $= 2(413) – 348 – 436 = +42$ kJ

    $\Delta H = 8\ D(C–H) + 1/2\ D(O=O) – D(C–C) – 6\ D(C–H) – 2\ D(O–H)$

        $= 2\ D(C–H) + 1/2\ D(O=O) – D(C–C) – 2\ D(O–H)$

        $= 2(413) + 1/2\ (495) – 348 – 2(463) = –200$ kJ

The fundamental difference in the two reactions is the formation of 1 mol of H–H bonds versus the formation of 2 mol of O–H bonds. The latter is much more exothermic, so the reaction involving oxygen is more exothermic.

8.97   (a) $\Delta H = 5\,D(C-H) + D(C-C) + D(C-O) + D(O-H) - 6\,D(C-H) - 2\,D(C-O)$

$= D(C-C) + D(O-H) - D(C-H) - D(C-O)$

$= 348\text{ kJ} + 463\text{ kJ} - 413\text{ kJ} - 358\text{ kJ}$

$\Delta H = +40$ kJ; ethanol has the lower enthalpy

(b) $\Delta H = 4\,D(C-H) + D(C-C) + 2\,D(C-O) - 4\,D(C-H) - D(C-C) - D(C=O)$

$= 2\,D(C-O) - D(C=O)$

$= 2(358\text{ kJ}) - 799\text{ kJ}$

$\Delta H = -83$ kJ; acetaldehyde has the lower enthalpy

(c) $\Delta H = 8\,D(C-H) + 4\,D(C-C) + D(C=C) - 8\,D(C-H) - 2\,D(C-C) - 2\,D(C=C)$

$= 2\,D(C-C) - D(C=C)$

$= 2(348\text{ kJ}) - 614\text{ kJ}$

$\Delta H = +82$ kJ; cyclopentene has the lower enthalpy

(d) $\Delta H = 3\,D(C-H) + D(C-N) + D(C \equiv N) - 3\,D(C-H) - D(C-C) - D(C \equiv N)$

$= D(C-N) - D(C-C)$

$= 293\text{ kJ} - 348\text{ kJ}$

$\Delta H = -55$ kJ; acetonitrile has the lower enthalpy

8.99   (a)

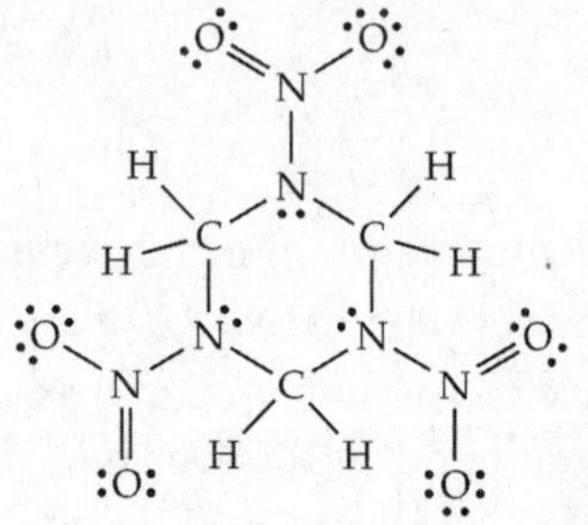

$C_3H_6N_6O_6$ $12 + 6 + 30 + 36 = 84$ e$^-$, 42 e$^-$ pairs

42 e$^-$ pairs – 24 shared e$^-$ pairs = 18 unshared (lone) e$^-$ pairs

Use unshared pairs to complete octets on terminal O atoms (15 unshared pairs) and ring N atoms (3 unshared pairs).

(b)   No C=N bonds in the 6-membered ring are possible, because all C octets are complete with 4 bonds to other atoms. N=N are possible, as shown below. There are 8 possibilities involving some combination of N–N and N=N groups [1 with 0 N=N, 3 with 1 N=N, 3 with 2N=N, 1 with 3N=N]. A resonance structure with 1 N=N is shown below.

Each terminal O=N–O group has two possible placements for the N=O. This generates 8 structures with 0 N=N groups (and 3 O = N–O groups), 12 with 1 N=N and 2 O=N–O, 6 with 2 N=N and 1 O=N–O, and 1 with 3 N=N and no O=N–O. This sums to a total of 27 resonance structures (that I can visualize). Can you find others?

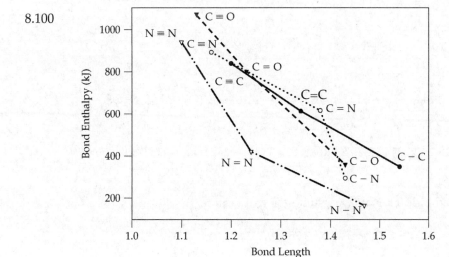

(c) $C_3H_6N_6O_6(s) \rightarrow 3\,CO(g) + 3\,N_2(g) + 3\,H_2O(g)$

(d)   The molecule contains N=O, N=N, C–H, C–N, N–O, and N–N bonds. According to Table 8.4, N–N bonds have the smallest bond enthalpy and are weakest.

(e)   Calculate the enthalpy of decomposition for the resonance structure drawn in part (a).

$\Delta H = 3\,D(N{=}O) + 3\,D(N{-}O) + 3\,D(N{-}N) + 6\,D(N{-}C) + 6\,D(C{-}H)$

$- 3\,D(C{\equiv}O) - 3\,D(N{\equiv}N) - 6\,D(O{-}H)$

$= 3(607) + 3(201) + 3(163) + 6(293) + 6(413) - 3(1072) - 3(941) - 6(463)$

$= -1668$ kJ/mol $C_3H_6N_6O_6$

$$5.0\text{ g } C_3H_6N_6O_6 \times \frac{1\text{ mol } C_3H_6N_6O_6}{222.1\text{ g } C_3H_6N_6O_6} \times \frac{-1668\text{ kJ}}{\text{mol } C_3H_6N_6O_6} = -37.55 = -38\text{ kJ}$$

Even though exchanging N=O and N–O bonds has no effect on the enthalpy calculation, structures with N=N and 2 N–O do have different enthalpy of decomposition. For the resonance structure with 3 N=N and 6 N–O bonds instead of 3 N–N, 3 N–O and 3 N=O, $\Delta H = -2121$ kJ/mol. The actual enthalpy of decomposition is probably somewhere between –1668 and –2121 kJ/mol. The enthalpy change for the decomposition of 5.0 g RDX is then in the range of –38 to –48 kJ.

8.100

(a)   The statement as written is false. When comparing the same pair of bonded atoms (C–N vs. C=N vs. C≡N), the shorter the bond the greater the bond energy. For different bonded atoms, all bets are off.

(b)   N–N < C–N < C–C < C–O. This is the order of decreasing bond enthalpy from Table 8.4. Note that it is not the order of increasing bond length.

(c)   To estimate the enthalpy of a carbon–carbon quadruple bond, we must estimate the bond length. The C=C is shorter than C–C by 0.20 Å, and C≡C is shorter than C=C by 0.14 Å. A lower limit for the length of the quadruple bond will be 1.06 Å. This a very short internuclear separation, so a better estimate might be 1.10 Å. Using a bond length range of 1.06 – 1.10 Å, the estimated bond enthalpy will be in the range 990–1050 kJ.

Integrative Exercises

8.102 (a) $Sr(s) \rightarrow Sr(g)$ $\qquad\qquad\qquad\quad$ $\Delta H_f^\circ\, Sr(g)\ [\Delta H_{sub}^\circ\, Sr(s)]$

$Sr(g) \rightarrow Sr^+(g) + 1\, e^-$ $\qquad\quad$ $I_1\, Sr$

$Sr^+(g) \rightarrow Sr^{2+}(g) + 1\, e^-$ $\qquad$ $I_2\, Sr$

$Cl_2(g) \rightarrow 2\, Cl(g)$ $\qquad\qquad\quad$ $2\, \Delta H_f^\circ\, Cl(g)\ [D(Cl_2)]$

$2\, Cl(g) + 2\, e^- \rightarrow 2\, Cl^-(g)$ $\qquad$ $2\, E_1\, Cl$

$\underline{SrCl_2(s) \rightarrow Sr(s) + Cl_2(g)} \qquad \underline{-\Delta H_f^\circ\, SrCl_2}$

$SrCl_2(s) \rightarrow Sr^{2+}(g) + 2\, Cl^-(g) \qquad \Delta H_{latt}$

(b) $\Delta H_f^\circ\, SrCl_2(s) = \Delta H_f^\circ\, Sr(g) + I_1(Sr) + I_2(Sr) + 2\, \Delta H_f^\circ\, Cl(g) + 2\, E(Cl) - \Delta H_{latt}\, SrCl_2$

$\Delta H_f^\circ\, SrCl_2(s) = 164.4\ kJ + 549\ kJ + 1064\ kJ + 2(121.7)\ kJ + 2(-349)\ kJ - 2127\ kJ$
$= -804\ kJ$

8.103   The pathway to the formation of K_2O can be written:

$2\, K(s) \rightarrow 2\, K(g)$ $\qquad\qquad\qquad$ $2\, \Delta H_f^\circ\, K(g)$

$2\, K(g) \rightarrow 2\, K^+(g) + 2\, e^-$ $\qquad\quad$ $2\, I_1(K)$

$1/2\, O_2(g) \rightarrow O(g)$ $\qquad\qquad\quad$ $\Delta H_f^\circ\, O(g)$

$O(g) + 1\, e^- \rightarrow O^-(g)$ $\qquad\qquad$ $E_1(O)$

$O^-(g) + 1\, e^- \rightarrow O^{2-}(g)$ $\qquad\quad$ $E_2(O)$

$\underline{2\, K^+(g) + O^{2-}(g) \rightarrow K_2O(s)} \qquad \underline{-\Delta H_{latt}\, K_2O(s)}$

$2\, K(s) + 1/2\, O_2(g) \rightarrow K_2O(s) \qquad \Delta H_f^\circ\, K_2O(s)$

$\Delta H_f^\circ\, K_2O(s) = 2\, \Delta H_f^\circ\, K(g) + 2\, I_1(K) + \Delta H_f^\circ\, O(g) + E_1(O) + E_2(O) - \Delta H_{latt}\, K_2O(s)$

$E_2(O) = \Delta H_f^\circ\, K_2O(s) + \Delta H_{latt}\, K_2O(s) - 2\, \Delta H_f^\circ\, K(g) - 2\, I_t(K) - \Delta H_f^\circ\, O(g) - E_1(O)$

$E_2(O) = -363.2\ kJ + 2238\ kJ - 2(89.99)\ kJ - 2(419)\ kJ - 247.5\ kJ - (-141)\ kJ$

$= +750\ kJ$

8.104   To calculate empirical formulas, assume 100 g of sample.

(a) $\dfrac{76.0\ g\ Ru}{101.07 g/mol} = 0.752\ mol\ Ru;\ 0.752/0.752 = 1\ Ru$

$\dfrac{24.0\ g\ O}{15.9994\ g/mol} = 1.50\ mol\ O;\ 1.50/0.762 = 2\ O$

The empirical formula of compound 1 is RuO_2.

(b) $\dfrac{61.2\,\text{g Ru}}{101.07\,\text{g / mol}} = 0.6055\ \text{mol Ru};\ 0.6055/0.6055 = 1\ \text{Ru}$

$\dfrac{38.8\ \text{g O}}{15.9994\,\text{g / mol}} = 2.425\ \text{mol O};\ 2.425/0.6055 = 4\ \text{O}$

The empirical formula of compound 2 is RuO_4.

(c)   The lower melting yellow compound is molecular. Substances with metals in high oxidation states are often molecular. RuO_4 contains Ru(VIII), whereas RuO_2 contains Ru(IV), so RuO_4 is more likely to be molecular. The yellow compound is RuO_4.

(d)   The very high melting black compound is ionic. The black compound is RuO_2.

(e)   Yellow RuO_4 is molecular.

(f)   Black RuO_2 is ionic.

8.105   (a)   Even though Cl has the greater (more negative) electron affinity, F has a much larger ionization energy, so the electronegativity of F is greater.

F: k(I–EA) = k(1681 – (–328)) = k(2009)

Cl: k(I–EA) = k(1251 – (–349)) = k(1600)

(b)   Electronegativity is the ability of an atom in a molecule to attract electrons to itself. It can be thought of as the ability to hold its own electrons (as measured by ionization energy) and the capacity to attract the electrons of other atoms (as measured by electron affinity). Thus, both properties are relevant to the concept of electronegativity.

(c)   EN = k(I – EA). For F: 4.0 = k(2009), k = 4.0/2009 = 2.0×10^{-3}

(d)   Cl: EN = 2.0×10^{-3}(1600) = 3.2

O: EN = 2.0×10^{-3}(1314 – (–141)) = 2.9

(e)   F: (I+EA)/2 = (1681 – 328)/2 = 676.5 = 677

To scale the value to 4.0 for F, 4.0 = k(677), k = 4.0/677 = 5.9×10^{-3}

Cl: 5.9×10^{-3}(1251 – 349)/2 = 2.7

Br: 5.9×10^{-3}(1140 – 325)/2 = 2.4

I: 5.9×10^{-3}(1008 – 295)/2 = 2.1

On this scale, the electronegativity of Br is 2.4.

8.106   (a)   Assume 100 g.

$14.52\ \text{g C} \times \dfrac{1\,\text{mol}}{12.011\,\text{g C}} = 1.209\ \text{mol C};\ 1.209/1.209 = 1$

$1.83\ \text{g H} \times \dfrac{1\,\text{mol}}{1.008\,\text{g H}} = 1.816\ \text{mol H};\ 1.816/1.209 = 1.5$

$$64.30 \text{ g Cl} \times \frac{1 \text{ mol}}{35.453 \text{ g Cl}} = 1.814 \text{ mol Cl}; \ 1.814 / 1.209 = 1.5$$

$$19.35 \text{ g O} \times \frac{1 \text{ mol}}{15.9994 \text{ g O}} = 1.209 \text{ mol O}; \ 1.209 / 1.209 = 1.0$$

Multiplying by 2 to obtain an integer ratio, the empirical formula is $C_2H_3Cl_3O_2$.

(b)   The empirical formula mass is $2(12.0) + 3(1.0) + 3(35.5) + 2(16) = 165.5$. The empirical formula is the molecular formula.

(c) $44 \ e^-$, $22 \ e^-$ pairs

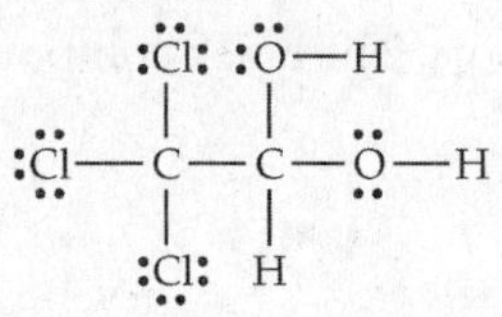

8.108   (a) C_2H_2: $10 \ e^-$, $5 \ e^-$ pair N_2: $10 \ e^-$, $5 \ e^-$ pair

$$H \!-\! C \!\equiv\! C \!-\! H \qquad\qquad :N \!\equiv\! N:$$

(b)   The enthalpy of formation for N_2 is 0 kJ/mol and for C_2H_2 is 226.77 kJ/mol. N_2 is an extremely stable, unreactive compound. Under appropriate conditions, it can be either oxidized or reduced. C_2H_2 is a reactive gas, used in combination with O_2 for welding and as starting material for organic synthesis.

(c) $2 N_2(g) + 5 O_2(g) \rightarrow 2 N_2O_5(g)$

$2 C_2H_2(g) + 5 O_2(g) \rightarrow 4 CO_2(g) + 2 H_2O(g)$

(d) $\Delta H^\circ_{rxn} (N_2) = 2 \ \Delta H^\circ_f \ N_2O_5(g) - 2 \ \Delta H^\circ_f \ N_2(g) - 5 \ \Delta H^\circ_f \ O_2(g)$

$$= 2(11.30) - 2(0) - 5(0) = 22.60 \text{ kJ}$$

$\Delta H^\circ_{ox} (N_2) = 11.30 \text{ kJ/ mol } N_2$

$\Delta H^\circ_{rxn} (C_2H_2) = 4 \ \Delta H^\circ_f \ CO_2(g) + 2 \ \Delta H^\circ_f \ H_2O(g) - 2 \ \Delta H^\circ_f \ C_2H_2(g) - 5 \ \Delta H^\circ_f \ O_2(g)$

$$= 4(-393.5 \text{ kJ}) + 2(-241.82 \text{ kJ}) - 2(226.77 \text{ kJ}) - 5(0) = -2511.18 \text{ kJ}$$

$\Delta H^\circ_{ox} (C_2H_2) = -1255.6 \text{ kJ/mol } C_2H_2$

(e) $N_2(g) + 3 H_2(g) \rightarrow 2 NH_3(g)$

$\Delta H^\circ_{rxn} (N_2) = 2 \ \Delta H^\circ_f \ NH_3(g) - \Delta H^\circ_f \ N_2(g) - 3 \ \Delta H^\circ_f \ H_2(g)$

$$= 2(-46.19) - (0) - 3(0) = -92.38 \text{ kJ}$$

$\Delta H^\circ_{rxn} (N_2) = -46.19 \text{ kJ/ mol } N_2$

$C_2H_2(g) + 3 H_2(g) \rightarrow 2 CH_4(g)$

$\Delta H^\circ_{rxn} (C_2H_2) = 2 \ \Delta H^\circ_f \ CH_4(g) - 2 \ \Delta H^\circ_f \ C_2H_2(g) - 3 \ \Delta H^\circ_f \ H_2(g)$

$$= 2(-679.9 \text{ kJ}) - 226.77 \text{ kJ} - 3(0) = -1586.6 \text{ kJ}$$

$\Delta H^\circ_{rxn} (C_2H_2) = -793.3 \text{ kJ/ mol } C_2H_2$

8.109   (a)   Assume 100 g of compound

$$69.6 \text{g S} \times \frac{1 \text{ mol S}}{32.07 \text{ g}} = 2.17 \text{ mol S}$$

$$30.4 \text{ g N} \times \frac{1 \text{ mol N}}{14.01 \text{ g}} = 2.17 \text{ mol N}$$

S and N are present in a 1:1 mol ratio, so the empirical formula is SN. The empirical formula mass is 46. MM/FW = 184.3/46 = 4 The molecular formula is S_4N_4.

(b)   44 e^-, 22 e^- pairs. Because of its small radius, N is unlikely to have an expanded octet. Begin with alternating S and N atoms in the ring. Try to satisfy the octet rule with single bonds and lone pairs. At least two double bonds somewhere in the ring are required.

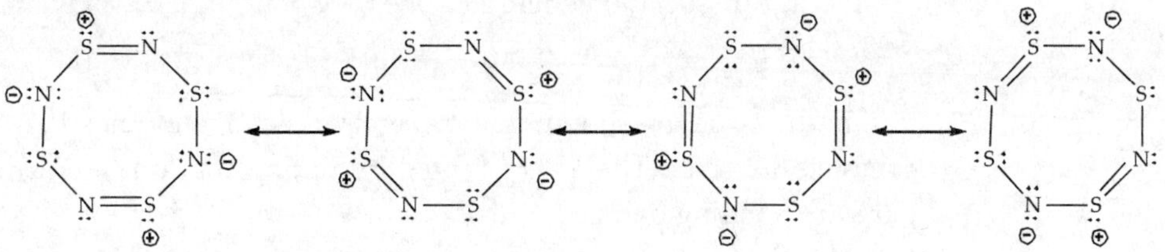

These structures carry formal charges on S and N atoms as shown. Other possibilities include:

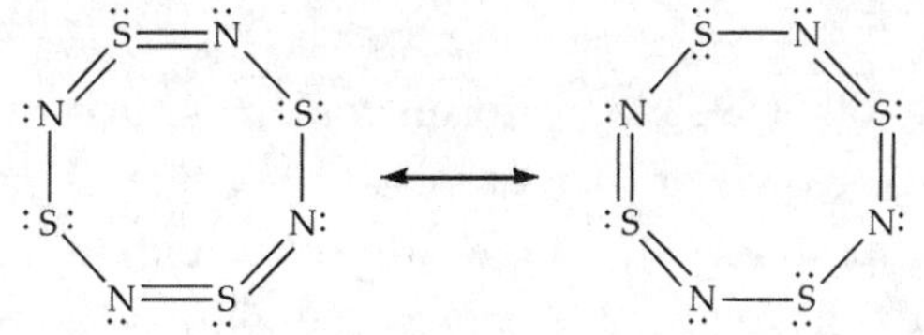

These structures have zero formal charges on all atoms and are likely to contribute to the true structure. Note that the S atoms that are shown with two double bonds are not necessarily linear because S has an expanded octet. Other resonance structures with four double bonds are:

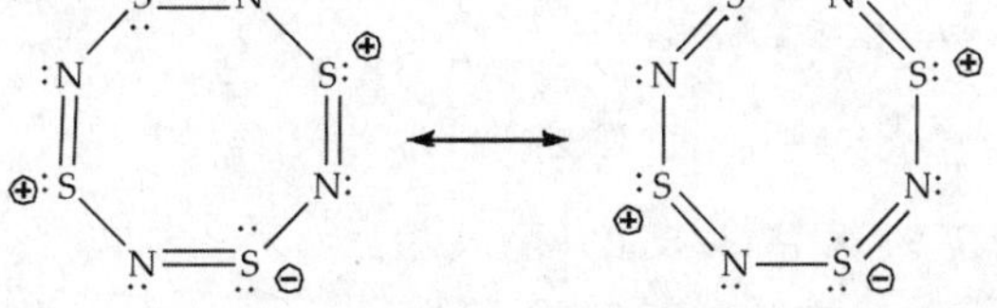

In either resonance structure, the two "extra" electron pairs can be placed on any pair of S atoms in ring, leading to a total of 10 resonance structures. The sulfur atoms alternately carry formal charges of +1 and –1. Without further structural information, it is not possible to eliminate any of the above structures. Clearly, the S_4N_4 molecule stretches the limits of the Lewis model of chemical bonding.

(c)   Each resonance structure has 8 total bonds and more than 8 but fewer than 16 bonding e^- pairs, so an "average" bond will be intermediate between a S–N single and double bond. We estimate an average S–N single bond length to be 1.77 Å (sum of bonding atomic radii from Figure 7.7). We do not have a direct value for a S–N double bond length. Comparing double and single bond lengths

135

for C–C (1.34 Å, 1.54 Å), N–N (1.24 Å, 1.47 Å), and O–O (1.21 Å, 1.48 Å) bonds from Table 8.5, we see that, on average, a double bond is approximately 0.23 Å shorter than a single bond. Applying this difference to the S–N single bond length, we estimate the S–N double bond length as 1.54 Å. Finally, the intermediate S–N bond length in S_4N_4 should be between these two values, approximately 1.60–165 Å. (The measured bond length is 1.62 Å.)

(d) $S_4N_4 \rightarrow 4\,S(g) + 4\,N(g)$

$\Delta H = 4\,\Delta H_f^\circ\,S(g) + 4\,\Delta H_f^\circ\,N(g) - \Delta H_f^\circ\,S_4N_4$
$\Delta H = 4(222.8\text{ kJ}) + 4(472.7\text{ kJ}) - 480\text{ kJ} = 2302\text{ kJ}$

This energy, 2302 kJ, represents the dissociation of 8 S–N bonds in the molecule; the average dissociation energy of one S–N bond in S_4N_4 is then 2302 kJ/8 bonds = 287.8 kJ.

8.110   (a)   Yes. In the structure shown in the exercise, each P atom needs 1 unshared pair to complete its octet. This is confirmed by noting that only 6 of the 10 valence e^- pairs are bonding pairs.

(b)   There are 6 P–P bonds in P_4.

(c)   20 e^-, 10 e^- pr

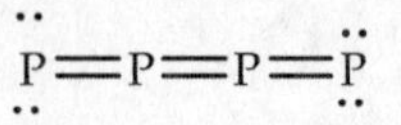

$$\ddot{P} = P = P = \ddot{P}$$

There are no other resonance forms for this structure. The octet rule is satisfied for all atoms. However, it requires P=P, which is uncommon because P has a covalent radius that is too large to accommodate the side-to-side π overlap of parallel p orbitals required for double bond formation.

(d)   From left to right, the formal charges are on the P atoms in the linear structure are −1, +1, +1, −1. In the tetrahedral structure, all formal charges are zero. Clearly the linear structure does not minimize formal charge and is probably less stable than the tetrahedral structure, owing to the difficulty of P=P bond formation (see above).

8.111   (a) $C_6H_6(g) \rightarrow 6\,H(g) + 6\,C(g)$

$\Delta H^\circ = 6\,\Delta H_f^\circ\,H(g) + 6\,\Delta H_f^\circ\,C(g) - \Delta H_f^\circ\,C_6H_6(g)$

$\Delta H^\circ = 6(217.94)\text{ kJ} + 6(718.4)\text{ kJ} - 82.9\text{ kJ} = 5535\text{ kJ}$

(b) $C_6H_6(g) \rightarrow 6\,CH(g)$

(c)

$C_6H_6(g) \rightarrow 6\,H(g) + 6\,C(g)$	ΔH°	5535 kJ
$6\,H(g) + 6\,C(g) \rightarrow 6\,CH(g)$	$-6\,D(C\text{–}H)$	$-6(413)$ kJ
$C_6H_6(g) \rightarrow 6\,CH(g)$		3057 kJ

3057 kJ is the energy required to break the six C–C bonds in $C_6H_6(g)$. The average bond dissociation energy for one carbon–carbon bond in $C_6H_6(g)$ is

$$\frac{3057\text{ kJ}}{6\,C\text{–}C\text{ bonds}} = 509.5\text{ kJ}.$$

(d)   The value of 509.5 kJ is between the average value for a C–C single bond (348 kJ) and a C=C double bond (614 kJ). It is somewhat greater than the average of these two values, indicating that the carbon–carbon bond in benzene is more like a carbon–carbon double bond than a carbon–carbon single bond.

8.113   (a) $Si(s) + O_2(g) \rightarrow SiO_2(s)$

$$\Delta H_{rxn}^{\circ} = \Delta H_f^{\circ}\ SiO_2(s) - \Delta H_f^{\circ}\ Si(s) - \Delta H_f^{\circ}\ O_2(g)$$

$$= -910.9 - (0) - (0) = -910.9\ kJ$$

$$1\ cm^3\ Si\ \times\ \frac{2.33\ g\ Si}{cm^3} \times \frac{1\ mol\ Si}{28.0855\ g\ Si}\ \times\ \frac{-910.9\ kJ}{mol\ Si} = -75.569 = -75.6\ kJ$$

The density for Si is from WebElements, 2013.

[The ΔH_{rxn}° could also be calculated from bond enthalpies in Table 8.4: 4 D(Si–Si) – 4 D(Si–O) = 4(226 kJ) – 4(368 kJ) = –568 kJ. This estimate does not take into account the extended structure of either Si or $SiO_2(s)$ and is a very lower limit. The Appendix C value of ΔH_f° for $SiO_2(s)$ used in the calculation above is for $SiO_2(s)$ in the form of quartz. This is probably an upper estimate.]

(b)   The ratio of bond enthalpies for the analogous bonds in carbon is

C=C/C–C = (614 kJ/mol)/(348 kJ/mol) = 1.76.

Applying this ratio to silicon, D(Si=Si) = 1.76 D(Si – Si) = 1.76(226) = 399 kJ/mol

9 Molecular Geometry and Bonding Theories

Visualizing Concepts

9.2   (a)   120°

      (b)   If the blue balloon expands, the angle between red and green balloons decreases.

      (c)   (ii)

9.4   (a) $4\,e^-$ domains

      (b)   The molecule has a nonzero dipole moment, because the C–H and C–F bond dipoles do not cancel each other.

      (c)   (ii)

9.6   (a)   (iii)

      (b) sp^3

9.7   (a)   Recall that π bonds require p atomic orbitals, so the maximum hybridization of a C atom involved in a double bond is sp^2 and in a triple bond is sp. There are 6 C atoms in the molecule. Starting on the left, the hybridizations are: $sp^2, sp^2, sp^3, sp, sp, sp^3$.

      (b)   All single bonds are σ bonds. Double and triple bonds each contain 1 σ bond. This molecule has 8 C–H σ bonds and 5 C–C σ bonds, for a total of 13 σ bonds.

      (c)   Double bonds have 1 π bond and triple bonds have 2 π bonds. This molecule has a total of 3 π bonds.

      (d)   Any central atom with sp^2 hybridization will have bond angles of 120° around it. The two left-most C atoms are sp^2 hybridized, so any angle with one of these C atoms central will be 120°. This amounts to 1 H–C–H, 4 H–C–C and 1 C–C–C angle.

9.8   (a)   (i)

      (b)   (iii)

9.9   (a) C_4H_4O (b) 26 valence e^- (c) sp^2 (d) $4\,e^-$ (e) (iii)

9.10   (a)   The lower-energy MO is σ_{1s}, the higher-energy MO is σ^*_{1s}.

      (b) H_2^+ (c) BO = ½   (d) σ_{1s} (the lowest energy available orbital)

9.12   (a)   The diagram has five electrons in MOs formed by 2p atomic orbitals. C has two 2p electrons, so X must have three 2p electrons. X is N.

      (b)   The molecule has an unpaired electron, so it is paramagnetic.

      (c)   Atom X is N, which is more electronegative than C. The atomic orbitals of the more electronegative N are slightly lower in energy than those of C. The lower-energy π_{2p} bonding molecular orbitals will have a greater contribution from the lower-energy N atomic orbitals. (Higher energy π^*_{2p} MOs will have a greater contribution from higher-energy C atomic orbitals.)

Molecular Shapes; the VSEPR Model (Sections 9.1 and 9.2)

9.14   (a)   In a symmetrical tetrahedron, the four bond angles are equal to each other, with values of 109.5°. The H–C–H angles in CH_4 and the O–Cl–O angles in ClO_4^- will have values close to 109.5°.

          (b)   'Planar' molecules are flat, so trigonal planar BF_3 is flat. In the trigonal pyramidal NH_3 molecule, the central N atom sits out of the plane of the three H atoms; this molecule is not flat.

9.16   (a)   three coplanar 120° angles

          (b)   four 109.5° angles

          (c)   90° angles in the equatorial square plane and between axial atoms and those in the square plane, 12 in all; 180° angles between atoms opposite each other, 3 in all

          (d)   one 180° angle

9.18   We expect the nonbonding electron domain in NH_3 to occupy a smaller volume than the one in PH_3. The electronegativity of N, 3.0, is larger than that of P, 2.1. The nonbonding electrons will be more strongly attracted to N than to P, and the volume of the domain will be smaller. This means that the charge density of the nonbonding domain in NH_3 will be greater and it will experience stronger repulsions than the nonbonding domain in PH_3.

9.20   Draw the Lewis structure of each molecule. If it has nonbonding electron pairs on the central atom, decide whether they will cause bond angles to deviate from ideal values for the particular electron-domain geometry.

          (a) H_2S, 8 valence e^-, 4 e^- pr, tetrahedral electron-domain geometry with 2 nonbonding electron pairs on S will cause the bond angle to deviate from ideal 109.5° angles

$$H \!-\! \overset{\displaystyle ..}{\underset{\displaystyle ..}{S}} \!-\! H$$

          (b) BCl_3, 24 valence e^-, 12 e^- pr, trigonal planar electron-domain geometry with zero nonbonding pairs on B. We confidently predict 120° angles.

$$:\!\ddot{C}l \!-\! B \!-\! \ddot{C}l\!:$$
$$|$$
$$:\!\ddot{C}l\!:$$

          (c) CH_3I, 14 valence e^-, 7 e^- pr, tetrahedral electron-domain geometry with zero nonbonding pairs on C. Because the bonding electron domains are not exactly the same, we predict some deviation from ideal 109.5° angles.

$$H$$
$$|$$
$$H \!-\! C \!-\! \ddot{I}\!:$$
$$|$$
$$H$$

          (d) CBr_4, 32 valence e^-, 16 e^- pr, tetrahedral electron-domain geometry with zero nonbonding pairs on C. We confidently predict 109.5° angles.

$$:\!\ddot{B}r\!:$$
$$|$$
$$:\!\ddot{B}r \!-\! C \!-\! \ddot{B}r\!:$$
$$|$$
$$:\!\ddot{B}r\!:$$

(e) $TeBr_4$, 34 valence e^-, 17 e^- pr, trigonal bipyramidal electron-domain geometry with one nonbonding pair on Te. The structure is similar to SF_4 shown in Sample Exercise 9.2. The bond angles will deviate from ideal values, but perhaps not as much as in SF_4. (Structure follows.)

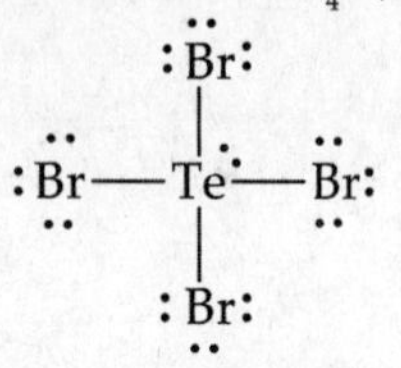

9.22   *Analyze/Plan.* See Table 9.1. *Solve.*

(a)   trigonal planar   (b)   tetrahedral

(c)   trigonal bipyramidal   (d)   octahedral

9.24

(a)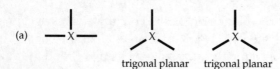

(b)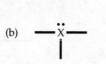

(c)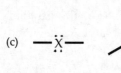

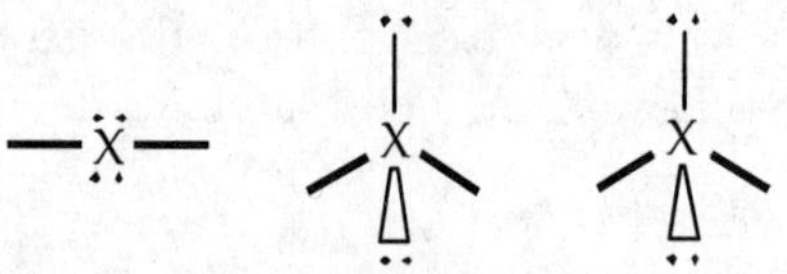

9.26   bent (b), linear (l), octahedral (oh), seesaw (ss), square pyramidal (sp), square planar (spl), tetrahedral (td), trigonal bipyramidal (tbp), trigonal planar (tr), trigonal pyramidal (tp), T-shaped (T)

Molecule or ion	Valence electrons	Lewis structure	Electron-domain Geometry	Molecular geometry
(a) AsF_3	26		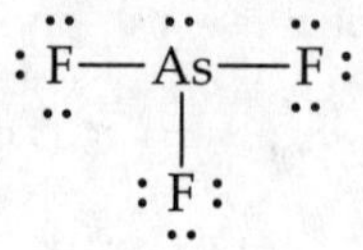 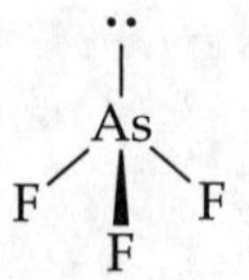 td	tp
(b) CH_3^+	6		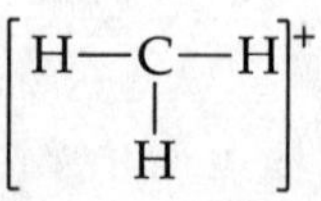 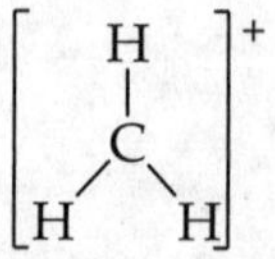 tr	tr

9.26   (Continued). bent (b), linear (l), octahedral (oh), seesaw (ss) square pyramidal (sp), square planar (spl), tetrahedral (td), trigonal bipyramidal (tbp), trigonal planar (tr), trigonal pyramidal (tp), T-shaped (T)

	Molecule or ion	Valence electrons	Lewis structure	Electron-domain geometry	Molecular geometry
(c)	BrF_3	28		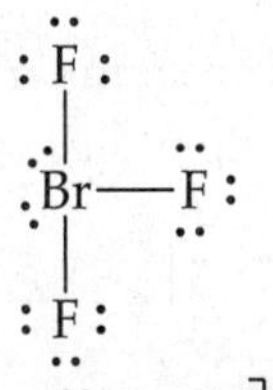 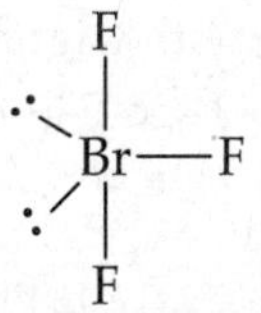 tbp	T
(d)	ClO_3^-	26		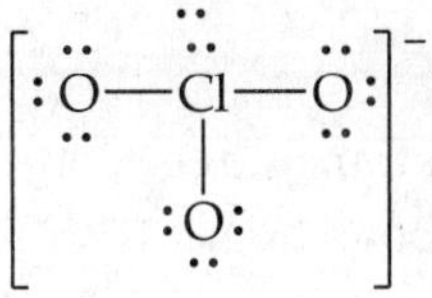 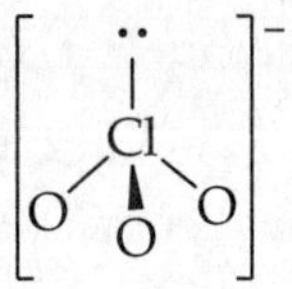 td	tp
(e)	XeF_2	22		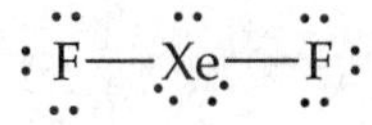 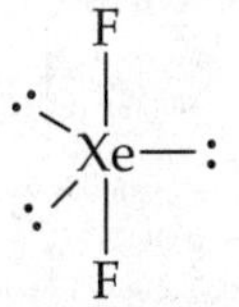 tbp	l
(f)	BrO_2^-	20	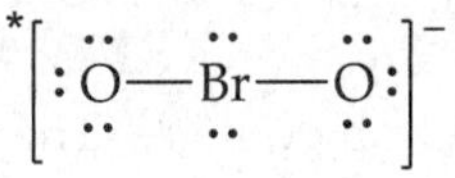 *| 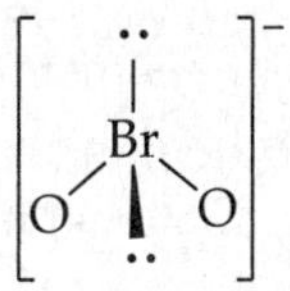	td	b

*More than one resonance structure is possible. All equivalent resonance structures predict the same molecular geometry.

9.28   (a)   Electron-domain geometries: (i), octahedral; (ii), tedrahedral; (iii), trigonal bipyramidal

(b)   nonbonding electron domains: (i), 2; (ii), 0; (iii), 1

(c)   S or Se. Shape (iii) has five electron domains, so A must be in or below the third row of the periodic table. This eliminates Be and C. Assuming each F atom has three nonbonding electron domains and forms only single bonds with A, A must have six valence electrons to produce these electron-domain and molecular geometries.

(d)   Xe. (See Table 9.3.) Assuming F behaves typically, A must be in or below the third row and have eight valence electrons. Only Xe fits this description. (Noble-gas elements above Xe have not been shown to form molecules of the type AF_4. See Section 7.8.)

9.30   (a)   1, less than 109.5°; 2, less than 120°

(b)   3, close to 109.5°; 4, slightly greater than 120°

(c)   5, less than 109.5°; 6, less than 109.5°

(d)   7, 180°; 8, close to 109.5°

9.32   *Analyze/Plan.* Given the formula of each molecule or ion, draw the correct Lewis structure and use principles of VSEPR to answer the question. *Solve.*

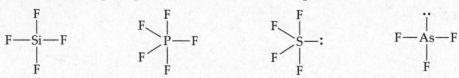

The three nonbonded electron pairs on each F atom have been omitted for clarity.

The two molecules with trigonal bipyramidal electron-domain geometry, PF_5 and SF_4, have more than one F–A–F bond angle.

9.34   (a) ClO_2^- 20 e⁻, 10 e⁻ pr

$$\left[:\ddot{O}—\ddot{C}l—\ddot{O}: \right]^-$$

4 e⁻ domains around Cl, tetrahedral e⁻ domain geometry, bent molecular geometry bond angle ≤109.5°

NO_2^- 18 e⁻, 9 e⁻ pr

$$\left[\ddot{O}=\ddot{N}—\ddot{O}: \right]^- \longleftrightarrow \left[:\ddot{O}—\ddot{N}=\ddot{O} \right]^-$$

3 e⁻ domains about N (both resonance structures), trigonal planar e⁻ domain geometry bent molecular geometry bond angle ≤120°

Both molecular geometries are described as "bent" because both molecules have two nonlinear bonding electron domains. The bond angles (the angle between the two bonding domains) in the two ions are different because the total number of electron domains, and thus the electron-domain geometries are different.

(b) XeF_2 22 e⁻, 11 e⁻ pr

$$:\ddot{F}—\ddot{X}e—\ddot{F}:$$

5 e⁻ domains around Xe, trigonal bipyramidal e⁻ domain geometry, linear molecular geometry

The question here really is: why do the three nonbonding domains all occupy the equatorial plane of the trigonal bipyramid? In a tbp, there are several different kinds of repulsions, bonding domain-bonding domain (bd-bd), bonding domain-nonbonding domain (bd-nd), and nonbonding domain-nonbonding domain (nd-nd). Each of these can have 90°, 120°, or 180° geometry. Because nonbonding domains occupy more space, 90° nd-nd repulsions are most significant and least desirable. The various electron domains arrange themselves to minimize these 90° nd-nd interactions. The arrangement shown above has no 90° nd-nd repulsions. An arrangement with one or two nonbonding domains in axial positions would lead to at least two 90° nd-nd repulsions, a less stable situation. (To convince yourself, tabulate the number and kinds of repulsions for each possible tbp arrangement of 2 bonding domains and 3 nonbonding domains.)

Shapes and Polarity of Polyatomic Molecules (Section 9.3)

9.36   For a polar A–X bond in an AX_3 molecule, as the X–A–X bond angle increases from 100° to 120°, the molecular dipole moment decreases. In a symmetrical AX_3 molecule with 120° bond angles, bond dipoles cancel and the molecule is nonpolar. As the bond angle decreases, the resultant of the three bond dipoles becomes larger, and the dipole moment increases.

9.38   (a)   If PH_3 is polar, it must have a measurable dipole moment. This means that the three P–H bond dipoles do not cancel. If PH_3 were planar, the P–H bond dipoles would cancel, and the molecule would be nonpolar. The measurable dipole moment of PH_3 is experimental evidence that the molecule cannot be planar.

      (b) O_3, 18 e⁻, 9 e⁻ pr;   :Ö=Ö—Ö: ⟷ :Ö—Ö=Ö:

          trigonal planar e⁻ domain geometry
          bent molecular geometry

          Because all atoms are the same, the individual bond dipoles are zero. However, the central O atom has a lone pair of electrons that cause an unequal electron (and charge) distribution in the molecule. This lone pair is the source of the dipole moment in O_3.

9.40   (a)   In Exercise 9.27, molecules (ii) and (iii) will have nonzero dipole moments. Molecule (i) has zero nonbonding electron pairs on A, and the 3 A–F dipoles are oriented so that the sum of their vectors is zero (the bond dipoles cancel). Molecules (ii) and (iii) have nonbonding electron pairs on A and their bond dipoles do not cancel. A nonbonding electron pair (or pairs) on a central atom almost guarantees at least a small molecular dipole moment, because no bond dipole exactly cancels a nonbonding pair. (Exceptions are molecular geometries with nonbonding electron domains 180° apart.)

      (b)   In Exercise 9.28, molecules (i) and (ii) have zero dipole moments and are nonpolar. AF_4 molecules will have a zero dipole moment if the 4 A–F bond dipoles are arranged (symmetrically) so that they cancel, and any nonbonding pairs are arranged so that they cancel.

9.42   (a)   Nonpolar, in a symmetrical tetrahedral structure (Figure 9.1) the bond dipoles cancel.

                 Cl
                 |
                 C
          Cl | Cl
                Cl

      (b)   Polar, there is an unequal charge distribution because of the nonbonded electron pair on N.

               N̈
         H | H
             H

(c)   Polar, there is an unequal charge distribution because of the nonbonded electron pair on S.

F
F—|
    S:
F—|
    F

(d)   Nonpolar, the bond dipoles and the nonbonded electron pairs cancel.

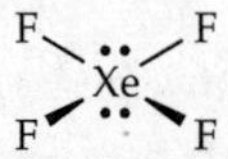

F   F
    Xe
F   F

(e)   Polar, the C–H and C–Br bond dipoles are not equal and do not cancel.

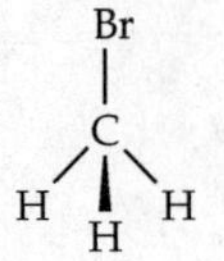

Br
|
C
H   H
H

(f)   Nonpolar, in a symmetrical trigonal planar structure, the bond dipoles cancel.

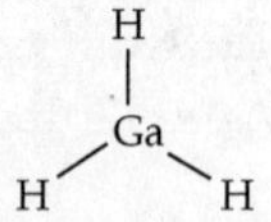

H
|
Ga
H   H

9.44   Each C–Cl bond is polar. The question is whether the vector sum of the C–Cl bond dipoles in each molecule will be nonzero. In the *ortho* and *meta* isomers, the C–Cl vectors are at 60° and 120° angles, respectively, and their resultant dipole moments are nonzero. In the *para* isomer, the C–Cl vectors are opposite, at an angle of 180°, with a resultant dipole moment of zero. The *ortho* and *meta* isomers are polar, the *para* isomer is nonpolar.

Orbital Overlap; Hybrid Orbitals (Sections 9.4 and 9.5)

9.46   (a)

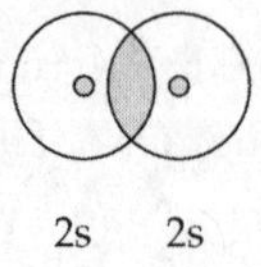

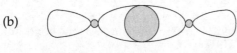

2s   2s

(b)

2p$_z$ 2p$_z$

(c)

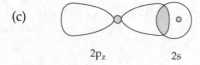

2p$_z$ 2s

9.48   By analogy to the H$_2$ molecule shown in Figure 9.14, as the distance between the atoms decreases, the overlap between their bonding orbitals increases. According to Figure 7.7, the bonding atomic radius for the halogens is in the order F < Cl < Br < I. The order of bond lengths in the molecules is I–F < I–Cl < I–Br < I–I. If the extent of orbital overlap increases as the distance between atoms decreases, I–F has the greatest overlap and I$_2$ the least. The order for extent of orbital overlap is I–I < I–Br < I–Cl < I–F.

9.50   (a)   S: $[Ne]3s^2 3p^4$

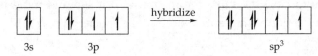

3s   3p   sp³

hybridize

        (b)   The hybrid orbitals are called sp³.

        (c)

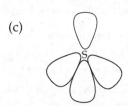

        (d)   The hybrid orbitals formed in (a) would not be appropriate for SF₄. There are five electron domains in SF₄, four bonding and one nonbonding. A set of four sp³ hybrid orbitals could not accommodate all the electron pairs around S.

9.52   (a)   32 e⁻, 16 e⁻ pairs

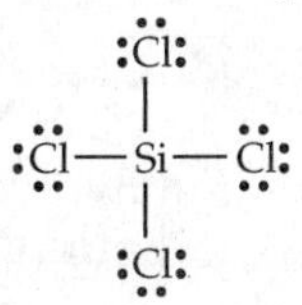

```
        :Cl:
          |
  :Cl—Si—Cl:
          |
        :Cl:
```

               4 e⁻ pairs around Si, tetrahedral e⁻ domain geometry, sp³ hybridization

        (b)   10 e⁻, 5 e⁻ pairs

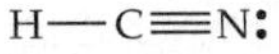

               H—C≡N:

               2 e⁻ domains around C, linear e⁻ domain geometry, sp hybridization

        (c)   24 e⁻, 12 e⁻ pairs

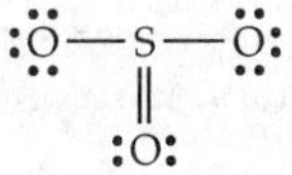

```
  :Ö—S—Ö:
        ‖
       :O:
```

               (other resonance structures are possible)
               3 e⁻ domains around S, trigonal planar e⁻ domain geometry, sp² hybridization

        (d)   20 e⁻, 10 e⁻ pairs

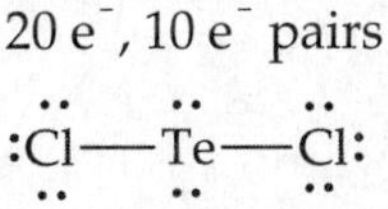

               :Cl—Te—Cl:

               4 e⁻ domains around Te, tetrahedral e⁻ domain geometry, sp³ hybridization

9.54   (a)   The three moieties, BH₄⁻, CH₄, and NH₄⁺, each have 8 valence e⁻, 4 e⁻ pairs, 4 bonding e⁻ domains, tetrahedral e⁻ domain and molecular geometry and sp³ hybridization at the central atom.

        (b)   The electronegativity of the central atoms decreases in the series N > C > B. The question is: where does the electronegativity of H lie in this series? By examination of electronegativity values in Figure 8.7, H is slightly less electronegative than C, and almost the same as B. The magnitude of the bond dipole decreases in the series N–H > C–H > B–H. The negative end of the dipole is toward N, C, and H, respectively.

(c) AlH_4^-, SiH_4, and PH_4^+. By the same arguments used in part (a), we expect these three moieties to have the same tetrahedral e⁻ domain and molecular geometry and sp^3 hybridization at the central atom as the species in part (a).

Multiple Bonds (Section 9.6)

9.56 (a) Two unhybridized p orbitals remain, and the atom can form two pi bonds.

(b) It would be much easier to twist or rotate around a single sigma bond. Sigma bonds are formed by end-to-end overlap of orbitals and the bonding electron density is symmetric about the internuclear axis. Rotating (twisting) around a sigma bond can be done without disrupting either the orbital overlap or bonding electron density, without breaking the bond.

The π part of a double bond is formed by side-to-side overlap of p atomic orbitals perpendicular to the internuclear axis. This π overlap locks the atoms into position and makes twisting difficult. Also, only a small twist (rotation) destroys overlap of the p orbitals and breaks the π bond.

9.58 (a) 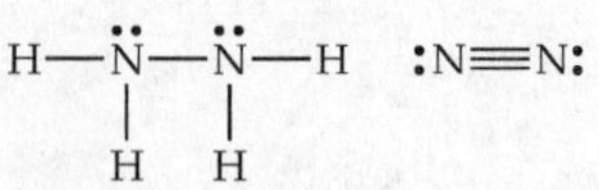 H—N̈—N̈—H :N≡N:
              | |
              H   H

(b) The N atoms in N_2H_4 are sp^3 hybridized; there are no unhybridized p orbitals available for π bonding. In N_2, the N atoms are sp hybridized, with two unhybridized p orbitals on each N atom available to form the two π bonds in the N≡N triple bond.

(c) The N–N triple bond in N_2 is significantly stronger than the N–N single bond in N_2H_4, because it consists of one σ and two π bonds, rather than a 'plain' sigma bond. Generally, bond strength increases as the extent of orbital overlap increases. The additional overlap from the two π bonds adds to the strength of the N–N bond in N_2.

9.60 (a) The C with a double bond to O has three electron domains and is sp^2 hybridized; the other three C atoms are sp^3 hybridized.

(b) $C_4H_8O_2$ has $4(4) + 8(1) + 2(6) = 36$ valence electrons.

(c) 13 pairs or 26 total valence electrons form σ bonds

(d) 1 pair or 2 total valence electrons form π bonds

(e) 4 pairs or 8 total valence electrons are nonbonding

9.62 (a) 1, ~120°; 2, ~120°; 3, less than 109.5°

(b) 1, sp^2; 2, sp^2; 3, sp^3

(c) 21 σ bonds

9.64 (a, b) 24 e⁻, 12 e⁻ pairs

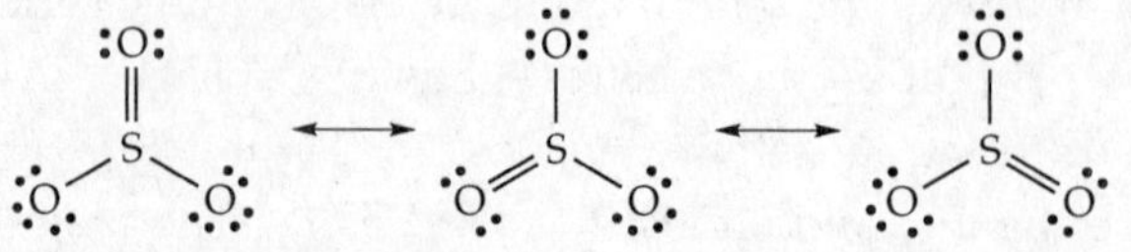

3 electron domains around S, trigonal planar electron-domain geometry, sp^2 hybrid orbitals

(c)   The multiple resonance structures indicate delocalized π bonding. All four atoms lie in the trigonal plane of the sp² hybrid orbitals. On each atom there is a p atomic orbital perpendicular to this plane in the correct orientation for π overlap. The resulting delocalized π electron cloud is Y-shaped (the shape of the molecule) and has electron density above and below the plane of the molecule.

9.66   (a)   The Lewis structure depicts an anion with a 1– charge. The chemical formula of the given structure is $C_3H_3O_2$. This grouping of atoms has 27 valence electrons, whereas the structure shown has 14 electron pairs or 28 electrons. This means that the structure is an anion with a 1– charge.

(b) sp^2

(c)   Yes, there is one other resonance structure.

(d)   The three C and two O atoms each have a p_π orbital.

(e)   There are six electrons in the π system of the molecule. If all the C and O atoms are sp² hybridized, there are seven bonding electron pairs and four nonbonding electron pairs in the σ system. This leaves three electron pairs or six electrons in the π system.

9.68   (a) $24 \, e^-$, $12 \, e^-$ pairs

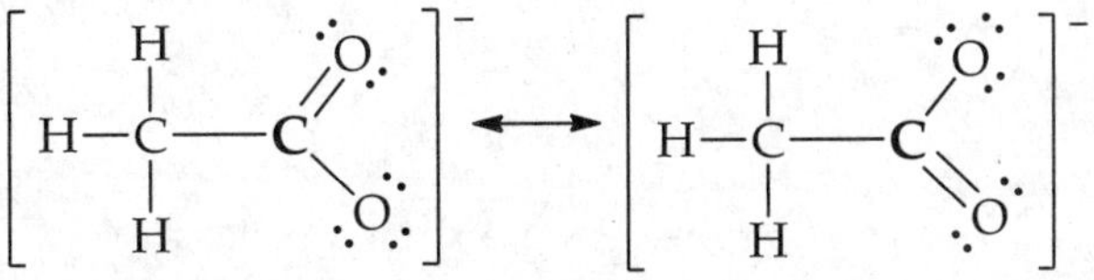

The designated C atom has 3 bonding e^- domains and sp² hybridization.

(b) $8 \, e^-$, $4 \, e^-$ pairs

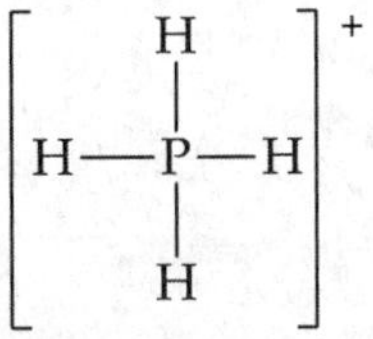

The P atom has 4 bonding e^- domains and sp³ hybridization

(c) $24 \, e^-$, $12 \, e^-$ pairs

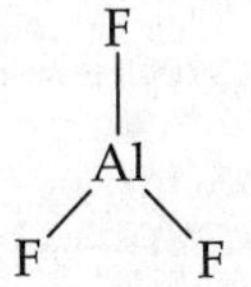

The Al atom has 3 bonding e^- domains and sp² hybridization

(d) $16 \, e^-$, $8 \, e^-$ pairs

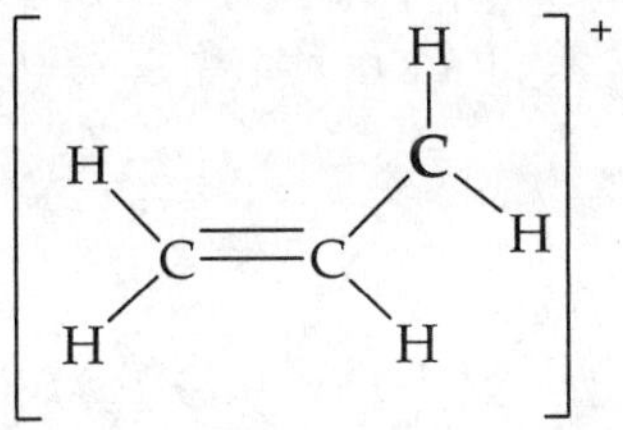

The designated C atom has 3 bonding e^- domains and sp² hybridization.

Molecular Orbitals and Period 2 Diatomic Molecules
(Sections 9.7 and 9.8)

9.70   (a)   An MO, because the AOs come from two different atoms.

       (b)   A hybrid orbital, because the AOs are on the same atom.

       (c)   Yes. The Pauli exclusion principle, that no two electrons can have the same four quantum numbers, means that an orbital can hold at most two electrons. (Because n, l, and m_l are the same for a particular orbital and m_s has only two possible values, an orbital can hold at most two electrons). This is true for atomic and molecular orbitals.

9.72   (a)

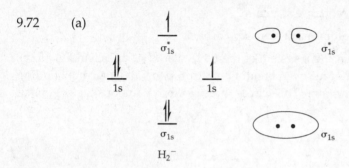

$$H_2^-$$

       (b) $\uparrow$ σ_{1s}^*

            $\uparrow\downarrow$ σ_{1s}

       (c)   Bond order = 1/2 (2 − 1) = ½

       (d)   If one electron moves from σ_{1s} to σ_{1s}^*, the bond order becomes −½. There is a net increase in energy relative to isolated H atoms, so the ion will decompose.

$$H_2^- \overset{h\nu}{\rightarrow} H + H^-.$$

       (e)   Statement (i) is true.

9.74   (a)   Zero

       (b)   The two π_{2p} molecular orbitals are degenerate; they have the same energy, but they have different spatial orientations 90° apart.

       (c)   In the bonding MO the electrons are stabilized by both nuclei. In an antibonding MO, the electrons are directed away from the nuclei, so the π_{2p} bonding MO is lower in energy than the π_{2p}^* antibonding MO.

9.76   (a) O_2^{2-} has a bond order of 1.0, whereas O_2^- has a bond order of 1.5. For the same bonded atoms, the greater the bond order the shorter the bond, so O_2^- has the shorter bond.

(b)   The two possible orbital energy level diagrams are:

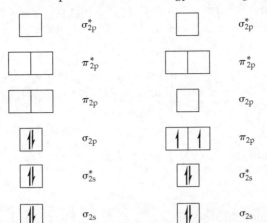

The magnetic properties of a molecule reveal whether it has unpaired electrons. If the σ_{2p} MOs are lower in energy, B_2 has no unpaired electrons. If the π_{2p} MOs are lower in energy than the σ_{2p} MO, there are two unpaired electrons. The magnetic properties of B_2 must indicate that it has unpaired electrons.

(c)   According to Figure 9.43, the two highest-energy electrons of O_2 are in antibonding π_{2p}^* MOs and O_2 has a bond order of 2.0. Removing these two electrons to form O_2^{2+} produces an ion with bond order 3.0. O_2^{2+} has a stronger O–O bond than O_2, because O_2^{2+} has a greater bond order.

9.78   (a)   Substances with unpaired electrons are attracted into a magnetic field. This property is called *paramagnetism*.

(b)   Weigh the substance normally and in a magnetic field, as shown in Figure 9.44. Paramagnetic substances appear to have a larger mass when weighed in a magnetic field.

(c)   See Figures 9.35 and 9.43. O_2^+, one unpaired electron; N_2^{2-}, two unpaired electrons; Li_2^+, one unpaired electron

9.80   Determine the number of "valence" (noncore) electrons in each molecule or ion. Use the homonuclear diatomic MO diagram from Figure 9.43 (shown below) to calculate bond order and magnetic properties of each species. The electronegativity difference between heteroatomics increases the energy difference between the 2s AO on one atom and the 2p AO on the other, rendering the "no interaction" MO diagram in Figure 9.43 appropriate.

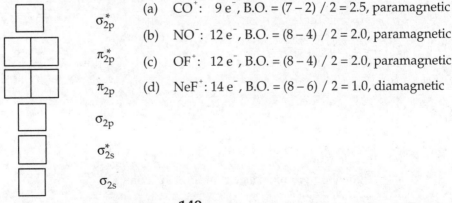

(a) CO^+: 9 e^-, B.O. = (7 – 2) / 2 = 2.5, paramagnetic

(b) NO^-: 12 e^-, B.O. = (8 – 4) / 2 = 2.0, paramagnetic

(c) OF^+: 12 e^-, B.O. = (8 – 4) / 2 = 2.0, paramagnetic

(d) NeF^+: 14 e^-, B.O. = (8 – 6) / 2 = 1.0, diamagnetic

9.82   (a)   Statement (ii) is the best explanation. The bond order of NO is $[1/2\,(8-3)] = 2.5$. The electron that is lost is in an antibonding molecular orbital, so the bond order in NO^+ is 3.0. The increase in bond order is the driving force for the formation of NO^+.

(b)   To form NO^-, an electron is added to an antibonding orbital, and the new bond order is $[1/2\,(8-4)] = 2$. The order of increasing bond order and bond strength is: $NO^- < NO < NO^+$. NO^- and NO are paramagnetic with two and one unpaired electrons, respectively. NO^+ is diamagnetic.

(c) NO^+ is isoelectronic with N_2, and NO^- is isoelectronic with O_2.

9.84   (a)   I: $5s, 5p_x, 5p_y, 5p_z$; Br: $4s, 4p_x, 4p_y, 4p_z$

(b)   By analogy to F_2, the BO of IBr will be 1.

(c)   I and Br have valence atomic orbitals with different principal quantum numbers. This means that the radial extensions (sizes) of the valence atomic orbital that contribute to the MO are different. The $n = 5$ valence AOs on I are larger than the $n = 4$ valence AOs on Br.

(d) σ_{np}^*

(e)   None

Additional Exercises

9.85   (a)   The physical basis of VSEPR is the electrostatic repulsion of like-charged particles, in this case groups or domains of electrons. That is, owing to electrostatic repulsion, electron domains will arrange themselves to be as far apart as possible.

(b)   The σ-bond electrons are localized in the region along the internuclear axes. The positions of the atoms and geometry of the molecule are thus closely tied to the locations of these electron pairs. Because the π-bond electrons are distributed above and below the plane that contains the σ bonds, these electron pairs do not, in effect, influence the geometry of the molecule. Thus, all σ- and π-bond electrons localized between two atoms are located in the same electron domain.

9.86   (a)   Two. If the electron-domain geometry is trigonal bipyramidal, there are five total electron domains around the central atom. An AB_3 molecule has three bonding domains, so there must be two nonbonding domains on A.

(b)   (iii) (See Table 9.3.)

9.88   (a) $40\ e^-$, $20\ e^-$ pairs

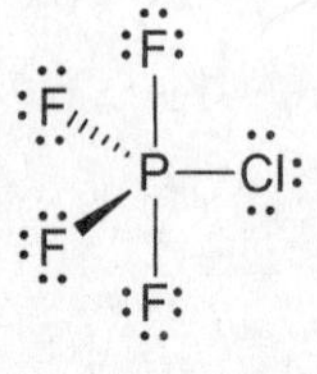

5 e^- domains
trigonal-bipyramidal electron-domain geometry

(b)   The greater the electronegativity of the terminal atom, the larger the negative charge centered on the atom, the smaller the effective size of the P–X bonding electron domain. A P–F bond will produce a smaller (and shorter) electron domain than a P–Cl bond.

(c)   The molecular geometry (shape) is also trigonal bipyramidal, because all five electron domains are bonding domains. Because we predicted the P–F electron domain to be smaller, the larger P–Cl bonding domain will occupy the equatorial plane of the molecule, minimizing the number of 90° P–Cl to P–F repulsions. This is the same argument that places a "larger" nonbonding domain in the equatorial position of a molecule like SF_4.

(d)   The molecular geometry is distorted from a perfect trigonal bipyramid because not all electron domains are alike. The 90° P–Cl to P–F repulsions will be greater than the 90° P–F to P–F repulsions, so the F(axial)–P–Cl angles will be greater than 90°. The equatorial F–P–F angles may distort slightly to "make room" for the axial F atoms that are "pushed away" from the equatorial Cl atom.

9.89   For any triangle, the law of cosines gives the length of side c as $c^2 = a^2 + b^2 - 2ab \cos\theta$.

Let the edge length of the cube (uy = vy = vz) = X

The length of the face diagonal (uv) is

$(uv)^2 = (uy)^2 + (vy)^2 - 2(uy)(vy) \cos 90$

$(uv)^2 = X^2 + X^2 - 2(X)(X) \cos 90$

$(uv)^2 = 2X^2; uv = \sqrt{2}X$

The length of the body diagonal (uz) is

$(uz)^2 = (vz^2) + (uv)^2 - 2(vz)(uv) \cos 90$

$(uz)^2 = X^2 + (\sqrt{2}X)^2 - 2(X)(\sqrt{2}X) \cos 90$

$(uz)^2 = 3X^2; uz = \sqrt{3}X$

For calculating the characteristic tetrahedral angle, the appropriate triangle has vertices u, v, and w. Theta, θ, is the angle formed by sides wu and wv and the hypotenuse is side uv.

$wu = wv = uz/2 = \sqrt{3}/2X; uv = \sqrt{2}X$

$(\sqrt{2}X)^2 = (\sqrt{3}/2X)^2 + (\sqrt{3}/2X)^2 - 2(\sqrt{3}/2X)(\sqrt{3}/2) \cos\theta$

$2X^2 = 3/4 X^2 + 3/4 X^2 - 3/2 X^2 \cos\theta$

$2X^2 = 3/2 X^2 - 3/2 X^2 \cos\theta$

$1/2 X^2 = -3/2 X^2 \cos\theta$

$\cos\theta = -(1/2 X^2) / (3/2 X^2) = -1/3 = -0.3333$

$\theta = 109.47°$

9.92   (a)

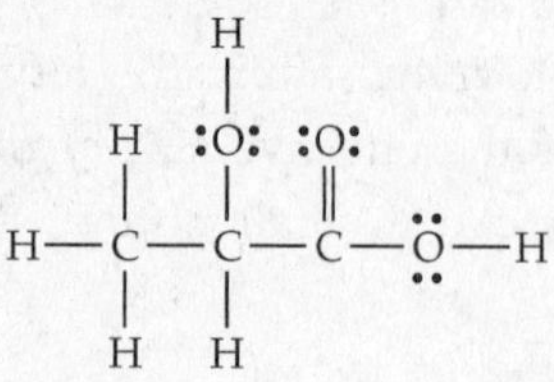

$$H-\overset{\overset{\displaystyle H}{|}}{\underset{\underset{\displaystyle H}{|}}{C}}-\overset{\overset{\displaystyle :O:}{\|}}{\underset{\underset{\displaystyle H}{|}}{C}}-\overset{\overset{\displaystyle :O:}{\|}}{C}-\overset{\cdot\cdot}{\underset{\cdot\cdot}{O}}-H$$

$3(4) + 3(6) + 6(1) = 36\,e^-$, $18\,e^-$ pr

(b)   There are 11 σ and 1 π bonds.

(c)   The C=O on the right-hand C atom is shortest. For the same bonded atoms, in this case C and O, the greater the bond order, the shorter the bond.

(d, e)   The right-most C has three e^- domains, so the hybridization is sp^2; bond angles about this C atom are approximately 120°. The middle and left-hand C atoms both have four bonding e^- domains and are sp^3 hybridized. Because the bonding domains about each C atom are not exactly the same, the bond angles will deviate somewhat from 109.5°; we expect larger deviations for the angles around the middle C atom.

9.93   (a)   Square pyramidal

(b)   Yes, there is one nonbonding electron domain on A. If there were only five bonding domains, the shape would be trigonal bipyramidal. With five bonding and one nonbonding electron domains, the molecule has octahedral domain geometry.

(c)   (iii). If the B atoms are halogens, each will have three nonbonding electron pairs; there are five bonding pairs, and A has one nonbonded pair, for a total of $[5(3) + 5 + 1] = 21\,e^-$ pairs and 42 electrons in the Lewis structure. If the five halogens contribute $35\,e^-$, A must contribute seven valence electrons. A is also a halogen.

9.94   (a)   The compound on the right has a dipole moment. In the square planar trans structure on the left, all equivalent bond dipoles can be oriented opposite each other, for a net dipole moment of zero.

(b)

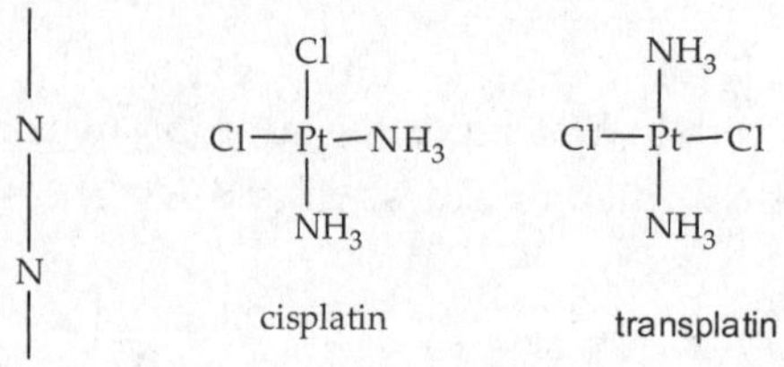

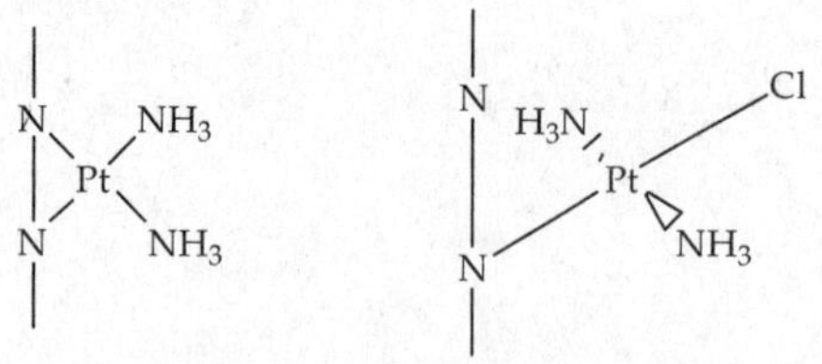

cisplatin   transplatin

The cis orientation of the Cl atoms in cisplatin means that when they leave, the Pt can bind two adjacent N sites on DNA. This "chelate" orientation (see Chapter 24) tightly binds cisplatin to DNA. Transplatin can bind only one DNA N atom at a time. Thus, to avoid bumping by transplatin NH_3 groups and DNA, the plane of transplatin must rotate away from the DNA backbone. This is a much looser bonding situation than for cisplatin.

9.95

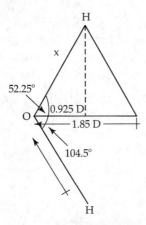

(a) The bond dipoles in H_2O lie along the O–H bonds with the positive end at H and the negative end at O. The dipole moment vector of the H_2O molecule is the resultant (vector sum) of the two bond dipoles. This vector bisects the H–O–H angle and has a magnitude of 1.85 D with the negative end pointing toward O.

(b) Because the dipole moment vector bisects the H–O–H bond angle, the angle between one H–O bond and the dipole moment vector is 1/2 the H–O–H bond angle, 52.25°. Dropping a perpendicular line from H to the dipole moment vector creates the right triangle pictured. If x = the magnitude of the O–H bond dipole, x cos (52.25) = 0.925 D. x = 1.51 D.

(c) The X–H bond dipoles (Table 8.3) and the electronegativity values of X (Figure 8.7) are

	Electronegativity	Bond dipole
F	4.0	1.82
O	3.5	1.51
Cl	3.0	1.08

Because the electronegativity of O is midway between the values for F and Cl, the O–H bond dipole should be approximately midway between the bond dipoles of HF and HCl. The value of the O–H bond dipole calculated in part (b) is consistent with this prediction.

9.96 (a) XeF_6 50 e^-, 25 e^- pairs

$$\begin{array}{c} :\!\ddot{F}\!: \;\; :\!\ddot{F}\!: \\ | \;\; \diagup \\ :\!\ddot{F}\!-\!Xe\!-\!\ddot{F}\!: \\ \diagup \;\; | \;\; \ddots \\ :\!\ddot{F} \;\; :\!\ddot{F}\!: \end{array}$$

(b) There are seven electron domains around Xe, and the maximum number of e^- domains in Table 9.3 is six.

(c) Tie seven balloons together and see what arrangement they adopt (seriously! see Figure 9.5). Alternatively, go to the chemical literature where VSEPR was first proposed and see if there is a preferred orientation for seven e^- domains.

(d) One of the seven e⁻ domains is a nonbonded domain. The question is whether it occupies an axial or equatorial position. The equatorial plane of a pentagonal bipyramid has F–Xe–F angles of 72°. Placing the nonbonded domain in the equatorial plane would create severe repulsions between it and the adjacent bonded domains. Thus, the nonbonded domain will reside in the axial position. The molecular structure is a pentagonal pyramid.

9.97 Statements (ii) and (iii) are true.

9.99 (a) $9\sigma, 3\pi$

(b) (i) 2, the second and third C atoms from the left

(ii) 2, the rightmost C atom and the N atom

(iii) 1, the leftmost C atom

9.100 (a) $16\,e^-, 8\,e^-$ pairs

$$\left[\overset{..}{\ddot{N}}=N=\overset{..}{\ddot{N}}\right]^- \longleftrightarrow \left[:N\equiv N-\overset{..}{\ddot{N}:}\right]^- \longleftrightarrow \left[:\overset{..}{\ddot{N}}-N\equiv N:\right]^-$$

(b) The observed bond length of 1.16 Å is intermediate between the values for N=N, 1.24 Å, and N ≡ N, 1.10 Å. This is consistent with the resonance structures, which indicate contribution from formally double and triple bonds to the true bonding picture in N_3^-.

(c) In each resonance structure, the central N has two electron domains, so it must be sp hybridized. It is difficult to predict the hybridization of terminal atoms in molecules where there are resonance structures because there are a different number of electron domains around the terminal atoms in each structure. Because the "true" electronic arrangement is a combination of all resonance structures, we will assume that the terminal N–N bonds have some triple bond character and that the terminal N atoms are sp hybridized. (There is no experimental measure of hybridization at terminal atoms, because there are no bond angles to observe.)

(d) In each resonance structure, N–N σ bonds are formed by sp hybrids and π bonds are formed by unhybridized p orbitals. Nonbonding e⁻ pairs can reside in

sp hybrids or p atomic orbitals.

(e) Recall that electrons in 2s orbitals are on the average closer to the nucleus than electrons in 2p orbitals. Because sp hybrids have greater s orbital character, it is reasonable to expect the radial extension of sp orbitals to be smaller than that of sp^2 or sp^3 orbitals and σ bonds formed by sp orbitals to be slightly shorter than those formed by other hybrid orbitals, assuming the same bonded atoms.

There are no solitary σ bonds in N_3^-. That is, the two σ bonds in N_3^- are each accompanied by at least one π bond between the bonding pair of atoms. Sigma bonds that are part of a double or triple bond must be shorter so that the p orbitals can overlap enough for the π bond to form. Thus, the observation is not applicable to this molecule. (Comparison of C–H bond lengths in C_2H_2, C_2H_4, C_2H_6, and related molecules would confirm or deny the observation.)

9.102 (a) Each C atom is surrounded by three electron domains (two single bonds and one double bond), so bond angles at each C atom will be approximately 120°.

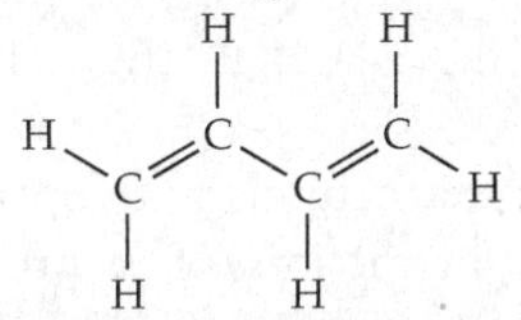

Because there is free rotation around the central C–C single bond, other conformations are possible.

(b) According to Table 8.5, the average C–C length is 1.54 Å, and the average C=C length is 1.34 Å. While the C=C bonds in butadiene appear "normal," the central C–C is significantly shorter than average. Examination of the bonding in butadiene reveals that each C atom is sp^2 hybridized and the π bonds are formed by the remaining unhybridized 2p orbital on each atom. If the central C–C bond is rotated so that all four C atoms are coplanar, the four 2p orbitals are parallel, and some delocalization of the π electrons occurs.

9.103 (a) $30\ e^-$, $15\ e^-$ pairs   (b)

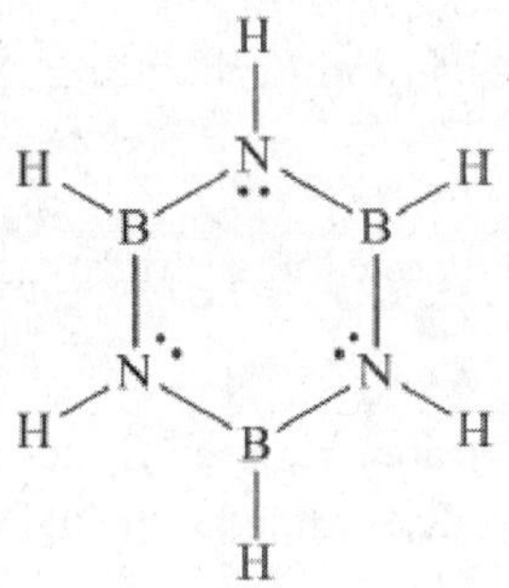

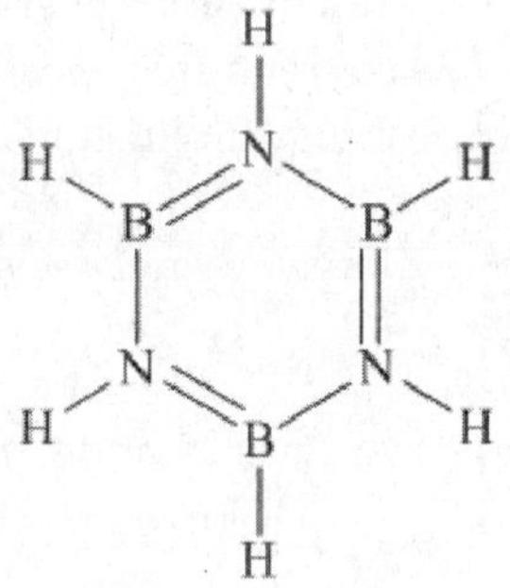

(c) In the Lewis structure in part (b), the N atoms have a +1 formal charge and the B atoms have a –1 formal charge. Because N is more electronegative than B, these formal charges do not seem favorable.

(d) The Lewis structure in part (b) has two resonance structures.

(e) In part (a), the B atoms are sp^2 hybridized and the N atoms are sp^3 hybridized. In part (b), both B and N are sp^2 hybridized. We would not expect the structure in part (a) to lead to a planar molecule, whereas the structure in (b) would be planar.

(f) The magnitude of the bond distance is between the values for single and double bonds, which favors multiple resonance structures with alternating single and double bonds, the structure in part (b). That the B–N bond lengths are identical also favors this structure.

(g) Six. There are 12 electron pairs in the σ system, which leaves 3 electron pairs in the π system.

9.105 (a) The orbital in the sketch is a σ antibonding MO.

(b) In H_2^-, there is one electron in the σ^* antibonding MO.

(c) In H_2^-, BO = ½.

(d) (iv). For the same two bonded atoms, the smaller the bond order, the weaker and longer the bond.

9.106   Use the MO diagrams in Figure 9.43 to calculate bond order, taking into account the correct number of electrons in each ion.

Ne_2 (BO = 0) < H_2^+ (BO = ½) < B_2 (BO = 1) < F_2^+ (BO = 1.5) < N_2^+ (BO = 2.5)

9.107   (a)   The diagram shows two s atomic orbitals with opposite phases. Because they are spherically symmetric, the interaction of s orbitals can only produce a σ molecular orbital. Because the two orbitals in the diagram have opposite phases, the interaction excludes electron density from the region between the nuclei. The resulting MO has a node between the two nuclei and is labeled σ_{2s}^*. The principal quantum number designation is arbitrary, because it defines only the size of the pertinent AOs and MOs. Shapes and phases of MOs depend only on these same characteristics of the interacting AOs.

           (b)   The diagram shows two p atomic orbitals with oppositely phased lobes pointing at each other. End-to-end overlap produces a σ-type MO; opposite phases mean a node between the nuclei and an antibonding MO. The interaction results in a σ_{2p}^* MO.

           (c)   The diagram shows parallel p atomic orbitals with like-phased lobes aligned. Side-to-side overlap produces a π-type MO; overlap of like-phased lobes concentrates electron density between the nuclei and a bonding MO. The interaction results in a π_{2p} MO.

9.109   We will refer to *azobenzene* (on the left) as A and *hydrazobenzene* (on the right) as H.

           (a)   A: sp^2; H: sp^3

           (b)   A: Each N and C atom has one unhybridized p orbital. H: Each C atom has one unhybridized p orbital, but the N atoms have no unhybridized p orbitals.

           (c)   A: ~120°; H: less than 109.5°

           (d)   Because all C and N atoms in A have unhybridized p orbitals, all can participate in delocalized π bonding. The delocalized π system extends over the entire molecule, including both benzene rings and the azo "bridge." In H, the N atoms have no unhybridized p orbitals, so they cannot participate in delocalized π bonding. Each of the benzene rings in H is delocalized, but the network cannot span the N atoms in the bridge.

           (e)   This is consistent with the answer to (d). In order for the unhybridized p orbitals in A to overlap, they must be parallel. This requires a planar σ-bond framework where all atoms in the molecule are coplanar.

           (f)   For a molecule to be useful in a solar energy conversion device, it must absorb visible light. This requires a HOMO-LUMO energy gap in the visible region. For organic molecules, the size of the gap is related to the number of conjugated π bonds; the more conjugated π bonds, the smaller the gap and the more likely the molecule is to be colored. Azobenzene has seven conjugated π bonds (π network delocalized over the entire molecule) and appears red-orange. Hydrazobenzene has only three conjugated π bonds (π network on benzene rings only) and appears white. Thus, the smaller HOMO-LUMO energy gap in A causes it to be both intensely colored and a more useful molecule for solar energy conversion.

9.110   (a)   H: $1s^1$; F: [He]$2s^2 2p^5$

When molecular orbitals are formed from atomic orbitals, the total number of orbitals is conserved. Because H and F have a total of five valence AOs $(H_{1s} + F_{2s} + 3F_{2p})$, the MO diagram for HF has five MOs.

(b)   H and F have a total of eight valence electrons. Because each MO can hold a maximum of two electrons, four of the five MOs would be occupied.

(c)

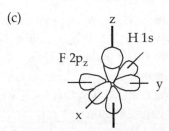

If H and F lie on the z-axis, then the $2p_z$ orbital of F will overlap with the 1s orbital of H.

(d)   Because F is more electronegative than H, the valence orbitals on F are at lower energy than those on H.

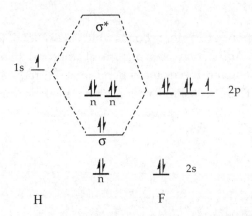

The HF MO diagram has 6 nonbonding, 2 bonding, and 0 antibonding electrons. The BO = [2 − 0]/2 = 1. (Nonbonding electrons do not "count" toward bond order.)

(e)   H—F̈:

In the Lewis structure for HF, the nonbonding electrons are on the (more electronegative) F atom, as they are in the MO diagram.

9.111   (a)   CO, 10 e⁻, 5 e⁻ pair

:C≡O:

(b)   The bond order for CO, as predicted by the MO diagram in Figure 9.46, is 1/2[8 − 2] = 3.0. A bond order of 3.0 agrees with the triple bond in the Lewis structure.

(c)   Applying the MO diagram in Figure 9.46 to the CO molecule, the highest-energy electrons would occupy the π_{2p} MOs. That is, π_{2p} would be the HOMO, highest occupied molecular orbital. If the true HOMO of CO is a σ-type MO, the order of the π_{2p} and σ_{2p} orbitals must be reversed. Figure 9.42 shows how the interaction of the 2s orbitals on one atom and the 2p orbitals on the other atom can affect the relative energies of the resulting MOs. This 2s–2p interaction in CO is significant enough so that the σ_{2p} MO is higher in energy than the π_{2p} MOs, and the σ_{2p} is the HOMO. This is the same energy order of MOs shown for large 2s–2p interaction in homonuclear diatomic molecules in Figure 9.43.

(d)   We expect the atomic orbitals of the more electronegative element to have lower energy than those of the less electronegative element. When atoms of the two elements combine, the lower-energy atomic orbitals make a greater contribution to the bonding MOs and the higher-energy atomic orbitals make a larger contribution to the antibonding orbitals. Thus, the π_{2p} bonding MOs will have a greater contribution from the more electronegative O atom.

9.112 (a)   The carbon–carbon bond order in ethylene is two. This means that there are two more bonding e^- pairs (4 bonding e^-) than antibonding e^- pairs. If one e^- is promoted form a bonding to an antibonding MO, there is one fewer bonding e^- and one more antibonding e^-; there is now an excess of two bonding e^- or one bonding e^- pair. The electron transition reduces the carbon–carbon bond order from two to one.

(b)   A bond order of one corresponds to a single bond, which is a σ bond; there is free rotation about a σ bond. Absorption of a photon of appropriate wavelength breaks the π bond in ethylene but leaves the σ bond intact; it reduces the bond order from two to one. There is facile rotation about the remaining carbon-carbon σ bond.

Integrative Exercises

9.113 (a)   Assume 100 g of compound

$$2.1\,g\,H \times \frac{1\,mol\,H}{1.008\,g\,H} = 2.1\,mol\,H; 2.1/2.1 = 1$$

$$29.8\,g\,N \times \frac{1\,mol\,N}{14.01\,g\,N} = 2.13\,mol\,N; 2.13/2.1 \approx 1$$

$$68.1\,g\,O \times \frac{1\,mol\,O}{16.00\,g\,O} = 4.26\,mol\,O; 4.26/2.1 \approx 2$$

The empirical formula is HNO_2; formula weight = 47. Because the approximate molar mass is 50, the molecular formula is HNO_2.

(b)   Assume N is central, because it is unusual for O to be central, and part (d) indicates as much. HNO_2: 18 valence e^-

$$\ddot{O}\!=\!\ddot{N}\!-\!\ddot{O}\!-\!H \longleftrightarrow :\!\ddot{O}\!-\!\ddot{N}\!=\!\ddot{O}\!-\!H$$
$$\qquad\qquad\qquad\qquad -1\quad 0\quad +1$$

The second resonance form is a minor contributor because of unfavorable formal charges.

(c)   The electron-domain geometry around N is trigonal planar with an O–N–O angle of approximately 120°. If the resonance structure on the right makes a significant contribution to the molecular structure, all four atoms would lie in a plane. If only the left structure contributes, the H could rotate in and out of the molecular plane. The relative contributions of the two resonance structures could be determined by measuring the O–N–O and N–O–H bond angles.

(d)   3 VSEPR e⁻ domains around N, sp^2 hybridization

(e)   3 σ, 1 π for both structures (or for H bound to N).

9.115 (a) PX_3, 26 valence e⁻, 13⁻ pairs

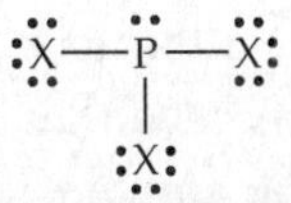

4 electron domains around P, tetrahedral e⁻ domain geometry,
If all bonding and nonbonding electron domains are the same size, perfect tetrahedral angles are 109.5°. If all bonding electron domains are the same size but the nonbonding domain is larger, bond angles are somewhat less than 109.5°.

(b)   As electronegativity increases (I < Br < Cl < F), the X–P–X angles decreases.

(c)   The greater the electronegativity of X, the larger the magnitude of negative charge centered on X. The more negative charge centered on X, the smaller the P–X bonding domains, the greater the effect of the nonbonding domain and the smaller the bond angle. Also, as the electronegativity of X decreases and the bonding domain size increases, the effect of the large nonbonding domain decreases.

(d) $PBrCl_4$, 40 valence electrons, 20 e⁻ pairs. The molecule will have trigonal-bipyramidal electron-domain geometry (similar to PCl_5 in Table 9.3). Based on the argument in part (c), the P–Br bond will have greater repulsions with P–Cl bonds than P–Cl bonds have with each other. Therefore, the Br will occupy an equatorial position in the trigonal bipyramid, so that the more unfavorable P–Br to P–Cl repulsions can be situated at larger angles in the equatorial plane.

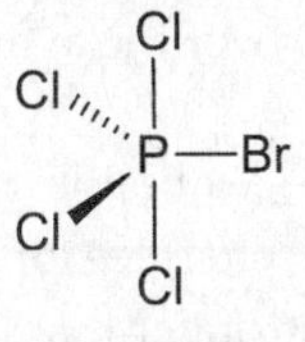

9.116 (a)   Three electron domains around each central C atom, sp^2 hybridization

(b)   A 180° rotation around the C=C double bond is required to convert the trans isomer into the cis isomer. A 90° rotation around the bond eliminates all overlap of the p orbitals that form the π bond and it is broken.

(c)   **average bond enthalpy**

    C=C   614 kJ/mol

    C–C   348 kJ/mol

The difference in these values, 266 kJ/mol, is the average bond enthalpy of a C–C π bond. This is the amount of energy required to break 1 mol of C–C π bonds. The energy per molecule is

$$266 \text{ kJ/mol} \times \frac{1000 \text{ J}}{1 \text{ kJ}} \times \frac{1 \text{ mol}}{6.022 \times 10^{23} \text{ molecules}} = 4.417 \times 10^{-19}$$

$$= 4.42 \times 10^{-19} \text{ J/molecule}$$

(d) $\lambda = \text{hc/E} = \dfrac{6.626 \times 10^{-34} \text{ J-s} \times 2.998 \times 10^{8} \text{ m/s}}{4.417 \times 10^{-19} \text{ J}} = 4.50 \times 10^{-7} \text{ m} = 450 \text{ nm}$

(e) Yes, 450 nm light is in the visible portion of the spectrum. A cis-trans isomerization in the retinal portion of the large molecule rhodopsin is the first step in a sequence of molecular transformations in the eye that leads to vision. The sequence of events enables the eye to detect visible photons, in other words, to see.

9.117 (a)   C $\equiv$ C   839 kJ/mol   (1 σ, 2 π)

 C=C   614 kJ/mol   (1 σ, 1 π)

 C–C   348 kJ/mol   (1 σ)

The contribution from 1 π-bond would be (614–348) 266 kJ/mol. From a second π bond, (839 – 614), 225 kJ/mol. An average π bond contribution would be (266 + 225)/2 = 246 kJ/mol.

This is $\dfrac{246 \text{ kJ/}\pi \text{ bond}}{348 \text{ kJ/}\sigma \text{ bond}} \times 100 = 71\%$ of the average enthalpy of a σ bond.

(b)   N $\equiv$ N   941 kJ/mol

 N=N   418 kJ/mol

 N–N   163 kJ/mol

first π = (418 – 163) = 255 kJ/mol

second π = (941 – 418) = 523 kJ/mol

average π-bond enthalpy = (255 + 523)/2 = 389 kJ/mol

This is $\dfrac{389 \text{ kJ/}\pi \text{ bond}}{163 \text{ kJ/}\sigma \text{ bond}} \times 100 = 240\%$ of the average enthalpy of a σ bond.

N–N σ bonds are weaker than C–C σ bonds, whereas N–N π bonds are stronger than C–C π bonds. The relative energies of C–C σ and π bonds are similar, whereas N–N π bonds are much stronger than N–N σ bonds.

(c) N_2H_4, 14 valence e⁻, 7 e⁻ pairs

$$\text{H}-\overset{\displaystyle ..}{\text{N}}-\overset{\displaystyle ..}{\text{N}}-\text{H}$$
$$\;\;\;\;\; | \;\;\; |$$
$$\;\;\;\;\; \text{H} \;\;\; \text{H}$$

4 electron domains around N, sp^3 hybridization

N_2H_2, 12 valence e⁻, 6 e⁻ pairs

$$\text{H}-\overset{\displaystyle ..}{\text{N}}=\overset{\displaystyle ..}{\text{N}}-\text{H}$$

3 electron domains around N, sp^2 hybridization

N_2, 10 valence e^-, 5 e^- pairs

:N≡N:

2 electron domains around N, sp hybridization

(d) In the three types of N–N bonds, each N atom has a nonbonding or lone pair of electrons. The lone pair to bond pair repulsions are minimized going from less than 109.5° to 120° to 180° bond angles, making the π bonds stronger relative to the σ bond. In the three types of C–C bonds, no lone-pair to bond-pair repulsions exist, and the σ and π bonds have more similar energies.

9.119   (a) $3d_{z^2}$

        (b)   Ignoring the donut of the d_{z^2} orbital

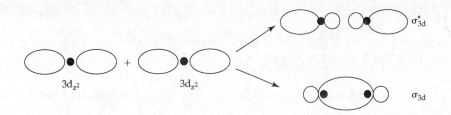

        (c)   A node is generated in σ_{3d}^* because antibonding MOs are formed when AO lobes with opposite phases interact. Electron density is excluded from the internuclear region and a node is formed in the MO.

        (d)   Sc: $[Ar]4s^2 3d^1$ Omitting the core electrons, there are six e^- in the energy-level diagram.

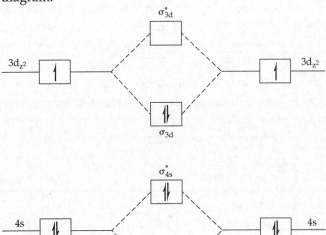

        (e)   The bond order in Sc_2 is $1/2\,(4-2) = 1.0$.

9.120 (a) The molecular and empirical formulas of the four molecules are:

benzene: molecular, C_6H_6; empirical, CH

napthalene: molecular, $C_{10}H_8$; empirical, C_5H_4

anthracene: molecular, $C_{14}H_{10}$; empirical, C_7H_5

tetracene: molecular, $C_{18}H_{12}$, empirical, C_3H_2

(b) Yes. Because the compounds all have different empirical formulas, combustion analysis could in principle be used to distinguish them. In practice, the mass % of C in the four compounds is not very different, so the data would have to be precise to at least 3 decimal places and 4 would be better.

(c) $C_{10}H_8(s) + 12O_2(g) \rightarrow 10CO_2(g) + 4H_2O(g)$

(d) $\Delta H_{comb} = 5D(C=C) + 6D(C-C) + 8D(C-H) + 12D(O=O) - 20D(C=O) - 8D(O-H)$

$= 5(614) + 6(348) + 8(413) + 12(495) - 20(799) - 8(463)$

$= -5282 \text{ kJ/mol } C_{10}H_8$

(e) Yes. For example, the resonance structures of naphthalene are:

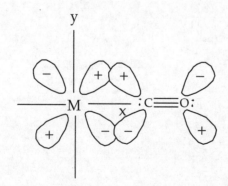

(f) Colored compounds absorb visible light and appear the color of the visible light that they reflect. Colorless compounds typically absorb shorter wavelength, higher-energy light. The energy of light absorbed corresponds to the energy gap between the HOMO and LUMO of the molecule. That tetracene absorbs longer wavelength, lower-energy visible light indicates that it has the smallest HOMO-LUMO energy gap of the four molecules. Tetracene also has the most conjugated double bonds of the four molecules. We might conclude that the more conjugated double bonds in an organic molecule, the smaller the HOMO-LUMO energy gap. More information about the absorption spectra of anthracene, naphthalene, and benzene is needed to confirm this conclusion.

9.121 (a, b)

(c)   The two lobes of a p AO have opposite phases. These are shown on the diagram as + and −. An antibonding MO is formed when p AOs with opposite phases interact.

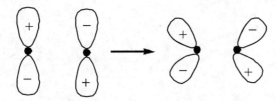

(d)   Note that the d_{xy} AO has lobes that lie between, not on, the x and y axes.

(e)   A π bond forms by overlap of orbitals on M and C. There is electron density above and below, but not along, the M–C axis.

(f)   According to Exercise 9.111, the HOMO of CO is a σ-type MO. So the appropriate MO diagram is shown on the left side of Figure 9.43. A lone CO molecule has 10 valence electrons, the HOMO is σ_{2p} and the bond order is 3.0. The LUMO is π_{2p}^{*}.

When M and CO interact as shown in the π_{2p}^{*} diagram, d-π back bonding causes the π_{2p}^{*} to become partially occupied. Electron density in the π_{2p}^{*} decreases electron density in the bonding molecular orbitals and decreases the BO of the bound CO. The strength of the C–O bond in a metal–CO complex decreases relative to the strength of the C–O bond in an isolated CO molecule.

10 Gases

Visualizing Concepts

10.2 (a) $V_1/T_1 = V_2/T_2$ (Charles' law)

$V_1/300 \text{ K} = V_2/500 \text{ K}$

$V_2 = 5/3 \, V_1$

(b) $P_1V_1 = P_2V_2$ (Boyle's law)

$1 \text{ atm} \times V_1 = 2 \text{ atm} \times V_2$

$V_2 = 1/2 \, V_1$

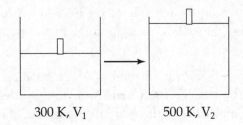

300 K, V_1 500 K, V_2

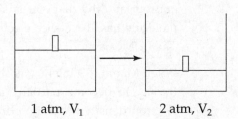

1 atm, V_1 2 atm, V_2

(c) $V_1/T_1 = V_2/T_2$ (Charles' law)

$V_1/300 \text{ K} = V_2/200 \text{ K}$

$V_2 = 2/3 \, V_1$

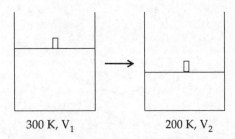

300 K, V_1 200 K, V_2

10.4 Statement (d) describes the volume change. At constant pressure and temperature, the container volume is directly proportional to the number of particles present (Avogadro's law). As the reaction proceeds, 3 gas molecules are converted to 2 gas molecules, so the number of particles and the container volume decrease by 33%.

10.6 Over time, the gases will mix perfectly. Each bulb will contain 4 blue and 3 red atoms. The "blue" gas has the greater partial pressure after mixing, because it has the greater number of particles (at the same T and V as the "red" gas.)

10.8

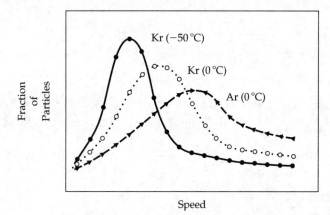

10.10   (a)   Total pressure is directly related to total number of particles (or total mol particles). P(ii) < P(i) = P(iii)

         (b)   Partial pressure of He is directly related to number of He atoms (yellow) or mol He atoms. P_{He}(iii) < P_{He}(ii) < P_{He}(i)

         (c)   Density is total mass of gas per unit volume. We can use the atomic or molar masses of He (4) and N_2 (28), as relative masses of the particles.

$$mass(i) = 5(4) + 2(28) = 76$$

$$mass(ii) = 3(4) + 1(28) = 40$$

$$mass(iii) = 2(4) + 5(28) = 148$$

Because the container volumes are equal, d(ii) < d(i) < d(iii).

         (d)   At the same temperature, all gases have the same average kinetic energy. The average kinetic energies of the particles in the three containers are equal.

10.12   (a)   The van der Waals constant a accounts for the influence of intermolecular forces in lowering the pressure of a real gas.

         (b)   According to the plot, Gas B is closest to ideal behavior, then Gas A and Gas C. Gas B will have the smallest a value, then A and C. From the values in Table 10.3, Gas B is N_2, a = 1.39 ; Gas A is CO_2, a = 3.59; Gas C = Cl_2, a = 6.49.

Gas Characteristics; Pressure (Sections 10.1 and 10.2)

10.14   (a)   Because gas molecules are far apart and in constant motion, the gas expands to fill the container. Attractive forces hold liquid molecules together and the volume of the liquid does not change.

         (b) H_2O and CCl_4 molecules are too dissimilar to displace each other and mix in the liquid state. All mixtures of gases are homogeneous. (See Solution 10.13 (c).)

         (c)   Because gas molecules are far apart, the mass present in 1 mL of a gas is very small. The mass of a gas present in 1 L is on the same order of magnitude as the mass of a liquid present in 1 mL.

10.16 $P = m \times a/A$; $1\,Pa = 1\,kg/m\text{-}s^2$; $A = 3.0\,cm \times 4.1\,cm \times 4 = 49.2 = 49\,cm^2$

$$\frac{262\,kg}{49.2\,cm^2} \times \frac{9.81\,m}{s^2} \times \frac{(100)^2\,cm^2}{1\,m^2} = 5.224 \times 10^5 \frac{kg}{m\text{-}s^2} = 5.2 \times 10^5\,Pa$$

10.18 Using the relationship derived in Solution 10.17 for two liquids under the influence of gravity, $(d \times h)_{1id} = (d \times h)_{Hg}$. At 749 torr, the height of an Hg barometer is 749 mm.

$$\frac{1.20\,g}{1\,mL} \times h_{1id} = \frac{13.6\,g}{1\,mL} \times 760\,mm; h_{1id} = \frac{13.6\,g/mL \times 749\,mm}{1.20\,g/mL} = 8.49 \times 10^3\,mm = 8.49\,m$$

10.20 (a) $0.912\,atm \times \dfrac{760\,torr}{1\,atm} = 693\,torr$

(b) $0.685\,bar \times \dfrac{1 \times 10^5\,Pa}{1\,bar} \times \dfrac{1\,kPa}{1 \times 10^3\,Pa} = 68.5\,kPa$

(c) $655\,mm\,Hg \times \dfrac{1\,atm}{760\,mm\,Hg} = 0.862\,atm$

(d) $1.323 \times 10^5\,Pa \times \dfrac{1\,atm}{1.01325 \times 10^5\,Pa} = 1.3057 = 1.306\,atm$

(e) $2.50\,atm \times \dfrac{14.70\,psi}{1\,atm} = 36.75 = 36.8\,psi$

10.22 $882\,mbar = 0.882\,bar$

(a) $0.882\,bar \times \dfrac{1 \times 10^5\,Pa}{1\,bar} \times \dfrac{1\,atm}{101,325\,Pa} = 0.8705 = 0.871\,atm$

(b) $0.882\,bar = 0.8705\,atm \times \dfrac{760\,torr}{1\,atm} = 661.55 = 662\,torr$

(c) $0.882\,bar = 661.55\,torr \times \dfrac{1\,mm\,Hg}{1\,torr} \times \dfrac{1\,cm\,Hg}{10\,mm\,Hg} \times \dfrac{1\,in.\,Hg}{2.54\,cm\,Hg} = 26.045 = 26.0\,in.\,Hg$

10.24 (a) The atmosphere is exerting 15.4 cm = 154 mm Hg (torr) more pressure than the gas.

$$P_{gas} = P_{atm} - 15.4\,torr = \left(0.985\,atm \times \frac{760\,torr}{1\,atm}\right) - 15.4\,torr = 733\,torr$$

(b) The gas is exerting 12.3 mm Hg (torr) more pressure than the atmosphere.

$$P_{gas} = P_{atm} + 12.3\,torr = \left(0.99\,atm \times \frac{760\,torr}{1\,atm}\right) + 12.3\,torr = 764.7\,torr = 7.6 \times 10^2\,torr$$

(Atmospheric pressure of 0.99 atm determines that the result has 2 sig figs.)

The Gas Laws (Section 10.3)

10.26 *Analyze.* Given: initial P, V, T. Find: final values of P, V, T for certain changes of condition. *Plan.* Select the appropriate gas law relationships from Section 10.3; solve for final conditions, paying attention to units. *Solve.*

(a) $P_1V_1 = P_2V_2$; the proportionality holds true for any pressure or volume units.

$P_1 = 752$ torr, $V_1 = 5.12$ L, $P_2 = 1.88$ atm

$$V_2 = \frac{P_1V_1}{P_2} = \frac{752 \text{ torr} \times 5.12 \text{ L}}{1.88 \text{ atm}} \times \frac{1 \text{ atm}}{760 \text{ torr}} = 2.69 \text{ L}$$

Check. As pressure increases, volume should decrease; our result agrees with this.

(b) $V_1/T_1 = V_2/T_2$; T must be in Kelvins for the relationship to be true.

$V_1 = 5.12$ L, $T_1 = 21\ °C = 294$ K, $T_2 = 175\ °C = 448$ K

$$V_2 = \frac{V_1T_2}{T_1} = \frac{5.12 \text{ L} \times 448 \text{ K}}{294 \text{ K}} = 7.80 \text{ L}$$

Check. As temperature increases, volume should increase; our result is consistent with this.

10.28   According to Avogadro's hypothesis, the mole ratios in the chemical equation will be volume ratios for the gases if they are at the same temperature and pressure.

$$N_2(g) + 3H_2(g) \rightarrow 2NH_3(g)$$

The volumes of H_2 and N_2 are in a stoichiometric $\dfrac{3.6 \text{ L}}{1.2 \text{ L}}$ or $\dfrac{3 \text{ vol } H_2}{1 \text{ vol } N_2}$ ratio, so either can be used to determine the volume of $NH_3(g)$ produced.

$$1.2 \text{ L } N_2 \times \frac{2 \text{ mol } NH_3}{1 \text{ mol } N_2} = 2.4 \text{ L } NH_3(g) \text{ produced.}$$

The Ideal-Gas Equation (Section 10.4)

(In *Solutions to Exercises*, the symbol for molar mass is MM.)

10.30   Assume 1 mole of Ne at STP. Sample volume = 22.4 L. Calculate the volume occupied by 1 mole of Ne atoms. $1 \text{ L} = 1 \text{ dm}^3$, $r = 0.69$ Å, $V = 4\pi r^3/3$

$$r = 0.69 \text{ Å} \times \frac{1 \times 10^{-10} \text{ m}}{1 \text{ Å}} \times \frac{10 \text{ dm}}{\text{m}} = 6.9 \times 10^{-10} \text{ dm}$$

$$V = \frac{4\pi r^3}{3}; \quad \frac{4\pi (6.9 \times 10^{-10} \text{ dm})^3}{3} \times \frac{6.022 \times 10^{23} \text{ Ne atoms}}{\text{mol}} = 8.287 \times 10^{-4} \text{ dm}^3 = 8.3 \times 10^{-4} \text{ L}$$

$$\frac{8.3 \times 10^{-4} \text{ L occupied by Ne atoms}}{22.4 \text{ L total gas volume}} = 1/27{,}000 \text{ of the total space occupied by Ne atoms}$$

10.32 $n = g/MM$; $PV = nRT = gRT/MM$; $MM = gRT/PV$.

2-L flask: $MM = 4.8 \text{ RT}/2.0(X) = 2.4 \text{ RT}/X$

3-L flask: $MM = 0.36 \text{ RT}/3.0 (0.1 \text{ X}) = 1.2 \text{ RT}/X$

The molar masses of the two gases are not equal. The gas in the 2-L flask has a molar mass that is twice as large as the gas in the 3-L flask.

10.34   *Analyze/Plan.* Follow the strategy for calculations involving many variables given in Section 10.4.   *Solve.*

(a) $n = 1.50$ mol, $P = 1.25$ atm, $T = -6\ °C = 267$ K

$$V = \frac{nRT}{P} = 1.50\ \text{mol} \times \frac{0.08206\ \text{L-atm}}{\text{mol-K}} \times \frac{267\ \text{K}}{1.25\ \text{atm}} = 26.3\ \text{L}$$

(b) $n = 3.33 \times 10^{-3}$ mol, $V = 478$ mL $= 0.478$ L

$$P = 750\ \text{torr} \times \frac{1\ \text{atm}}{760\ \text{torr}} = 0.9868 = 0.987\ \text{atm}$$

$$T = \frac{PV}{nR} = 0.9868\ \text{atm} \times \frac{0.478\ \text{L}}{3.33 \times 10^{-3}\ \text{mol}} \times \frac{1\ \text{mol-K}}{0.08206\ \text{L-atm}} = 1726 = 1.73 \times 10^3\ \text{K}$$

(c) $n = 0.00245$ mol, $V = 413$ mL $= 0.413$ L, $T = 138\ °C = 411$ K

$$P = \frac{nRT}{V}; = 0.00245\ \text{mol} \times \frac{0.08206\ \text{L-atm}}{\text{mol-K}} \times \frac{411\ \text{K}}{0.413\ \text{L}} = 0.200\ \text{atm}$$

(d) $V = 126.5$ L, $T = 54\ °C = 327$ K,

$$P = 11.25\ \text{kPa} \times \frac{1\ \text{atm}}{101.325\ \text{kPa}} = 0.11103 = 0.1110\ \text{atm}$$

$$n = \frac{PV}{RT} = 0.11103\ \text{atm} \times \frac{\text{mol-K}}{0.08206\ \text{L-atm}} \times \frac{126.5\ \text{L}}{327\ \text{K}} = 0.523\ \text{mol}$$

10.36   Find the volume of the tube in cm^3; $1\ \text{cm}^3 = 1$ mL.

$r = d/2 = 2.5\ \text{cm}/2 = 1.25 = 1.3\ \text{cm}$; $h = 5.5\ \text{m} = 5.5 \times 10^2$ cm

$V = \pi r^2 h = 3.14159 \times (1.25\ \text{cm})^2 \times (5.5 \times 10^2\ \text{cm}) = 2.700 \times 10^3\ \text{cm}^3 = 2.7\ \text{L}$

$$PV = \frac{g}{MM}RT;\ g = \frac{MM \times PV}{RT};\ P = 1.78\ \text{torr} \times \frac{1\ \text{atm}}{760\ \text{torr}} = 2.342 \times 10^{-3} = 2.34 \times 10^{-3}\ \text{atm}$$

$$g = \frac{20.18\ \text{g Ne}}{1\ \text{mol Ne}} \times \frac{\text{mol-K}}{0.08206\ \text{L-atm}} \times \frac{2.342 \times 10^{-3}\ \text{atm} \times 2.700\ \text{L}}{308\ \text{K}} = 5.049 \times 10^{-3} = 5.0 \times 10^{-3}\ \text{g Ne}$$

10.38   (a) $P_{O_3} = 3.0 \times 10^{-3}$ atm; $T = 250$ K; $V = 1$ L (exact)

$$\text{\# of } O_3 \text{ molecules} = \frac{PV}{RT} \times 6.022 \times 10^{23}$$

$$\text{\#} = \frac{3.0 \times 10^{-3}\ \text{atm} \times 1\ \text{L}}{250\ \text{K}} \times \frac{\text{mol-K}}{0.08206\ \text{L-atm}} \times \frac{6.022 \times 10^{23}\ \text{molecules}}{\text{mol}}$$

$$= 8.8 \times 10^{19}\ O_3 \text{ molecules}$$

(b) $\text{\# of } CO_2 \text{ molecules} = \dfrac{PV}{RT} \times 6.022 \times 10^{23} \times 0.0004$

$$\text{\#} = \frac{1.0\ \text{atm} \times 2.0\ \text{L}}{300\ \text{K}} \times \frac{\text{mol-K}}{0.08206\ \text{L-atm}} \times \frac{6.022 \times 10^{23}\ \text{molecules}}{\text{mol}} \times 0.0004$$

$$= 1.957 \times 10^{19} = 2 \times 10^{19}\ CO_2 \text{ molecules}$$

10.40   (a) $V = 0.250$ L, $T = 23\ °C = 296$ K, $n = 2.30\ \text{g } C_3H_8 \times \dfrac{1\ \text{mol } C_3H_8}{44.1\ \text{g } C_3H_8} = 0.052154$

$$= 0.0522\ \text{mol}$$

$$P = \frac{nRT}{V} = 0.052154\ \text{mol} \times \frac{0.08206\ \text{L-atm}}{\text{mol-K}} \times \frac{296\ \text{K}}{0.250\ \text{L}} = 5.07\ \text{atm}$$

(b)   STP = 1.00 atm, 273 K

$$V = \frac{nRT}{P} = 0.052154 \, \text{mol} \times \frac{0.08206 \, \text{L-atm}}{\text{mol-K}} \times \frac{273 \, \text{K}}{1.00 \, \text{atm}} = 1.1684 = 1.17 \, \text{L}$$

(c)   °C = 5/9 (°F − 32°); K = °C + 273.15 = 5/9 (130 °F − 32°) + 273.15 = 327.59 = 328 K

$$P = \frac{nRT}{V} = 0.052154 \, \text{mol} \times \frac{0.08206 \, \text{L-atm}}{\text{mol-K}} \times \frac{327.59 \, \text{K}}{0.250 \, \text{L}} = 5.608 = 5.61 \, \text{atm}$$

10.42   Calculate the mass of He that will produce a pressure of 75 atm in the cylinder, then subtract that mass from 5.225 g He to calculate the mass of He that must be released.

$$PV = \frac{g}{MM}RT; \quad g = \frac{MM \times PV}{RT}$$

$$g = \frac{4.0026 \, \text{g He}}{1 \, \text{mol He}} \times \frac{\text{mol-K}}{0.08206 \, \text{L-atm}} \times \frac{75 \, \text{atm} \times 0.334 \, \text{L}}{296 \, \text{K}} = 4.1279 = 4.1 \, \text{g He remain}$$

5.225 g He initial − 4.1279 g He remain = 1.0971 = 1.1 g He must be released

10.44   T = 23 °C = 296 K,   P = 16,500 kPa $\times \dfrac{1 \, \text{atm}}{101.325 \, \text{kPa}} = 162.84 = 163 \, \text{atm}$

$$V = 55.0 \, \text{gal} \times \frac{3.7854 \, \text{L}}{1 \, \text{gal}} = 208.20 = 208 \, \text{L}$$

(a) $g = \dfrac{MM \times PV}{RT}; g = \dfrac{32.0 \, \text{g O}_2}{1 \, \text{mol O}_2} \times \dfrac{\text{mol-K}}{0.08206 \, \text{L-atm}} \times \dfrac{162.84 \, \text{atm}}{296 \, \text{K}} \times 208.20 \, \text{L}$

$$= 4.4665 \times 10^4 \, \text{g O}_2 = 44.7 \, \text{kg O}_2$$

(b) $V_2 = \dfrac{P_1 V_1 T_2}{T_1 P_2} = \dfrac{16,500 \, \text{kPa} \times 208.20 \, \text{L} \times 273 \, \text{K}}{296 \, \text{K} \times 101.325 \, \text{kPa}} = 3.13 \times 10^4 \, \text{L}$

(c) $T_2 = \dfrac{P_2 T_1}{P_1} = \dfrac{150.0 \, \text{atm} \times 296 \, \text{K}}{16,500 \, \text{kPa}} \times \dfrac{101.325 \, \text{kPa}}{1 \, \text{atm}} = 272.7 = 273 \, \text{K}$

(d) $P_2 = \dfrac{P_1 V_1 T_2}{V_2 T_1} = \dfrac{16,500 \, \text{kPa} \times 208.20 \, \text{L} \times 297 \, \text{K}}{55.0 \, \text{L} \times 296 \, \text{K}} = 62,671 = 6.27 \times 10^4 \, \text{kPa}$

10.46   Change mass to kg; 1 hr = 60 min; pay attention to units.

(a) $185 \, \text{lb} \times \dfrac{1 \, \text{kg}}{2.2046 \, \text{lb}} \times \dfrac{47.5 \, \text{mL O}_2}{\text{kg-min}} \times 60 \, \text{min} = 2.39 \times 10^5 \, \text{mL}$

(b) $165 \, \text{lb} \times \dfrac{1 \, \text{kg}}{2.2046 \, \text{lb}} \times \dfrac{65.0 \, \text{mL O}_2}{\text{kg-min}} \times 60 \, \text{min} = 2.92 \times 10^5 \, \text{mL}$

Further Applications of the Ideal-Gas Equation (Section 10.5)

10.48 $CO_2 < SO_2 < HBr$. For gases at the same conditions, density is directly proportional to molar mass. The order of increasing molar mass is the order of increasing density. CO_2, 44 g/mol $< SO_2$, 64 g/mol $< HBr$, 81 g/mol.

10.50   (b) Xe atoms have a higher mass than N_2 molecules. Because both gases at STP have the same number of molecules per unit volume, the Xe gas must be denser.

10.52   (a) $d = \dfrac{MM \times P}{RT}$; $MM = 146.1$ g/mol, $T = 21\,°C = 294$ K, $P = 707$ torr

$$d = \frac{146.1\,g}{1\,mol} \times \frac{mol\text{-}K}{0.08206\,L\text{-}atm} \times \frac{707\,torr}{294\,K} \times \frac{1\,atm}{760\,torr} = 5.63\,g/L$$

   (b) $MM = \dfrac{dRT}{P} = \dfrac{7.135\,g}{1\,L} \times \dfrac{0.08206\,L\text{-}atm}{mol\text{-}K} \times \dfrac{285\,K}{743\,torr} \times \dfrac{760\,torr}{1\,atm} = 171\,g/mol$

10.54 $MM = \dfrac{gRT}{PV} = \dfrac{0.846\,g}{0.354\,L} \times \dfrac{0.08206\,L\text{-}atm}{mol\text{-}K} \times \dfrac{373\,K}{752\,torr} \times \dfrac{760\,torr}{1\,atm} = 73.9\,g/mol$

10.56 $n_{H_2} = \dfrac{P_{H_2}V}{RT}$; $P = 825$ torr $\times \dfrac{1\,atm}{760\,torr} = 1.0855 = 1.09$ atm; $T = 273 + 21\,°C = 294$ K

$$n_{H_2} = 1.0855\,atm \times \frac{mol\text{-}K}{0.08206\,L\text{-}atm} \times \frac{145\,L}{294\,K} = 6.5242 = 6.52\,mol\,H_2$$

$$6.5242\,mol\,H_2 \times \frac{1\,mol\,CaH_2}{2\,mol\,H_2} \times \frac{42.10\,g\,CaH_2}{1\,mol\,CaH_2} = 137.34 = 137\,g\,CaH_2$$

10.58   Follow the logic in Sample Exercise 10.9. The $H_2(g)$ will be used in a balloon, which operates at atmospheric pressure. Because atmospheric pressure is not explicitly given, assume 1 atm (infinite sig figs).

$$n = \frac{PV}{RT} = 1\,atm \times \frac{3.1150 \times 10^4\,L}{295\,K} \times \frac{mol\text{-}K}{0.08206\,L\text{-}atm} = 1.28678 \times 10^3 = 1.29 \times 10^3\,mol\,H_2$$

From the balanced equation, 1 mol of Fe produces 1 mol of H_2, so 1.29×10^3 mol Fe are required.

$$1.28678 \times 10^3\,mol\,Fe \times \frac{55.845\,g\,Fe}{mol\,Fe} \times \frac{1\,kg}{1000\,g} = 71.86 = 71.9\,kg\,Fe$$

10.60   The gas sample is a mixture of $C_2H_2(g)$ and $H_2O(g)$. Find the partial pressure of C_2H_2, then moles CaC_2 and C_2H_2.

$P_t = 753$ torr $= P_{C_2H_2} + P_{H_2O}$. P_{H_2O} at $23\,°C = 21.07$ torr

$$P_{C_2H_2} = (753\,torr - 21.07\,torr) \times \frac{1\,atm}{760\,torr} = 0.96307 = 0.963\,atm$$

$$1.524\,g\,CaC_2 \times \frac{1\,mol\,CaC_2}{64.10\,g} \times \frac{1\,mol\,C_2H_2}{1\,mol\,CaC_2} = 0.023775 = 0.02378\,mol\,C_2H_2$$

$$V = 0.023775\,mol \times \frac{0.08206\,L\text{-}atm}{mol\text{-}K} \times \frac{296\,K}{0.96307\,atm} = 0.600\,L\,C_2H_2$$

Partial Pressures (Section 10.6)

10.62   (a)   The partial pressure of gas A is **not affected** by the addition of gas C. The partial pressure of A depends only on moles of A, volume of container, and conditions; none of these factors changes when gas C is added.

   (b)   The total pressure in the vessel **increases** when gas C is added, because the total number of moles of gas increases.

(c)   The mole fraction of gas B **decreases** when gas C is added. The moles of gas B stay the same, but the total moles increase, so the mole fraction of B (n_B/n_t) decreases.

10.64   Given mass, V and T of O_2 and He, find the partial pressure of each gas. Sum to find the total pressure in the tank.

V = 10.0 L; T = 19 °C; 19 + 273 = 292 K

$$n_{O_2} = 51.2 \text{ g O}_2 \times \frac{1 \text{ mol O}_2}{31.999 \text{ g O}_2} = 1.600 = 1.60 \text{ mol O}_2$$

$$n_{He} = 32.6 \text{ g He} \times \frac{1 \text{ mol He}}{4.0026 \text{ g He}} = 8.1447 = 8.14 \text{ mol He}$$

$$P_{O_2} = 1.600 \text{ mol} \times \frac{0.08206 \text{ L-atm}}{\text{mol-K}} \times \frac{292 \text{ K}}{10.0 \text{ L}} = 3.8338 = 3.84 \text{ atm}$$

$$P_{He} = 8.1447 \text{ mol} \times \frac{0.08206 \text{ L-atm}}{\text{mol-K}} \times \frac{292 \text{ K}}{10.0 \text{ L}} = 19.5159 = 19.5 \text{ atm}$$

$$P_t = 3.8338 + 19.5159 = 23.3497 = 23.3 \text{ atm}$$

10.66 $X_{Xe} = 4/100 = 0.04$; $X_{Ne} = X_{He} = (1 - 0.04)/2 = 0.48$

$$V_t = 0.900 \text{ mm} \times 0.300 \text{ mm} \times 10.0 \text{ mm} \times \frac{1 \text{ cm}^3}{10^3 \text{ mm}^3} \times \frac{1 \text{ L}}{1000 \text{ cm}^3} = 2.70 \times 10^{-6} \text{ L}$$

$$P_t = 500 \text{ torr} \times \frac{1 \text{ atm}}{760 \text{ torr}} = 0.657895 = 0.658 \text{ atm}$$

$$n_t = \frac{PV}{RT} = 0.657895 \text{ atm} \times \frac{\text{mol-K}}{0.08206 \text{ L-atm}} \times \frac{2.70 \times 10^{-6} \text{ L}}{298 \text{ K}} \times \frac{6.022 \times 10^{23} \text{ atoms}}{\text{mol}}$$

$$= 4.3743 \times 10^{16} = 4.37 \times 10^{16} \text{ total atoms}$$

Xe atoms = $X_{Xe} \times$ total atoms = $0.04(4.3743 \times 10^{16}) = 1.75 \times 10^{15} = 2 \times 10^{15}$ Xe atoms

Ne atoms = He atoms = $0.48(4.3743 \times 10^{16}) = 2.10 \times 10^{16} = 2.1 \times 10^{16}$ Ne and He atoms

Assumptions: To calculate total moles of gas and total atoms, we assumed a reasonable room temperature. Because '4% Xe' was not defined, we conveniently assumed mole percent. The 1:1 relationship of Ne to He is assumed to be by volume and not by mass.

10.68 $V \text{ C}_2\text{H}_5\text{OC}_2\text{H}_5(l) \rightarrow \text{mass C}_2\text{H}_5\text{OC}_2\text{H}_5 \rightarrow \text{mol C}_2\text{H}_5\text{OC}_2\text{H}_5 \rightarrow P_{\text{C}_2\text{H}_5\text{OC}_2\text{H}_5} \text{ (at V, T)}$

$P_t = P_{N_2} + P_{O_2} + P_{\text{C}_2\text{H}_5\text{OC}_2\text{H}_5}$; T = 273.15 + 35.0 °C = 308.15 = 308.2 K

(a) $5.00 \text{ mL C}_2\text{H}_5\text{OC}_2\text{H}_5 \times \frac{0.7134 \text{ g C}_2\text{H}_5\text{OC}_2\text{H}_5}{\text{mL}} \times \frac{1 \text{ mol C}_2\text{H}_5\text{OC}_2\text{H}_5}{74.12 \text{ g C}_2\text{H}_5\text{OC}_2\text{H}_5}$

$$= 0.048125 = 0.0481 \text{ mol C}_2\text{H}_5\text{OC}_2\text{H}_5$$

$$P = \frac{nRT}{V} = 0.048125 \text{ mol} \times \frac{308.15}{6.00 \text{ L}} \times \frac{0.08206 \text{ L-atm}}{\text{mol-K}} = 0.20282 = 0.203 \text{ atm}$$

(b) $P_t = P_{N_2} + P_{O_2} + P_{\text{C}_2\text{H}_5\text{OC}_2\text{H}_5} = 0.751 \text{ atm} + 0.208 \text{ atm} + 0.203 \text{ atm} = 1.162 \text{ atm}$

10.70 $T = 320\,^{\circ}C + 273 = 593\ K;\quad P_t\,(593\ K) = P_{N_2} + P_{O_2}$

(i) Use Amonton's law (see Solution 10.27) to calculate P_{N_2} at 593 K.

(ii) Use stoichiometry to calculate mol O_2 produced by decomposing 5.15 g Ag_2O.

(iii) Use the ideal-gas law to calculate P_{O_2} at 593 K. Sum the pressures.

(i) $P_1 = 760\ torr = 1\ atm;\quad T_1 = 32\,^{\circ}C + 273 = 305\ K;\quad T_2 = 320\,^{\circ}C + 273 = 593\ K$

$$\frac{P_1}{T_1} = \frac{P_2}{T_2}\ \text{or}\ P_2 = \frac{P_1 T_2}{T_1} = \frac{1.00\ atm\ \times\ 593\ K}{305\ K} = 1.9443 = 1.94\ atm$$

(ii) $2\ Ag_2O(s) \rightarrow 4\ Ag(s) + O_2(g)$

$$5.15\ g\ Ag_2O \times \frac{1\ mol\ Ag_2O}{231.74\ g\ Ag_2O} \times \frac{1\ mol\ O_2}{2\ mol\ Ag_2O} = 0.01111 = 0.0111\ mol\ O_2$$

(iii) $V = 0.0750\ L,\ T = 593\ K,\ n = 0.0111$

$$P = \frac{nRT}{V};\ P = \frac{0.01111\ mol}{0.0750\ L} \times \frac{0.08206\ L\text{-}atm}{mol\text{-}K} \times 593\ K = 4.876 = 7.209 = 7.21\ atm$$

$$P_t\,(593\ K) = P_{N_2} + P_{O_2} = 1.94\ atm + 7.21\ atm = 9.15\ atm$$

10.72 (a) $n_{O_2} = 15.08\ g\ O_2 \times \dfrac{1\ mol}{31.999\ g} = 0.4713\ mol;\ n_{N_2} = 8.17\ g\ N_2 \times \dfrac{1\ mol}{28.02\ g} = 0.292\ mol$

$n_{H_2} = 2.64\ g\ H_2 \times \dfrac{1\ mol}{2.016\ g} = 1.31\ mol;\ n_t = 0.4713 + 0.292 + 1.31 = 2.07\ mol$

$X_{O_2} = \dfrac{n_{O_2}}{n_t} = \dfrac{0.4713}{2.07} = 0.228;\ X_{N_2} = \dfrac{n_{N_2}}{n_t} = \dfrac{0.292}{2.07} = 0.141$

$X_{H_2} = \dfrac{1.31}{2.07} = 0.633$

(b) $P_{O_2} = n \times \dfrac{RT}{V};\ P_{O_2} = 0.4713\ mol \times \dfrac{0.08206\ L\text{-}atm}{mol\text{-}K} \times \dfrac{288.15\ K}{15.50\ L} = 0.7190\ atm$

$P_{N_2} = 0.292\ mol \times \dfrac{0.08206\ L\text{-}atm}{mol\text{-}K} \times \dfrac{288.15\ K}{15.50\ L} = 0.445\ atm$

$P_{H_2} = 1.31\ mol \times \dfrac{0.08206\ L\text{-}atm}{mol\text{-}K} \times \dfrac{288.15\ K}{15.50\ L} = 2.00\ atm$

10.74 Calculate the pressure of the gas in the second vessel directly from mass and conditions using the ideal-gas equation.

(a) $P_{SO_2} = \dfrac{gRT}{M\,V} = \dfrac{3.00\ g\ SO_2}{64.07\ g\ SO_2/mol} \times \dfrac{0.08206\ L\text{-}atm}{mol\text{-}K} \times \dfrac{299\ K}{10.0\ L} = 0.11489 = 0.115\ atm$

(b) $P_{N_2} = \dfrac{gRT}{M\,V} = \dfrac{2.35\ g\ N_2}{28.01\ g\ N_2/mol} \times \dfrac{0.08206\ L\text{-}atm}{mol\text{-}K} \times \dfrac{299\ K}{10.0\ L} = 0.20585 = 0.206\ atm$

(c) $P_t = P_{SO_2} + P_{N_2} = 0.11489\ atm + 0.20585\ atm = 0.321\ atm$

Kinetic-Molecular Theory of Gases; Effusion and Diffusion
(Sections 10.7 and 10.8)

10.76   (a)   False. The average kinetic energy per molecule in a collection of gas molecules is the same for all gases at the same temperature.

        (b)   True.

        (c)   False. The molecules in a gas sample at a given temperature exhibit a distribution of kinetic energies.

        (d)   True.

        (e)   False. Gas molecules at the same temperature exhibit a distribution of speeds.

10.78   Newton's model provides no explanation of the effect of a change in temperature on the pressure of a gas at constant volume or on the volume of a gas at constant pressure. On the other hand, the assumption that the average kinetic energy of gas molecules increases with increasing temperature explains Charles' law, that an increase in temperature requires an increase in volume to maintain constant pressure.

10.80   The gas undergoes a chemical reaction that has fewer gas particles in products than in reactants. Mass is conserved when a chemical reaction occurs, so the mass of (flask + contents) remains constant. Pressure is directly proportional to number of particles, so pressure decreases as the number of gaseous particles decreases. One simple example of such a reaction is the dimerization of NO_2: $2\,NO_2(g) \rightarrow N_2O_4$.

10.82   (a)   They have the same number of molecules (equal volumes of gases at the same temperature and pressure contain equal numbers of molecules).

        (b) N_2 is more dense because it has the larger molar mass. Because the volumes of the samples and the number of molecules are equal, the gas with the larger molar mass will have the greater density.

        (c)   The average kinetic energies are equal (statement 5, section 10.7).

        (d) CH_4 will effuse faster. The lighter the gas molecules, the faster they will effuse (Graham's law).

10.84   (a)   *Plan.* The greater the molecular (and molar) mass, the smaller the rms and average speeds of the molecules. Calculate the molar mass of each gas, and place them in decreasing order of mass and increasing order of rms and average speed.

           Solve. CO = 28 g/mol; SF_6 = 146 g/mol; H_2S = 34 g/mol; Cl_2 = 71 g/mol; HBr = 81 g/mol. In order of increasing speed (and decreasing molar mass):

           $SF_6 < HBr < Cl_2 < H_2S < CO$

        (b)   *Plan.* Follow the logic of Sample Exercise 10.13. *Solve.*

$$CO: u_{rms} = \sqrt{\frac{3RT}{MM}} = \left(\frac{3 \times 8.314\,kg\text{-}m^2/s^2\text{-}mol\text{-}K \times 300\,K}{28.0 \times 10^{-3}\,kg/mol}\right)^{1/2} = 5.17 \times 10^2\,m/s$$

$$Cl_2: u_{rms} = \left(\frac{3 \times 8.314\,kg\text{-}m^2/s^2\text{-}mol\text{-}K \times 300\,K}{70.9 \times 10^{-3}\,kg/mol}\right)^{1/2} = 3.25 \times 10^2\,m/s$$

As expected, the lighter molecule moves at the greater speed.

(c)   *Plan.* From Equations [10.20] and [10.21], we see that the ratio of most probable speed to rms speed is $(2/3)^{1/2}$. Use this ratio and the results from part (b) to calculate most probable speeds.   *Solve.*

$$CO: u_{mp} = (2/3)^{1/2}(5.17 \times 10^2 \text{ m/s}) = 422 \text{ m/s}$$

$$Cl_2: u_{mp} = (2/3)^{1/2}(3.25 \times 10^2 \text{ m/s}) = 265 \text{ m/s}$$

The lighter molecule, CO, has the greater most probable speed. Note that the most probable speed is less than the rms speed, as shown on Figure 10.13(b).

10.86   Write each proportionality relationship as an equation, then combine them to obtain a formula for mean free path.

The operational symbols and units are: mean free path, λ, meters (m); temperature, T, kelvins (K); pressure, P, atmospheres (atm); diameter of a gas molecule, d, meters (m), constant, R_{mfp}.

$\lambda = \text{constant} \times T$; $\lambda = \text{constant}/P$; $\lambda = \text{constant}/d^2$

$$\text{Combining}: \lambda = \frac{R_{mfp} \times T}{P \times d^2}$$

The units of R_{mfp} are chosen and arranged so that they cancel the units of measurement, leaving an appropriate length unit for λ.

With the units defined above, R_{mfp} will have units of $\dfrac{\text{atm} - \text{m}^3}{K}$.

(Note that $1 \text{ m}^3 = 10^3 \text{ dm}^3 = 1000$ L. Substituting, R_{mfp} would have units of $\dfrac{\text{atm} - L}{K}$, with the factor of 1000 incorporated into the value of R_{mfp}.)

10.88 $\dfrac{\text{rate}^{235}U}{\text{rate}^{238}U} = \sqrt{\dfrac{238.05}{235.04}} = \sqrt{1.0128} = 1.0064$

There is a slightly greater rate enhancement for $^{235}U(g)$ atoms than $^{235}UF_6(g)$ molecules (1.0043), because ^{235}U is a greater percentage (100%) of the mass of the diffusing particles than in $^{235}UF_6$ molecules. The masses of the isotopes were taken from *The Handbook of Chemistry and Physics*.

10.90   The time required is proportional to the reciprocal of the effusion rate.

$$\frac{\text{rate}(X)}{\text{rate}(O_2)} = \frac{105 \text{ s}}{31 \text{ s}} = \left[\frac{32 g\, O_2}{MM_x}\right]^{1/2}; MM_x = 32 g\, O_2 \times \left[\frac{105}{31}\right]^2 = 370 g/\text{mol (two sig figs)}$$

Nonideal-Gas Behavior (Section 10.9)

10.92   Ideal-gas behavior is most likely to occur at high temperature and low pressure, so the atmosphere on Mercury is more likely to obey the ideal-gas law. The higher temperature on Mercury means that the kinetic energies of the molecules will be larger relative to intermolecular attractive forces. Further, the gravitational attractive forces on Mercury are lower because the planet has a much smaller mass. This means that for the same column mass of gas (Figure 10.1), atmospheric pressure on Mercury will be lower.

10.94   The constant a is a measure of the strength of intermolecular attractions among gas molecules; b is a measure of molecular volume. Both increase with increasing molecular mass and structural complexity.

10.96   *Analyze.* Conditions and amount of $CCl_4(g)$ are given. *Plan.* Use ideal-gas equation and van der Waals equation to calculate pressure of gas at these conditions.   *Solve.*

(a) $P = 1.00 \text{ mol} \times \dfrac{0.08206 \text{ L-atm}}{\text{mol-K}} \times \dfrac{313 \text{ K}}{33.3 \text{ L}} = 0.771 \text{ atm}$

(b) $P = \dfrac{nRT}{V - nb} - \dfrac{an^2}{V^2} = \dfrac{1.00 \times 0.08206 \times 313}{33.3 - (1.00 \times 0.1383)} - \dfrac{20.4(1.00)^2}{(33.3)^2} = 0.756 \text{ atm}$

   Check. The van der Waals result indicates that the real pressure will be less than the ideal pressure. That is, intermolecular forces reduce the effective number of particles and the real pressure. This is reasonable for 1 mole of gas at relatively low temperature and pressure.

(c)   According to Table 10.3, CCl_4 has larger a and b values. That is, CCl_4 experiences stronger intermolecular attractions and has a larger molecular volume than Cl_2 does. CCl_4 will deviate more from ideal behavior at these conditions than Cl_2 will.

10.98   From section 7.3, the nonbonding or *van der Waals* radius is half of the shortest internuclear distance when two nonbonding atoms collide. So, radii calculated from the van der Waals equation are nonbonding radii. According to the kinetic-molecular theory, ideal-gas particles undergo perfectly elastic, billiard-ball collisions, in keeping with the definition of nonbonding radii.

Also, from the results of Exercise 10.97, the atomic radius calculated from the van der Waals b value is twice as large as the bonding atomic radius from Figure 7.7. Nonbonding radii are larger than bonding radii because no lasting penetration of electron clouds occurs during a nonbonding collision.

Additional Exercises

10.100 $PV = nRT$, $n = PV/RT$. Because RT is constant, n is proportional to PV.

   Total available n = $(15.0 \text{ L} \times 1.00 \times 10^2 \text{ atm}) - (15.0 \text{ L} \times 1.00 \text{ atm}) = 1485$

$$= 1.49 \times 10^3 \text{ L-atm}$$

   Each balloon holds 2.00 L × 1.00 atm = 2.00 L-atm

   $\dfrac{1485 \text{ L-atm available}}{2.00 \text{ L-atm/balloon}} = 742.5 = 742 \text{ balloons}$

   (Only 742 balloons can be filled completely, with a bit of He left over.)

10.102   (a)   Change mass CO_2 to mol CO_2.   P = 1.00 atm, T = 27 °C = 300 K.

   $6 \times 10^6 \text{ tons } CO_2 \times \dfrac{2000 \text{ lb}}{\text{ton}} \times \dfrac{453.6 \text{ g}}{\text{lb}} \times \dfrac{1 \text{ mol } CO_2}{44.01 \text{ g } CO_2} = 1.237 \times 10^{11} = 1 \times 10^{11} \text{ mol}$

$$V = \frac{nRT}{P} = 1.237 \times 10^{11} \text{ mol} \times \frac{300 \text{ K}}{1.00 \text{ atm}} \times \frac{0.08206 \text{ L-atm}}{\text{mol-K}} = 3.045 \times 10^{12} = 3 \times 10^{12} \text{ L}$$

(b) $1.237 \times 10^{11} \text{ mol CO}_2 \times \frac{44.01 \text{ g CO}_2}{\text{mol CO}_2} \times \frac{1 \text{ cm}^3}{1.2 \text{ g}} \times \frac{1 \text{ L}}{1000 \text{ cm}^3} = 4.536 \times 10^9 = 5 \times 10^9 \text{ L}$

(c) $n = 1.237 \times 10^{11} \text{ mol, } P = 90 \text{ atm, } T = 36 \,^\circ\text{C} = 309 \text{ K}$

$$V = \frac{nRT}{P} = 1.237 \times 10^{11} \text{ mol} \times \frac{309 \text{ K}}{90 \text{ atm}} \times \frac{0.08206 \text{ L-atm}}{\text{mol-K}} = 3.485 \times 10^{10} = 3 \times 10^{10} \text{ L}$$

10.104   Vol of room $= 12 \text{ ft} \times 20 \text{ ft} \times 9 \text{ ft} \times \frac{12^3 \text{ in.}^3}{1 \text{ ft}^3} \times \frac{2.54^3 \text{ cm}^3}{1^3 \text{ in.}^3} \times \frac{1 \text{ L}}{1000 \text{ cm}^3} = 61{,}164 = 6 \times 10^4 \text{ L}$

Calculate the **total** moles of gas in the laboratory at the conditions given.

$$n_t = \frac{PV}{RT} = 1.00 \text{ atm} \times \frac{\text{mol-K}}{0.08206 \text{ L-atm}} \times \frac{61{,}164 \text{ L}}{297 \text{ K}} = 2510 = 3 \times 10^3 \text{ mol gas}$$

A $Ni(CO)_4$ concentration of 1 part in 10^9 means 1 mol $Ni(CO)_4$ in 1×10^9 total moles of gas.

$$\frac{x \text{ mol Ni(CO)}_4}{2.510 \times 10^3 \text{ mol gas}} = \frac{1}{1 \times 10^9} = 2.510 \times 10^{-6} = 3 \times 10^{-6} \text{ mol Ni(CO)}_4$$

$$2.510 \times 10^{-6} \text{ mol Ni(CO)}_4 \times \frac{170.74 \text{g Ni(CO)}_4}{1 \text{ mol Ni(CO)}_4} = 4.286 \times 10^{-3} = 4 \times 10^{-3} \text{ g} = 4 \text{ mg Ni(CO)}_4$$

10.106   It is simplest to calculate the partial pressure of each gas as it expands into the total volume, then sum the partial pressures.

$$P_2 = P_1 V_1 / V_2; \; P_{N_2} = 265 \text{ torr} (1.0 \text{ L}/2.5 \text{ L}) = 106 = 1.1 \times 10^2 \text{ torr}$$

$$P_{Ne} = 800 \text{ torr} (1.0 \text{ L}/2.5 \text{ L}) = 320 = 3.2 \times 10^2 \text{ torr}$$

$$P_{H_2} = 532 \text{ torr} (0.5 \text{ L}/2.5 \text{ L}) = 106 = 1.1 \times 10^2 \text{ torr}$$

$$P_t = P_{N_2} + P_{Ne} + P_{H_2} = (106 + 320 + 106) \text{ torr} = 532 = 5.3 \times 10^2 \text{ torr}$$

10.108   (a)   Pressure percent = mole percent. Change pressure/mole percents to mole fraction. Partial pressure of each gas is mole fraction (X) times total pressure. $P_x = X_x P_t$

$P_{N_2} = 0.748(0.985 \text{ atm}) = 0.737 \text{ atm}; \;\; P_{O_2} = 0.153(0.985 \text{ atm}) = 0.151 \text{ atm}$

$P_{CO_2} = 0.037(0.985 \text{ atm}) = 0.03645 = 0.036 \text{ atm}$

$P_{H_2O} = 0.062(0.985 \text{ atm}) = 0.06107 = 0.061 \text{ atm}$

(b) $PV = nRT, n = PV/RT; \; P = 0.036 \text{ atm}, V = 0.455 \text{ L}, T = 37 \,^\circ\text{C} = 310 \text{ K}$

$$n = 0.03645 \text{ atm} \times \frac{0.455 \text{ L}}{310 \text{ K}} \times \frac{\text{mol-K}}{0.08206 \text{ L-atm}} = 6.520 \times 10^{-4} = 6.5 \times 10^{-4} \text{ mol}$$

(c) $C_6H_{12}O_6 + 6 \, O_2 \rightarrow 6 \, CO_2 + 6 \, H_2O$

$$6.520 \times 10^{-4} \text{ mol CO}_2 \times \frac{1 \text{ mol C}_6\text{H}_{12}\text{O}_6}{6 \text{ mol CO}_2} \times \frac{180.15 \text{ g C}_6\text{H}_{12}\text{O}_6}{1 \text{ mol C}_6\text{H}_{12}\text{O}_6} = 0.01958$$

$$= 0.020 \text{ g C}_6\text{H}_{12}\text{O}_6$$

10.110   At constant temperature, an ideal gas at a certain pressure and volume, P_1V_1, expands into a larger volume and lower pressure, P_2V_2. This is a Boyle's law problem.

Let $V_1 = x$ L, $V_2 = 0.800$ L $+ x$. Change 1.50 atm to torr, so the pressure units cancel.

1.50 atm $\times$ 760 torr/atm = 1140 torr

$P_1V_1 = P_2V_2$.   1140 torr (x L) = 695 torr [(0.800 + x) L]. Units cancel.

1140 x = 556 + 695 x; 445 x = 556; x = 1.25 L

Check. 1140 (1.25) = 695 (2.05); 1425 = 1425. The algebra is correct.

10.112   Calculate the number of moles of Ar in the vessel:

n = (339.854 − 337.428)/39.948 = 0.060729 = 0.06073 mol

The total number of moles of the mixed gas is the same (Avogadro's law). Thus, the average atomic weight is (339.076 − 337.428)/0.060729 = 27.137 = 27.14. Let the mole fraction of Ne be *X*. Then,

X (20.183) + (1 − *X*) (39.948) = 27.137; 12.811 = 19.765 *X*; *X* = 0.6482

Neon is thus 64.82 mole percent of the mixture.

10.114   (a)   Assumption 3 states that attractive and repulsive forces between molecules are negligible. All gases in the list are nonpolar. The largest and most structurally complex molecule, SF_6, is most likely to depart from this assumption.

           (b)   The monatomic gas Ne is smallest and least structurally complex, so it will behave most like an ideal gas.

           (c)   Root-mean-square speed is inversely related to molecular mass. The lightest gas, CH_4, has the highest rms speed.

           (d)   The heaviest and most structurally complex is SF_6. Also, S and F have larger atomic radii than C and H; this means that S–F bonds will be longer than C–H bonds and the volume of SF_6 will be greater than that of CH_4. It is reasonable to assume that SF_6 will occupy the greatest molecular volume relative to total volume. A quantitative measure is the *b* value in Table 10.3, with units of L/mol. Unfortunately, SF_6 does not appear in Table 10.3.

           (e)   Average kinetic energy is only related to absolute (K) temperature. At the same temperature, they all have the same average kinetic-molecular energy.

           (f)   Rate of effusion is inversely related to molecular mass. The lighter the molecule, the faster it effuses. Ne and CH_4 have smaller molecular masses and effuse faster than N_2.

           (g)   If SF_6 occupies the greatest molecular volume [see part (d)], we expect it to have the largest van der Waals *b* parameter.

10.116   The larger and heavier the particle, in this case a single atom, the more likely it is to deviate from ideal behavior. Other than Rn, Xe is the largest (atomic radius = 1.40 Å), heaviest (molar mass = 131.3 g/mol) and most dense (5.90 g/L) noble gas. Its susceptibility to intermolecular interactions is also demonstrated by its ability to form compounds like XeF_4.

10.118 (a) $120.00 \text{ kg N}_2(g) \times \dfrac{1000 \text{ g}}{1 \text{ kg}} \times \dfrac{1 \text{ mol N}_2}{28.0135 \text{ g N}} = 4283.6 \text{ mol N}_2$

$P = \dfrac{nRT}{V} = 4283.6 \text{ mol} \times \dfrac{0.08206 \text{ L-atm}}{\text{mol-K}} \times \dfrac{553 \text{ K}}{1100.0 \text{ L}} = 176.72 = 177 \text{ atm}$

(b) Rearranging Equation [10.25] to isolate P, $P = \dfrac{nRT}{V - nb} - \dfrac{n^2 a}{V^2}$

$P = \dfrac{(4283.6 \text{ mol})(0.08206 \text{ L-atm/mol-K})(553 \text{ K})}{1100.0 \text{ L} - (4283.6 \text{ mol})(0.0391 \text{ L/mol})} - \dfrac{(4283.6 \text{ mol})^2 (1.39 \text{ L}^2\text{-atm/mol}^2)}{(1100.0 \text{ L})^2}$

$P = \dfrac{194{,}388 \text{ L-atm}}{1100.0 \text{ L} - 167.5 \text{ L}} - 21.1 \text{ atm} = 208.5 \text{ atm} - 21.1 \text{ atm} = 187.4 \text{ atm}$

(c) The pressure corrected for the real volume of the N_2 molecules is 208.5 atm, 31.8 atm higher than the ideal pressure of 176.7 atm. The 21.1 atm correction for intermolecular forces reduces the calculated pressure somewhat, but the "real" pressure is still higher than the ideal pressure. The correction for the real volume of molecules dominates. Even though the value of b is small, the number of moles of N_2 is large enough so that the molecular volume correction is larger than the attractive forces correction.

Integrative Exercises

10.120 *Plan.* Write the balanced equation for the combustion of methanol. Because amounts of both reactants are given, determine the limiting reactant. Use mole ratios to calculate moles H_2O produced, based on the amount of limiting reactant. Change moles to grams H_2O, then use density to calculate volume of $H_2O(l)$ produced. Assume the condensed $H_2O(l)$ is at 25 °C, where density = 0.99707 g/mol. *Solve.*

methanol = $CH_3OH(l)$. $2 CH_3OH(l) + 3 O_2(g) \rightarrow 2 CO_2(g) + 4 H_2O(g)$

$25.0 \text{ mL CH}_3\text{OH} \times \dfrac{0.850 \text{ g CH}_3\text{OH}}{\text{mL}} \times \dfrac{1 \text{ mol CH}_3\text{OH}}{32.04 \text{ g}} = 0.6632 = 0.663 \text{ mol CH}_3\text{OH}$

$\text{mol O}_2 = n = \dfrac{PV}{RT} = 1.00 \text{ atm} \times \dfrac{12.5 \text{ L}}{273 \text{ K}} \times \dfrac{\text{mol-K}}{0.08206 \text{ L-atm}} = 0.5580 = 0.558 \text{ mol O}_2$

$0.558 \text{ mol O}_2 \times \dfrac{2 \text{ mol CH}_3\text{OH}}{3 \text{ mol O}_2} = 0.372 \text{ mol CH}_3\text{OH}$

0.558 mol O_2 can react with only 0.372 mol CH_3OH, so O_2 is the limiting reactant. Note that a large volume of $O_2(g)$ is required to completely react with a relatively small volume of $CH_3OH(l)$.

$0.558 \text{ mol O}_2 \times \dfrac{4 \text{ mol H}_2\text{O}}{3 \text{ mol O}_2} \times \dfrac{18.02 \text{ g H}_2\text{O}}{1 \text{ mol H}_2\text{O}} \times \dfrac{1 \text{ mL H}_2\text{O}}{0.99707 \text{ g H}_2\text{O}} = 13.446 = 13.4 \text{ mL H}_2\text{O}$

10.122 (a) *Plan.* Use the ideal-gas law to calculate the moles CO_2 that react.

Solve. P(reacted) = P(initial) – P(final), at constant V, T. Because both CaO and BaO react with CO_2 in a 1:1 mole ratio, mol CaO + mol BaO = mol CO_2. Use molar masses to calculate % CaO in sample.

$$P(\text{reacted}) = 730\,\text{torr} - 150\,\text{torr} = 580\,\text{torr};\ 580\,\text{torr} \times \frac{1\,\text{atm}}{760\,\text{torr}} = 0.76316 = 0.763\,\text{atm}$$

$$n = \frac{PV}{RT} = 0.76316\,\text{atm} \times \frac{1.0\,\text{L}}{298\,\text{K}} \times \frac{\text{mol-K}}{0.08206\,\text{L-atm}} = 0.03121 = 0.0312\,\text{mol CO}_2$$

(b)   *Plan.* Use the stoichiometry of the reaction and definition of moles to calculate the mass and Mass % of CaO.

Solve. $CaO(s) + CO_2(s) \rightarrow CaCO_3(s).$ $BaO(s) + CO_2(g) \rightarrow BaCO_3(s)$

mol CO_2 reacted = mol CaO + mol BaO

Let x = g CaO, 4.00 – x = g BaO

$$0.03121 = \frac{x}{56.08} + \frac{4.00 - x}{153.3}$$

$$0.03121(56.08)(153.3) = 153.3x + 56.08(4.00 - x)$$

$$268.3 = (153.3x - 56.08x) + 224.3$$

$$43.98 = 97.22x,\ x = 0.452 = 0.45\,\text{g CaO}$$

$$\frac{0.452\,\text{g CaO}}{4.00\,\text{g sample}} \times 100 = 11.3 = 11\%\ \text{CaO}$$

(By strict sig fig rules, the result has 2 sig figs, because 268 – 224 = 44 has 0 decimal places and 2 sig figs.)

10.124 $n = \dfrac{PV}{RT} = 1.00\,\text{atm} \times \dfrac{\text{mol-K}}{0.08206\,\text{L-atm}} \times \dfrac{2.7 \times 10^{12}\,\text{L}}{273\,\text{K}} = 1.205 \times 10^{11} = 1.2 \times 10^{11}\,\text{mol CH}_4$

$$CH_4(g) + 2\,O_2(g) \rightarrow CO_2(g) + 2\,H_2O(l) \quad \Delta H^\circ = -890.4\,\text{kJ}$$

(At STP, H_2O is in the liquid state.)

$$\Delta H^\circ_{\text{rxn}} = \Delta H^\circ_f CO_2(g) + 2\Delta H^\circ_f H_2O(l) - \Delta H^\circ_f CH_4(g) - 2\Delta H^\circ_f O_2(g)$$

$$\Delta H^\circ_{\text{rxn}} = -393.5\,\text{kJ} + 2(-285.83\,\text{kJ}) - (-74.8\,\text{kJ}) - 0 = -890.4\,\text{kJ}$$

$$\frac{-890.4\,\text{kJ}}{1\,\text{mol CH}_4} \times 1.205 \times 10^{11}\,\text{mol CH}_4 = -1.073 \times 10^{14} = -1.1 \times 10^{14}\,\text{kJ}$$

The negative sign indicates heat evolved by the combustion reaction.

10.126   (a) $\text{ft}^3\,CH_4 \rightarrow L\,CH_4 \rightarrow \text{mol}\,CH_4 \rightarrow \text{mol}\,CH_3OH \rightarrow g\,CH_3OH \rightarrow L\,CH_3OH$

$$10.7 \times 10^9\,\text{ft}^3\,CH_4 \times \frac{1\,\text{yd}^3}{3^3\,\text{ft}^3} \times \frac{1\,\text{m}^3}{(1.0936)^3\,\text{yd}^3} \times \frac{1\,\text{L}}{1 \times 10^{-3}\,\text{m}^3} = 3.03001 \times 10^{11}$$

$$= 3.03 \times 10^{11}\,\text{L CH}_4$$

$$n = \frac{PV}{RT} = \frac{3.03 \times 10^{11}\,\text{L} \times 1.00\,\text{atm}}{298\,\text{K}} \times \frac{\text{mol-K}}{0.08206\,\text{L-atm}} = 1.2391 \times 10^{10}$$

$$= 1.24 \times 10^{10}\,\text{mol CH}_4$$

1 mol CH_4 = 1 mol CH_3OH

$$1.2391 \times 10^{10} \text{ mol } CH_3OH \times \frac{32.04 \text{ g } CH_3OH}{\text{mol } CH_3OH} \times \frac{1 \text{ mL } CH_3OH}{0.791 \text{ g}} \times \frac{1 \text{ L}}{1000 \text{ mL}}$$

$$= 5.0189 \times 10^8 = 5.02 \times 10^8 \text{ L } CH_3OH$$

(b) $CH_4(g) + 2 O_2(g) \rightarrow CO_2(g) + 2 H_2O(l)$ $\Delta H° = -890.4$ kJ

$\Delta H° = \Delta H_f° \, CO_2(g) + 2 \, \Delta H_f° \, H_2O(l) - \Delta H_f° \, CH_4(g) - 2 \, \Delta H_f° \, O_2(g)$

$\Delta H° = -393.5 \text{ kJ} + 2(-285.83 \text{ kJ}) - (-74.8 \text{ kJ}) - 0 = -890.4 \text{ kJ}$

$$1.2391 \times 10^{10} \text{ mol } CH_4 \times \frac{-890.4 \text{ kJ}}{1 \text{ mol } CH_4} = -1.10 \times 10^{13} \text{ kJ}$$

$CH_3OH(l) + 3/2 \, O_2(g) \rightarrow CO_2(g) + 2 H_2O(l)$ $\Delta H° = -726.6$ kJ

$\Delta H° = \Delta H_f° \, CO_2(g) + 2 \, \Delta H_f° \, H_2O(l) - \Delta H_f° \, CH_3OH(l) - 3/2 \, \Delta H_f° \, O_2(g)$

$\quad\quad = -393.5 \text{ kJ} + 2(-285.83 \text{ kJ}) - (-238.6 \text{ kJ}) - 0 = -726.6 \text{ kJ}$

$$1.2391 \times 10^{10} \text{ mol } CH_3OH \times \frac{-726.6 \text{ kJ}}{1 \text{ mol } CH_3OH} = -9.00 \times 10^{12} \text{ kJ}$$

(c)   Assume a volume of 1.00 L of each liquid.

$$1.00 \text{ L } CH_4(l) \times \frac{466 \text{ g}}{1 \text{ L}} \times \frac{1 \text{ mol}}{16.04 \text{ g}} \times \frac{-890.4 \text{ kJ}}{\text{mol } CH_4} = -2.59 \times 10^4 \text{ kJ/ L } CH_4$$

$$1.00 \text{ L } CH_3OH \times \frac{791 \text{ g}}{1 \text{ L}} \times \frac{1 \text{ mol}}{32.04 \text{ g}} \times \frac{-726.6 \text{ kJ}}{\text{mol } CH_3OH} = -1.79 \times 10^4 \text{ kJ/ L } CH_3OH$$

Clearly $CH_4(l)$ has the higher enthalpy of combustion per unit volume.

10.128   (a) $MgCO_3(s) + 2HCl(aq) \rightarrow MgCl_2(aq) + H_2O(l) + CO_2(g)$

$CaCO_3(s) + 2HCl(aq) \rightarrow CaCl_2(aq) + H_2O(l) + CO_2(g)$

(b) $n = \dfrac{PV}{RT} = 743 \text{ torr} \times \dfrac{1 \text{ atm}}{760 \text{ torr}} \times \dfrac{\text{mol - K}}{0.08206 \text{ L - atm}} \times \dfrac{1.72 \text{ L}}{301 \text{ K}}$

$$= 0.06808 = 0.0681 \text{ mol } CO_2$$

(c) $x = \text{g } MgCO_3, \; y = \text{g } CaCO_3, \; x + y = 6.53 \text{ g}$

$\text{mol } MgCO_3 + \text{mol } CaCO_3 = \text{mol } CO_2 \text{ total}$

$$\frac{x}{84.32} + \frac{y}{100.09} = 0.06808; \; y = 6.53 - x$$

$$\frac{x}{84.32} + \frac{6.53 - x}{100.09} = 0.06808$$

$100.09x - 84.32x + 84.32(6.53) = 0.06808 \, (84.32)(100.09)$

$15.77x + 550.610 = 574.549; \; x = 1.52 \text{ g } MgCO_3$

$$\text{mass \% } MgCO_3 = \frac{1.52 \text{ g } MgCO_3}{6.53 \text{ g sample}} \times 100 = 23.3\%$$

[By strict sig fig rules, the answer has 2 sig figs: 15.77x + 551 (3 digits from 6.53) = 575; 575 − 551 = 24 (no decimal places, 2 sig figs) leads to 1.5 g $MgCO_3$ and 23% $MgCO_3$]

11 Liquids and Intermolecular Forces

Visualizing Concepts

In this chapter we will use the temperature units °C and K interchangeably when designating specific heats and *changes* in temperature.

11.2   (a)   (i)   Hydrogen bonding; H–F interactions qualify for this narrowly defined interaction.

              (ii)   London dispersion forces, the only intermolecular forces between nonpolar F_2 molecules.

              (iii)   Ion-dipole forces between Na^+ cation and the negative end of a polar covalent water molecule.

              (iv)   Dipole-dipole forces between oppositely charged portions of two polar covalent SO_2 molecules.

        (b)   London dispersion forces in (ii) are probably the weakest.

11.3   (a)   The viscosity of glycerol will be greater than that of 1-propanol.

        (b)   Viscosity is the resistance of a substance to flow. The stronger the intermolecular forces in a liquid, the greater its viscosity. Hydrogen bonding is the predominant force for both molecules. Glycerol has three times as many –OH groups and many more hydrogen-bonding interactions than 1-propanol, so it experiences stronger intermolecular forces and greater viscosity. (Both molecules have the same carbon-chain length, so dispersion forces are similar.)

11.5   (a)   385 mm Hg. Find 30 °C on the horizontal axis, and follow a vertical line from this point to its intersection with the red vapor pressure curve. Follow a horizontal line from the intersection to the vertical axis and read the vapor pressure.

        (b)   22 °C. Reverse the procedure outlined in part (a). Find 300 torr on the vertical axis, follow it to the curve and down to the value on the horizontal axis.

        (c)   47 °C. The normal boiling point of a liquid is the temperature at which its vapor pressure is 1 atm, or 760 mm Hg. A vapor pressure of 1 atm is very near the top of this diagram, at approximately 47 °C.

11.7   (a)   360 K, the normal boiling point; 260 K, normal freezing point. The left-most line is the freezing/melting curve, the right-most line is the condensation/boiling curve. The normal boiling and freezing points are the temperatures of boiling and freezing at 1 atm pressure.

     (b)   The material is solid in the left-most green (or pale blue) zone, liquid in the blue zone, and gas in the tan zone. (i) gas; (ii) solid; (iii) liquid.

     (c)   The triple point, where all three phases are in equilibrium, is the point where the three lines on the phase diagram meet. For this substance, the triple point is approximately 185 K at 0.45 atm.

11.8   (a)   The substance is in a liquid crystalline state at temperatures T_1 and T_2. At T_1, the molecules are aligned in layers and the long molecular axes are perpendicular to the layer planes; this describes a smectic A phase. At T_2, the long molecular axes are aligned but the ends are not aligned; this describes a nematic phase.

     (b) T_3 is the highest temperature. The molecular arrangement in this phase has the least order, so it represents the highest temperature. (The molecules are closely packed, but not aligned in any way; this describes an ordinary liquid phase.)

Molecular Comparisons of Gases, Liquids, and Solids (Section 11.1)

11.10   (a)   In solids, particles are in essentially fixed positions relative to each other, so the average energy of attraction is stronger than average kinetic energy. In liquids, particles are close together but moving relative to each other. The average attractive energy and average kinetic energy are approximately balanced. In gases, particles are far apart and in constant, random motion. Average kinetic energy is much greater than average energy of attraction.

     (b)   As the temperature of a substance is increased, the average kinetic energy of the particles increases. In a collection of particles (molecules), the state is determined by the strength of interparticle forces relative to the average kinetic energy of the particles. As the average kinetic energy increases, more particles are able to overcome intermolecular attractive forces and move to a less ordered state, from solid to liquid to gas.

     (c)   If a gas is placed under very high pressure, the particles undergo many collisions with the container and with each other. The large number of particle-particle collisions increases the likelihood that intermolecular attractions will cause the molecules to coalesce (liquefy).

11.12   (a) $Ar < CCl_4 < Si$. Intermolecular energy of attraction increases from gas to liquid to solid. (See Solution 11.10.)

     (b) $Ar < CCl_4 < Si$. Boiling point increases as intermolecular energy of attraction increases. The greater the intermolecular attractive energy among particles, the greater the kinetic energy and temperature required to overcome this attractive energy and produce the less ordered gaseous state.

11.14   (a)   The average distance between molecules is greater in the liquid state. Density is the ratio of the mass of a substance to the volume it occupies. For the same substance in different states, mass will be the same. The smaller the density, the greater the volume occupied, and the greater the distance between molecules. The liquid at 130° has the lower density (1.08 g/cm^3), so the average distance between molecules is greater.

(b)   As the temperature of a substance increases, the average kinetic energy and speed of the molecules increases. At the melting point (mp) the molecules, on average, have enough kinetic energy to break away from the very orderly array that was present in the solid. As the translational motion of the molecules increases, the occupied volume increases and the density decreases. Thus, the solid density, 1.266 g/cm^3 at 15 °C, is greater than the liquid density, 1.08 g/cm^3 at 130 °C.

Intermolecular Forces (Section 11.2)

11.16   (a)   *Intra*molecular interactions are generally stronger than *inter*molecular interactions. That is, interactions within molecules, such as chemical bonds, are stronger than interactions between molecules.

        (b)   Intermolecular interactions are broken when a liquid is converted to a gas.

11.18   (a) CH_3OH experiences hydrogen bonding, but CH_3SH does not.

        (b)   Both gases are influenced by London dispersion forces. The heavier the gas particles, the stronger the London dispersion forces. The heavier Xe is a liquid at the specified conditions, whereas the lighter Ar is a gas.

        (c)   Both gases are influenced by London dispersion forces. The larger, diatomic Cl_2 molecules are more polarizable, experience stronger dispersion forces, and have the higher boiling point.

        (d)   Acetone and 2-methylpropane are molecules with similar molar masses and London dispersion forces. Acetone also experiences dipole-dipole forces and has the higher boiling point.

11.20   (a)   True. A more polarizable molecule can develop a larger transient dipole, increasing the strength of electrostatic attractions and dispersion forces among molecules.

        (b)   False. The noble gases are all monoatomic. Going down the family, the atomic radius and the size of the electron cloud increase. The larger the electron cloud, the more polarizable the atom, the stronger the London dispersion forces, and the higher the boiling point. (In general, strength of forces and boiling point vary in the same direction, not opposite directions.)

        (c)   False. Generally, dipole-dipole forces are stronger than dispersion forces for molecules of similar size and mass. The size of the molecule and the magnitude of its dipole moment (if there is one) determine the relative magnitudes of dispersion and dipole-dipole forces.

        (d)   True. For molecules with similar molecular weights and elemental composition, linear molecules have the possibility for greater contact along and around their surfaces than spherical molecules. Their electron clouds are thus more polarizable, and dispersion forces are greater.

11.22   For molecules with similar structures, the strength of dispersion forces increases with molecular size (molecular weight and number of electrons in the molecule).

        (a) Br_2

(b) $CH_3CH_2CH_2CH_2CH_2SH$

(c) $CH_3CH_2CH_2Cl$. These two molecules have the same molecular formula and molecular weight (C_3H_7Cl, molecular weight = 78.5 amu), so the shapes of the molecules determine which has the stronger dispersion forces. According to Figure 11.6, the cylindrical (not branched) molecule will have stronger dispersion forces.

11.24   Both molecules experience hydrogen bonding through their –OH groups and dispersion forces between their hydrocarbon portions. The position of the –OH group in isopropyl alcohol shields it somewhat from approach by other molecules and slightly decreases the extent of hydrogen bonding. Also, isopropyl alcohol is less rod-like (it has a shorter chain) than propyl alcohol, so dispersion forces are weaker. Because hydrogen bonding and dispersion forces are weaker in isopropyl alcohol, it has the lower boiling point.

11.26   (a)   HF has the higher boiling point because hydrogen bonding is stronger than dipole-dipole forces.

(b) $CHBr_3$ has the higher boiling point because it has the higher molar mass, which leads to greater polarizability and stronger dispersion forces.

(c)   ICl has the higher boiling point because it is a polar molecule. For molecules with similar structures and molar masses, dipole-dipole forces are stronger than dispersion forces.

11.28   (a) C_4H_{10}. Both molecules experience dispersion forces. C_4H_{10} has the higher boiling point due to greater molar mass and similar strength of forces.

(b) $CH_3CH_2CH_2CH_2OH$ has the higher boiling point due to the influence of hydrogen bonding.

(c) SO_3. This is a tough call; SO_2 has dipole-dipole forces; SO_3 has dispersion forces but a larger molecular weight. The relative strength of dispersion and dipole-dipole forces depends on the mass and shape of the molecules. SO_3 molecules have greater molecular weight and are planar, so alignment is facile and dispersion forces are strong; SO_3 has the higher boiling point (confirmed by CRC *Handbook of Chemistry and Physics*).

(d) Cl_2CO has the higher boiling point due to greater molecular weight and stronger dispersion forces. (Note that H_2CO does not have hydrogen bonding, because the H atoms are bound to C, not to O.)

11.30   (a)   In the solid state, NH_3 molecules are arranged so as to form the maximum number of hydrogen bonds. At the melting point, the average kinetic energy of the molecules is large enough so that they are free to move relative to each other. As they move, old hydrogen bonds break and new ones form, but the strict relative order required for maximum hydrogen bonding is no longer present.

(b)   In the liquid state, molecules are moving relative to one another while touching, which makes some hydrogen bonding possible. When molecules achieve enough kinetic energy to vaporize, the distance between them increases beyond the point where hydrogen bonds can form.

11.32 The longer the alkyl side chain of the 1-alkyl-3-methylimidazolium cation, the more irregular the shape of the cation. Particles with irregular shapes are more difficult to pack into solids, so the melting point of the salt decreases as the length of the alkyl group and irregularity increases. (This is called a steric effect.)

Select Properties of Liquids (Section 11.3)

11.34 The order of increasing strength of intermolecular forces is also the order of increasing viscosity and surface tension

(a) $CH_3CH_2CH_3 < CH_2Cl_2 < CH_3CH_2OH$

(b) $CH_3CH_2CH_3 < CH_2Cl_2 < CH_3CH_2OH$

(c) $CH_3CH_2CH_3 < CH_2Cl_2 < CH_3CH_2OH$

11.36 (a)

$$H-\ddot{N}-\ddot{N}-H \qquad H-\ddot{O}-\ddot{O}-H \qquad H-\ddot{O}-H$$
$$\quad\ \ |\quad\ |$$
$$\quad\ \ H\ \ H$$

(b) All have bonds (N–H or O–H, respectively) capable of forming hydrogen bonds. Hydrogen bonding is the strongest intermolecular interaction between neutral molecules and leads to very strong cohesive forces in liquids. The stronger the cohesive forces in a liquid, the greater the surface tension.

11.38 (a) For molecules with similar shapes, viscosity usually decreases with decreasing molecular weight. Because n-pentane has one fewer carbon atom and a shorter chain than n-hexane, the molecules are slightly more free to move around each other and n-pentane will have the smaller viscosity.

(b) According to Figure 11.6, neopentane is roughly spherical, whereas n-pentane is cylindrical or rod shaped. The spherical neopentane has weaker dispersion forces and the molecules are more free to tumble, so it will have the smaller viscosity.

Phase Changes (Section 11.4)

11.40 (a) condensation, exothermic

(b) sublimation, endothermic

(c) vaporization (evaporation), endothermic

(d) freezing, exothermic

11.42 (a) Liquid ethyl chloride at room temperature is far above its boiling point. When the liquid contacts the room temperature surface, heat sufficient to vaporize the liquid is transferred from the surface to the ethyl chloride, and the heat content of the molecules increases. At constant atmospheric pressure, $\Delta H = q$, so the heat content and the enthalpy content of $C_2H_5Cl(g)$ are higher than that of $C_2H_5Cl(l)$. This indicates that the specific heat of the gas is less than that of the liquid, because the heat content of the gas starts at a higher level.

(b)   Liquid C_2H_5Cl is vaporized (boiled), $C_2H_5Cl(g)$ is warmed to the final temperature, and the solid surface is cooled to the final temperature. The enthalpy of vaporization (ΔH_{vap}) of $C_2H_5Cl(l)$, the specific heat of $C_2H_5Cl(g)$, and the specific heat of the solid surface must be considered.

11.44   Energy released when 200 g of H_2O is cooled from 15 °C to 0 °C:

$$\frac{4.184\,J}{g\text{-}K} \times 200\,g\,H_2O \times 15\,°C = 12.55 \times 10^3\,J = 13\,kJ$$

Energy released when 200 g of H_2O is frozen (there is no change in temperature during a change of state):

$$\frac{334\,J}{g} \times 200\,g\,H_2O = 6.68 \times 10^4\,J = 66.8\,kJ$$

Total energy released = 12.55 kJ + 66.8 kJ = 79.35 = 79.4 kJ

Mass of freon that will absorb 79.4 kJ when vaporized:

$$79.35\,kJ \times \frac{1 \times 10^3\,J}{1\,kJ} \times \frac{1\,g\,CCl_2F_2}{289\,J} = 275\,g\,CCl_2F_2$$

11.46   Consider the process in steps, using the appropriate thermochemical constant.

Heat the liquid from 10.00 °C to 47.6 °C, $\Delta T = 37.6\,°C = 37.6\,K$, using the specific heat of the liquid.

$$35.0\,g\,C_2Cl_3F_3 \times \frac{0.91\,J}{g\text{-}K} \times 37.6\,K \times \frac{1\,kJ}{1000\,J} = 1.1976 = 1.2\ kJ$$

Boil the liquid at 47.6 °C (320.6 K), using the enthalpy of vaporization.

$$35.0\,g\,C_2Cl_3F_3 \times \frac{1\,mol\,C_2Cl_3F_3}{187.4\,g\,C_2Cl_3F_3} \times \frac{27.49\,kJ}{mol} = 5.1342 = 5.13\ kJ$$

Heat the gas from 47.6 °C to 105.00 °C, $\Delta T = 57.4\,°C = 57.4\,K$, using the specific heat of the gas.

$$35.0\,g\,C_2Cl_3F_3 \times \frac{0.67\,J}{g\text{-}K} \times 57.4\,K \times \frac{1\,kJ}{1000\,J} = 1.3460 = 1.3\ kJ$$

The total energy required is 1.1967 kJ + 5.1342 kJ + 1.3460 kJ = 7.6778 = 7.7 kJ.

11.48   (a) CCl_3F, CCl_2F_2, and $CClF_3$ are polar molecules that experience dipole-dipole and London dispersion forces with like molecules. CF_4 is a nonpolar compound that experiences only dispersion forces.

(b)   According to Solution 11.47(b), the higher the critical temperature, the stronger the intermolecular attractive forces of a substance. Therefore, the strength of intermolecular attraction increases moving from right to left across the series and as molecular weight increases. $CF_4 < CClF_3 < CCl_2F_2 < CCl_3F$.

(c)   The increasing intermolecular attraction with increasing molecular weight indicates that the critical temperature and pressure of CCl_4 will be greater than that of CCl_3F. Looking at the numerical values in the series, an increase of 88 K in critical temperature and 3.1 atm in critical pressure to the corresponding values for CCl_3F seem reasonable.

Physical Property	CCl_3F	CCl_4(predicted)	CCl_4 (CRC)
Critical Temperature (K)	471	557	556.6
Critical Pressure (atm)	43.5	46.6	44.6

The predicted values for CCl_4 are in very good agreement with literature values. The key concept is that dispersion, not dipole-dipole, forces dominate the physical properties in the series.

Vapor Pressure (Section 11.5)

11.50   A normal boiling point of 56 °C places the liquid-vapor curve for acetone between the curves for diethyl ether and ethanol on Figure 11.25. Following a vertical line of increasing vapor pressure at 25 °C, we first cross the ethanol curve, then the (virtual) acetone curve. This means that, at 25 °C, the vapor pressure of acetone is higher than the vapor pressure of ethanol. (The lower boiling point of acetone is a strong indicator that it will have a higher vapor pressure than ethanol at a given temperature.)

11.52   (a)   False. The heavier (and larger) CBr_4 has stronger dispersion forces, a higher boiling point, lower vapor pressure, and is less volatile.

(b)   True.

(c)   False.

(d)   False.

11.54   (a)   On a humid day, there are more gaseous water molecules in the air and more are recaptured by the surface of the liquid, making evaporation slower.

(b)   At high altitude, atmospheric pressure is lower and water boils at a lower temperature. This lower boiling temperature at high altitude means that cooking an egg takes longer.

11.56   (a)

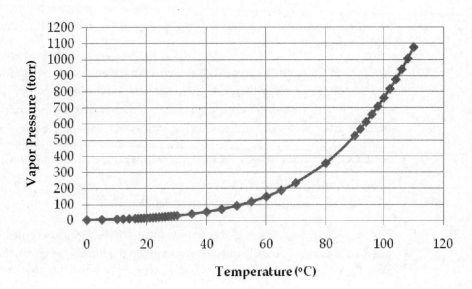

A plot of vapor pressure versus temperature data for H_2O from Appendix B is shown here. The vapor pressure of water at body temperature, 37 °C, is approximately 50 torr.

(b) The data point at 760.0 torr, 100 °C is the normal boiling point of H_2O. This is the temperature at which the vapor pressure of H_2O is equal to a pressure of 1 atm or 760 torr.

(c) At an external (atmospheric) pressure of 633 torr, the boiling point of H_2O is approximately 96 °C.

(d) At an external pressure of 774 torr, the boiling point of water is approximately 100.5 °C.

(e) Follow the logic in Sample Exercise 10.13 to calculate rms speeds at the two temperatures. The rms speed is one way to represent the "average" speed of a large collection of particles.

$u = (3RT/MM)^{1/2}$. MM = 18.0 g/mol = 18.0×10^{-3} kg/mol

R = 8.314 kg - m^2/s^2 - mol - K. At T = 96 °C = 369 K,

$$u = \left(\frac{3(8.314 \text{ kg - m}^2/s^2 \text{ - mol - K}) \, 369 \text{ K}}{18.0 \times 10^{-3} \text{ kg/mol}} \right)^{1/2} = 715 \text{ m/s}$$

At T = 100.5 °C = 373.6 K,

$$u = \left(\frac{3(8.314 \text{ kg - m}^2/s^2 \text{ - mol - K}) \, 373.6 \text{ K}}{18.0 \times 10^{-3} \text{ kg/mol}} \right)^{1/2} = 719.50 = 720 \text{ m/s}$$

The difference in the two rms speeds is less than 1% of the average of the two values. Given the precision of estimating boiling temperatures from the plot, the two rms speeds are essentially equal.

Phase Diagrams (Section 11.6)

11.58 (a) The *triple point* on a phase diagram represents the temperature and pressure at which the gas, liquid, and solid phases are in equilibrium.

(b) No. A phase diagram represents a closed system, one where no matter can escape and no substance other than the one under consideration is present; air cannot be present in the system. Even if air is excluded, at 1 atm of external pressure, the triple point of water is inaccessible, regardless of temperature [see Figure 11.28].

11.60 (a) Solid CO_2 sublimes to form $CO_2(g)$ at a temperature of about –60 °C.

(b) Solid CO_2 melts to form $CO_2(l)$ at a temperature of about –55 °C. The $CO_2(l)$ boils when the temperature reaches approximately –45 °C.

11.62 (a) The normal boiling point is the temperature where liquid becomes gas at 1 atm pressure. Moving vertically down from 1 atm on the liquid-gas line to the temperature axis, the normal boiling point is approximately 27 to 28 K, or –246 °C to –245 °C.

(b) The much higher critical temperature and pressure of Ar (150.9 K, 48 atm) compared with those of Ne (25 K, 0.43 atm), indicate that Ar experiences much stronger intermolecular forces than Ne.

11.64   The density of Ga(s), 5.91 g/cm^3, is less than the density of Ga(l), 6.1 g/cm^3, just above the melting temperature. "Typically" the density of a solid is greater than the density of its liquid. Gallium is then an atypical substance, like water, where the solid state is denser and more compact than the liquid. This results in a backward sloping solid-liquid line on the phase diagram for water, and we also expect to see this unusual feature on the diagram for gallium.

Liquid Crystals (Section 11.7)

11.66   Reinitzer observed that cholesteryl benzoate has a phase that exhibits properties intermediate between those of the solid and liquid phases. This "liquid-crystalline" phase, formed by melting at 145 °C, is viscous and opaque; its viscosity decreases on heating and it becomes clear at 179 °C.

11.68   (a)   The molecule has one double bond (and one triple bond).

        (b)   The "LCD molecule" is long relative to its thickness. It has one C=C and one C≡N group that promote rigidity and polarizability along the length of the molecule. The C≡N group also provides dipole-dipole interactions that encourage alignment. Unlike the molecules in Figure 11.33, which contain planar phenyl rings, the LCD molecule contains nonaromatic, nonplanar six-membered rings. These rings are subject to substantial London dispersion effects. They probably contribute to specific physical properties such as the liquid crystal temperature range that make this molecule particularly functional in LCD displays.

11.70   A nematic phase is composed of sheets of molecules aligned along their lengths, but with no additional order within the sheet or between sheets. A cholesteric phase also contains this kind of sheet, but with some ordering between sheets. In a cholesteric phase, there is a characteristic angle between molecules in one sheet and those in an adjacent sheet. That is, one sheet of molecules is twisted at some characteristic angle relative to the next, producing a "screw" axis perpendicular to the sheets.

11.72   In the nematic phase, molecules are aligned in one dimension, the long dimension of the molecule. In a smectic phase (A or C), molecules are aligned in two dimensions. Not only are the long directions of the molecules aligned, but the ends are also aligned. The molecules are organized into layers; the height of the layer is related to the length of the molecule.

Additional Exercises

11.74

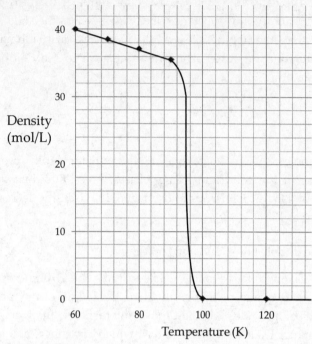

The normal boiling point of O_2 is between 90 and 100 K, approximately 95 K.

11.75   (a)   Correct.

(b)   The lower boiling liquid must experience less total intermolecular forces.

(c)   If both liquids are structurally similar nonpolar molecules, the lower boiling liquid has a lower molecular weight than the higher boiling liquid.

(d)   Correct.

(e)   At their boiling points, both liquids have vapor pressures of 760 mm Hg.

11.77   (a)   Increase

(b)   Decrease

(c)   The magnitude of the dispersion forces increases from to CH_2F_2 to CH_2Cl_2 to CH_2Br_2 because polarizability increases going from fluorine to chlorine to bromine. Also, the dipole-dipole contribution decreases in the series because molecular polarity (dipole moment) decreases.

11.79   The GC base pair, with more hydrogen bonds, is more stable to heating. To break up a base pair by heating, sufficient thermal energy must be added to break the existing hydrogen bonds. With 50% more hydrogen bonds, the GC pair is definitely more stable (harder to break apart) than the AT pair.

11.80   The two O–H groups in ethylene glycol are involved in many hydrogen-bonding interactions, leading to its high boiling point and viscosity, relative to pentane, which experiences only dispersion forces.

11.82   Ionic liquids are the liquid phase of ionic compounds. Upon melting, the ions are free to move relative to one another. The ion-ion interparticle attractive forces at work in an ionic liquid are extremely strong relative to dispersion, dipole-dipole, and even hydrogen-bonding forces operating in most molecular solvents. These powerful ion-ion forces must be broken for an ion to escape to the vapor phase. In the distribution of particle energies at room temperature, very few ions have sufficient kinetic energy to escape these interactions and move to the vapor phase. With few particles in the vapor phase, the vapor pressures of ionic liquids are extremely low.

11.84   (a)   If the Clausius-Clapeyron equation is obeyed, a graph of ln P versus 1/T (K^{-1}) should be linear. Here are the data in a form for graphing.

T (K)	1/T	P (torr)	ln P
280.0	3.571×10^{-3}	32.42	3.479
300.0	3.333×10^{-3}	92.47	4.527
320.0	3.125×10^{-3}	225.1	5.417
330.0	3.030×10^{-3}	334.4	5.812
340.0	2.941×10^{-3}	482.9	6.180

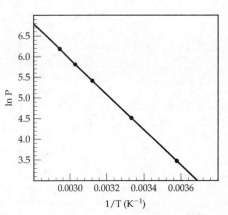

According to the graph, the Clausius-Clapeyron equation is obeyed, to a first approximation.

$$\Delta H_{vap} = -\text{slope} \times R; \ \text{slope} = \frac{3.479 - 6.180}{(3.571 - 2.941) \times 10^{-3}} = -\frac{2.701}{0.630 \times 10^{-3}} = -4.29 \times 10^3$$

$$\Delta H_{vap} = -(-4.29 \times 10^3) \times 8.314 \ \text{J/K - mol} = 35.7 \ \text{kJ/mol}$$

(b)   The normal boiling point is the temperature at which the vapor pressure of the liquid equals atmospheric pressure, 760 torr. From the graph,

ln 760 = 6.63, 1/T for this vapor pressure = 2.828×10^{-3}; T = 353.6 K

11.85   (a)   The Clausius-Clapeyron equation is $\ln P = \dfrac{-\Delta H_{vap}}{RT} + C$.

For two vapor pressures, P_1 and P_2, measured at corresponding temperatures T_1 and T_2, the relationship is

$$\ln P_1 - \ln P_2 = \left(\frac{-\Delta H_{vap}}{RT_1} + C \right) - \left(\frac{-\Delta H_{vap}}{RT_2} + C \right)$$

$$\ln P_1 - \ln P_2 = \frac{-\Delta H_{vap}}{R} \left(\frac{1}{T_1} - \frac{1}{T_2} \right) + C - C; \ \ln \frac{P_1}{P_2} = \frac{-\Delta H_{vap}}{R} \left(\frac{1}{T_1} - \frac{1}{T_2} \right)$$

(b) P_1 = 13.95 torr, T_1 = 298 K; P_2 = 144.78 torr, T_2 = 348 K

$$\ln \frac{13.95}{144.78} = \frac{-\Delta H_{vap}}{8.314 \ \text{J/mol - K}} \left(\frac{1}{298} - \frac{1}{348} \right)$$

$$-2.33974 \, (8.314 \ \text{J/mol - K}) = -\Delta H_{vap} \, (4.821 \times 10^{-4} / \text{K})$$

$$\Delta H_{vap} = 4.035 \times 10^4 = 4.0 \times 10^4 \ \text{J/mol} = 40 \ \text{kJ/mol}$$

$[(1/T_1) - (1/T_2)]$ has 2 sig figs and so does the result.

(c) The normal boiling point of a liquid is the temperature at which the vapor pressure of the liquid is 760 torr.

P_1 = 144.78 torr, T_1 = 348 K; P_2 = 760 torr, T_2 = bp of octane

$$\ln\left(\frac{144.78}{760.0}\right) = \frac{-4.035 \times 10^4 \text{ J/mol}}{8.314 \text{ J/mol-K}}\left(\frac{1}{348 \text{ K}} - \frac{1}{T_2}\right)$$

$$\frac{-1.6581}{-4.8533 \times 10^3} = 2.874 \times 10^{-3} - \frac{1}{T_2}; \quad \frac{1}{T_2} = 2.874 \times 10^{-3} - 3.416 \times 10^{-4}$$

$$1/T_2 = 2.532 \times 10^{-3} = 2.53 \times 10^{-3}; \quad T_2 = 395 \text{ K } (122 \,^{\circ}\text{C})$$

From the plot of boiling point versus number of carbon atoms in Solution 11.81, we read an approximate boiling point for octane of 130 °C. These two temperatures are close, but do differ by more than 5%. Considering experimental uncertainties in the vapor pressure (vp) data, and the empirical nature of the plot, the two values are surprisingly close. The literature boiling point of octane, 126 °C, is exactly midway between our two estimates.

(d) P_1 = vp of octane at –30 °C, T_1 = 243 K; P_2 = 144.78 torr, T_2 = 348 K

$$\ln\frac{P_1}{144.78 \text{ } torr} = \frac{-4.035 \times 10^4 \text{ J/mol}}{8.314 \text{ J/mol-K}}\left(\frac{1}{243} - \frac{1}{348}\right)$$

$$\ln\frac{P_1}{144.78 \text{ } torr} = \frac{-4.035 \times 10^4 \text{ J/mol}}{8.314 \text{ J/mol-K}} \times 1.242 \times 10^{-3} = -6.026 = -6.03$$

$$\frac{P_1}{144.78 \text{ torr}} = e^{-6.026}; \quad P_1 = 0.002415(144.78) = 0.3496 = 0.35 \text{ torr}$$

[This result has 2 sig figs because (ln = –6.03) has 2 decimal places. In a ln or log, the places left of the decimal show order of magnitude, and places right of the decimal show sig figs in the real number.] The result, 0.35 torr at –30 °C, is reasonable, because we expect vapor pressure to decrease as temperature decreases, and we are approaching the freezing point of octane, –57 °C.

11.86 Physical data for the two compounds from the *Handbook of Chemistry and Physics*:

	MM	dipole moment	boiling point
CH_2Cl_2	85 g/mol	1.60 D	40.0 °C
CH_3I	142 g/mol	1.62 D	42.4 °C

(a) The two substances have very similar molecular structures; each is an unsymmetrical tetrahedron with a single central carbon atom and no hydrogen bonding. Because the structures are very similar, the magnitudes of the dipole-dipole forces should be similar. This is verified by their very similar dipole moments. The heavier compound, CH_3I, will have slightly stronger London dispersion forces. Because the nature and magnitude of the intermolecular forces in the two compounds are nearly the same, it is very difficult to predict which will be more volatile [or which will have the higher boiling point as in part (b)].

(b)   Given the structural similarities discussed in part (a), one would expect the boiling points to be very similar, and they are. Based on its larger molar mass (and dipole-dipole forces being essentially equal) one might predict that CH_3I would have a slightly higher boiling point; this is verified by the known boiling points.

(c)   According to Equation [11.1], $\ln P = \dfrac{-\Delta H_{vap}}{RT} + C$

A plot of ln P versus 1/T for each compound is linear. Because the order of volatility changes with temperature for the two compounds, the two lines must cross at some temperature; the slopes of the two lines, ΔH_{vap} for the two compounds, and the y-intercepts, C, must be different.

(d)

CH_2Cl_2			**CH_3I**		
ln P	**T (K)**	**1/T**	**ln P**	**T (K)**	**1/T**
2.303	229.9	4.351×10^{-3}	2.303	227.4	4.398×10^{-3}
3.689	250.9	3.986×10^{-3}	3.689	249.0	4.016×10^{-3}
4.605	266.9	3.747×10^{-3}	4.605	266.2	3.757×10^{-3}
5.991	297.3	3.364×10^{-3}	5.991	298.5	3.350×10^{-3}

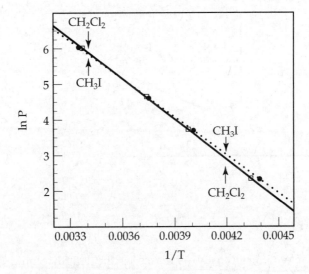

For CH_2Cl_2, $-\Delta H_{vap}/R$ = slope = $\dfrac{(5.991 - 2.303)}{(3.364 \times 10^{-3} - 4.350 \times 10^{-3})} = \dfrac{-3.688}{0.987 \times 10^{-3}}$

$$= -3.74 \times 10^{3} = -\Delta H_{vap}/R$$

$\Delta H_{vap} = 8.314\,(3.74 \times 10^{3}) = 3.107 \times 10^{4}$ J/mol = 31.1 kJ/mol

For CH_3I, $-\Delta H_{vap}/R$ = slope = $\dfrac{(5.991 - 2.303)}{(3.350 \times 10^{-3} - 4.398 \times 10^{-3})} = \dfrac{-3.688}{1.048 \times 10^{-3}} = -3.519 \times 10^{3}$

$$= -\Delta H_{vap}/R$$

$\Delta H_{vap} = 8.314\,(3.519 \times 10^{3}) = 2.926 \times 10^{4}$ J/mol = 29.3 kJ/mol

11.87   The normal melting point (nmp) and normal boiling point (nbp) are at a pressure of
1 atm. The triple point (tp) occurs at 1000 Pa. Change this pressure to atm.

$$1000 \times \frac{1 \text{ atm}}{1.01325 \times 10^5 \text{ Pa}} = 0.009869 = 9.87 \times 10^{-3} = 1 \times 10^{-2} \text{ atm}$$

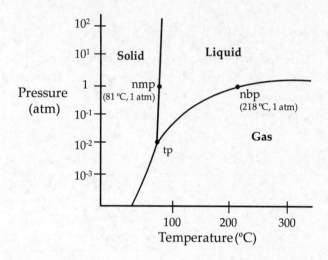

With no information about the critical point of naphthalene, we cannot denote it or the
supercritical fluid region.

11.89

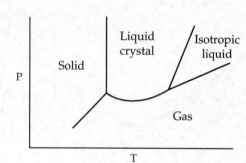

Integrative Exercises

11.90   (a)   In Table 11.5, viscosity increases as the length of the carbon chain increases.
Longer molecular chains become increasingly entangled, increasing resistance to
flow.

(b)   Whereas viscosity depends on molecular chain length in a critical way, surface
tension depends on the strengths of intermolecular interactions between molecules. These dispersion forces do not increase as rapidly with increasing chain
length and molecular weight as viscosity does.

(c)   The –OH group in n-octyl alcohol gives rise to hydrogen bonding among
molecules, which increases molecular entanglement and leads to greater viscosity
and higher boiling point.

11.91   (a)   24 valence e^-, 12 e^- pairs

$$
\begin{array}{ccccc}
\text{H} & & :\!\text{O}: & & \text{H} \\
| & & \| & & | \\
\text{H}-\!\!\!\!& \text{C}-\!\!\!\!& \text{C}-\!\!\!\!& \text{C}-\!\!\!\!& \text{H} \\
| & & & & | \\
\text{H} & & & & \text{H}
\end{array}
$$

The geometry around the central C atom is trigonal planar, and around the two terminal C atoms, tetrahedral.

      (b)   Polar. The C=O bond is quite polar and the dipoles in the trigonal plane around the central C atom do not cancel.

      (c)   Dipole-dipole and London dispersion forces

      (d)   Because the molecular weights of acetone and 1-propanol are similar, the strength of the London dispersion forces in the two compounds is also similar. The big difference is that 1-propanol has hydrogen bonding, whereas acetone does not. These relatively strong attractive forces lead to the higher boiling point for 1-propanol.

11.93   Ethanol will evaporate until its vapor fills the flask at a pressure of 40.0 torr. Calculate the mass of 2.00 L of ethanol vapor at 19 $^{\circ}$C and a pressure of 40.0 torr. Subtract this mass from the original 1.00 g to find the mass of liquid ethanol remaining.

$$
T = 19\,^{\circ}\text{C} + 273.15 = 292 \text{ K}; \; MM = 46.07 \text{ g/mol}; \; 40.0 \text{ torr} \times \frac{1 \text{ atm}}{760 \text{ torr}} = 0.052632 = 0.0526 \text{ atm}
$$

$$
g = \frac{MM \times PV}{RT}; \; g = \frac{46.07 \text{ g ethanol}}{1 \text{ mol ethanol}} \times \frac{\text{mol-K}}{0.08206 \text{ L-atm}} \times \frac{0.052632 \text{ atm}}{292 \text{ K}} \times 2.0 \text{ L} = 0.202 \text{ g vapor}
$$

g ethanol liquid = 1.00 g – 0.202 g vapor = 0.798 = 0.80 g

11.94   (a)   For butane to be stored as a liquid at temperatures above its boiling point (–5 °C), the pressure in the tank must be greater than atmospheric pressure. In terms of the phase diagram of butane, the pressure must be high enough so that, at tank conditions, the butane is "above" the gas-liquid line and in the liquid region of the diagram.

The pressure of a gas is described by the ideal-gas law as P = nRT/V; pressure is directly proportional to moles of gas. The more moles of gas present in the tank the greater the pressure, until sufficient pressure is achieved for the gas to liquefy. At the point where liquid and gas are in equilibrium and temperature is constant, liquid will vaporize or condense to maintain the equilibrium vapor pressure. That is, as long as some liquid is present, the gas pressure in the tank will be constant.

      (b)   If butane gas escapes the tank, butane liquid will vaporize (evaporate) to maintain the equilibrium vapor pressure. Vaporization is an endothermic process, so the butane will absorb heat from the surroundings. The temperature of the tank and the liquid butane will decrease.

(c) $250 \, g \, C_4H_{10} \times \dfrac{1 \, mol \, C_4H_{10}}{58.12 \, g \, C_4H_{10}} \times \dfrac{21.3 \, kJ}{mol} = 91.6 \, kJ$

$V = \dfrac{nRT}{P} = 250 \, g \times \dfrac{1 \, mol}{58.12 \, g} \times \dfrac{0.08206 \, L\text{-}atm}{mol\text{-}K} \times \dfrac{308 \, K}{755 \, torr} \times \dfrac{760 \, torr}{1 \, atm} = 109.44 = 109 \, L$

11.95   *Plan.*

(i)   Using thermochemical data from Appendix B, calculate the energy (enthalpy) required to melt and heat the H_2O.

(ii)   Using Hess's Law, calculate the enthalpy of combustion, ΔH_{comb}, for C_3H_8.

(iii)   Solve the stoichiometry problem.

Solve.

(i)   Heat $H_2O(s)$ from $-20 \, ^\circ C$ to $0.0 \, ^\circ C$; $5.50 \times 10^3 \, g \, H_2O \times \dfrac{2.092 \, J}{g\text{-}^\circ C} \times 20 \, ^\circ C = 2.301 \times 10^2$

$= 2.3 \times 10^2 \, kJ$

Melt $H_2O(s)$; $5.50 \times 10^3 \, g \, H_2O \times \dfrac{6.008 \, kJ}{mol \, H_2O} \times \dfrac{1 \, mol \, H_2O}{18.02 \, g \, H_2O} = 1834 = 1.83 \times 10^3 \, kJ$

Heat $H_2O(l)$ from $0 \, ^\circ C$ to $75 \, ^\circ C$; $5.50 \times 10^3 \, g \, H_2O \times \dfrac{4.184 \, J}{g\text{-}^\circ C} \times 75 \, ^\circ C = 1726 = 1.7 \times 10^3 \, kJ$

Total energy $= 230.1 \, kJ + 1834 \, kJ + 1726 \, kJ = 3790 = 3.8 \times 10^3 \, kJ$

(The result is significant to 100 kJ, limited by $1.7 \times 10^3 \, kJ$)

(ii) $C_3H_8(g) + 5O_2(g) \rightarrow 3CO_2(g) + 4H_2O(l)$

Assume that one product is $H_2O(l)$, because this leads to a more negative ΔH_{comb} and fewer grams of $C_3H_8(g)$ required.

$\Delta H_{comb} = 3\Delta H_f^\circ \, CO_2(g) + 4\Delta H_f^\circ \, H_2O(l) - \Delta H_f^\circ \, C_3H_8(g) - 5\Delta H_f^\circ \, O_2(g)$

$= 3(-393.5 \, kJ) + 4(-285.83 \, kJ) - (-103.85 \, kJ) - 5(0) = -2219.97 = -2220 \, kJ$

(iii) $3.790 \times 10^3 \, kJ \, required \times \dfrac{1 \, mol \, C_3H_8}{2219.97 \, kJ} \times \dfrac{44.096 \, g \, C_3H_8}{1 \, mol \, C_3H_8} = 75 \, g \, C_3H_8$

$(3.8 \times 10^3 \, kJ$ required has 2 sig figs and so does the result)

11.97   *Plan.* Relative humidity and vp of H_2O at given $T \rightarrow P_{H_2O} \rightarrow$ ideal-gas law $\rightarrow$ mol $H_2O(g) \rightarrow H_2O$ molecules. Change $^\circ F \rightarrow ^\circ C$, volume of room from $ft^3 \rightarrow L$.

Solve. $^\circ C = 5/9 \, (^\circ F - 32)$; $^\circ C = 5/9 \, (68 \, ^\circ F - 32) = 20 \, ^\circ C$;

$rh = (P_{H_2O}$ in air $/$ vp of $H_2O) \times 100$

From Appendix B, vp of H_2O at $20 \, ^\circ C = 17.54 \, torr$

P_{H_2O} in air $= rh \times$ vp of $H_2O/100 = 58 \times 17.54 \, torr/100 = 10.173 = 10 \, torr$

$$V = 12 \text{ ft} \times 10 \text{ ft} \times 8 \text{ ft} \times \frac{12^3 \text{ in}^3}{\text{ft}^3} \times \frac{2.54^3 \text{ cm}^3}{\text{in}^3} \times \frac{1 \text{ L}}{1000 \text{ cm}^3} = 2.718 \times 10^4 = 3 \times 10^4 \text{ L}$$

(The result has 1 sig fig, as does the measurement 8 ft.)

$PV = nRT$; $n = PV/RT$

$$n = 10.173 \text{ torr} \times \frac{1 \text{ atm}}{760 \text{ torr}} \times \frac{\text{mol - K}}{0.08206 \text{ L - atm}} \times \frac{2.718 \times 10^4 \text{ L}}{293 \text{ K}} = 15.13 = 2 \times 10^1 \text{ mol H}_2\text{O}$$

$$15.13 \text{ mol H}_2\text{O} \times \frac{6.022 \times 10^{23} \text{ molecules}}{1 \text{ mol}} = 9.112 \times 10^{24} = 9 \times 10^{24} \text{ H}_2\text{O molecules}$$

12 Solids and Modern Materials

Visualizing Concepts

12.2   When choosing a unit cell, remember that the environment of each lattice point must be identical and that unit cells must *tile* to generate the complete two-dimensional lattice. For a given structure, there are often several ways to draw a unit cell. We will select the unit cell with higher symmetry (more $90°$ or $120°$ angles) and smaller area ($a \times b$).

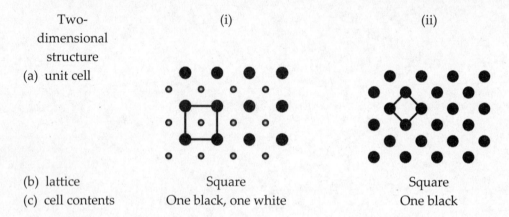

	(i)	(ii)
Two-dimensional structure (a) unit cell		
(b) lattice	Square	Square
(c) cell contents	One black, one white	One black

12.4   Arrangement (b) is more stable. In (a), the cation is so small that its neighboring anions are nearly touching. Very close contacts among like-charged particles produce strong electrostatic repulsions and an unstable arrangement.

12.6   (a)   Band A is the valence band.

    (b)   Band B is the conduction band.

    (c)   Band A (the valence band) consists of bonding molecular orbitals (MOs).

    (d)   This is the electronic structure of a p-type doped semiconductor. The electronic structure shows a few empty MOs or "positive" holes in the valence band. This fits the description of a p-type doped semiconductor.

    (e)   The dopant is Ga. Of the three elements listed, only Ga has fewer valence electrons than Ge, the requirement for a p-type dopant.

12.8   The smaller the nanocrystals, the greater the band gap, E_g, and the shorter the wavelength of emitted light. According to Figure 6.4, the shortest visible wavelengths appear violet or purple and the longest appear red.

    (a)   The 4.0 nm nanocrystals will have the smallest E_g, and emit the longest wavelength. That describes vial 2, the red one.

    (b)   The 2.8 nm nanocrystals will have the largest E_g, and emit the shortest wavelength. That describes vial 1, the green one.

(c)   The band gap of the 100 nm CdTe crystals is 1.5 eV.

$$\lambda = \frac{hc}{E} = \frac{6.626 \times 10^{-34} \text{ J-s} \times 2.998 \times 10^8 \text{ m}}{\text{s}} \times \frac{1}{1.5 \text{ eV}} \times \frac{1 \text{ eV}}{1.602 \times 10^{-19} \text{ J}} = 8.27 \times 10^{-7} \text{ m}$$

$$\upsilon = \frac{c}{\lambda} = \frac{2.998 \times 10^8 \text{ m/s}}{8.27 \times 10^{-7} \text{ m}} = 3.63 \times 10^{14} \text{ s}^{-1}$$

The visible portion of the electromagnetic spectrum has wavelengths up to 750 nm (7.50×10^{-7} m). The 100 nm CdTe crystals emit a wavelength longer than this, 827 nm (8.27×10^{-7} m). This light is in the IR portion of the spectrum and is not visible to the human eye.

Classification of Solids (Section 12.1)

12.10   (a)   Covalent-network solid. According to Figure 12.1, diamond and silicon are covalent-network solids. Individual atoms are bound into a three-dimensional network by strong covalent (single) bonds.

        (b)   Covalent-network solid. The physical properties could describe a covalent-network or ionic solid. Silicon and oxygen are both nonmetals, so their bonding is likely to be covalent.

12.12   (a)   metallic

        (b)   molecular or metallic (physical properties of metals vary widely)

        (c)   covalent-network or ionic

        (d)   covalent-network

12.14   (a)   InAs—covalent-network (InAs is a compound semiconductor, with an average of four electrons per atom and somewhat polar covalent bonds.)

        (b)   MgO—ionic crystal (metal and nonmetal)

        (c)   HgS—ionic crystal (metal and nonmetal)

        (d)   In—metallic (metal, Figure 7.13)

        (e)   HBr—molecular (two nonmetals)

12.16   According to Figure 12.1, the solid could be either ionic with low water solubility or covalent-network. Because of the extremely high sublimation temperature, it is probably covalent-network.

Structures of Solids (Section 12.2)

12.18   Statement (e) is the best explanation for the difference in density. In amorphous silica (SiO_2) the regular structure of quartz is disrupted; the loose, disordered structure, has many vacant "pockets" throughout. There are fewer SiO_2 groups per volume in the amorphous solid; the packing is less efficient and less dense.

12.20

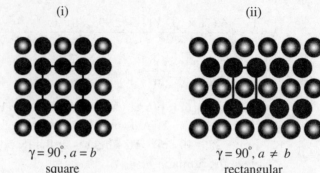

Two-dimensional structure
(a) unit cell

(i)   (ii)

(b) γ, a, b   γ = 90°, a = b   γ = 90°, a ≠ b
(c) lattice type   square   rectangular

12.22   Tetragonal, $a = b \neq c$, $\alpha = \beta = \gamma = 90°$. In the new lattice, one of the edge lengths is shorter than (not the same as) the other two, and all angles remain 90°.

12.24   (c). If all three lattice vectors have the same length, $a = b = c$. This is characteristic of the rhombohedral as well as the cubic lattice.

12.26   (d). A face-centered cubic lattice is composed of face-centered cubic unit cells. A unit cell contains the minimum number of atoms when it has atoms only at the lattice points. A face-centered cubic unit cell like this is shown in Figure 12.12(c). There is one atom centered on each face ($6 \times 1/2$) and one at each corner ($8 \times 1/8$) for a total of 4 atoms in the unit cell. (Only metallic elements have face-centered cubic lattices and unit cells.)

12.28   (a) $a = b = 3.98$ Å. $c = 3.47$ Å $\neq a$ or b. $\alpha = \beta = \gamma = 90°$. This is a tetragonal unit cell and crystal lattice.

(b)   There is one Sr atom totally inside the cell, Fe atoms at each corner ($8 \times 1/8$), and O atoms centered on eight of the unit cell edges ($8 \times 1/4$). The unit cell contains 1 Sr, 1 Fe, and 2 O atoms. The empirical formula is $SrFeO_2$.

Metallic Solids (Section 12.3)

12.30   Metallic: (b) NiCo alloy and (c) W. The lattices of these substances are composed of neutral metal atoms. Delocalization of valence electrons produces metallic properties.

Not metallic: (d) Ge is a metalloid, not a metal. (a) $TiCl_4$ and (e) ScN are ionic compounds; in ionic compounds, electrons are localized on the individual ions, precluding metallic properties.

12.32   (a)   The density of a crystalline solid is (unit cell mass/unit cell volume). Solve for unit cell volume, then use geometry and the properties of a body-centered cubic unit cell to calculate the atomic radius of sodium. There are 2 Na atoms in each body-centered cubic unit cell (Figure 12.12).

$$V = \frac{\text{unit cell mass}}{\rho} = \frac{2\,\text{Na atoms} \times 22.99\,\text{g Na}}{6.022 \times 10^{23}\,\text{Na atoms}} \times \frac{cm^3}{0.97\,\text{g}} \times \frac{1\,\text{Å}^3}{(10^{-8}\,\text{cm})^3} = 78.71 = 79\,\text{Å}^3$$

For a cubic unit cell, $V = a^3$. We need the relationship between atomic radius and unit cell edge length for a body-centered cubic unit cell. In a body-centered cubic metal structure, the atoms touch along the body diagonal, d_2. Then, $d_2 = 4$ r. From the Pythagorean theorem, $d_2 = \sqrt{3}\ a$. (See Sample Exercise 12.2.)

$$a = (V)^{1/3} = (78.71)^{1/3} = 4.2857 = 4.3 \text{ Å}.$$

$$d_2 = 4\,r_{Na}\ ;\quad d_2 = \sqrt{3}\,a;\quad r_{Na} = \sqrt{3}\,a/4 = \frac{\sqrt{3} \times 4.2857\ \text{Å}}{4} = 1.8557 = 1.9\ \text{Å}$$

(b)   A cubic close-packed metal structure has a face-centered cubic unit cell; there are 4 atoms in each unit cell and atoms touch along the face diagonal, d_1. Then, $d_1 = 4\,r$. From the Pythagorean theorem, $d_1 = \sqrt{2}\,a$.

$$d_1 = 4\,r_{Na}\ ;\quad d_1 = \sqrt{2}\,a;\quad a = 4\,r_{Na}\ /\sqrt{2} = \frac{4 \times 1.8557\ \text{Å}}{\sqrt{2}} = 5.2489 = 5.2\ \text{Å} = 5.2 \times 10^{-8}\,\text{cm}$$

$$\rho = \frac{4\ \text{Na atoms}}{(5.2489 \times 10^{-8}\ \text{cm})^3} \times \frac{22.99\ \text{g Na}}{6.022 \times 10^{23}\ \text{Na atoms}} = 1.056 = 1.1\,\text{g/cm}^3$$

Sodium metal with a cubic close-packed structure would not float on water.

12.34   (a)   In a body-centered cubic unit cell, there is one atom totally inside the unit cell (1×1) and one atom at each of the eight corners ($8 \times 1/8$), for a total of 2 Ca atoms in each unit cell.

   (b)   Because atoms are only at lattice points in a body-centered metal structure, each metal atom has an equivalent environment. That is, atoms at the corner of the cell and the interior atom must have equivalent environments. Consider the Ca atom at the middle of the unit cell. It has eight nearest neighbors, the eight Ca atoms at the corners of the cell. Interior atoms in adjacent unit cells are farther from the reference Ca atom and are not "nearest." (For a corner Ca atom, the eight nearest neighbors are the interior atoms in the eight unit cells that include 1/8 of that corner atom.)

   (c)   In a body-centered cubic metal structure, the atoms touch along the the body diagonal, d_2. Then, $d_2 = 4\,r$. From the Pythagorean theorem, $d_2 = \sqrt{3}\,a$. (See Sample Exercise 12.2.)

$$4\,r_{Ca} = \sqrt{3}\,a;\quad a = 4\,r_{Ca}\ /\sqrt{3} = \frac{4 \times 1.97\ \text{Å}}{\sqrt{3}} = 4.5495 = 4.55\ \text{Å}$$

   (d) $$\rho = \frac{2\ \text{Ca atoms}}{(4.5495 \times 10^{-8}\ \text{cm})^3} \times \frac{40.08\ \text{g Ca}}{6.022 \times 10^{23}\ \text{Ca atoms}} = 1.4136 = 1.41\,\text{g/cm}^3$$

12.36   *Analyze.* Given the atomic arrangement, length of the cubic unit cell edge, and density of the solid, calculate the atomic weight of the element. *Plan.* If we calculate the mass of a single unit cell, and determine the number of atoms in one unit cell, we can calculate the mass of a single atom and of a mole of atoms.   *Solve.*

The volume of the unit cell is $(2.86 \times 10^{-8}\ \text{cm})^3$. The mass of the unit cell is:

$$\frac{7.92\ \text{g}}{\text{cm}^3} \times \frac{(2.86 \times 10^{-8})^3\ \text{cm}^3}{\text{unit cell}} = 1.853 \times 10^{-22}\ \text{g/unit cell}$$

There are two atoms of the element present in the body-centered cubic unit cell. Thus the atomic weight is:

$$\frac{1.853 \times 10^{-22}\ \text{g}}{\text{unit cell}} \times \frac{1\ \text{unit cell}}{2\ \text{atoms}} \times \frac{6.022 \times 10^{23}\ \text{atoms}}{1\ \text{mol}} = 55.8\ \text{g/mol}$$

Check. The result is a reasonable atomic weight and the units are correct. The element could be iron.

12.38   (a)   False. Substitutional and interstitial alloys are both solution alloys.

         (b)   True

         (c)   True

12.40   (a) $Cu_{0.66}Zn_{0.34}$, substitutional alloy; similar atomic radii, substantial amounts of both components

         (b) Ag_3Sn, intermetallic compound; set stoichiometric ratio of components

         (c) $Ti_{0.99}O_{0.01}$, interstitial alloy; very different atomic radii, tiny amount of smaller component

12.42   (a)   True

         (b)   True

         (c)   False. In stainless steel, the chromium atoms replace iron atoms in the structure. (The atomic radii of Fe and Cr are 1.25 Å and 1.27 Å, respectively. Metal atoms with similar radii form substitutional alloys.)

12.44   Purple gold is $AuAl_2$. The composition is not variable, because it is a compound, not an alloy, which is a mixture. Because of their ordered structures, intermetallic compounds tend to be higher melting and more brittle, but less malleable and ductile than alloys. These properties make them hard to manipulate into shapes and are the reason purple gold is not used to make rings or necklaces.

Metallic Bonding (Section 12.4)

12.46   The part of the bar sitting in the dark will feel hot. The structure of the metal and its delocalized electrons extend over the whole bar. The high thermal conductivity of the metal facilitates heat transfer from the part of the bar in the sun to the part of the bar in the shade.

12.48

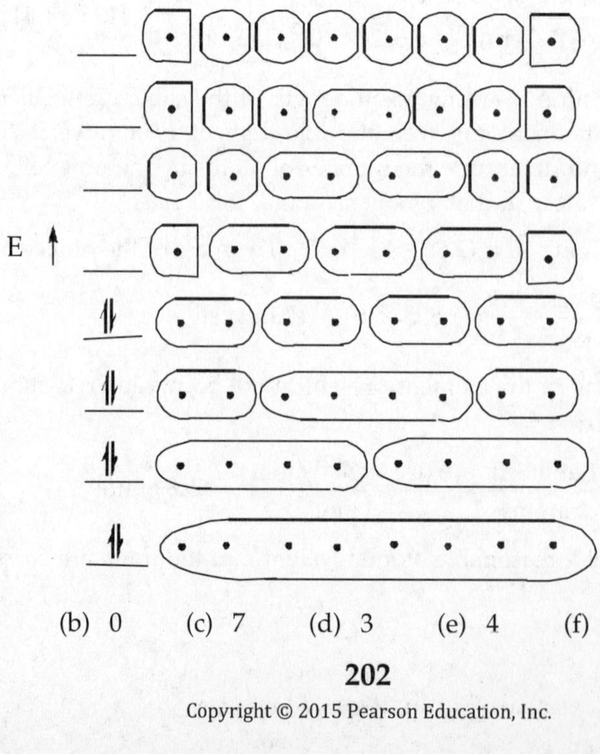

         (a) 8   (b) 0   (c) 7   (d) 3   (e) 4   (f) smaller

12.50   Statement (d) is false and does not follow from the fact that alkali metals have relatively weak metal-metal bonds. We expect strong metal-metal bonds to be present in metals with high melting points (Figure 12.21).

12.52   In each group, choose the metal that has the number of valence electrons closest to six.

   (a)   Re   (b)   Mo   (c)   Ru

Ionic and Molecular Solids (Sections 12.5 and 12.6)

12.54   (a)   Ti: 8 corners $\times$ 1/8 sphere/corner + [1 center $\times$ 1 sphere/center] = 2 Ti atoms

      O: 4 faces $\times$ 1/2 sphere/face + [2 interior $\times$ 1 sphere/interior] = 4 O atoms

      Formula: TiO_2

   (b)   From inspection of the unit cell diagram, the Ti at the center of the cell is bound to six (6) O atoms. Each totally interior O atom is bound to three titanium atoms. The coordination number of Ti is 6; the coordination number of O is 3.

12.56   Calculate the mass of a single unit cell and then use density to find the volume of a single unit cell. The edge length is the cube root of the volume of a cubic cell. From the previous exercise, there are four PbSe units in a NaCl-type unit cell. The unit cell edge length is designated a.

$$8.27 \text{ g/ cm}^3 = \frac{4 \text{ PbSe units}}{a^3} \times \frac{286.2 \text{ g}}{6.022 \times 10^{23} \text{ PbSe units}} \times \left(\frac{1 \text{ Å}}{1 \times 10^{-8} \text{ cm}}\right)^3$$

$a^3 = 229.87 \text{ Å}^3, a = 6.13 \text{ Å}$

12.58   (a)   Each cell edge goes through two half Rb^+ ions (at the corners) and one full I^- ion (centered on the edge). The length of an edge, a, is then

      $a = 2r_{Rb} + 2r_I = 2(1.66 \text{ Å}) + 2(2.06 \text{ Å}) = 7.44 \text{ Å}$

   (b) $\text{density} = \dfrac{4 \text{ RbI units}}{(7.44 \text{ Å})^3} \times \dfrac{212.37 \text{ g}}{6.022 \times 10^{23} \text{ RbI units}} \times \left(\dfrac{1 \text{ Å}}{1 \times 10^{-8} \text{ cm}}\right)^3 = 3.43 \text{ g/cm}^3$

   (c)   In the CsCl-type structure, I^- anions sit at the corners of the cube and an Rb^+ cation sits completely inside the unit cell (at or near the body center). Assume that the anions and cations touch along the body diagonal (bd), so that

      bd $= 2r_{Rb} + 2r_I = 2(1.66 \text{ Å}) + 2(2.06 \text{ Å}) = 7.44 \text{ Å}$. (This is the edge length of the NaCl-type structure.)

      From Solution 12.34, the relationship between the body diagonal and edge of a cube is: bd $= \sqrt{3} \times a$; $a = bd/\sqrt{3}$. $a = 7.44 \text{ Å}/\sqrt{3} = 4.2955 = 4.30 \text{ Å}$

   (d)   There is one RbI unit ($8 \times 1/8$ I^- anions and 1 Rb^+ cation) in the CsCl-type unit cell.

      $\text{density} = \dfrac{1 \text{ RbI unit}}{(4.2955 \text{ Å})^3} \times \dfrac{212.37 \text{ g}}{6.022 \times 10^{23} \text{ RbI units}} \times \left(\dfrac{1 \text{ Å}}{1 \times 10^{-8} \text{ cm}}\right)^3 = 4.45 \text{ g/cm}^3$

      The density of the CsCl-type structure is greater than the density of the NaCl-type structure, owing to the much smaller unit cell volume.

12.60   (a)   In CaF_2 the ionic radii are very similar, Ca^{2+} (r = 1.14 Å) and F^- (r = 1.19 Å). In ZnF_2 the cation radius is smaller and the ionic radii are more different, Zn^{2+} (r = 0.88 Å) and F^- (r = 1.19 Å). Cations in both structures in the exercise are shown with equal radii, so direct inspections does not answer the question. We can, however, refer to the trend that, for compounds with the same cation/anion ratio, as cation size decreases, coordination number of the anion decreases. The anion coordination number (CN) for the rutile structure is 3 and for the fluorite structure is 4. The compound with the smaller cation, ZnF_2, will adopt the rutile structure and the compound with the larger cation, CaF_2, will adopt the fluorite structure. (Detailed analysis of coordination numbers follows.)

   (b)   Rutile (top) structure, cation (blue) coordination number (CN) = 6, anion (green) CN = 3. [The blue cation in the interior or the cell is coordinated to the six (green) anions associated with the same cell; either of the green interior anions is associated with the triangle of blue cations located inside and at the two nearest corners of the cell.]

         Fluorite (bottom) structure, cation (blue) CN = 8, anion (green) CN = 4. [A blue cation at one of the face centers is coordinated to the four nearest green anions inside the cell and to four identical anions in an adjacent cell that also contains the cation. Any of the green interior anions is coordinated to a tetrahedron of blue cations located at one corner and the three nearest face centers.]

12.62 $$\frac{\text{\# of cations per formula unit}}{\text{\# of anions per formula unit}} = \frac{\text{anion coordination number}}{\text{cation coordination number}}$$

$$\text{cation CN} = \frac{\text{anion CN} \times \text{\# of anions per formula unit}}{\text{\# of cations per formula unit}}$$

   (a) AlF_3:   1 cation, 3 anions, anion CN = 2; cation CN = $(2 \times 3)/1 = 6$

   (b) Al_2O_3:   2 cations, 3 anions, anion CN = 6; cation CN = $(6 \times 3)/2 = 9$

   (c)   AlN: 1 cation, 1 anion, anion CN = 4; cation CN = $(4 \times 1)/1 = 4$

12.64   (a)   False. Melting point is a bulk property, not a molecular property.

   (b)   True. Strengths of intermolecular forces determine properties of the bulk material like melting point.

Covalent-Network Solids (Section 12.7)

12.66   The extended network of localized covalent bonds in a covalent-network solid produces solids that are inflexible, hard, and high-melting. The delocalized nature of metallic bonding leads to flexibility, because atoms can move relative to one another without "breaking" bonds. Metals have a wide range of hardness and melting point, depending on the occupancy of the bands.

   (a)   Ductility, metallic solid

   (b)   Hardness, covalent-network solid and metallic solid, depending on the strength of metallic bonding

   (c)   High melting point, covalent-network solid and metallic solid, depending on the strength of metallic bonding

12.68   (a)   InP. P is in the same family and higher than As.

        (b)   AlP. Al and P are horizontally separated, resulting in greater bond polarity, and Al and P are in the row above Ge. Band gap values in Table 12.4 confirm this order.

        (c)   AgI. The four elements are in the same row; Ag and I are farther apart than Cd and Te.

12.70   p-type semiconductors have a slight electron deficit. If the dopant replaces As, Group 5A, it should have fewer than five valence electrons. The dopant will be a 4A element, probably Si or Ge. Si would be closer to As in bonding atomic radius.

12.72   (a)   From Table 12.4, the band gap of CdTe is 1.50 eV (or 145 kJ/mol).

        (b) $\lambda = \dfrac{hc}{E} = \dfrac{6.626 \times 10^{-34} \text{ J-s} \times 2.998 \times 10^8 \text{ m}}{s} \times \dfrac{1}{1.50 \text{ eV}} \times \dfrac{1 \text{ eV}}{1.602 \times 10^{-19} \text{ J}}$

$$= 8.267 \times 10^{-7} = 8.27 \times 10^{-7} \text{ m}$$

        (c, d)   CdTe can absorb energies ≥ 1.50 eV or wavelengths $\leq 8.27 \times 10^{-7}$ m. A wavelength of 8.27×10^{-7} m corresponds to 827 nm. CdTe absorbs a smaller range of wavelengths and thus a smaller portion of the solar spectrum than Si.

12.74   The band gap, ΔE, of GaAs is 1.43 eV.

$$\lambda = \frac{hc}{E} = \frac{6.626 \times 10^{-34} \text{ J-s} \times 2.998 \times 10^8 \text{ m}}{s} \times \frac{1}{1.43 \text{ eV}} \times \frac{1 \text{ eV}}{1.602 \times 10^{-19} \text{ J}} = 8.67 \times 10^{-7} \text{ m (867 nm)}$$

    The visible portion of the electromagnetic spectrum has wavelengths up to 750 nm (7.50×10^{-7} m). GaAs emits longer, 867 nm, light in the IR portion of the spectrum.

12.76   Reverse the logic in Solution 12.75. Given λ, calculate E_g. Then solve for x assuming the value of E_g is a linear combination of the stoichiometric contributions of GaP and GaAs. That is, $2.26x + 1.43(1 - x) = E_g$.

$$E_g = \frac{hc}{\lambda} = \frac{6.626 \times 10^{-34} \text{ J-S}}{6.60 \times 10^{-7} \text{ m}} \times \frac{2.998 \times 10^8 \text{ m}}{s} \times \frac{1 \text{ eV}}{1.602 \times 10^{-19} \text{ J}} = 1.8788 = 1.88 \text{ eV}$$

$2.26x + 1.43(1 - x) = 1.8788$; $2.26x - 1.43x + 1.43 = 1.8788$

$0.83x = 0.4488$; $x = 0.5407 = 0.54$

Check. From Solution 12.73, an E_g of 1.85 V corresponds to x = 0.5. E_g =1.88 eV should have a very similar composition, and x = 0.54 is very close to x = 0.5. The P/As composition is very sensitive to wavelength, and provides a useful mechanism to precisely tune the wavelength of the diode.

Polymers (Section 12.8)

12.78   *n*-decane does not have a sufficiently high chain length or molecular mass to be considered a polymer.

12.80 True. In an addition polymerization, all atoms present in the monomer are present in the polymer. (We are assuming 100% yield. However, there is usually some unreacted monomer as well as new polymer.)

12.82

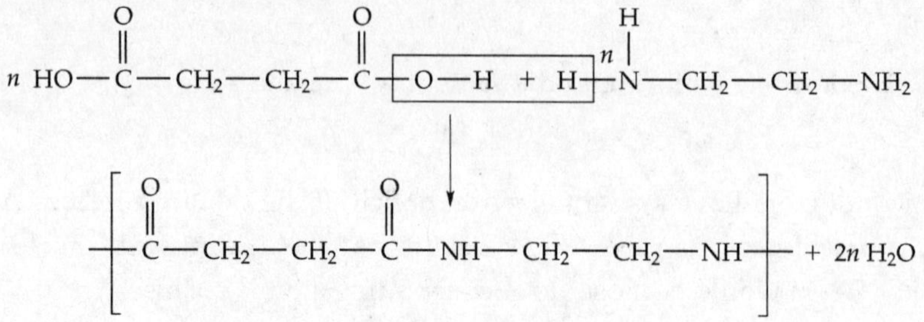

$$n \text{ HO—C—CH}_2\text{—CH}_2\text{—C} \boxed{\text{—O—H } + \text{ H}} \text{—N—CH}_2\text{—CH}_2\text{—NH}_2$$

$$\left[\text{—C—CH}_2\text{—CH}_2\text{—C—NH—CH}_2\text{—CH}_2\text{—NH—} \right] + 2n \text{ H}_2\text{O}$$

12.84 (a) By analogy to polyisoprene, Figure 12.41,

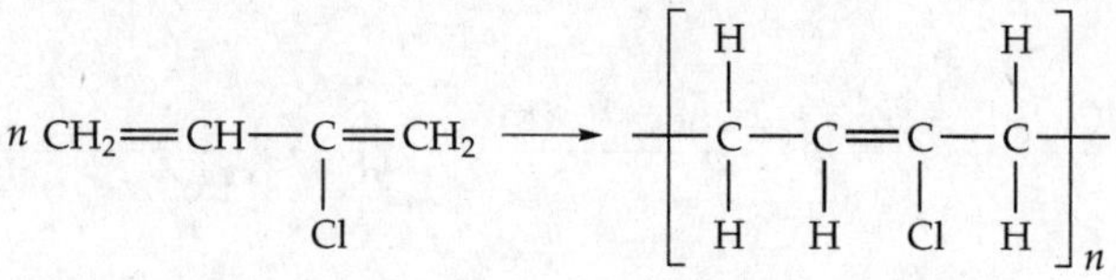

$$n \text{ CH}_2\text{=CH—C=CH}_2 \text{ (Cl)} \longrightarrow \left[\text{—C—C=C—C—} \right]_n$$

(b) 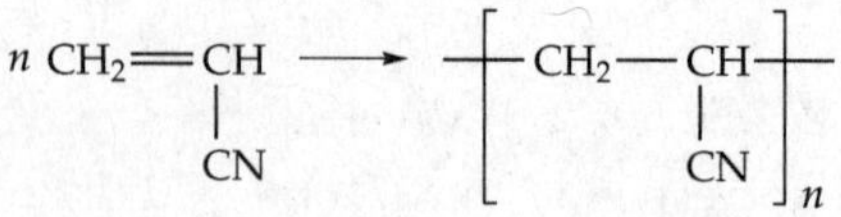
$$n \text{ CH}_2\text{=CH (CN)} \longrightarrow \left[\text{—CH}_2\text{—CH—} \right]_n \text{ (CN)}$$

12.86 (a) 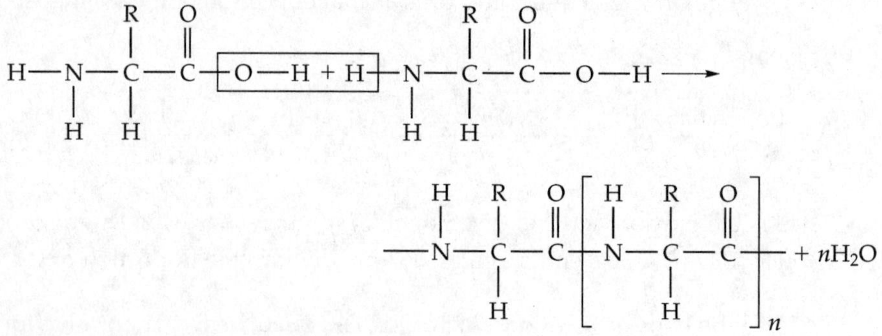
$$\text{H—N—C—C} \boxed{\text{—O—H + H}} \text{—N—C—C—O—H} \longrightarrow$$

$$\text{—N—C—C} \left[\text{—N—C—C—} \right]_n + n\text{H}_2\text{O}$$

(b) 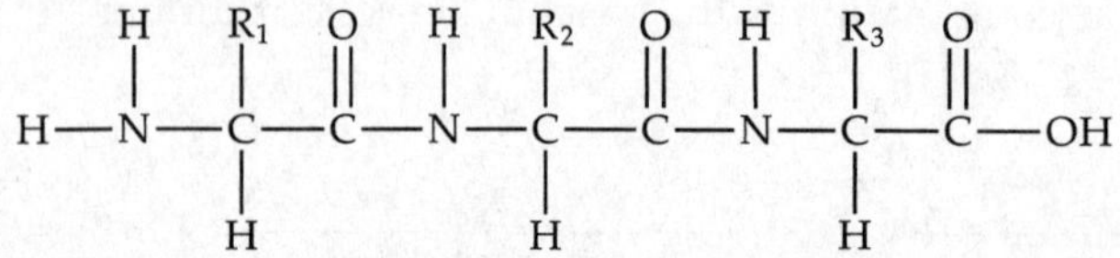
$$\text{H—N—C—C—N—C—C—N—C—C—OH}$$

In the peptide shown here, the N terminus amino acid contains R1 and the C terminus amino acid contains R3.

(c) Six (R1–R2–R3, R2–R3–R1, R3–R1–R2, R1–R3–R2, R3–R2–R1, R2–R1–R3)

12.88 At the molecular level, the longer, unbranched chains of HDPE fit closer together and have more crystalline (ordered, aligned) regions than the shorter, branched chains of LDPE. Closer packing leads to higher density.

12.90 (a) True

(b) True

(c) True

Nanomaterials (Section 12.9)

12.92   (a)   Calculate the wavelength of light that corresponds to 2.4 eV, then look at a visible spectrum such as Figure 6.4 to find the color that corresponds to this wavelength.

$$\lambda = \frac{hc}{E} = \frac{6.626 \times 10^{-34} \text{ J-s} \times 3.00 \times 10^{8} \text{ m}}{\text{s}} \times \frac{1}{2.4 \text{ eV}} \times \frac{1 \text{ eV}}{1.602 \times 10^{-19} \text{ J}}$$

$$= 5.167 \times 10^{-7} = 5.2 \times 10^{-7} \text{ m (520 nm)}$$

The emitted 520 nm light is green.

     (b)   Yes. The wavelength of blue light is shorter than the wavelength of green light. Emitting a shorter wavelength requires a larger band gap. As particle size decreases, band gap increases. Appropriately sized CdS quantum dots, far smaller than a large CdS crystal, would have a band gap greater than 2.4 eV and could emit shorter wavelength blue light.

     (c)   No. The wavelength of red light is greater than the wavelength of green light. The large CdS crystal has plenty of AOs contributing to the MO scheme to ensure maximum delocalization. A bigger crystal will not reduce the size of the band gap. The 2.4 eV band gap of the large CdS crystal represents the minimum energy and maximum wavelength of light that can be emitted by this material.

12.94   True. Blue light has short wavelengths, corresponding to a relatively large band gap. As particle size decreases, band gap increases and wavelength decreases. We could begin with a semiconductor with a smaller band gap and make it a nanoparticle to increase E_g and decrease wavelength. (Nanoparticle size becomes one more way to tune the properties of semiconductors.)

12.96   There are four CdTe formula units in each cubic unit cell. Calculate the number of unit cells contained in a 5.00 nm cubic crystal. The volume of one unit cell is

$$(6.49)^3 \text{ Å}^3 \times \frac{1 \text{ nm}^3}{10^3 \text{ Å}^3} = 0.2734 = 0.273 \text{ nm}^3$$

The volume of the 5.00 nm crystal is $(5.00)^3 \text{ nm}^3 = 125 \text{ nm}^3$

$$\frac{125 \text{ nm}^3 / \text{ crystal}}{0.2734 \text{ nm}^3 / \text{ unit cell}} \times \frac{4 \text{ CdTe units}}{\text{unit cell}} = 1829 = 1.83 \times 10^3 \text{ CdTe units}$$

That is, 1829 Cd atoms and 1829 Te atoms.

12.98   Statement (e) is correct. Statements (b) and (c) both support the notion that buckyballs are discrete molecules and not extended materials.

Additional Exercises

12.99   According to Figure 12.6, a tetragonal unit cell has a square base, with the third lattice vector perpendicular to the base but with a different length. In other words, $a = b \neq c$, $\alpha = \beta = \gamma = 90°$. To create a face-centered tetragonal unit cell, add a lattice point in the center of each face (both square and rectangular faces). Draw a second face-centered tetragonal unit cell above or below the first one.

The square base of a new tetragonal unit cell can be drawn by connecting the face centers of the four rectangular faces. Connecting these four face centers with those of the adjacent (old) tetragonal cell creates the new tetragonal unit cell. The lattice point at the center of the old square face becomes the body center of the new unit cell.

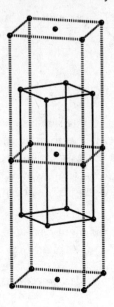

12.100   Qualitatively from Figure 12.12, a face-centered cubic structure has a greater portion of its volume occupied by metal atoms and less "empty" space than a body-centered structure. The face-centered structure will have the greater density.

Quantitatively, use the atomic radius of iron, 1.25 Å from Figure 7.7, to estimate the unit cell edge length in each of the structures, then calculate the estimated density of both structures. Recall that a face-centered cubic structure has 4 atoms per unit cell, and a body-centered structure has 2.

Face-centered cubic: $4\,r_{Fe} = \sqrt{2}\,a$; $a = 4\,r_{Fe}/\sqrt{2} = \dfrac{4 \times 1.25\,\text{Å}}{\sqrt{2}} = 3.5355 = 3.54\,\text{Å}$

$$\rho = \frac{4\,\text{Fe atoms}}{(3.5355 \times 10^{-8}\,\text{cm})^3} \times \frac{55.845\,\text{g Fe}}{6.022 \times 10^{23}\,\text{Fe atoms}} = 8.3934 = 8.39\,\text{g/cm}^3$$

Body-centered cubic: $4\,r_{Fe} = \sqrt{3}\,a$; $a = 4\,r_{Fe}/\sqrt{3} = \dfrac{4 \times 1.25\,\text{Å}}{\sqrt{3}} = 2.8868 = 2.89\,\text{Å}$

$$\rho = \frac{2\,\text{Fe atoms}}{(2.8868 \times 10^{-8}\,\text{cm})^3} \times \frac{55.845\,\text{g Ca}}{6.022 \times 10^{23}\,\text{Ca atoms}} = 7.7098 = 7.71\,\text{g/cm}^3$$

The face-centered cubic structure has the greater density.

12.101   The metallic properties of malleability, ductility, and high electrical and thermal conductivity are results of the delocalization of valence electrons throughout the lattice. Delocalization occurs because metal atom valence orbitals of nearest-neighbor atoms interact to produce nearly continuous molecular orbital energy bands. When C atoms are introduced into the metal lattice, their valence orbitals do not have the same energies as metal orbitals, and the interaction is different. This causes a discontinuity in the band structure and limits delocalization of electrons. The properties of the carbon-infused metal begin to resemble those of a covalent-network lattice with localized electrons. The substance is harder and less conductive than the pure metal.

12.102   Density is the mass of the unit cell contents divided by the unit cell volume [(edge length)^3]. Refer to Figure 12.12(c) to determining the number of Ni atoms and Figure 12.17 for the number of Ni$_3$Al units in each unit cell.

Ni: There are 4 Ni atoms in each face-centered cubic unit cell (8 × 1/8 at the corners, 6 × 1/2 on the face centers).

$$\text{density} = \frac{4 \text{ Ni atoms}}{(3.53 \text{ Å})^3} \times \frac{58.6934 \text{ g Ni}}{6.022 \times 10^{23} \text{ Ni atoms}} \times \left(\frac{\text{Å}}{1 \times 10^{-8} \text{ cm}}\right)^3 = 8.86 \text{ g/cm}^3$$

Ni$_3$Al: There is 1 Ni$_3$Al unit in each cubic unit cell. According to Figure 12.17, Ni is at the face centers (6 × 1/2 = 3 Ni atoms) and Al is at the corners (8 × 1/8 = 1 Al atom); the stoichiometry is correct.

$$\text{density} = \frac{1 \text{ Ni}_3\text{Al unit}}{(3.56 \text{ Å})^3} \times \frac{203.062 \text{ g Ni}_3\text{Al}}{6.022 \times 10^{23} \text{ Ni atoms}} \times \left(\frac{\text{Å}}{1 \times 10^{-8} \text{ cm}}\right)^3 = 7.47 \text{ g/cm}^3$$

The density of the Ni$_3$Al alloy (intermetallic compound) is ~85% of the density of pure Ni. The sizes of the two unit cells are very similar. In Ni$_3$Al, one out of every four Ni atoms is replaced with an Al atom. The mass of an Al atom is (~27 amu) is about half that of a Ni atom (~59 amu); the mass of the unit cell contents of Ni$_3$Al is ~7/8 (87.5%) that of Ni, and the densities show the same relationship.

12.104   (a)   CsCl, primitive cubic lattice (Figure 12.25)

(b)   Au, face-centered cubic lattice (Figure 12.18)

(c)   NaCl, face-centered cubic lattice (Figure 12.25)

(d)   Po, primitive cubic lattice, rare for metals (Section 12.3)

(e)   ZnS, face-centered cubic lattice (Figure 12.25)

12.106   Aluminum, density = 2.70 g/mL, is indeed more dense than silicon, density = 2.33 g/mL. This difference is significant, but not as large as one might expect relative to the difference in electrical conductivity. It is true that under very high pressure, about 12 GPa or 120,000 atm, the structure of silicon changes from the diamond structure to a close-packed structure. The conductivity also changes from that of a semiconductor to that of a metal. The structure change can be detected by monitoring the conductivity (or resistivity) of the material.

In terms of atomic properties, aluminum adopts a metallic (close-packed) structure because it is electron deficient. With only three valence electrons, it is difficult for an aluminum atom to acquire a complete octet by covalent bonding. A close-packed structure (face-centered cubic) provides each aluminum atom with twelve nearest neighbors and the possibility of electron delocalization to satisfy its bonding needs. On the other hand, each silicon atom has four valence electrons and can complete its octet by forming four bonds in a covalent-network structure with localized bonding.

12.108   Semiconductors have a filled valence band and an empty conduction bond, separated by a characteristic difference in energy, the band gap, E$_g$. When a semiconductor is heated, more electrons have sufficient energy to jump the band gap, and conductivity increases. Metals have a partially filled continuous energy band. Heating a metal increases the average kinetic energy of the metal atoms, usually through increased vibrations within the lattice. The greater vibrational energy of the atoms leads to imperfections in the lattice and discontinuities in the energy band. Thermal vibrations create barriers to electron delocalization and reduce the conductivity of the metal.

12.109   (a)   There is a Re atom (gray sphere) at each corner of the unit cell: $(8 \times 1/8 = 1)$. There is an O atom (red sphere) in the middle of each cell edge: $(12 \times 1/4 = 3)$. There are 1 Re and 3 O atoms per unit cell, for an empirical formula of ReO_3.

      (b)   Each cell edge goes through two half Re atoms (at the corners) and one full O atom (centered on the edge). The length of an edge, a, is then

$$a = 2r_{Re} + 2r_O = 2(0.70 \text{ Å}) + 2(1.26 \text{ Å}) = 3.92 \text{ Å}$$

      (c)   The density of a crystalline solid is the mass of the unit cell contents divided by the unit cell volume. There is one ReO_3 unit in each primitive cubic unit cell. The unit cell volume is a^3, $(3.92 \text{ Å})^3$.

$$\frac{1 \, ReO_3 \text{ unit}}{(3.92 \text{ Å})^3} \times \frac{1 \text{ Å}^3}{(10^{-8} \text{ cm})^3} \times \frac{234.205 \text{ g } ReO_3}{6.022 \times 10^{23} \, ReO_3 \text{ units}} = 6.46 \text{ g/cm}^3$$

Check. The units of density are correct.

12.110   (a)

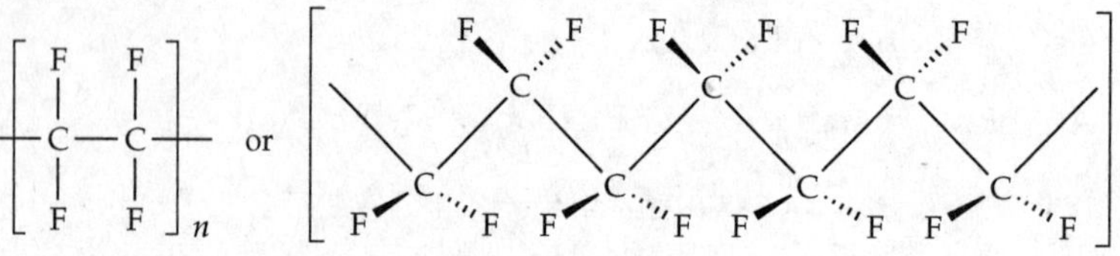

      (b)   Teflon^{TM} is formed by addition polymerization.

12.111

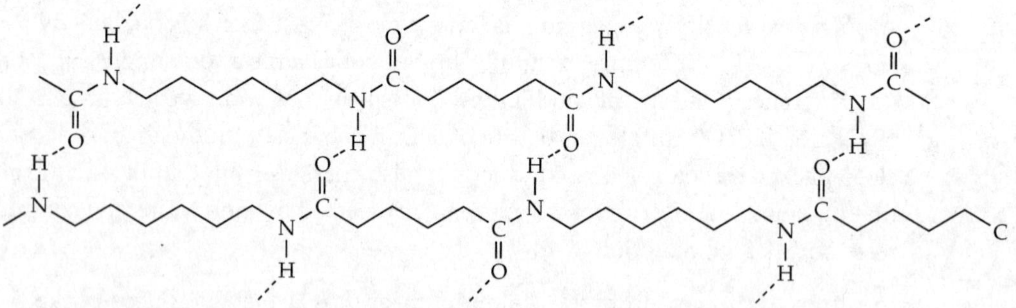

Hydrogen bonding occurs between — amide groups of adjacent chains.

12.112   X-ray *diffraction* is the phenomenon that enables us to measure interatomic distances in crystals. Diffraction is most efficient when the wavelength of light is similar to the size of the object (e.g., the slit) doing the diffracting. Interatomic distances are on the order of 1-10 Å, and the wavelengths of X-rays are also in this range. Visible light has wavelengths of 400-700 nm, or 4000-7000 Å, too long to be diffracted efficiently by atoms (electrons) in crystals.

12.114   Germanium is in the same family but below Si on the periodic chart. This means that Ge will probably have bonding characteristics and crystal structure similar to those of Si. Because Ge has a larger bonding atomic radius than Si, we expect a larger unit cell and d-spacing for Ge. In Bragg's law, $n\lambda = 2d\sin\theta$, d and $\sin\theta$ are inversely proportional. That is, the larger the d-spacing, the smaller the value of $\sin\theta$ and θ. In a diffraction experiment, we expect a Ge crystal to diffract X-rays at a smaller θ-angle than a Si crystal, assuming the X-rays have the same wavelength.

12.115   (a)   Both diamond (d = 3.5 g/cm^3) and graphite (d = 2.3 g/cm^3) are covalent-network solids with efficient packing arrangements in the solid state; there is relatively little empty space in their respective crystal lattices. Diamond, with bonded C–C distances of 1.54 Å in all directions, is more dense than graphite, with shorter C–C distances within carbon sheets but longer 3.35 Å separations between sheets (Figure 12.29). Buckminsterfullerene has much more empty space, both inside each C_{60} "ball" and between balls, than either diamond or graphite, so its density will be considerably less than 2.3 g/cm^3.

(b)   In a face-centered cubic unit cell, there are 4 complete C_{60} units.

$$\frac{4\,C_{60}\text{ units}}{(14.2\text{ Å})^3} \times \frac{720.66\text{ g}}{6.022 \times 10^{23}\,C_{60}\text{ units}} \times \left(\frac{1\text{ Å}}{1 \times 10^{-8}\text{ cm}}\right)^3 = 1.67\text{ g/cm}^3$$

(1.67 g/cm^3 is the smallest density of the three allotropes: diamond, graphite, and buckminsterfullerene.)

12.116   (b)   Increase. Promotion of an electron from the valence band to the conduction band generates a mobile electron in the conduction band and a mobile hole in the valence band. Both phenomena promote delocalization and increase the conductivity of the semiconductor.

Integrative Exercises

12.117   The karat scale is based on mass%, not mol%. In each case, change mol% to mass % Au. Then, karat = mass fraction Au × 24. Determine color using mass% and Figure 12.18.

(a)   Assume 0.50 mol Ag and 0.50 mol Au.

$$0.50\text{ mol Ag} \times \frac{107.87\text{ g Ag}}{\text{mol Ag}} = 53.935 = 54\text{ g Ag}$$

$$0.50\text{ mol Au} \times \frac{196.97\text{ g Au}}{\text{mol Au}} = 98.485 = 98\text{ g Au}$$

$$\text{mass \% Au} = \frac{98.485\text{ g Au}}{(98.485 + 53.935)\text{ g total}} \times 100 = 64.61 = 65\text{ \% Au}$$

karat = 0.6461 × 24 = 15.5064 = 16 karat Au

On Figure 12.18, the Au/Ag alloy line is labeled in terms of mass% Ag. For an alloy that is 65% Au and 35% Ag, the color is greenish yellow.

(b)   Assume 0.50 mol Cu and 0.50 mol Au.

$$0.50 \text{ mol Cu} \times \frac{63.546 \text{ g Cu}}{\text{mol Cu}} = 31.773 = 32 \text{ g Cu}$$

$$0.50 \text{ mol Au} = 98.485 = 98 \text{ g Au}$$

$$\text{mass\% Au} = \frac{98.485 \text{ g Au}}{(98.485 + 31.773) \text{ g total}} \times 100 = 75.61 = 76 \% \text{ Au}$$

$$\text{karat} = 0.7561 \times 24 = 18.15 = 18 \text{ karat Au}$$

On Figure 12.18, for an alloy that is 76% Au and 24% Cu, the color is reddish gold.

12.118   *Analyze.* Given: mass % of Al, Mg, O; density, unit cell edge length. Find: number of each type of atom. *Plan.* We are not given the type of cubic unit cell, primitive, body centered, face centered. So we must calculate the number of formula units in the unit cell, using density, cell volume, and formula weight. Begin by determining the empirical formula and formula weight from mass % data. *Solve.* Assume 100 g spinel.

$$37.9 \text{ g Al} \times \frac{1 \text{ mol Al}}{26.98 \text{ g Al}} = 1.405 \text{ mol Al}; 1.405/0.7036 \approx 2$$

$$17.1 \text{ g Mg} \times \frac{1 \text{ mol Mg}}{24.305 \text{ g Mg}} = 0.7036 \text{ mol Mg}; 0.7036/0.7036 = 1$$

$$45.0 \text{ g O} \times \frac{1 \text{ mol O}}{16.00 \text{ g Al}} = 2.813 \text{ mol O}; 2.813/0.7036 \approx 4$$

The empirical formula is Al_2MgO_4; formula weight = 142.3 g/mol

Calculate the number of formula units per unit cell.

$$8.09 \text{ Å} = 8.09 \times 10^{-10} \text{ m} = 8.09 \times 10^{-8} \text{ cm}; V = (8.09 \times 10^{-8})^3 \text{ cm}^3$$

$$\frac{3.57 \text{ g}}{\text{cm}^3} \times (8.09 \times 10^{-8})^3 \text{ cm}^3 \times \frac{1 \text{ mol}}{142.3 \text{ g}} \times \frac{6.022 \times 10^{23} \text{ units}}{\text{mol}} = 7.999 = 8$$

There are 8 formula units per unit cell, for a total of 16 Al atoms, 8 Mg atoms, and 32 O atoms.

[The relationship between density (d), unit cell volume (V), number of formula units (Z), formula weight (FW), and Avogadro's number (N) is a useful one. It can be rearranged to calculate any single variable, knowing values for the others. For densities in g/cm^3 and unit cell volumes in cm^3 the relationship is Z = (N × d × V)/FW.]

12.120   (a)

HDPE

$$\Delta H = D(C{=}C) - 2D(C{-}C) = 614 - 2(348) = -82 \text{ kJ/mol C}_2\text{H}_4$$

(b) $(n+1)$ HOOC——$(CH_2)_6$——COOH + $(n+1)$ H_2N——$(CH_2)_6$——NH_2 $\longrightarrow$

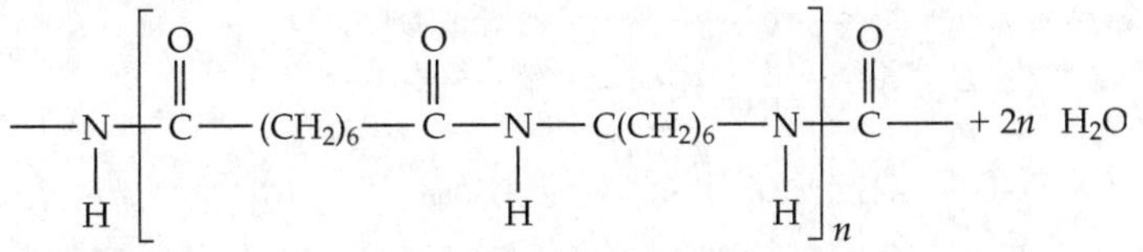

Nylon 6,6

$\Delta H = 2D(C-O) + 2D(N-H) - 2D(C-N) - 2D(H-O)$

$\Delta H = 2(358) + 2(391) - 2(293) - 2(463) = -14$ kJ/mol

(This is −14 kJ/mol of either reactant.)

(c) $(n+1)$ HOOC——⬡——COOH + $(n+1)$ HO——CH_2——CH_2——OH $\longrightarrow$

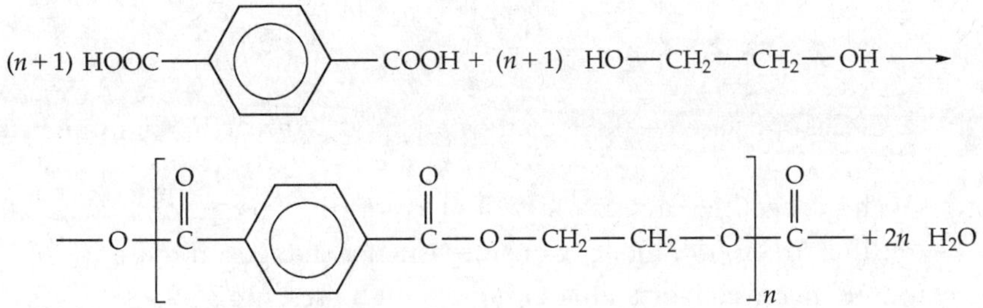

PET

$\Delta H = 2D(C-O) + 2D(O-H) - 2D(C-O) - 2D(O-H) = 0$ kJ

12.121   (a) sp^3 hybrid orbitals at C, 109° bond angles around C

(b)

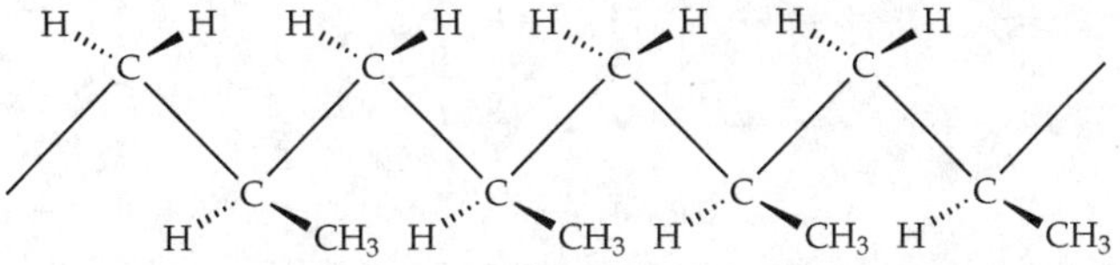

isotactic

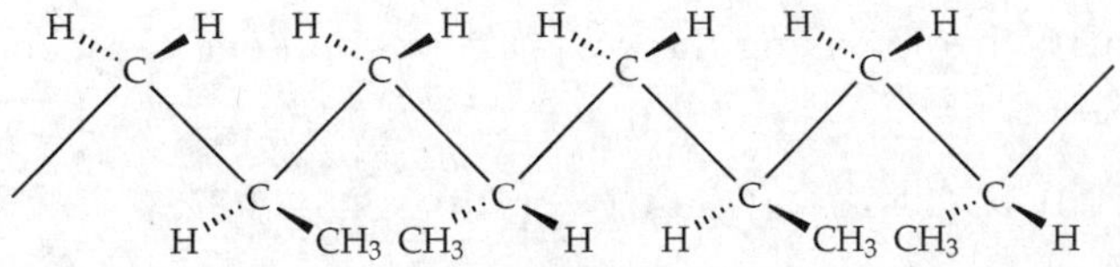

syndiotactic

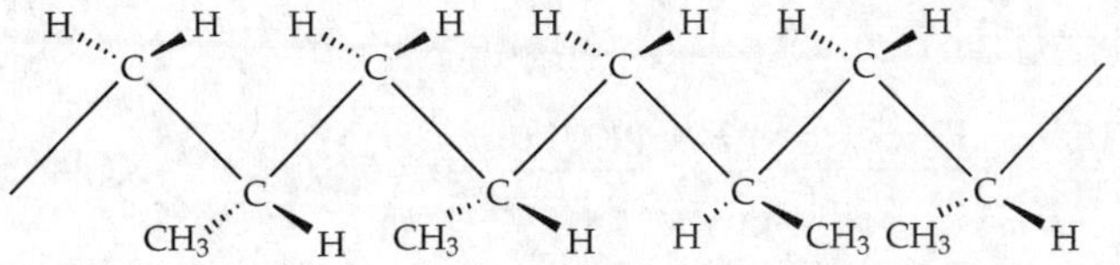

atactic

Isotactic polypropylene has the highest degree of crystallinity and highest melting point. The regular shape of the polymer backbone allows for close, orderly (almost zipper-like) contact between chains. This maximizes dispersion forces between chains and produces higher order (crystallinity) and melting point. Atactic polypropylene has the least order and the lowest melting point.

(c)   Cotton, with $-\overset{\displaystyle |}{\underset{\displaystyle OH}{C}}-$ groups and polyester, with $-\overset{\displaystyle O}{\overset{\displaystyle \|}{C}}-O-C$ groups

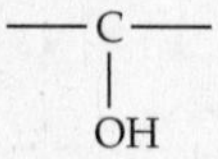

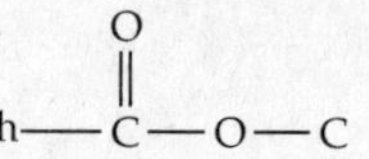

Both participate in hydrogen-bonding interactions with H_2O molecules. These are strong intermolecular forces that hold the "moisture" at the surface of the fabric next to the skin. Polypropylene has no strong interactions with water, and capillary action "wicks" the moisture away from the skin.

12.122   (a)

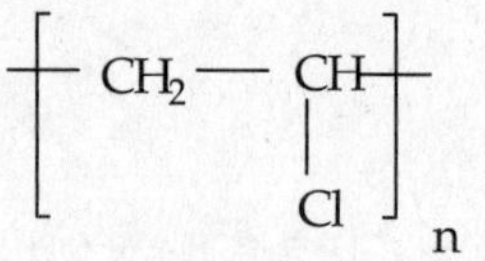

$$\begin{bmatrix} CH_2 - \overset{\displaystyle |}{\underset{\displaystyle Cl}{CH}} \end{bmatrix}_n$$

C—Cl	328 kJ/mol	← lowest
C—C	348 kJ/mol	
C—H	413 kJ/mol	

(b)   C–Cl bonds are weakest, so they are most likely to break upon heating.

(c)   The repeating unit in polyvinyl chloride consists of two C atoms, each in a different environment. Consider the net changes in these two C atoms when the polymer is converted to diamond a high pressure.

Diamond is a covalent-network structure where each C atom is tetrahedrally bound to four other C atoms [Figure 12.29(a)].

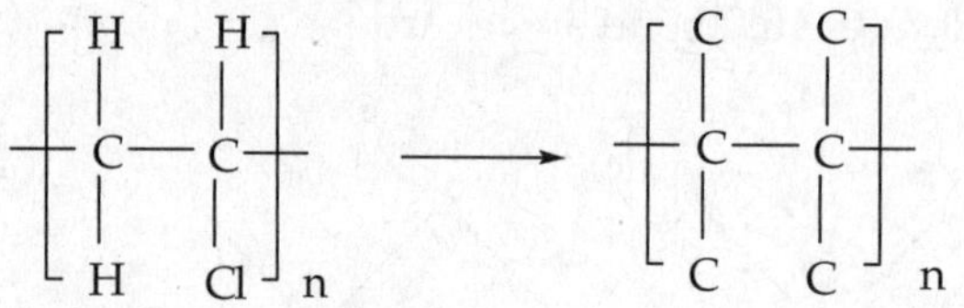

$$\begin{bmatrix} \overset{\displaystyle H}{\underset{\displaystyle H}{C}} - \overset{\displaystyle H}{\underset{\displaystyle Cl}{C}} \end{bmatrix}_n \longrightarrow \begin{bmatrix} \overset{\displaystyle C}{\underset{\displaystyle C}{C}} - \overset{\displaystyle C}{\underset{\displaystyle C}{C}} \end{bmatrix}_n$$

Assume that there is no net change to the C–C bonds in the structure, even though they may be broken and reformed. The net change to the 2-C vinyl chloride unit is then breaking three C–H bonds and one C–Cl bond, and making four C–C bonds.

$\Delta H = 3D(C–H) – D(C–Cl) – 4D(C–C) = 3(413) + 328 – 4(348)$

$= 523$ kJ/vinyl chloride unit

12.124   Avogadro's number is the number of KCl formula units in 74.55 g of KCl.

$$74.55 \text{ g KCl} \times \frac{1 \text{ cm}^3}{1.984 \text{ g}} \times \frac{(1 \times 10^8 \text{ Å})^3}{1 \text{ cm}^3} \times \frac{4 \text{ KCl units}}{6.28^3 \text{ Å}^3} = 6.07 \times 10^{23} \text{ KCl formula units}$$

13 Properties of Solutions

Visualizing Concepts

13.2   (a)   The oxygen atom of the water molecule is associated with the cation. In the polar water molecule, the partial negative charge is localized on the oxygen atom. The electrostatic attractive interaction is between the positive charge of the cation and the partially negative charge of the oxygen atom.

        (b)   Statement (c). The smaller ionic radius of lithium ion means that the positive charge is localized over a smaller volume; it is essentially a point charge. This increases the strength of the interaction between lithium cation and each individual water molecule.

13.4   Yes. Gases are always miscible with each other because they are in constant, random motion and gas particles are far apart relative to each other. Interparticle forces among gas molecules are too weak and to restrain the particles from mixing.

13.5   Diagram (b) is the best representation of a saturated solution. There is some undissolved solid with particles that are close together and ordered, in contact with a solution containing mobile, separated solute particles. As much solute has dissolved as can dissolve, leaving some undissolved solid in contact with the saturated solution.

13.6   (a)   The solubility of the gases in water increases in the order Ar, $1.50 \times 10^{-3}\ M$ < Kr, $2.79 \times 10^{-3}\ M$ < Xe, $5 \times 10^{-3}\ M$

        (b)   As the polarizability of the gas atoms increases, their water solubility increases. As the molar mass of the ideal gas increases, atomic size increases and the electron cloud is less tightly held by the nucleus, causing the cloud to be more polarizable. The greater the polarizability, the stronger the dispersion forces between the gas atoms and water, the more likely the gas atom is to stay dissolved rather than escape the solution, the greater the solubility of the gas.

13.8   The second friend is correct. As pressure decreases, gas solubility decreases. The first bubbles were from air gases dissolved at atmospheric pressure but not dissolved at lower pressure. The second batch of bubbles is from the water boiling. As the pressure in the vessel decreases, eventually the vapor pressure of water equals external pressure, and the water boils.

13.10   (a)   The blue line represents the solution. According to Raoult's law, the presence of a nonvolatile solute lowers the vapor pressure of a volatile solvent. At any given temperature, the blue line has a lower vapor pressure and represents the solution.

(b)   The boiling point of a liquid is the temperature at which the vapor pressure of the liquid is equal to atmospheric pressure. Assuming atmospheric pressure of 1.0 atm, the boiling point of the solvent (red line) is approximately 64 °C. The boiling point of the solution is approximately 70 °C.

13.12   A detergent for solubilizing large hydrophobic proteins (or any other large nonpolar solute, such as greasy dirt) needs a hydrophobic part to interact with the solute, and a hydrophilic part to interact with water. In n-octyl glucoside, the eight-carbon n-octyl chain has strong dispersion interactions with the hydrophobic (nonpolar) protein. The –OH groups on the glucoside (sugar) ring form strong hydrogen bonds with water. This causes the glucoside to dissolve, dragging the hydrophobic protein along with it.

The Solution Process (Section 13.1)

13.14   (a)   False. NaCl is more soluble in water than it is in benzene because the enthalpy of mixing for NaCl and water is much more exothermic. Ion-dipole interactions between NaCl and water are much stronger than the interactions between ionic NaCl and nonpolar benzene. (And, water is more dense than benzene.)

         (b)   True

         (c)   True

13.16   From weakest to strongest solvent-solute interactions:

         (b), dispersion forces < (c), hydrogen bonding < (a), ion-dipole

13.18   (a)   This solution process is endothermic. The enthalpy of the solution is greater than the enthalpy of the unmixed solute plus solvent.

         (b)   The solution forms because the favorable entropy of mixing outweighs the increase in enthalpy by the solution.

13.20 ΔH_{mix} is much more negative (exothermic) than $\Delta H_{solvent}$ or ΔH_{solute}. Both $\Delta H_{solvent}$ and ΔH_{solute} will be endothermic, because separating solvent molecules or solute ions requires energy. Because ΔH_{soln} is exothermic, ΔH_{mix} must be exothermic, and not just more negative than the other two, but more negative than the sum of the other two. This is not surprising, because ΔH_{mix} involves formation of many ion-dipole interactions, strong interparticle forces.

13.22   KBr is quite soluble in water because of the sizeable increase in disorder of the system (ordered KBr lattice → freely moving hydrated ions) associated with the dissolving process. An increase in disorder or randomness in a process tends to make that process spontaneous.

Saturated Solutions; Factors Affecting Solubility (Sections 13.2 and 13.3)

13.24   (a) $\dfrac{1.22 \text{ mol } MnSO_4 \cdot H_2O}{1 \text{ L soln}} \times \dfrac{169.0 \text{ g } MnSO_4 \cdot H_2O}{1 \text{ mol}} \times 0.100 \text{ L}$

                                                     $= 20.6 \text{ g } MnSO_4 \cdot H_2O/100 \text{ mL}$

         The 1.22 *M* solution is unsaturated.

(b)   Add a known mass, say 5.0 g, of $MnSO_4 \cdot H_2O$, to the unknown solution. If the solid dissolves, the solution is unsaturated. If there is undissolved $MnSO_4 \cdot H_2O$, filter the solution and weigh the solid. If there is less than 5.0 g of solid, some of the added $MnSO_4 \cdot H_2O$, dissolved and the unknown solution is unsaturated. If there is exactly 5.0 g, no additional solid dissolved and the unknown is saturated. If there is more than 5.0 g, excess solute has precipitated and the solution is supersaturated.

13.26   (a)   at 30 °C, $\dfrac{10 \text{ g } KClO_3}{100 \text{ g } H_2O} \times 250 \text{ g } H_2O = 25 \text{ g } KClO_3$

(b) $\dfrac{66 \text{ g } Pb(NO_3)_2}{100 \text{ g } H_2O} \times 250 \text{ g } H_2O = 165 = 1.7 \times 10^2 \text{ g } Pb(NO_3)_2$

(c) $\dfrac{3 \text{ g } Ce_2(SO_4)_3}{100 \text{ g } H_2O} \times 250 \text{ g } H_2O = 7.5 = 8 \text{ g } Ce_2(SO_4)_3$

13.28   The most likely reason is statement (b), although statement (c) also contributes.

Many substances are called "oil," but they typically contain molecules composed mostly of carbon and hydrogen with fairly high molecular weights. Both statements contribute to the fact that oil molecules are nonpolar and experience strong dispersion forces. The properties of water are dominated by its strong hydrogen bonding. Oil and water have very dissimilar intermolecular interactions and the liquids are not miscible.

There are examples of high molecular weight compounds (e.g., sugars and proteins) with many hydrogen-bonding interactions that are soluble in water, so statement (b) is the better answer.

13.30   We expect alanine to be more soluble in water than hexane. Alanine has a –COOH and a –NH$_2$ group available to form hydrogen bonds with water molecules. Although there are some potential dispersion forces between the terminal –CH$_3$ group of alanine and hexane molecules, we expect the hydrogen bonding between alanine and water to be stronger. Stronger intermolecular attractive forces between alanine and water lead to a more negative ΔH_{mix} and more negative (smaller positive) ΔH_{soln} for water than for hexane.

13.32   The red part of the molecule, a carboxyl group able to form hydrogen bonds, contributes to its water solubility. The gray and white parts, a phenyl ring and several –CH$_3$ groups, form a large nonpolar area that contributes to its water insolubility.

13.34   *Analyze/Plan.* Water, H_2O, is a polar solvent that forms hydrogen bonds with other H_2O molecules. The more soluble solute in each case will have intermolecular interactions that are most similar to the hydrogen bonding in H_2O. *Solve.*

(a)   Glucose, $C_6H_{12}O_6$, is more soluble because it is capable of hydrogen bonding (Figure 13.10). Nonpolar C_6H_{12} is capable only of dispersion interactions and does not have strong intermolecular interactions with polar (hydrogen bonding) H_2O.

(b)   Ionic sodium propionate, CH_3CH_2COONa, is more soluble. Sodium propionate is a crystalline solid, whereas propionic acid is a liquid. The increase in disorder or entropy when an ionic solid dissolves leads to significant water solubility, despite the strong ion-ion forces (large ΔH_{solute}) present in the solute (see Solution 13.22).

(c)   HCl is more soluble because it is a strong electrolyte and completely ionized in water. Ionization leads to ion-dipole solute-solvent interactions, and an increase in disorder. CH_3CH_2Cl is a molecular solute capable of relatively weak dipole-dipole solute-solvent interactions and is much less soluble in water.

13.36   Pressure has an effect on O_2 solubility in water because, at constant temperature and volume, pressure is directly related to the amount of O_2 available to dissolve. The greater the partial pressure of O_2 above water, the more O_2 molecules are available for dissolution, and the more molecules that strike the surface of the liquid.

          Pressure does not affect the amount or physical properties of NaCl, or ionic solids in general, so it has little influence on the dissolving of NaCl in water.

13.38 $650 \text{ torr} \times \dfrac{1\,\text{atm}}{760\,\text{torr}} = 0.855\,\text{atm}; P_{O_2} = \chi_{O_2}(P_t) = 0.21(0.855\,\text{atm}) = 0.1796 = 0.18\,\text{atm}$

          $S_{O_2} = kP_{O_2} = \dfrac{1.38 \times 10^{-3}\,\text{mol}}{\text{L-atm}} \times 0.1796\,\text{atm} = 2.5 \times 10^{-4}\,M$

Concentrations of Solutions (Section 13.4)

13.40   (a) $\text{mass \%} = \dfrac{\text{mass solute}}{\text{total mass solution}} \times 100$

               $\text{mass solute} = 0.035\,\text{mol}\,I_2 \times \dfrac{253.8\,\text{g}\,I_2}{1\,\text{mol}\,I_2} = 8.883 = 8.9\,\text{g}\,I_2$

               $\text{mass \%}\,I_2 = \dfrac{8.883\,\text{g}\,I_2}{8.883\,\text{g}\,I_2 + 125\,\text{g}\,CCl_4} \times 100 = 6.635 = 6.6\%\,I_2$

   (b) $\text{ppm} = \dfrac{\text{mass solute}}{\text{total mass solution}} \times 10^6 = \dfrac{0.0079\,\text{g}\,Sr^{2+}}{1 \times 10^3\,\text{g}\,H_2O} \times 10^6 = 7.9\,\text{ppm}\,Sr^{2+}$

13.42   (a) $\dfrac{20.8\,\text{g}\,C_6H_5OH}{94.11\,\text{g/mol}} = 0.2210 = 0.221\,\text{mol}\,C_6H_5OH$

               $\dfrac{425\,\text{g}\,CH_3CH_2OH}{46.07\,\text{g/mol}} = 9.2251 = 9.23\,\text{mol}\,CH_3CH_2OH$

               $X_{C_6H_5OH} = \dfrac{0.2210}{0.2210 + 9.2251} = 0.02340 = 0.0234$

   (b) $\text{mass \%} = \dfrac{20.8\,\text{g}\,C_6H_5OH}{20.8\,\text{g}\,C_6H_5OH + 425\,\text{g}\,CH_3CH_2OH} \times 100 = 4.67\%\,C_6H_5OH$

   (c) $m = \dfrac{0.2210\,\text{mol}\,C_6H_5OH}{0.425\,\text{kg}\,CH_3CH_2OH} = 0.5200 = 0.520\,m\,C_6H_5OH$

13.44   (a) $M = \dfrac{\text{mol solute}}{\text{L soln}}; \dfrac{15.0\,\text{g}\,Al_2(SO_4)_3}{0.250\,\text{L soln}} \times \dfrac{1\,\text{mol}\,Al_2(SO_4)_3}{342.2\,\text{g}\,Al_2(SO_4)_3} = 0.175\,M\,Al_2(SO_4)_3$

   (b) $\dfrac{5.25\,\text{g}\,Mn(NO_3)_2 \cdot 2H_2O}{0.175\,\text{L soln}} \times \dfrac{1\,\text{mol}\,Mn(NO_3)_2 \cdot 2H_2O}{215.0\,\text{g}\,Mn(NO_3)_2 \cdot 2H_2O} = 0.140\,M\,Mn(NO_3)_2$

   (c) $M_c \times L_c = M_d \times L_d; 9.00\,M\,H_2SO_4 \times 0.0350\,\text{L} = ?M\,H_2SO_4 \times 0.500\,\text{L}$
          500 mL of $0.630\,M\,H_2SO_4$

13.46   (a) $16.0 \text{ mol H}_2\text{O} \times \dfrac{18.02 \text{ g H}_2\text{O}}{1 \text{ mol H}_2\text{O}} = 288.3 \text{ g H}_2\text{O} = 0.288 \text{ kg H}_2\text{O}$

$m = \dfrac{1.12 \text{ mol KCl}}{0.2883 \text{ kg H}_2\text{O}} = 3.8846 = 3.88 \; m \text{ KCl}$

    (b) $m = \dfrac{\text{mol solute}}{\text{kg solute}}$; $\text{mol S}_8 = m \times \text{kg C}_{10}\text{H}_8 = 0.12 \; m \times 0.1000 \text{ kg C}_{10}\text{H}_8 = 0.012 \text{ mol}$

$0.012 \text{ mol S}_8 \times \dfrac{256.5 \text{ g S}_8}{1 \text{ mol S}_8} = 3.078 = 3.1 \text{ g S}_8$

13.48   (a) $\text{mass \%} = \dfrac{\text{mass C}_6\text{H}_8\text{O}_6}{\text{total mass solution}} \times 100$;

$\dfrac{80.5 \text{ g C}_6\text{H}_8\text{O}_6}{80.5 \text{ g C}_6\text{H}_8\text{O}_6 + 210 \text{ g H}_2\text{O}} \times 100 = 27.71 = 27.7\% \text{ C}_6\text{H}_8\text{O}_6$

    (b) $\text{mol C}_6\text{H}_8\text{O}_6 = \dfrac{80.5 \text{ g C}_6\text{H}_8\text{O}_6}{176.1 \text{ g/mol}} = 0.4571 = 0.457 \text{ mol C}_6\text{H}_8\text{O}_6$

$\text{mol H}_2\text{O} = \dfrac{210 \text{ g H}_2\text{O}}{18.02 \text{ g/mol}} = 11.654 = 11.7 \text{ mol H}_2\text{O}$

$X_{\text{C}_6\text{H}_8\text{O}_6} = \dfrac{0.4571 \text{ mol C}_6\text{H}_8\text{O}_6}{0.4571 \text{ mol C}_6\text{H}_8\text{O}_6 + 11.654 \text{ mol H}_2\text{O}} = 0.0377$

    (c) $m = \dfrac{0.4571 \text{ mol C}_6\text{H}_8\text{O}_6}{0.210 \text{ kg H}_2\text{O}} = 2.18 \; m \text{ C}_6\text{H}_8\text{O}_6$

    (d) $M = \dfrac{\text{mol C}_6\text{H}_8\text{O}_6}{\text{L solution}}$; $290.5 \text{ g soln} \times \dfrac{1 \text{ mL}}{1.22 \text{ g}} \times \dfrac{1 \text{ L}}{1000 \text{ mL}} = 0.2381 = 0.238 \text{ L}$

$M = \dfrac{0.4571 \text{ mol C}_6\text{H}_8\text{O}_6}{0.2381 \text{ L soln}} = 1.92 \; M \text{ C}_6\text{H}_8\text{O}_6$

13.50   Given: $8.10 \text{ g C}_4\text{H}_4\text{S}$, 1.065 g/mL; $250.0 \text{ mL C}_7\text{H}_8$, 0.867 g/mL

    (a) $\text{mol C}_4\text{H}_4\text{S} = 8.10 \text{ g C}_4\text{H}_4\text{S} \times \dfrac{1 \text{ mol C}_4\text{H}_4\text{S}}{84.15 \text{ g C}_4\text{H}_4\text{S}} = 0.09626 = 0.0963 \text{ mol C}_4\text{H}_4\text{S}$

$\text{mol C}_7\text{H}_8 = \dfrac{0.867 \text{ g}}{1 \text{ mL}} \times 250.0 \text{ mL} \times \dfrac{1 \text{ mol C}_7\text{H}_8}{92.14 \text{ g C}_7\text{H}_8} = 2.352 = 2.35 \text{ mol}$

$X_{\text{C}_4\text{H}_4\text{S}} = \dfrac{0.09626 \text{ mol C}_4\text{H}_4\text{S}}{0.09626 \text{ mol C}_4\text{H}_4\text{S} + 2.352 \text{ mol C}_7\text{H}_8} = 0.03932 = 0.0393$

    (b) $m_{\text{C}_4\text{H}_4\text{S}} = \dfrac{\text{mol C}_4\text{H}_4\text{S}}{\text{kg C}_7\text{H}_8}$; $250.0 \text{ mL} \times \dfrac{0.867 \text{ g}}{1 \text{ mL}} \times \dfrac{1 \text{ kg}}{1000 \text{ g}} = 0.2168 = 0.217 \text{ kg C}_7\text{H}_8$

$m_{\text{C}_4\text{H}_4\text{S}} = \dfrac{0.09626 \text{ mol C}_4\text{H}_4\text{S}}{0.2168 \text{ kg C}_7\text{H}_8} = 0.444 \; m \text{ C}_4\text{H}_4\text{S}$

(c) $8.10 \, g \, C_4H_4S \times \dfrac{1 \, mL}{1.065 \, g} = 7.606 = 7.61 \, mL \, C_4H_4S;$

$V_{soln} = 7.61 \, mL \, C_4H_4S + 250.0 \, mL \, C_7H_8 = 257.6 \, mL$

$M_{C_4H_4S} = \dfrac{0.09626 \, mol \, C_4H_4S}{0.2576 \, L \, soln} = 0.374 \, M \, C_4H_4S$

13.52   (a) $\dfrac{1.50 \, mol \, HNO_3}{1 \, L \, soln} \times 0.255 \, L = 0.3825 = 0.383 \, mol \, HNO_3$

        (b)   Assume that for dilute aqueous solutions, the mass of the solvent is the mass of solution.

$\dfrac{1.50 \, mol \, NaCl}{1 \, kg \, H_2O} = \dfrac{x \, mol}{50.0 \times 10^{-6} \, kg}; x = 7.50 \times 10^{-5} \, mol \, NaCl$

        (c) $\dfrac{1.50 \, g \, C_{12}H_{22}O_{11}}{100 \, g \, soln} = \dfrac{x \, g \, C_{12}H_{22}O_{11}}{75.0 \, g \, soln}; x = 1.125 = 1.13 \, g \, C_{12}H_{22}O_{11}$

$1.125 \, g \, C_{12}H_{22}O_{11} \times \dfrac{1 \, mol \, C_{12}H_{22}O_{11}}{342.3 \, g \, C_{12}H_{22}O_{11}} = 3.287 \times 10^{-3} = 3.29 \times 10^{-3} \, mol \, C_{12}H_{22}O_{11}$

13.54   (a) $\dfrac{0.110 \, mol \, (NH_4)_2SO_4}{1 \, L \, soln} \times 1.50 \, L \times \dfrac{132.2 \, g \, (NH_4)_2SO_4}{1 \, mol \, (NH_4)_2SO_4} = 21.81 = 21.8 \, g \, (NH_4)_2SO_4$

Weigh 21.8 g $(NH_4)_2SO_4$, dissolve in a small amount of water, continue adding water with thorough mixing up to a total solution volume of 1.50 L.

        (b)   Determine the mass fraction of Na_2CO_3 in the solution:

$\dfrac{0.65 \, mol \, Na_2CO_3}{1000 \, g \, H_2O} \times \dfrac{106.0 \, g \, Na_2CO_3}{1 \, mol \, Na_2CO_3} = 68.9 \, g = \dfrac{69 \, g \, Na_2CO_3}{1000 \, g \, H_2O}$

$\text{mass fraction} = \dfrac{68.9 \, g \, Na_2CO_3}{1000 \, g \, H_2O + 68.9 \, g \, Na_2CO_3} = 0.06446 = 0.064$

In 225 g of solution, there are 0.06446(225) = 14.503 = 15 g Na_2CO_3.

Weigh out 15 g Na_2CO_3 and dissolve it in 225 – 15 = 210 g H_2O to make exactly 225 g of solution. (210 g H_2O/0.997 g H_2O/mL @ 25 °C = 211 mL H_2O)

[Carrying 3 sig figs, weigh 14.5 g Na_2CO_3 and dissolve it in 225 – 14.5 = 210.5 g H_2O. This produces a solution that is much closer to 0.65 *m*.]

        (c) $1.20 \, L \times \dfrac{1000 \, mL}{1 \, L} \times \dfrac{1.16 \, g}{1 \, mL} = 1392 \, g \, \text{solution}; \, 0.150(1392 \, g \, soln) = 209 \, g \, Pb(NO_3)_2$

Weigh 209 g $Pb(NO_3)_2$ and add (1392 – 209) = 1183 g H_2O to make exactly (1392 = 1.39×10^3) g or 1.20 L of solution.

(1183 g H_2O/0.997 g/mL @ 25 °C = 1187 mL H_2O)

        (d)   Calculate the mol HCl necessary to neutralize 5.5 g $Ba(OH)_2$.

$Ba(OH)_2(s) + 2HCl(aq) \rightarrow BaCl_2(aq) + 2H_2O(l)$

$5.5 \, g \, Ba(OH)_2 + \dfrac{1 \, mol \, Ba(OH)_2}{171 \, g \, Ba(OH)_2} \times \dfrac{2 \, mol \, HCl}{1 \, mol \, Ba(OH)_2} = 0.0643 = 0.064 \, mol \, HCl$

$$M = \frac{\text{mol}}{\text{L}}; \text{L} = \frac{\text{mol}}{M} = \frac{0.0643\,\text{mol HCl}}{0.50\,M\,\text{HCl}} = 0.1287 = 0.13\,\text{L} = 130\,\text{mL}$$

130 mL of 0.50 M HCl are needed.

$$M_c \times L_c = M_d \times L_d;\ 6.0\,M \times L_c = 0.50\,M \times 0.1287\,\text{L};\ L_c = 0.01072\,\text{L} = 11\,\text{mL}$$

Using a pipette, measure exactly 11 mL of 6.0 M HCl and dilute with water to a total volume of 130 mL.

13.56 *Analyze/Plan.* Assume 1.00 L of solution. Calculate mass of 1 L of solution using density. Calculate mass of NH_3 using mass %, then mol NH_3 in 1.00 L. *Solve.*

$$1.00\,\text{L soln} \times \frac{1000\,\text{mL}}{1\,\text{L}} \times \frac{0.90\,\text{g soln}}{1\,\text{mL soln}} = 9.0 \times 10^2\,\text{g soln/L}$$

$$\frac{900\,\text{g soln}}{1.00\,\text{L soln}} \times \frac{28\,\text{g }NH_3}{100\,\text{g soln}} \times \frac{1\,\text{mol }NH_3}{17.03\,\text{g }NH_3} = 14.80 = 15\,\text{mol }NH_3/\text{L soln} = 15\,M\,NH_3$$

13.58 (a)
$$\frac{0.0500\,\text{mol }C_8H_{10}N_4O_2}{1\,\text{kg }CHCl_3} \times \frac{194.2\,\text{g }C_8H_{10}N_4O_2}{1\,\text{mol }C_8H_{10}N_4O_2} = 9.7100$$

$$= 9.71\,\text{g }C_8H_{10}N_4O_2/\text{kg }CHCl_3$$

$$\frac{9.710\,\text{g }C_8H_{10}N_4O_2}{9.710\,\text{g }C_8H_{10}N_4O_2 + 1000.00\,\text{g }CHCl_3} \times 100 = 0.9617 = 0.962\%\,C_8H_{10}N_4O_2\,\text{by mass}$$

(b)
$$1000\,\text{g }CHCl_3 \times \frac{1\,\text{mol }CHCl_3}{119.4\,CHCl_3} = 8.375 = 8.38\,\text{mol }CHCl_3$$

$$X_{C_8H_{10}N_4O_2} = \frac{0.0500}{0.0500 + 8.375} = 0.00593$$

13.60 (a) For gases at the same temperature and pressure, volume % = mol %. The volume and mol % of CO_2 in this breathing air is 4.0%.

(b) $P_{CO_2} = X_{CO_2} \times P_t = 0.040\,(1\,\text{atm}) = 0.040\,\text{atm}$

$$M_{CO_2} = \frac{P_{CO_2}}{RT} = \frac{0.040\,\text{atm}}{310\,\text{K}} \times \frac{\text{mol - K}}{0.08206\,\text{L - atm}} = 1.6 \times 10^{-3}\,M$$

Colligative Properties (Section 13.5)

13.62 (a) False. Adding solvent decreases the molality of the solution and elevates the freezing point of the solution.

(b) False. The solid that forms is nearly pure solvent.

(c) False. The more concentrated the solution, the lower the freezing point.

(d) True

(e) True

13.64 (a) An *ideal solution* is a solution that obeys Raoult's law.

(b) *Analyze/Plan.* Calculate the vapor pressure predicted by Raoult's law and compare it to the experimental vapor pressure. Assume ethylene glycol (eg) is the solute. *Solve.*

$X_{H_2O} = X_{eg} = 0.500; \quad P_A = X_A P_A{}^\circ = 0.500(149) \text{ torr} = 74.5 \text{ torr}$

The experimental vapor pressure (P_A), 67 torr, is less than the value predicted by Raoult's law for an ideal solution. The solution is not ideal.

Check. An ethylene glycol-water solution has extensive hydrogen bonding, which causes deviation from ideal behavior. We expect the experimental vapor pressure to be less than the ideal value and it is.

13.66 (a) H_2O vapor pressure will be determined by the mole fraction of H_2O in the solution. The vapor pressure of pure H_2O at 343 K (70 °C) = 233.7 torr.

$$\frac{28.5 \text{ g } C_3H_8O_3}{92.10 \text{ g/mol}} = 0.3094 = 0.309 \text{ mol}; \quad \frac{125 \text{ g } H_2O}{18.02 \text{ g/mol}} = 6.937 = 6.94 \text{ mol}$$

$$P_{H_2O} = \frac{6.937 \text{ mol } H_2O}{6.937 + 0.309} \times 233.7 \text{ torr} = 223.7 = 224 \text{ torr}$$

(b) Calculate X_B by vapor pressure lowering; $X_B = \Delta P_A / P_A{}^\circ$. [See Solution 13.65(b).] Given moles solvent, calculate moles solute from the definition of mole fraction.

$$X_{C_2H_6O_2} = \frac{10.0 \text{ torr}}{100 \text{ torr}} = 0.100$$

$$\frac{1.00 \times 10^3 \text{ g } C_2H_5OH}{46.07 \text{ g/mol}} = 21.71 = 21.7 \text{ mol } C_2H_5OH; \text{ let } y = \text{mol } C_2H_6O_2$$

$$X_{C_2H_6O_2} = \frac{y \text{ mol } C_2H_6O_2}{y \text{ mol } C_2H_6O_2 + 21.71 \text{ mol } C_2H_5OH} = 0.100 = \frac{y}{y + 21.71}$$

0.100 y + 2.171 = y; 0.900 y = 2.171; y = 2.412 = 2.41 mol $C_2H_6O_2$

$$2.412 \text{ mol } C_2H_6O_2 \times \frac{62.07 \text{ g}}{1 \text{ mol}} = 150 \text{ g } C_2H_6O_2$$

13.68 (a) Because C_6H_6 and C_7H_8 form an ideal solution, we can use Raoult's law. Because both components are volatile, both contribute to the total vapor pressure of 35 torr.

$$P_t = P_{C_6H_6} + P_{C_7H_8}; \quad P_{C_6H_6} = X_{C_6H_6} P_{C_6H_6}^\circ; \quad P_{C_7H_8} = X_{C_7H_8} P_{C_7H_8}^\circ$$

$$X_{C_7H_8} = 1 - X_{C_6H_6}; \quad P_T = X_{C_6H_6} P_{C_6H_6}^\circ + (1 - X_{C_6H_6}) P_{C_7H_8}^\circ$$

$$35 \text{ torr} = X_{C_6H_6}(75 \text{ torr}) + (1 - X_{C_6H_6})22 \text{ torr}$$

$$13 \text{ torr} = 53 \text{ torr } (X_{C_6H_6}); \quad X_{C_6H_6} = \frac{13 \text{ torr}}{53 \text{ torr}} = 0.2453 = 0.25; \quad X_{C_7H_8} = 0.7547 = 0.75$$

(b) $P_{C_6H_6} = 0.2453(75 \text{ torr}) = 18.4 \text{ torr}; \quad P_{C_7H_8} = 0.7547(22 \text{ torr}) = 16.6 \text{ torr}$

In the vapor, $\quad X_{C_6H_6} = \dfrac{P_{C_6H_6}}{P_t} = \dfrac{18.4 \text{ torr}}{18.4 \text{ torr} + 16.6 \text{ torr}} = 0.53; \quad X_{C_7H_8} = 0.47$

13.70 *Analyze/Plan.* ΔT_b depends on mol dissolved particles. Assume 100 g of each solution, calculate mol solute and mol dissolved particles. Glucose and sucrose are molecular solutes, but $NaNO_3$ dissociates into 2 mol particles per mol solute. *Solve.*

10% by mass means 10 g solute in 100 g solution. If we have 10 g of each solute, the one with the smallest molar mass will have the largest mol solute. The molar masses are: glucose, 180.2 g/mol; sucrose, 342.3 g/mol; $NaNO_3$, 85.0 g/mol. $NaNO_3$ has most mol solute, and twice as many dissolved particles, so it will have the highest boiling point. Sucrose has least mol solute and lowest boiling point. Glucose is intermediate.

In order of increasing boiling point: 10% sucrose < 10% glucose < 10% $NaNO_3$.

13.72 0.030 *m* phenol > 0.040 *m* glycerin = 0.020 *m* KBr. Phenol is very slightly ionized in water, but not enough to match the number of particles in a 0.040 *m* glycerin solution. Assuming the ideal van't Hoff factor of 2.00, the KBr solution is 0.040 *m* in particles, so it has the same freezing point as 0.040 *m* glycerin, which is a nonelectrolyte. (The measured van't Hoff factor for 0.040 *m* KBr will be slightly less than 2. We expect the measured order of freezing points to be 0.030 *m* phenol > 0.020 *m* KBr > 0.040 *m* glycerin.)

13.74 $\Delta T = K(m)$; first calculate the **molality** of the solute particles.

(a) 0.25 *m*

(b) $\dfrac{20.0\,g\,C_{10}H_{22}}{0.0500\,kg\,CHCl_3} \times \dfrac{1\,mol\,C_{10}H_{22}}{142.3\,g\,C_{10}H_{22}} = 2.811 = 2.81\ m$

(c) $3.50\,g\,NaOH \times \dfrac{1\,mol\,NaOH}{40.00\,g\,NaOH} \times \dfrac{2\,mol\,particles}{1\,mol\,NaOH} = 0.1750 = 0.175\,mol\,particles$

$m = \dfrac{0.1750\,mol\,NaCl}{0.175\,kg\,H_2O} = 1.000 = 1.00\ m$

(d) $m = \dfrac{0.45\,mol\,eg + 2(0.15)\,mol\,KBr}{0.150\,kg\,H_2O} = \dfrac{0.75\,mol\,particles}{0.150\,kg\,H_2O} = 5.0\ m$

Then, fp $= T_f - K_f(m)$; bp $= T_b + K_b(m)$; T in °C

	m	T_f	- $K_f(m)$	fp	T_b	+ $K_b(m)$	bp
(a)	0.25	−114.6	−1.99(0.25) = −0.50	−115.1	78.4	1.22(0.25) = 0.31	78.7
(b)	2.81	−63.5	−4.68(2.81) = −13.2	−76.7	61.2	3.63(2.81) = 10.2	71.4
(c)	1.00	0.0	−1.86(1.00) = −1.86	−1.9	100.0	0.51(1.00) = 0.51	100.5
(d)	5.0	0.0	−1.86(5.0) = −9.3	−9.3	100.0	0.51(5.0) = 2.6	102.6

13.76 Use ΔT_b = find *m* of aqueous solution, and then use *m* to calculate ΔT_f and freezing point. $K_b = 0.51, K_f = 1.86$.

bp = 105.0 °C; ΔT_b = 105.0 °C − 100.0 °C = 5.0 °C

$\Delta T_b = K_b(m)$; $m = \dfrac{\Delta T_b}{K_b} = \dfrac{5.0\,°C}{0.51} = 9.804 = 9.8\ m$

$\Delta T_f = 1.86\,°C/m \times 9.804\,m = 18.24 = 18\ °C$; freezing point = 0.0 °C − 18.24 °C = −18 °C

13.78 $\Pi = MRT$; $T = 20\ °C + 273 = 293\ K$

$$M\ (\text{of ions}) = \frac{\text{mol NaCl} \times 2}{\text{L soln}} = \frac{3.4\ \text{g NaCl}}{1\ \text{L soln}} \times \frac{1\ \text{mol NaCl}}{58.4\ \text{g NaCl}} \times \frac{2\ \text{mol ions}}{1\ \text{mol NaCl}} = 0.116 = 0.12\ M$$

$$\Pi = \frac{0.116\ \text{mol}}{L} \times \frac{0.08206\ \text{L - atm}}{\text{mol - K}} \times 293\ K = 2.8\ \text{atm}$$

13.80 $\Delta T_f = 5.5 - 4.1 = 1.4$; $m = \dfrac{\Delta T_f}{K_f} = \dfrac{1.4}{5.12} = 0.273 = 0.27\ m$

$$\text{MM lauryl alcohol} = \frac{\text{g lauryl alcohol}}{m\ \times\ \text{kg } C_6H_6} = \frac{5.00\ \text{g lauryl alcohol}}{0.273\ \times\ 0.100\ \text{kg } C_6H_6}$$

$$= 1.8 \times 10^2\ \text{g/ mol lauryl alcohol}$$

13.82 $M = P/RT = \dfrac{0.605\ \text{atm}}{298\ K} \times \dfrac{\text{mol - K}}{0.08206\ \text{L - atm}} = 0.02474 = 0.0247\ M$

$$\text{MM} = \frac{g}{M \times L} \ = \ \frac{2.35\ g}{0.02474\ M \times 0.250\ L} = 380\ \text{g/ mol}$$

13.84   If these were ideal solutions, they would have equal ion concentrations and equal ΔT_f values. Data in Table 13.4 indicates that the van't Hoff factors (i) for both salts are less than the ideal values. For 0.030 m NaCl, i is between 1.87 and 1.94, about 1.92. For 0.020 m K_2SO_4, i is between 2.32 and 2.70, about 2.62. From Equation 13.15,

ΔT_f (measured) = $i \times \Delta T_f$ (calculated for nonelectrolyte)

NaCl: ΔT_f (measured) = $1.92 \times 0.030\ m \times 1.86\ °C/m = 0.11\ °C$

K_2SO_4: ΔT_f (measured) = $2.62 \times 0.020\ m \times 1.86°C/m = 0.097\ °C$

0.030 m NaCl would have the larger ΔT_f.

The deviations from ideal behavior are due to ionpairing in the two electrolyte solutions. K_2SO_4 has more extensive ionpairing and a larger deviation from ideality because of the higher charge on SO_4^{2-} relative to Cl^-.

Colloids (Section 13.6)

13.86   The best answer is (b) emulsion (Table 13.5).

13.88   The presence of aerosols in the atmosphere decreases the amount of sunlight that arrives at Earth's surface, compared to an "aerosol-free" atmosphere. All colloids scatter light (the Tyndall effect). Aerosols in the atmosphere scatter the incoming sunlight. Although some of this scattered light will eventually reach Earth, some will not.

13.90   (a)   The nonpolar hydrophobic tails of soap particles (the hydrocarbon chain of stearate ions) establish attractive intermolecular dispersion forces with the nonpolar oil molecules, whereas the charged hydrophilic head of the soap particles interacts with H_2O to keep the oil molecules suspended. (This is the mechanism by which laundry detergents remove greasy dirt from clothes.)

(b)   Electrolytes from the acid neutralize surface charges of the suspended particles in milk, causing the colloid to coagulate. (It is probably also true that H^+ from the acid disrupts the hydrogen-bonding pattern of the protein, causing the protein to denature and precipitate.)

Additional Exercises

13.92 In this equilibrium system, molecules move from the surface of the solid into solution, whereas molecules in solution are deposited on the surface of the solid. As molecules leave the surface of the small particles of powder, the reverse process preferentially deposits other molecules on the surface of a single crystal. Eventually, all molecules that were present in the 50 g of powder are deposited on the surface of a 50 g crystal; this can only happen if the dissolution and deposition processes are ongoing.

13.93 Assume that the density of the solution is 1.00 g/mL.

(a) $4\,\text{ppm}\,O_2 = \dfrac{4\,\text{mg}\,O_2}{1\,\text{kg soln}} = \dfrac{4\times10^{-3}\,g\,O_2}{1\,L\,\text{soln}} \times \dfrac{1\,\text{mol}\,O_2}{32.0\,g\,O_2} = 1.25\times10^{-4} = 1\times10^{-4}\,M$

(b) $S_{O_2} = kP_{O_2};\; P_{O_2} = S_{O_2}/k = \dfrac{1.25\times10^{-4}\,\text{mol}}{L} \times \dfrac{L\text{-atm}}{1.71\times10^{-3}\,\text{mol}} = 0.0731 = 0.07\,\text{atm}$

$0.0731\,\text{atm} \times \dfrac{760\,\text{mm Hg}}{1\,\text{atm}} = 55.6 = 60\,\text{mm Hg}$

13.95 0.10% by mass means 0.10 g glucose/100 g blood.

(a) $\text{ppm glucose} = \dfrac{g\,\text{glucose}}{g\,\text{solution}} \times 10^6 = \dfrac{0.10\,g\,\text{glucose}}{100\,g\,\text{blood}} \times 10^6 = 1000\,\text{ppm glucose}$

(b) m = mol glucose/kg solvent. Assume that the mixture of nonglucose components is the "solvent."

mass solvent = 100 g blood – 0.10 g glucose = 99.9 g solvent = 0.0999 kg solvent

$\text{mol glucose} = 0.10\,g \times \dfrac{1\,\text{mol}}{180.2\,g\,C_6H_{12}O_6} = 5.55\times10^{-4} = 5.6\times10^{-4}\,\text{mol glucose}$

$m = \dfrac{5.55\times10^{-4}\,\text{mol glucose}}{0.0999\,\text{kg solvent}} = 5.6\times10^{-3}\,m\,\text{glucose}$

(c) To calculate molarity, solution volume must be known. The density of blood is needed to relate mass and volume.

13.96 *Analyze.* Given 13 ppt Au in seawater, find grams of Au in 1.0×10^3 gal seawater. The definition of ppt is (mass solute/mass solution) $\times 10^{12}$. *Plan.* Assume seawater is a dilute aqueous solution with a density of 1.00 g/mL. Use the definition of ppt to calculate g Au. *Solve.*

$\dfrac{13\,g\,Au}{1\times10^{12}\,g\,\text{soln}} \times \dfrac{1.0\,g\,\text{soln}}{mL\,\text{soln}} \times \dfrac{1000\,mL}{1\,L} \times \dfrac{3.7854\,L}{\text{gal}} \times 1.0\times10^3\,\text{gal soln} = 4.9\times10^{-5}\,g\,Au$

13.97 *Analyze.* The definition of ppb is (mass solute/mass solution) $\times 10^9$. *Plan.* Use the definition to get g Pb and g solution. Change g Pb to mol Pb, g solution to L solution, calculate molarity. *Solve.*

(a) $9.0\,\text{ppb} = \dfrac{9.0\,g\,Pb}{1\times10^9\,g\,\text{soln}} \times 10^9$

For dilute aqueous solutions (drinking water) assume that the density of the solution is the density of H_2O.

$\dfrac{9.0\,g\,Pb}{1\times10^9\,g\,\text{soln}} \times \dfrac{1.0\,g\,\text{soln}}{mL\,\text{soln}} \times \dfrac{1000\,mL}{1\,L} \times \dfrac{1\,\text{mol}\,Pb}{207.2\,g\,Pb} = 4.34\times10^{-8}\,M = 4.3\times10^{-8}\,M$

(b)   Change 60 m³ H_2O to cm³ (mL) H_2O to g H_2O (or g soln).

$$60\ m^3 \times \frac{100^3\ cm^3}{m^3} \times \frac{1\,g\ H_2O}{cm^3\ H_2O} = 6.0 \times 10^7\ g\ H_2O\ or\ soln$$

$$\frac{9.0\ g\ Pb}{1 \times 10^9\ g\ soln} \times 6.0 \times 10^7\ g\ soln = 0.54\ g\ Pb$$

13.99   Mole fraction ethyl alcohol, $X_{C_2H_5OH} = \dfrac{P_{C_2H_5OH}}{P^{\circ}_{C_2H_5OH}} = \dfrac{8\ torr}{100\ torr} = 0.08$

$$\frac{620 \times 10^3\ g\ C_{24}H_{50}}{338.6\ g/mol} = 1.83 \times 10^3\ mol\ C_{24}H_{50};\quad let\ y = mol\ C_2H_5OH$$

$$X_{C_2H_5OH} = 0.08 = \frac{y}{y + 1.83 \times 10^3};\ \ 0.92\,y = 146.4;\ \ y = 1.6 \times 10^2\ mol\ C_2H_5OH$$

(Strictly speaking, y should have 1 sig fig because 0.08 has 1 sig fig, but this severely limits the calculation.)

$$1.6 \times 10^2\ mol\ C_2H_5OH \times \frac{46\,g\ C_2H_5OH}{1\,mol} = 7.4 \times 10^3\ g\ or\ 7.4\ kg\ C_2H_5OH$$

13.101   (a)   The solvent vapor pressure over each solution is determined by the total particle concentrations present in the solutions. When the particle concentrations are equal, the vapor pressures will be equal and equilibrium established. The particle concentration of the nonelectrolyte is just 0.050 *M*, the ion concentration of the NaCl is 2 × 0.035 *M* = 0.070 *M*. Solvent will diffuse from the less concentrated nonelectrolyte solution. The level of the NaCl solution will rise, and the level of the nonelectrolyte solution will fall.

         (b)   Let x = volume of solvent transferred

$$\frac{0.050\ M \times 30.0\ mL}{(30.0 - x)\ mL} = \frac{0.070\ M \times 30.0\ mL}{(30.0 + x)\ mL};\,1.5(30.0 + x) = 2.1(30.0 - x)$$

               45 + 1.5 x = 63 − 2.1 x;   3.6 x = 18;   x = 5.0 = 5 mL transferred

               The volume in the nonelectrolyte beaker is (30.0 − 5.0) = 25.0 mL; in the NaCl beaker (30.0 + 5.0) = 35.0 mL.

13.102   If pure ethylene glycol is used in a radiator, it freezes at its regular (not depressed) freezing point, −11.5 °C. In the recommended 30% solution, water is the solvent and ethylene glycol the solute; the freezing point of water, 0.0 °C, is depressed by the nonvolatile solute ethylene glycol. On the day when the car owner's radiator froze, the temperature must have been between the freezing point of pure ethylene glycol and the depressed freezing point of the 30% solution. The depressed freezing point of the 30% solution must have been lower than the atmospheric temperature.

13.104   (a) $K_b = \dfrac{\Delta T_b}{m}$; $\Delta T_b = 47.46\ °C - 46.30\ °C = 1.16\ °C$

$$m = \frac{mol\ solute}{kg\ CS_2} = \frac{0.250\ mol}{400.0\ mL\ CS_2} \times \frac{1\,mL\ CS_2}{1.261\,g\ CS_2} \times \frac{1000\,g}{1\,kg} = 0.4956 = 0.496\ m$$

$$K_b = \frac{1.16\ ^\circ C}{0.4956\ m} = 2.34\ ^\circ C/\ m$$

(b) $$m = \frac{\Delta T_b}{K_b} = \frac{(47.08 - 46.30)\ ^\circ C}{2.34\ ^\circ C/\ m} = 0.333 = 0.33\ m$$

$$m = \frac{\text{mol unknown}}{\text{kg } CS_2};\quad m \times \text{kg } CS_2 = \frac{\text{g unknown}}{\text{MM unknown}};\quad MM = \frac{\text{g unknown}}{m \times \text{kg } CS_2}$$

$$50.0\ \text{mL } CS_2 \times \frac{1.261\ \text{g } CS_2}{1\ \text{mL}} \times \frac{1\ \text{kg}}{1000\ \text{g}} = 0.06305 = 0.0631\ \text{kg } CS_2$$

$$MM = \frac{5.39\ \text{g unknown}}{0.333\ m \times 0.06305\ \text{kg } CS_2} = 257 = 2.6 \times 10^2\ \text{g/mol}$$

13.105 $$M = \frac{\Pi}{RT} = \frac{57.1\ \text{torr}}{298\ K} \times \frac{1\ \text{atm}}{760\ \text{torr}} \times \frac{\text{mol-K}}{0.08206\ \text{L-atm}} = 3.072 \times 10^{-3} = 3.07 \times 10^{-3}\ M$$

$$\frac{0.036\ \text{g solute}}{100\ \text{g } H_2O} \times \frac{1000\ \text{g } H_2O}{1\ \text{kg } H_2O} = 0.36\ \text{g solute/kg } H_2O$$

Assuming molarity and molality are the same in this dilute solution, we can then say 0.36 g solute = 3.072×10^{-3} mol; MM = 117 g/mol. Because the salt is completely ionized, the formula weight of the lithium salt is **twice** this calculated value, or **234 g/mol**. The organic portion, $C_nH_{2n+1}O_2^-$, has a formula weight of 234 – 7 = 227 g. Subtracting 32 for the oxygens, and 1 to make the formula C_nH_{2n}, we have C_nH_{2n}, MM = 194 g/mol. Because each CH_2 unit has a mass of 14, n ≈ 194/14 ≈ 14. The formula for our salt is $LiC_{14}H_{29}O_2$.

Integrative Exercises

13.107   (a) $$\frac{0.015\ \text{g } N_2}{1\ \text{L blood}} \times \frac{1\ \text{mol } N_2}{28.01\ \text{g } N_2} = 5.355 \times 10^{-4} = 5.4 \times 10^{-4}\ \text{mol } N_2/\text{L blood}$$

      (b)   At 100 ft, the partial pressure of N_2 in air is 0.78 (4.0 atm) = 3.12 atm. This is just four times the partial pressure of N_2 at 1.0 atm air pressure. According to Henry's law, $S_g = kP_g$, a fourfold increase in P_g results in a fourfold increase in S_g, the solubility of the gas. Thus, the solubility of N_2 at 100 ft is $4(5.355 \times 10^{-4}\ M)$ = $2.142 \times 10^{-3} = 2.1 \times 10^{-3}\ M$.

      (c)   If the diver suddenly surfaces, the amount of N_2/L blood released is the difference in the solubilities at the two depths:

$$(2.142 \times 10^{-3}\ \text{mol/L} - 5.355 \times 10^{-4}\ \text{mol/L}) = 1.607 \times 10^{-3} = 1.6 \times 10^{-3}\ \text{mol } N_2/\text{L blood}.$$

At surface conditions of 1.0 atm external pressure and 37 °C = 310 K,

$$V = \frac{nRT}{P} = 1.607 \times 10^{-3}\ \text{mol} \times \frac{310\ K}{1.0\ \text{atm}} \times \frac{0.08206\ \text{L-atm}}{\text{mol-K}} = 0.041\ L$$

That is, 41 mL of tiny N_2 bubbles are released from each liter of blood.

13.108   The stronger the intermolecular forces, the higher the heat (enthalpy) of vaporization.

      (a)   None of the substances are capable of hydrogen bonding in the pure liquid, and they have similar molar masses. All intermolecular forces are van der Waals forces, dipole-dipole, and dispersion forces. In decreasing order of strength of forces: acetone > acetaldehyde > ethylene oxide > cyclopropane

The first three compounds have dipole-dipole and dispersion forces, the last only dispersion forces.

(b) The order of solubility in hexane should be the reverse of the order above. The least polar substance, cyclopropane, will be most soluble in hexane. Ethanol, CH_3CH_2OH, is capable of hydrogen bonding with the three polar compounds. Thus, acetaldehyde, acetone, and ethylene oxide should be more soluble than cyclopropane, but without further information we cannot distinguish among the polar molecules.

13.110   (a) $Zn(s) + H_2SO_4(aq) \rightarrow ZnSO_4(aq) + H_2(g)$

$$2.050\,g\,Zn \times \frac{1\,mol\,Zn}{65.39\,g\,Zn} = 0.03135\,mol\,Zn$$

$1.00\,M\,H_2SO_4 \times 0.0150\,L = 0.0150\,mol\,H_2SO_4$

Because Zn and H_2SO_4 react in a 1:1 mole ratio, H_2SO_4 is the limiting reactant; 0.0150 mol of $H_2(g)$ are produced.

(b) $P = \dfrac{nRT}{V} = \dfrac{0.0150\,mol}{0.122\,L} \times \dfrac{0.08206\,L\text{-}atm}{mol\text{-}K} \times 298\,K = 3.0066 = 3.01\,atm$

(c) $S_{H_2} = kP_{H_2} = \dfrac{7.8 \times 10^{-4}\,mol}{L\text{-}atm} \times 3.0066\,atm = 0.002345 = 2.3 \times 10^{-3}\,M$

$$\frac{0.002345\,mol\,H_2}{L\,soln} \times 0.0150\,L = 3.518 \times 10^{-5} = 3.5 \times 10^{-5}\,mol\,dissolved\,H_2$$

$$\frac{3.5 \times 10^{-5}\,mol\,dissolved\,H_2}{0.0150\,mol\,H_2\,produced} \times 100 = 0.23\%\,dissolved\,H_2$$

This is approximately 2.3 parts per thousand; for every 10,000 H_2 molecules, 23 are dissolved. It was reasonable to ignore dissolved $H_2(g)$ in part (b).

13.111   (a) $\dfrac{1.3 \times 10^{-3}\,mol\,CH_4}{L\,soln} \times 4.0\,L = 5.2 \times 10^{-3}\,mol\,CH_4$

$$V = \frac{nRT}{P} = \frac{5.2 \times 10^{-3}\,mol \times 298\,K}{1.0\,atm} \times \frac{0.08206\,L\text{-}atm}{mol\text{-}K} = 0.13\,L$$

(b) All three hydrocarbons are nonpolar; they have zero net dipole moment. In CH_4 and C_2H_6, the C atoms are tetrahedral and all bonds are σ bonds. C_2H_6 has a higher molar mass than CH_4, which leads to stronger dispersion forces and greater water solubility. In C_2H_4, the C atoms are trigonal planar and the π electron cloud is symmetric above and below the plane that contains all the atoms. This planar arrangement facilitates contact between molecules, leading to stronger dispersion forces. The π cloud in C_2H_4 is an area of concentrated electron density that experiences attractive forces with the positive ends of H_2O molecules. These forces increase the solubility of C_2H_4 relative to the other hydrocarbons.

(c)   The molecules have similar molar masses. NO is most soluble because it is polar. The triple bond in N_2 is shorter than the double bond in O_2. It is more difficult for H_2O molecules to surround the smaller N_2 molecules, so they are less soluble than O_2 molecules.

(d) H_2S and SO_2 are polar molecules capable of hydrogen bonding with water. Hydrogen bonding is the strongest force between neutral molecules and causes the much greater solubility. H_2S is weakly acidic in water. SO_2 reacts with water to form H_2SO_3, a weak acid. The large solubility of SO_2 is a sure sign that a chemical process has occurred.

(e) N_2 and C_2H_4. N_2 is too small to be easily hydrated, so C_2H_4 is more soluble in H_2O.

      NO (31) and C_2H_6 (30). The structures of these two molecules are very different, yet they have similar solubilities. NO is slightly polar, but too small to be easily hydrated. The larger C_2H_6 is nonpolar, but more polarizable (stronger dispersion forces).

      NO (31) and O_2 (32). The slightly polar NO is more soluble than the slightly larger (longer O=O bond than N $\equiv$ O bond) but nonpolar O_2.

13.112   The resulting solution is very dilute, so assume ideal behavior. Assume the amount of water consumed in the reaction is negligible. Ignore the solubility of $H_2(g)$ in the solution (see Solution 3.110).

$$1.0 \text{ mm}^3 \times \frac{0.535 \text{ g}}{\text{cm}^3} \times \frac{1^3 \text{ cm}^3}{10^3 \text{ mm}^3} = 5.35 \times 10^{-4} = 5.4 \times 10^{-4} \text{ g Li}$$

$$5.35 \times 10^{-4} \text{ g Li} \times \frac{1 \text{ mol Li}}{6.941 \text{ g Li}} = 7.708 \times 10^{-5} = 7.7 \times 10^{-5} \text{ mol Li}$$

mol Li = mol LiOH; 2 mol ions per mol LiOH

$$7.708 \times 10^{-5} \text{ mol Li} = 7.708 \times 10^{-5} \text{ mol LiOH} = 1.542 \times 10^{-4} \text{ mol ions} = 1.5 \times 10^{-4} \text{ mol ions}$$

$$m = \frac{1.542 \times 10^{-4} \text{ mol ions}}{0.500 \text{ L } H_2O} \times \frac{1 \text{ L}}{1000 \text{ mL}} \times \frac{1 \text{ mL } H_2O}{0.997 \text{ g } H_2O} \times \frac{1000 \text{ g}}{1 \text{ kg}} = 3.092 \times 10^{-4} = 3.1 \times 10^{-4} \, m$$

$$\Delta T_f = K_f \, m = -1.86(3.092 \times 10^{-4}) = -5.8 \times 10^{-4} \, ^\circ C; \quad T_f = 0.00000 - 0.00058 = -0.00058 \, ^\circ C$$

The freezing point of the LiOH(aq) solution is essentially zero.

13.114   (a)   True solutions do not scatter light, colloids do. Below the critical micelle concentration, cmc, the mixture of solvent and surfactant is a true solution. Above the cmc, the mixture is a colloid. The micelles are too large to be perfectly mixed in the solvent. They are suspended in the solvent, resulting in a colloid that scatters light.

      (b)   Surfactant monomers are anions; the "head" carries a negative charge. Below the cmc, each monomer is an independent particle. Above the cmc, many monomers aggregate into one micelle, drastically reducing the effective number of particles "in solution." (A micelle does have a greater negative charge than a monomer.) This dramatically changes the ionic conductivity.

(c)   The interior of a micelle is a hydrophobic environment.   If a dye molecule becomes entrapped in a micelle, it will fluoresce.   In the absence of micelles, the dye will not fluoresce in an aqueous solution.

At low sodium stearate concentrations, the fluorescent intensity will be low. As the surfactant concentration increases, fluorescent intensity increases gradually until the cmc is reached (assuming a few micelles form at concentrations below cmc).   At the cmc, there is a large increase in fluorescent intensity.   At concentrations greater than the cmc, fluorescent intensity remains high and probably increases gradually.

14 Chemical Kinetics

Visualizing Concepts

14.2   *Analyze/Plan.* Given the plot of [X] vs time, answer questions about reaction speed and rate. Consider the definitions of average reaction rate and instantaneous rate. *Solve.*

(a)   True. X is a product, because its concentration increases with time.

(b)   False. The speed of a reaction is its rate, or how quickly the concentration of a reactant or product changes over time. This graph shows how [X] increases over time. The rate at any particular time, the instantaneous rate, is the slope of the tangent to the curve at that time. Visualizing the tangents at points 0, 1, 2, and 3, we see that the slopes of these lines are decreasing with time. That is, the rate of reaction is decreasing; the reaction is slowing down as time progresses.

(c)   True. The average rate of reaction between any two points on the graph is the slope of the line connecting the two points. Points 1 and 2 are earlier in the reaction when more reactants are available, so the average rate of formation of products is greater. As reactants are used up, the rate of X production decreases, and the average rate between points 2 and 3 is smaller.

(d)   False. The graph shows the build-up of [X] as the reaction progresses. [X] will not decrease once it reaches its maximum. (In the case of a chemical equilibrium, [X] will increase until reaching its equilibrium concentration, but it will not decrease after equilibrium is established.)

14.4   *Analyze/Plan.* Given a plot of increase in [M] over time, answer questions about reaction rate and progress. Consider the definition of reaction rate. *Solve.*

(a)   Yes. The plot of [M] versus time from t = 0 to t = 15 is a straight line, so [M] increases at a constant rate and the reaction occurs at a constant rate. The rate is zero after t = 15 min.

(b)   Yes. [M] does not change after 15 min. This means that no more M is being produced and the reaction is no longer occurring.

(c)   Statement (ii) is correct. After t = 15 min, 0.00 mol K and 0.20 mol L remain in solution. When an additional 0.20 mol K is added at 30 min, the second half of the reaction occurs. The plot of [M] versus t looks like the first half of the reaction, from t = 0 to t = 15 min.

14.5   *Analyze.* Given three mixtures and the order of reaction in each reactant, determine which mixture will have the fastest initial rate.

Plan. Write the rate law. Count the number of reactant molecules in each container. The three containers have equal volumes and total numbers of molecules. Use the molecule count as a measure of concentration of NO and O_2. Calculate the initial rate for each container and compare.

Solve. Rate = $k[NO]^2[O_2]$; rate is proportional to $[NO]^2[O_2]$

Container	[NO]	[O_2]	$[NO]^2[O_2] \propto$ rate
(1)	5	4	100
(2)	7	2	98
(3)	3	6	54

The relative rates in containers (1) and (2) are very similar, with (1) having the slightly faster initial rate.

14.6   *Plan.* For a first-order reaction, a plot of ln[A] versus time is linear, as shown in the diagram. The slope is –k, and the intercept is $[A]_0$. According to the Arrhenius equation 14.19, k increases with increasing temperature. *Solve.*

(a)   Graphs 1 and 2 have the same slope and thus the same rate constant, k. These experiments are done at the same temperature. The y-intercepts of the two graphs are different; the experiments had different initial concentrations of A.

(b)   Graphs 2 and 3 have the same y-intercept and thus the same starting concentration of A. The slopes of the two graphs are different, so their rate constants are different and they occur at different temperatures. Graph 3, with the smaller slope and k value will occur at the lower temperature.

14.7   *Analyze.* Given concentrations of reactants and products at two times, as represented in the diagram, find $t_{1/2}$ for this first-order reaction.

Plan. For a first-order reaction, $t_{1/2} = 0.693/k$; $t_{1/2}$ depends only on k. Use Equation 14.12 to solve for k. *Solve.*

(a)   Because reactants and products are in the same container, use number of particles as a measure of concentration. The red dots are reactant A, and the blue are product B. $[A]_0 = 16$, $[A]_{30} = 4$, t = 30 min.

$$\ln\frac{[A]_t}{[A]_0} = -kt. \quad \ln(4/16) = -k(30 \text{ min}); \quad \frac{-1.3863}{-30\text{min}} = k;$$

$k = 0.046210 = 0.0462 \text{ min}^{-1}$

$t_{1/2} = 0.693/k = 0.693/0.046210 = 15 \text{ min}$

By examination, $[A]_0 = 16$, $[A]_{30} = 4$. After 1 half-life, [A] = 8; after a second half-life, [A] = 4. Thirty minutes represents exactly 2 half-lives, so $t_{1/2} = 15$ min. [This is more straightforward than the calculation, but a less general method.]

(b)   After 4 half-lives, $[A]_t = [A]_0 \times 1/2 \times 1/2 \times 1/2 \times 1/2 = [A]_0/16$. In general, after n half-lives, $[A] = [A]_0/2^n$.

14.8   (a)   Plot (v) is zero order.

(b)   Plot (i) is first order.

(c)   Plot (iii) is second order.

14.10   *Analyze/Plan.* On a plot of ln k versus 1/T, the slope is $-E_a/R$ and the y-intercept is ln A, where E_a is activation energy and A is the frequency factor.   *Solve.*

(a)   Blue. The magnitude of the slope of the blue line is greater than that of the red line.

(b)   Red. The y-intercept of the red line is greater than that of the blue line.

14.11   (a)   False. The red pathway is slower, because it has the greater activation energy, E_a.

(b)   True. For both reactions, the difference in potential energy between the products and the activated complex is greater than the difference between the reactants and the activated complex.

(c)   True. ΔE is the difference between the energy of the reactants and the energy of the products.

14.13   This is the profile of a two-step mechanism, A → B and B → C. There is one intermediate, B. Because there are two energy maxima, there are two transition states. The B → C step is faster, because its activation energy is smaller. For the overall reaction A → C, ΔE is negative, because the potential energy of the products is lower than the potential energy of the reactants.

14.14   The most likely transition state shows the relative geometry of both reactants and products. It is reasonable to assume that multiple bonds, with greater total bond energy, remain intact at the expense of single bonds. In the black-and-white diagram below, open circles represent the red balls and closed circles represent the blue.

14.15   (a) $A_2 + AB + AC \rightarrow BA_2 + A + AC$

$BA_2 + A + AC \rightarrow A_2 + BA_2 + C$

net: $AB + AC \rightarrow BA_2 + C$

(b)   A is the intermediate; it is produced and consumed.

(c) A_2 is the catalyst; it is consumed and reproduced.

Reaction Rates (Sections 14.1 and 14.2)

14.18   (a)   *M/s*

(b)   The hotter the oven, the faster the cake bakes. Milk sours faster in hot weather than cool weather.

(c)   The *average rate* is the rate over a period of time, whereas the *instantaneous rate* is the rate at a particular time.

14.20

Time (s)	Mol A	(a) Mol B	Δ Mol A	(b) Rate –(Δ mol A/s)
0	0.100	0.000		
40	0.067	0.033	−0.033	8.3×10^{-4}
80	0.045	0.055	−0.022	5.5×10^{-4}
120	0.030	0.070	−0.015	3.8×10^{-4}
160	0.020	0.080	−0.010	2.5×10^{-4}

(c) (ii) The volume of the container must be known to report the rate in units of concentration (mol/L) per time.

14.22

(a)

Time (min)	Time Interval (min)	Concentration (M)	Δ M	Rate (M/s)
0.0		1.85		
54.0	54.0	1.58	−0.27	8.3×10^{-5}
107.0	53.0	1.36	−0.22	6.9×10^{-5}
215.0	108	1.02	−0.34	5.2×10^{-5}
430.0	215	0.580	−0.44	3.4×10^{-5}

(b) $-\dfrac{\Delta M}{\Delta t} = -\dfrac{(1.85-0.580)\,M}{(430-0)\,min} \times \dfrac{1\,min}{60\,s} = 4.9225 \times 10^{-5} = 4.92 \times 10^{-5}\ M/s$

(c) $-\dfrac{\Delta M}{\Delta t} = -\dfrac{(1.02-1.58)\,M}{(215-54)\,min} \times \dfrac{1\,min}{60\,s} = 5.8 \times 10^{-5}\ M/s$

$-\dfrac{\Delta M}{\Delta t} = -\dfrac{(0.580-1.36)\,M}{(430-107)\,min} \times \dfrac{1\,min}{60\,s} = 4.0 \times 10^{-5}\ M/s$

The average rate between t = 54.0 and t = 215.0 min is greater. In general, the rate of a reaction decreases over time.

(d) From the slopes of the lines in the figure at the right, the rates are: at 75.0 min, $4.2 \times 10^{-3}\,M/min$, or $7.0 \times 10^{-5}\,M/s$; at 250 min, $2.1 \times 10^{-3}\,M/min$ or $3.5 \times 10^{-5}\,M/s$.

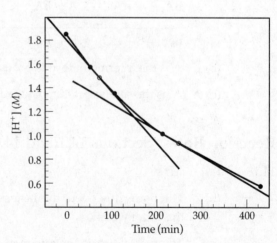

14.24   (a)   Rate $= -\Delta[H_2O]/2\Delta t = \Delta[H_2]/2\Delta t = \Delta[O_2]/\Delta t$

       (b)   Rate $= -\Delta[SO_2]/2\Delta t = -\Delta[O_2]/\Delta t = \Delta[SO_3]/2\Delta t$

       (c)   Rate $= -\Delta[NO]/2\Delta t = -\Delta[H_2]/2\Delta t = \Delta[N_2]/\Delta t = \Delta[H_2O]/2\Delta t$

       (d)   Rate $= -\Delta[N_2]/\Delta t = -\Delta[H_2]/2\Delta t = \Delta[N_2H_4]/\Delta t$

14.26   (a) $-\Delta[C_2H_4]/\Delta t = \Delta[CO_2]/2\Delta t = \Delta[H_2O]/2\Delta t$

            $-2\Delta[C_2H_4]/\Delta t = \Delta[CO_2]/\Delta t = \Delta[H_2O]/\Delta t$

            C_2H_4 is burning, $-\Delta[C_2H_4]/\Delta t = 0.036 \; M/s$

            CO_2 and H_2O are produced, at twice the rate that C_2H_4 is consumed.

            $\Delta[CO_2]/\Delta t = \Delta[H_2O]/\Delta t = 2(0.036) \; M/s = 0.072 \; M/s$

       (b)   In this reaction, pressure is a measure of concentration.

            $-\Delta[N_2H_4]/\Delta t = -\Delta[H_2]/\Delta t = \Delta[NH_3]/2\Delta t$

            N_2H_4 is consumed, $-\Delta[N_2H_4]/\Delta t = 74 \; torr/h$

            H_2 is consumed, $-\Delta[H_2]/\Delta t = 74 \; torr/h$

            NH_3 is produced at twice the rate that N_2H_4 and H_2 are consumed,

            $\Delta[NH_3]/\Delta t = -2\Delta[N_2H_4]/\Delta t = 2(74) \; torr/h = 148 \; torr/h$

            $\Delta P_T/\Delta t = (+148 \; torr/h - 74 \; torr/h - 74 \; torr/h) = 0 \; torr/h$

Rate Laws (Section 14.3)

14.28   (a)   Rate $= k[A][C]^2$

       (b)   Rate is proportional to [A], rate doubles

       (c)   Rate is not affected by [B], no change

       (d)   Rate changes as $[C]^2$, rate increases by a factor of 3^2 or 9

       (e)   Rate increases by a factor of $(3)(3)^2 = 27$

       (f)   Rate decreases by a factor of $(1/2)(1/2)^2 = 1/8$

14.30   (a)   Rate $= k[H_2][NO]^2$

       (b)   Rate $= (6.0 \times 10^4 \; M^{-2} s^{-1})(0.035 \; M)^2 (0.015 \; M) = 1.1 \; M/s$

       (c)   Rate $= (6.0 \times 10^4 \; M^{-2} s^{-1})(0.10 \; M)^2 (0.010 \; M) = 6.0 \; M/s$

       (d)   Rate $= (6.0 \times 10^4 \; M^{-2} s^{-1})(0.010 \; M)^2 (0.030 \; M) = 0.18 \; M/s$

14.32   (a, b)   Rate $= k[C_2H_5Br][OH^-]$; $k = \dfrac{rate}{[C_2H_5Br][OH^-]}$

            at 298 K, $k = \dfrac{1.7 \times 10^{-7} \; M/s}{[0.0477 \; M][0.100 \; M]} = 3.6 \times 10^{-5} \; M^{-1}s^{-1}$

(c) Adding an equal volume of ethyl alcohol reduces both $[C_2H_5Br]$ and $[OH^-]$ by a factor of two. New rate $= (1/2)(1/2) = 1/4$ of old rate.

14.34 (a) From the data given, when $[ClO_2]$ increases by a factor of 3 (experiment 2 to experiment 1), the rate increases by a factor of 9. When $[OH^-]$ increases by a factor of 3 (experiment 2 to experiment 3), the rate increases by a factor of 3. The reaction is second order in $[ClO_2]$ and first order in $[OH^-]$. Rate $= k[ClO_2]^2[OH^-]$.

(b) Using data from experiment 2:

$$k = \frac{rate}{[ClO_2]^2[OH^-]} = \frac{0.00276 \ M/s}{(0.020 \ M)^2(0.030 \ M)} = 2.3 \times 10^2 \ M^{-2}s^{-1}$$

(c) Rate $= 2.3 \times 10^2 \ M^{-2}s^{-1}(0.100 \ M)^2(0.050 \ M) = 0.115 = 0.12 \ M/s$

14.36 *Analyze/Plan.* Follow the logic in Sample Exercise 14.6 to deduce the rate law. Rearrange the rate law to solve for k and deduce units. Calculate a k value for each set of concentrations and then average the three values. *Solve.*

(a) Doubling [NO] while holding $[O_2]$ constant increases the rate by a factor of 4 (experiments 1 and 2). Doubling $[O_2]$ while holding [NO] constant doubles the rate (experiments 2 and 3). The reaction is second order in [NO] and first order in $[O_2]$. Rate $= k[NO]^2[O_2]$.

(b, c) From experiment 1: $k_1 = \dfrac{1.41 \times 10^{-2} \ M/s}{(0.0126 \ M)^2(0.0125 \ M)} = 7105 = 7.11 \times 10^3 \ M^{-2}s^{-1}$

$k_2 = 0.113/(0.0252)^2(0.0250) = 7118 = 7.12 \times 10^3 \ M^{-2}s^{-1}$

$k_3 = 5.64 \times 10^{-2}/(0.0252)^2(0.125) = 7105 = 7.11 \times 10^3 \ M^{-2}s^{-1}$

$k_{avg} = (7105 + 7118 + 7105)/3 = 7109 = 7.11 \times 10^3 \ M^{-2}s^{-1}$

(d) Rate $= 7.109 \times 10^3 \ M^{-2}s^{-1}(0.0750 \ M)^2(0.0100 \ M) = 0.3999 = 0.400 \ M/s$.

(e) The data are given in terms of the disappearance of NO. Use Equation 14.4 to relate the disappearance of NO to the disappearance of O_2.

$-\Delta[NO]/2\Delta t = -[O_2]/\Delta t$

For the concentrations given in part (d), $\Delta[NO]/\Delta t = 0.400 \ M/s$.

$\Delta[O_2]/\Delta t = \Delta[NO]/2\Delta t = 0.400 \ M/s/2 = 0.200 \ M/s$

14.38 (a) Increasing $[S_2O_8^{2-}]$ by a factor of 1.5 while holding $[I^-]$ constant increases the rate by a factor of 1.5 (Experiments 1 and 2). Doubling $[S_2O_8^{2-}]$ and increasing $[I^-]$ by a factor of 1.5 triples the rate ($2 \times 1.5 = 3$, experiments 1 and 3). Thus the reaction is first order in both $[S_2O_8^{2-}]$ and $[I^-]$; rate $= k[S_2O_8^{2-}][I^-]$.

(b) $k = rate/[S_2O_8^{2-}][I^-]$

$k_1 = 2.6 \times 10^{-6} \ M/s/(0.018 \ M)(0.036 \ M) = 4.01 \times 10^{-3} = 4.0 \times 10^{-3} \ M^{-1}s^{-1}$

$k_2 = 3.9 \times 10^{-6}/(0.027)(0.036) = 4.01 \times 10^{-3} = 4.01 \times 10^{-3} = 4.0 \times 10^{-3} \ M^{-1}s^{-1}$

$k_3 = 7.8 \times 10^{-6} / (0.036)(0.054) = 4.01 \times 10^{-3} = 4.01 \times 10^{-3} = 4.0 \times 10^{-3}\ M^{-1}\,s^{-1}$

$k_4 = 1.4 \times 10^{-5} / (0.050)(0.072) = 3.89 \times 10^{-3} = 3.9 \times 10^{-3}\ M^{-1}\,s^{-1}$

$k_{avg} = 3.98 \times 10^{-3} = 4.0 \times 10^{-3}\ M^{-1}\,s^{-1}$

(c) $-\Delta[S_2O_8^{2-}]/\Delta t = -\Delta[I^-]/3\Delta t$; the rate of disappearance of $S_2O_8^{2-}$ is one-third the rate of disappearance of I^-.

(d)   Note that the data are given in terms of disappearance of $S_2O_8^{2-}$.

$$\frac{-\Delta[I^-]}{\Delta t} = \frac{-3\Delta[S_2O_8^{2-}]}{\Delta t} = 3(3.98 \times 10^{-3}\ M^{-1}s^{-1})(0.025\ M)(0.050\ M) = 1.5 \times 10^{-5}\ M/s$$

Change of Concentration with Time (Section 14.4)

14.40   (a)   A graph of $1/[A]$ versus time yields a straight line for a second-order reaction.

        (b)   On a graph of $1/[A]$ versus time, the slope of the straight line is the rate constant, k.

        (c)   The half-life of a first-order reaction is independent of $[A]_0$, $t_{1/2} = 0.693/k$. Whereas, the half-life of a second-order reaction does depend on $[A]_0$, $t_{1/2} = 1/k[A]_0$.

14.42   (a)   For a first-order reaction, $t_{1/2} = 0.693/k$.

            $t_{1/2} = 0.693/0.271\ s^{-1} = 2.5572 = 2.56\ s$

        (b)   For a first-order reaction, $\ln[A]_t - \ln[A]_0 = -kt$. $\ln[A]_t = -kt + \ln[A]_0$

            $[A]_0 = 0.050\ M\ I_2$, $t = 5.12\ s$, $k = 0.271\ s^{-1}$

            $\ln[I_2] = -0.271\ s^{-1}(5.12\ s) + \ln(0.050)$

            $\ln[I_2] = -4.3833$, $[I_2] = 0.0125\ M$

            Check. 5.12 s is 2 half-lives. $[I_2]$ should be reduced by a factor of 4, and it is.

14.44   (a)   Using Equation 14.13 for a first-order reaction: $\ln[A]_t = -kt + \ln[A]_0$

            5.0 min = 300 s; $[N_2O_5]_0 = (0.0250\ mol/2.0\ L) = 0.0125 = 0.013\ M$

            $\ln[N_2O_5]_{300} = -(6.82 \times 10^{-3}\ s^{-1})(300\ s) + \ln(0.0125)$

            $\ln[N_2O_5]_{300} = -2.0460 + (-4.3820) = -6.4280 = -6.43$

            $[N_2O_5]_{300} = 1.616 \times 10^{-3} = 1.6 \times 10^{-3}\ M$

            mol $N_2O_5 = 1.616 \times 10^{-3}\ M \times 2.0\ L = 3.2 \times 10^{-3}\ mol$

        (b) $[N_2O_5]_t = 0.010\ mol/2.0\ L = 0.0050\ M$; $[N_2O_5]_0 = 0.0125\ M$

            $\ln(0.0050) = -(6.82 \times 10^{-3}\ s^{-1})(t) + \ln(0.0125)$

$$t = \frac{-[\ln(0.0050) - \ln(0.0125)]}{(6.82 \times 10^{-3}\,s^{-1})} = 134.35 = 1.3 \times 10^2\ s \times \frac{1\ min}{60\ s} = 2.24 = 2.2\ min$$

        (c) $t_{1/2} = 0.693/k = 0.693/6.82 \times 10^{-3}\ s^{-1} = 101.6 = 102\ s$ or 1.69 min

14.46

Time (s)	P_{CH_3NC}	$\ln P_{CH_3NC}$
0	502	6.219
2,000	335	5.814
5,000	180	5.193
8,000	95.5	4.559
12,000	41.7	3.731
15,000	22.4	3.109

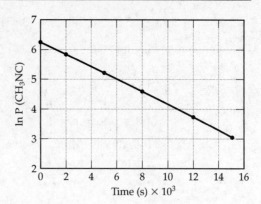

A graph of ln P vs t is linear with a slope of -2.08×10^{-4} s^{-1}. The rate constant, k, = $-$slope = 2.08×10^{-4} s^{-1}. Half-life = $t_{1/2}$ = $0.693/k = 3.33 \times 10^3$ s.

14.48   (a)   Make both first- and second-order plots to see which is linear. Moles is a satisfactory concentration unit, because volume is constant.

Time (s)	mol A	ln (mol A)	1/mol A
0	0.1000	−2.303	10.00
40	0.067	−2.70	14.9
80	0.045	−3.10	22.2
120	0.030	−3.51	33.3
160	0.020	−3.91	50.0

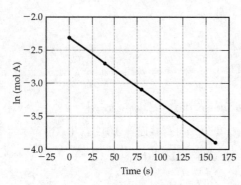 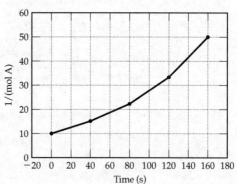

The plot of ln (mol A) vs time is linear, so the reaction is first order in A.

(b)   k = −slope = − [−3.91 − (−2.70)]/120 = 0.010083 = 0.0101 s^{-1}
(The best fit to this line yields the same value for the slope, 0.01006 = 0.0101 s^{-1})

(c) $t_{1/2} = 0.693/k = 0.693/0.010083$ s^{-1} = 68.7 s

14.50   (a)   Make both first- and second-order plots to see which is linear.

Time (min)	$[C_{12}H_{22}O_{11}](M)$	$\ln[C_{12}H_{22}O_{11}]$	$1/[C_{12}H_{22}O_{11}]$
0	0.316	−1.152	3.16
39	0.274	−1.295	3.65
80	0.238	−1.435	4.20
140	0.190	−1.661	5.26
210	0.146	−1.924	6.85

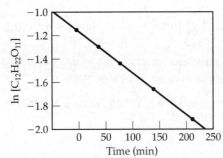 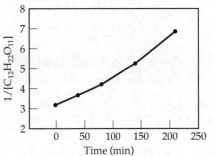

The plot of $\ln [C_{12}H_{22}O_{11}]$ is linear, so the reaction is first order in $C_{12}H_{22}O_{11}$.

  (b) $k = -\text{slope} = -[-1.924 - (-1.295)] / 171 \text{ min} = 3.68 \times 10^{-3} \text{ min}^{-1}$

       (The slope of the best-fit line is $-3.67 \times 10^{-3} \text{ min}^{-1}$.)

  (c)   For a reaction zero order in sucrose, the rate does not change as [sucrose] changes. A plot of [sucrose] versus time is linear with negative slope, until all reactant is consumed. $[\text{sucrose}]_t = -kt + [\text{sucrose}]_0$

       @39 min, $[\text{sucrose}] = -3.68 \times 10^{-3} \text{ min}^{-1} (39 \text{ min}) + 0.316 \, M = 0.17 \, M$

       @80 min, $[\text{sucrose}] = -3.68 \times 10^{-3} \text{ min}^{-1} (80 \text{ min}) + 0.316 \, M = 0.022 \, M$

       @140 min, $[\text{sucrose}] = -3.68 \times 10^{-3} \text{ min}^{-1} (140 \text{ min}) + 0.316 \, M = 0 \, M$

       @210 min, $[\text{sucrose}] = 0 \, M$. All sucrose is consumed at $(0.316/3.68 \times 10^{-3} \text{ min}^{-1}) = 85.9 \text{ min}$.

Temperature and Rate (Section 14.5)

14.52   (a)   The orientation factor is less important in $H + Cl \rightarrow HCl$, because the reactants are monatomic and spherical (nondirectional); all collision orientations are equally effective.

  (b)   The kinetic-molecular theory tells us that at some temperature T, there will be a distribution of molecular speeds and kinetic energies, and that the average kinetic energy of the sample is proportional to temperature. That is, as temperature of the sample increases, the average speed and kinetic energy of the molecules increases. At higher temperatures, there will be more molecular collisions (owing to greater speeds) and more energetic collisions (owing to greater kinetic energies). Overall there will be more collisions that have sufficient energy to form an activated complex, and the reaction rate will be greater.

14.54   (a) $f = e^{-E_a/RT}$ $E_a = 160$ kJ/mol $= 1.60 \times 10^5$ J/mol, T = 500 K

$$-E_a/RT = -\frac{1.60 \times 10^5 \text{ J/mol}}{500 \text{ K}} \times \frac{\text{mol - K}}{8.314 \text{ J}} = -38.489 = -38.5$$

$$f = e^{-38.489} = 1.924 \times 10^{-17} = 2 \times 10^{-17}$$

(b) $-E_a/RT = -\dfrac{1.60 \times 10^5 \text{ J/mol}}{520 \text{ K}} \times \dfrac{\text{mol - K}}{8.314 \text{ J}} = -37.009 = -37.0$

$$f = e^{-37.009} = 8.45712 \times 10^{-17} = 8.46 \times 10^{-17}$$

$$\frac{f \text{ at } 520 \text{ K}}{f \text{ at } 500 \text{ K}} = \frac{8.46 \times 10^{-17}}{1.92 \times 10^{-17}} = 4.41$$

An increase of 20 K means that 4.41 times more molecules have this energy.

14.56   *Analyze/Plan.* Use the definitions of activation energy ($E_{max} - E_{react}$) and ΔE ($E_{prod} - E_{react}$) to sketch the graph and calculate E_a for the reverse reaction. *Solve.*

(a)   (b) E_a(reverse) = 18 kJ

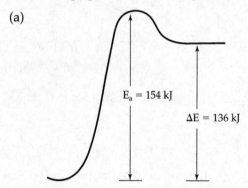

$E_a = 154$ kJ

$\Delta E = 136$ kJ

14.58   (a)   False. If you measure the rate constant for a reaction at different temperatures, you can calculate the overall activation energy, E_a, for the reaction.

(b)   False. Exothermic reactions are not necessarily faster than endothermic reactions. (The rate of a reaction is not determined by the overall enthalpy change going from reactants to products.)

(c)   False. If you double the temperature for a reaction, there is no change to the activation energy, E_a.

14.60 E_a for the reverse reaction is:

(a) 45 – (–25) = 70 kJ   (b) 35 – (–10) = 45 kJ   (c) 55 – 10 = 45 kJ

Based on the magnitude of E_a, the reverse of reactions (b) and (c) occur at the same rate, which is faster than the reverse of reaction (a).

14.62 $T_1 = 737\,°C + 273 = 1010$ K, $k_1 = 0.0796$ $M^{-1}s^{-1}$;

$T_2 = 947\,°C + 273 = 1220$ K, $k_2 = 0.0815$ $M^{-1}s^{-1}$

$$\ln\left(\frac{k_1}{k_2}\right) = \frac{E_a}{R}\left(\frac{1}{T_2} - \frac{1}{T_1}\right)$$

$$\ln\left(\frac{0.0796}{0.0815}\right) = \frac{E_a}{8.314 \text{ J/mol}}\left(\frac{1}{1220} - \frac{1}{1010}\right)$$

$$-0.023589 = \frac{E_a\ (-1.704 \times 10^{-4})}{8.314 \text{ J/mol}}$$

$$E_a = \frac{8.314\ (-0.023589)\ \text{J/mol}}{(-1.704 \times 10^{-4})} = 1.151 \times 10^3 \text{ J/mol} = 1.15 \text{ kJ/mol}$$

14.64

k	ln k	T(K)	$1/T(\times 10^3)$
0.028	−3.58	600	1.67
0.22	−1.51	650	1.54
1.3	0.26	700	1.43
6.0	1.79	750	1.33
23	3.14	800	1.25

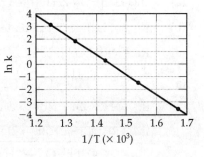

Using the relationship $\ln k = \ln A - E_a/RT$, the slope, $-15.94 \times 10^3 = -16 \times 10^3$, is $-E_a/R$. $E_a = 15.94 \times 10^3 \times 8.314$ J/mol $= 1.3 \times 10^2$ kJ/mol. To calculate A, we will use the rate data at 700 K. From the equation given above, $0.262 = \ln A - 15.94 \times 10^3/700$; $\ln A = 0.262 + 22.771$. $A = 1.0 \times 10^{10}$.

Reaction Mechanisms (Section 14.6)

14.66   (a)   The *molecularity* of a process indicates the number of molecules that participate as reactants in the process. A unimolecular process has one reactant molecule, a bimolecular process has two reactant molecules and a termolecular process has three reactant molecules.

           (b)   Termolecular processes are rare because it is highly unlikely that three molecules will simultaneously collide with the correct energy and orientation to form an activated complex.

           (c)   An *intermediate* is a substance that is produced and then consumed during a chemical reaction. It does not appear in the balanced equation for the overall reaction.

           (d)   A *transition state* is a high energy complex formed when one or more reactants collide and distort in a way that can lead to formation of product(s). An *intermediate* is the product of an early elementary reaction in a multistep reaction mechanism. A transition state occurs at an energy maximum or peak of a reaction profile as in Figure 14.20. An intermediate exists at an energy minimum or trough of a reaction profile. Every reaction, single- or multi-step, has a transition state. Only multistep reactions have intermediates.

14.68   (a)   bimolecular, rate $= k[NO]^2$

           (b)   unimolecular, rate $= k[C_3H_6]$

           (c)   unimolecular, rate $= k[SO_3]$

14.70   (a)   Two elementary reactions; two energy maxima

          (b)   One intermediate; one energy minimum between reactants and products

          (c)   The second step is rate-limiting; second energy maximum and E_a is larger.

          (d)   For the overall reaction, ΔE is negative, because the potential energy of the products is lower than the potential energy of the reactants.

14.72   (a) $2H_2O_2(aq) \rightarrow 2H_2O(l) + O_2(g)$

          (b) $IO^-(aq)$ is the intermediate.

          (c)   Rate = $k[H_2O_2][I^-]$

14.74   (a)   i. $HBr + O_2 \rightarrow HOOBr$

                  ii. $HOOBr + HBr \rightarrow 2HOBr$

                  iii. $\underline{2HOBr + 2HBr \rightarrow 2H_2O + 2Br_2}$

                         $4HBr + O_2 \rightarrow 2H_2O + 2Br_2$

          (b)   The observed rate law is: rate = $k[HBr][O_2]$, the rate law for the first elementary step. The first step must be rate-determining.

          (c)   HOOBr and HOBr are both intermediates; HOOBr is produced in i and consumed in ii and HOBr is produced in ii and consumed in iii.

          (d)   Because the first step is rate-determining, it is possible that neither of the intermediates accumulates enough to be detected. This does not disprove the mechanism, but indicates that steps ii and iii are very fast, relative to step i.

Catalysis (Section 14.7)

14.76   (a)   The smaller the particle size of a solid catalyst, the greater the surface area. The greater the surface area, the more active sites and the greater the increase in reaction rate.

          (b)   Adsorption is the binding of reactants onto the surface of the heterogeneous catalyst. It is usually the first step in the catalyzed reaction.

14.78   For an acid-catalyzed reaction in solution, H^+ is a homogeneous catalyst. It is consumed and then regenerated during the reaction. (This assumes that H^+ is present in excess and that H^+ is not a reactant, that the reactants are neither acids nor bases.) The $[H^+]$ is a maximum at $t = 0$ and when the reaction is complete.

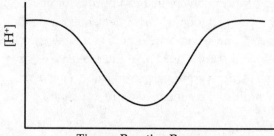

Time or Reaction Progress

14.80   (a) $2[NO(g) + N_2O(g) \rightarrow N_2(g) + NO_2(g)]$

$$\frac{2NO_2(g) \rightarrow 2NO(g) + O_2(g)}{2N_2O(g) \rightarrow 2N_2(g) + O_2(g)}$$

(b)   NO serves as a catalyst in this reaction. It is present when the reaction sequence begins and after the last step is completed.

(c)   No. The proposed mechanism cannot be ruled out if there is no build-up of NO_2. In this reaction, NO_2 functions as an intermediate; it is produced and then consumed during the reaction. If there is no measurable build-up of NO_2, the first step is slow relative to the second. As soon as NO_2 is produced by the slow first step, it is consumed by the faster second step.

14.82   (a)   Catalytic converters are heterogeneous catalysts that adsorb gaseous CO and hydrocarbons and speed up their oxidation to $CO_2(g)$ and $H_2O(g)$. They also adsorb nitrogen oxides, NO_x, and speed up their reduction to $N_2(g)$ and $O_2(g)$. If a catalytic converter is working effectively, the exhaust gas should have very small amounts of the undesirable gases CO, $(NO)_x$, and hydrocarbons.

(b)   The high temperatures could increase the rate of the desired catalytic reactions given in part (a). It could also increase the rate of undesirable reactions such as corrosion, which decrease the lifetime of the catalytic converter.

(c)   The rate of flow of exhaust gases over the converter will determine the rate of adsorption of CO, $(NO)_x$, and hydrocarbons onto the catalyst and thus the rate of conversion to desired products. Too fast an exhaust flow leads to less than maximum adsorption. A very slow flow leads to back pressure and potential damage to the exhaust system. Clearly the flow rate must be adjusted to balance chemical and mechanical efficiency of the catalytic converter.

14.84   Just as the π electrons in C_2H_4 are attracted to the surface of a hydrogenation catalyst, the nonbonding electron density on S causes compounds of S to be attracted to these same surfaces. Strong interactions could cause the sulfur compounds to be permanently attached to the surface, blocking active sites and reducing adsorption of alkenes for hydrogenation.

14.86   (a) $(NH_2)_2C=O(aq) + H_2O(l) \rightarrow CO_2(g) + 2\,NH_3(aq)$

(b)   From Solution 14.85, $E_a - E_{ac} = RT \ln(k_c/k)$. $k_c = 3.4 \times 10^4\,s^{-1}$, $k = 4.15 \times 10^{-5}\,s^{-1}$, $T = 100\,°C = 373\,K$.

$$E_a - E_{ac} = \frac{8.314\,J}{mol\text{-}K} \times 373\,K \times \ln\frac{3.4 \times 10^4}{4.15 \times 10^{-5}} = 63,647\,J = 63.647\,kJ = 64\,kJ$$

(c)   Because reaction rate always increases with increasing temperature, we expect the rate of the catalyzed reaction to be significantly greater at 100 °C than at 21 °C.

(d)   The 64 kJ difference between activation energies for the catalyzed and uncatalyzed reaction calculated in part (b) is a minimum difference. Because we expect the value of k_c to be significantly greater than $3.4 \times 10^4\,s^{-1}$, the difference in activation energies for the catalyzed and uncatalyzed reactions will be greater than 64 kJ.

14.88   Let k and E_a equal the rate constant and activation energy for the uncatalyzed reaction. Let k_c and E_{ac} equal the rate constant and activation energy of the catalyzed reaction. A is the same for the uncatalyzed and catalyzed reactions. $k_c/k = 1 \times 10^5$, T = 37 °C = 310 K.

According to Equation 14.20, $\ln k = -E_a/RT + \ln A$. Subtracting ln k from ln k_c

$$\ln k_c - \ln k = \left[\frac{-E_{ac}}{RT}\right] + \ln A - \left[\frac{-E_a}{RT}\right] - \ln A$$

$$\ln (k_c/k) = \frac{E_a - E_{ac}}{RT}; \quad E_a - E_{ac} = RT \ln (k_c/k)$$

$$E_a - E_{ac} = \frac{8.314\,J}{mol\text{-}K} \times 310\,K \times \ln(1 \times 10^5) = 2.966 \times 10^4\,J = 29.66\,kJ = 3 \times 10^1\,kJ$$

The enzyme must lower the activation energy by 30 kJ to achieve a 1×10^5-fold increase in reaction rate.

Additional Exercises

14.89   (a)   False. If this is not an elementary reaction, we need more information to write the correct rate law.

(b)   True. If the reaction is elementary, the rate law in part (a) is correct, and the reaction is second order overall.

(c)   False. If the reaction is elementary, the reverse reaction is second order.

(d)   False. The relationship between the forward and reverse activation energies depends on whether ΔE for the reaction is positive or negative.

14.90 $$\text{Rate} = \frac{-\Delta[H_2S]}{\Delta t} = \frac{\Delta[Cl^-]}{2\Delta t} = k[H_2S][Cl_2]$$

$$\frac{-\Delta[H_2S]}{\Delta t} = -(3.5 \times 10^{-2}\,M^{-1}s^{-1})(2.0 \times 10^{-4}\,M)(0.025\,M) = -1.75 \times 10^{-7} = -1.8 \times 10^{-7}\,M/s$$

$$\frac{\Delta[Cl^-]}{\Delta t} = \frac{-2\Delta[H_2S]}{\Delta t} = -2(-1.75 \times 10^{-7}\,M/s) = 3.5 \times 10^{-7}\,M/s$$

14.92   *Analyze/Plan.* Using the relationship rate = $k[A]^x$, determine the value of x that produces a rate law to match the described situation. *Solve.*

(a)   x = 0. The rate of reaction does not depend on $[A]_0$, so the reaction is zero-order in A.

(b)   x = 2. When $[A]_0$ increases by a factor of 3, rate increases by a factor of $(3)^2 = 9$.

(c)   x = 3. When $[A]_0$ increases by a factor of 2, rate increases by a factor of $(2)^3 = 8$.

14.93   (a)   The rate increases by a factor of nine when $[C_2O_4^{2-}]$ triples (compare experiments 1 and 2). The rate doubles when $[HgCl_2]$ doubles (compare experiments 2 and 3). The apparent rate law is: Rate = $k[HgCl_2][C_2O_4^{2-}]^2$

(b) $k = \dfrac{\text{rate}}{[HgCl_2][C_2O_4^{2-}]^2}$ Using the data for Experiment 1,

$k = \dfrac{(3.2 \times 10^{-5} \ M/s)}{[0.164 \ M][0.15 \ M]^2} = 8.672 \times 10^{-3} = 8.7 \times 10^{-3} \ M^{-2}s^{-1}$

(c)   Rate $= (8.672 \times 10^{-3} \ M^{-2}s^{-1})(0.100 \ M)(0.25 \ M)^2 = 5.4 \times 10^{-5} \ M/s$

14.94   (a)   Compare experiments 2 and 3, where the $[X]_0$ is the same and $[Z]_0$ increases by a factor of 1.5. The rate increases by a factor or 2.25, or $(1.5)^2$. The reaction is second order in $[Z]$. Next compare experiments 1 and 2, where both $[X]_0$ and $[Z]_0$ increase by a factor of two. The rate increases by a factor of eight. We know that doubling $[Z]_0$ increases rate by a factor of four; doubling $[X]_0$ then increases rate by a factor of two. The reaction is first order in $[X]_0$. The apparent rate law is: Rate $= k[X][Z]^2$

Check. Compare experiments 1 and 3. Here $[X]_0$ doubles and $[Z]_0$ triples. If our rate law is correct, the rate should increase by a factor of $(2)(3)^2 = 18$. This is indeed the relationship between the rates of Experiments 1 and 3.

(b) $k = \dfrac{\text{rate}}{[X][Z]^2}$. Using the data for Experiment 2,

$k = \dfrac{(3.2 \times 10^2 \ M/s)}{[0.50 \ M][0.50 \ M]^2} = 2.560 \times 10^3 = 2.6 \times 10^3 \ M^{-2}s^{-1}$

The same value is obtained using data from the other two experiments.

(c)   Rate $= (2.56 \times 10^3 \ M^{-2}s^{-1})(0.75 \ M)(1.25 \ M)^2 = 3.0 \times 10^3 \ M/s$

14.96   For a first-order reaction, $t_{1/2} = 0.633/k$. For a second-order reaction, $t_{1/2} = 1/k[A]_0$. Half-life is constant over the course of the reaction for first-order reactions. Although second-order half-life does not appear to depend on t, the value of "$[A]_0$" does change over the course of the reaction, and $t_{1/2}$ increases with time. (For a zero reaction, $t_{1/2}$ decreases with time.)

The rate law for reaction (1) must be first order, because that is the only reaction type that has a constant half-life. Reaction (2), where $t_{1/2}$ increases with time, is second order.

14.97   *Analyze/Plan.* Given k and $[A]_0$, use the integrated form of the first-order rate law to calculate $[A]$ at $t = 660$ s. Then, Rate $= k[A]$.

For a first-order reaction, $\ln[A]_t - \ln[A]_0 = -kt$. $\ln[A]_t = -kt + \ln[A]_0$

$[A]_0 = 2.5 \times 10^{-2} \ M, t = 660 \ s, k = 3.2 \times 10^{-3} \ s^{-1}$

$\ln[A] = -3.2 \times 10^{-3} \ s^{-1} (660 \ s) + \ln(0.025)$

$\ln[A] = -5.8009, \ [A] = 3.025 \times 10^{-3} \ M = 3.0 \times 10^{-3} \ M$

Rate $= k[A] = (3.2 \times 10^{-3} \ s^{-1})(3.025 \times 10^{-3} \ M) = 9.7 \times 10^{-6} \ M/s$

14.98   (a) $t_{1/2} = 0.693/k = 0.693/7.0 \times 10^{-4} \ s^{-1} = 990 = 9.9 \times 10^2 \ s$

(b) $k = \dfrac{0.693}{t_{1/2}} = \dfrac{0.693}{56.3 \ min} \times \dfrac{1 \ min}{60 \ s} = 2.05 \times 10^{-4} \ s^{-1}$

14.100   (a) $k = (8.56 \times 10^{-5} \; M/s)/(0.200 \; M) = 4.28 \times 10^{-4} \; s^{-1}$

           (b) $\ln[\text{urea}] = -(4.28 \times 10^{-4} \; s^{-1} \times 4.00 \times 10^{3} \; s) + \ln(0.500)$

               $\ln[\text{urea}] = -1.712 - 0.693 = -2.405 = -2.41; [\text{urea}] = 0.0903 = 0.090 \; M$

           (c) $t_{1/2} = 0.693/k = 0.693/4.28 \times 10^{-4} \; s^{-1} = 1.62 \times 10^{3} \; s$

14.101   (a) $A = \varepsilon bc$, Equation 14.5. $A = 0.605, \varepsilon = 5.60 \times 10^{3} \; cm^{-1} \; M^{-1}, b = 1.00 \; cm$

$$c = \frac{A}{\varepsilon b} = \frac{0.605}{(5.60 \times 10^{3} \; cm^{-1} \; M^{-1})(1.00 \; cm)} = 1.080 \times 10^{-4} = 1.08 \times 10^{-4} \; M$$

           (b)   Calculate $[c]_t$ using Beer's law. We calculated $[c]_0$ in part (a). Use Equation 14.13 to calculate k.

$$A_{30} = \varepsilon bc_{30}; c_{30} = \frac{A_{30}}{\varepsilon b} = \frac{0.250}{(5.60 \times 10^{3} \; cm^{-1} \; M^{-1})(1.00 \; cm)} = 4.464 \times 10^{-5} \; M$$

$$\ln[c]_t = -kt + \ln[c]_0; \quad \frac{\ln[c]_0 - \ln[c]_t}{t} = k; \quad t = 30 \; \text{min} \times \frac{60 \; s}{\text{min}} = 1800 \; s$$

$$k = (\ln(1.080 \times 10^{-4}) - \ln(4.464 \times 10^{-5}))/1800 \; s = 4.910 \times 10^{-4} = 4.91 \times 10^{-4} \; s^{-1}$$

           (c)   For a first-order reaction, $t_{1/2} = 0.693/k$.

               $t_{1/2} = 0.693/4.910 \times 10^{-4} \; s^{-1} = 1.411 \times 10^{3} = 1.41 \times 10^{3} \; s = 23.5 \; \text{min}$

           (d) $A_t = 0.100$; calculate c_t using Beer's law, then t from the first-order integrated rate equation.

$$c_t = \frac{A}{\varepsilon b} = \frac{0.100}{(5.60 \times 10^{3} \; cm^{-1} \; M^{-1})(1.00 \; cm)} = 1.786 \times 10^{-5} = 1.79 \times 10^{-5} \; M$$

$$t = \frac{\ln[c]_0 - \ln[c]_t}{k} = \frac{\ln(1.080 \times 10^{-4}) - \ln(1.786 \times 10^{-5})}{4.910 \times 10^{-4} \; s^{-1}}$$

               $t = 3.666 \times 10^{3} = 3.67 \times 10^{3} \; s = 61.1 \; \text{min}$

14.102   Calculate [dye] at each time, using Beer's law, $A = \varepsilon bc$; calculate ln[dye] and 1/[dye] and plot these quantities vs time in two separate graphs. The straight-line plot indicates the order of reaction with respect to [dye].

$$A_0 = \varepsilon bc_0; c_0 = \frac{A_0}{\varepsilon b} = \frac{1.254}{(4.7 \times 10^{4} \; cm^{-1} \; M^{-1})(1.00 \; cm)} = 2.668 \times 10^{-5} = 2.7 \times 10^{-5} \; M$$

Time (min)	A at 608 nm	[dye]	ln [dye]	1/[dye]
0	1.254	2.7×10^{-5}	−10.53	3.7×10^{4}
30	0.941	2.0×10^{-5}	−10.82	5.0×10^{4}
60	0.752	1.6×10^{-5}	−11.04	6.3×10^{4}
90	0.672	1.4×10^{-5}	−11.16	7.0×10^{4}
120	0.545	1.2×10^{-5}	−11.36	8.6×10^{4}

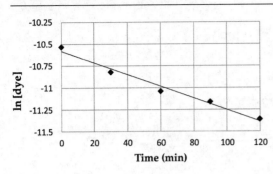

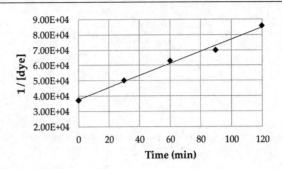

Although the graphs are not absolutely definitive, the plot of $1/[dye]$ vs time appears to be more linear. (The data point at $t = 90$ min us "out of line" in both plots and is suspect. More precision and accuracy in the experimental data would be helpful.) Assuming the reaction is second order with respect to the dye, the rate law is: Rate $= k[dye]^2$.

$$k = slope = (8.6 \times 10^4 - 3.7 \times 10^4) \, M^{-1}/(120-0)min = 4.1 \times 10^2 \, M^{-1} \, min^{-1}$$

(The best-fit slope and k value is $3.9 \times 10^2 \, M^{-1} \, min^{-1}$.)

14.104

ln k	1/T
–24.17	3.33×10^{-3}
–20.72	3.13×10^{-3}
–17.32	2.94×10^{-3}
–15.24	2.82×10^{-3}

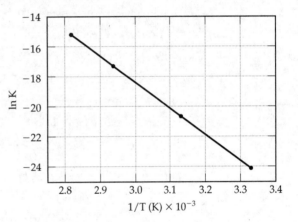

The calculated slope is -1.751×10^4. The activation energy E_a, equals $-(slope) \times (8.314 \, J/mol)$. Thus, $E_a = 1.8 \times 10^4 (8.314) = 1.5 \times 10^5$ $J/mol = 1.5 \times 10^2 \, kJ/mol$. (The best-fit slope is $-1.76 \times 10^4 = -1.8 \times 10^4$ and the value of E_a is 1.5×10^2 kJ/mol.)

14.105   *Analyze.* Given the time required to reach the "sour" point, at two different temperatures, estimate the activation energy for the reaction.

Plan/Solve. The warmer reaction at 28 °C is faster than the cooler reaction at 5 °C. The ratio of the rates of reaction is the inverse ratio of times.

$$\frac{rate_{28}}{rate_5} = \frac{k_{28}[sour \, milk]_{28}}{k_5[sour \, milk]_5} = \frac{t_5}{t_{28}} = \frac{48 \, hr}{4 \, hr} = 12$$

Assume the concentration of sour milk at 28 °C is the same as the concentration of sour milk at 5 °C. That is, $[sour \, milk]_{28} = [sour \, milk]_5$. Then, the ratio of rate constants, k_{28}/k_5, equals 12.

$T_1 = 28 \, °C = 301 \, K$; $T_2 = 5 \, °C = 278 \, K$; $k_1/k_2 = 12$

$$\ln\left(\frac{k_1}{k_2}\right) = \frac{E_a}{R}\left[\frac{1}{T_2} - \frac{1}{T_1}\right]; \quad \ln(12) = \frac{E_a}{8.314 \, J/mol}\left[\frac{1}{278} - \frac{1}{301}\right]$$

$$E_a = \frac{\ln(12)(8.314 \text{ J}/\text{mol})}{2.749 \times 10^{-4}} = 7.516 \times 10^4 \text{ J} = 75 \text{ kJ}/\text{mol}$$

14.106   (a) $T_1 = 77\ °\text{F}$; $°\text{C} = 5/9\ (°\text{F} - 32) = 5/9\ (77 - 32) = 25\ °\text{C} = 298 \text{ K}$

$T_2 = 59\ °\text{F}$; $°\text{C} = 5/9\ (59{-}32) = 15\ °\text{C} = 288 \text{ K}$; $k_1/k_2 = 6$

$$\ln\left(\frac{k_1}{k_2}\right) = \frac{E_a}{R}\left[\frac{1}{T_2} - \frac{1}{T_1}\right]; \quad \ln(6) = \frac{E_a}{8.314 \text{ J}/\text{mol}}\left[\frac{1}{288} - \frac{1}{298}\right]$$

$$E_a = \frac{\ln(6)(8.314 \text{ J}/\text{mol})}{1.165 \times 10^{-4}} = 1.28 \times 10^5 \text{ J} = 1.3 \times 10^2 \text{ kJ}/\text{mol}$$

$T_1 = 77\ °\text{F} = 25\ °\text{C} = 298 \text{ K}$; $T_2 = 41\ °\text{F} = 5\ °\text{C} = 278 \text{ K}$, $k_1/k_2 = 40$

$$\ln(40) = \frac{E_a}{8.314 \text{ J}/\text{mol}}\left[\frac{1}{278} - \frac{1}{298}\right]; \quad E_a = \frac{\ln(40)(8.314 \text{ J}/\text{mol})}{2.414 \times 10^{-4}}$$

$$E_a = 1.27 \times 10^5 \text{ J} = 1.3 \times 10^2 \text{ kJ}/\text{mol}$$

The values are amazingly consistent, considering the precision of the data.

(b)   For a first-order reaction, $t_{1/2} = 0.693/k$, $k = 0.693/t_{1/2}$

k_1 at $298 \text{ K} = 0.693/2.7 \text{ yr} = 0.257 = 0.26 \text{ yr}^{-1}$

$T_1 = 298 \text{ K}$, $T_2 = 273 - 15\ °\text{C} = 258 \text{ K}$

$$\ln\left(\frac{0.257}{k_2}\right) = \frac{1.27 \times 10^5 \text{ J}}{8.314 \text{ J}/\text{mol}}\left[\frac{1}{258} - \frac{1}{298}\right] = 7.94727 = 7.95$$

$0.257/k_2 = e^{7.94727} = 2.828 \times 10^3$; $k_2 = 0.257/2.828 \times 10^3 = 9.088 \times 10^{-5} = 9.1 \times 10^{-5} \text{ yr}^{-1}$

$t_{1/2} = 0.693/k = 0.693/9.088 \times 10^{-5} = 7.625 \times 10^3 \text{ yr} = 7.6 \times 10^3 \text{ yr}$

14.108   (a)

$$\text{Cl}(g) + \text{O}_3(g) \rightarrow \text{ClO}(g) + \text{O}_2(g)$$
$$\text{ClO}(g) + \text{O}(g) \rightarrow \text{Cl}(g) + \text{O}_2(g)$$

$$\text{Cl}(g) + \text{O}_3(g) + \text{ClO}(g) + \text{O}(g) \rightarrow \text{ClO}(g) + \text{O}_2(g) + \text{Cl}(g) + \text{O}_2(g)$$
$$\text{O}_3(g) + \text{O}(g) \rightarrow 2\text{O}_2(g)$$

(b) $\text{Cl}(g)$ is the catalyst. It is consumed in the first step and reproduced in the second.

(c) $\text{ClO}(g)$ is the intermediate. It is produced in the first step and consumed in the second.

14.109   (a)

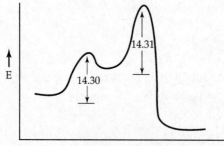

14.30 = E_a for reaction [14.30]

14.31 = E_a for reaction [14.31]

(b)   The fact that Br_2 builds up during the reaction tells us that the appearance of Br_2 (reaction [14.30]) is faster than the disappearance of Br_2 (reaction [14.31]). This is the reason that E_a of [14.31] in the energy profile above is larger than E_a of [14.30].

14.111   (a)   Rate $= k_1[A][B]$

(b)   Follow the logic in Sample Exercise 14.15.

$$Rate = k_2[A][X]; \; k_1[A][B] = k_{-1}[C][X]; \; [X] = \frac{k_1}{k_{-1}}\frac{[A][B]}{[C]}$$

Substituting for [X] in the rate expression,

$$Rate = k_2[A]\frac{k_1}{k_{-1}}\frac{[A][B]}{[C]} = k\frac{[A]^2[B]}{[C]}$$

(c)   (iii) The result of part (b) might be surprising because [C] appears in the rate law, and [C] has a negative reaction order.

14.112   (a) $(CH_3)_3AuPH_3 \rightarrow C_2H_6 + (CH_3)AuPH_3$

(b) $(CH_3)_3Au$, $(CH_3)Au$, and PH_3 are intermediates.

(c)   Reaction 1 is unimolecular, Reaction 2 is unimolecular, Reaction 3 is bimolecular.

(d)   Reaction 2, the slow one, is rate determining.

(e)   If Reaction 2 is rate determining, Rate $= k_2[(CH_3)_3Au]$.

$(CH_3)_3Au$ is an intermediate formed in Reaction 1, an equilibrium. By definition, the rates of the forward and reverse processes in Reaction 1 are equal:

$k_1[(CH_3)_3AuPH_3] = k_{-1}[(CH_3)_3Au][PH_3]$; solving for $[(CH_3)_3Au]$,

$$[(CH_3)_3Au] = \frac{k_1[(CH_3)_3AuPH_3]}{k_{-1}[PH_3]}$$

Substituting into the rate law

$$Rate = \left(\frac{k_2k_1}{k_{-1}}\right)\frac{[(CH_3)_3AuPH_3]}{[PH_3]} = \frac{k[(CH_3)_3AuPH_3]}{[PH_3]}$$

(f)   The rate is inversely proportional to $[PH_3]$, so adding PH_3 to the $(CH_3)_3AuPH_3$ solution would decrease the rate of the reaction.

14.113   *Analyze/Plan.* Use the structure and unit cell edge of Pt, along with the formulas for volume and surface area of a sphere, to calculate the number of Pt atoms in a 2-nm sphere and on the surface of a 2-nm sphere.

(a)   For a Pt sphere with a 2.0 nm diameter, radius = 1.0 nm.

$$V = 4/3\pi r^3 = \frac{4\pi(1.0 \text{ nm})^3}{3} \times \frac{10^3 \text{ Å}^3}{1^3 \text{ nm}^3} = 4.188879 \times 10^3 = 4.2 \times 10^3 \text{ Å}^3$$

In a face-centered cubic metal structure, there are 4 metal atoms per unit cell. The volume of the unit cell is $(3.924 \text{ Å})^3 = 60.42 \text{ Å}^3$

$$\frac{4 \text{ Pt atoms}}{60.42 \text{ Å}^3} \times 4.1889 \times 10^3 \text{ Å}^3 = 277.3 = 2.8 \times 10^2 \text{ Pt atoms in a 2.0 - nm sphere}$$

(b)   Assume that the "footprint" of an atom is its cross-sectional area, the area of a circle with the radius of the atom. The area of this circle is πr^2. The diameter, d, of a Pt atom is 2.8 Å, so r = d/2 = 1.4 Å. The footprint of the Pt atom is then

$\pi(1.4 \text{ Å})^2 = 6.1575 = 6.2 \text{ Å}^2$

The surface area of the 2.0-nm sphere is

$$4\pi r^2 = 4\pi (1.0 \text{ nm})^2 \times \frac{10^2 \text{ Å}^2}{1^2 \text{ nm}^2} = 12.56637 \times 10^2 = 1.3 \times 10^3 \text{ Å}^2$$

$$\frac{1 \text{ Pt atoms}}{6.1575 \text{ Å}^2} \times 1.2566 \times 10^3 \text{ Å}^2 = 204.1 = 2.0 \times 10^2 \text{ surface Pt atoms on a 2.0-nm sphere.}$$

(c) $\dfrac{204 \text{ surface Pt atoms}}{277 \text{ total Pt atoms}} \times 100 = 74\% \text{ Pt atoms on the surface}$

(d)   For a 5.0-nm Pt sphere, radius = 2.5 nm

$$V = 4/3\pi r^3 = \frac{4\pi (2.50 \text{ nm})^3}{3} \times \frac{10^3 \text{ Å}^3}{1^3 \text{ nm}^3} = 65.4498 \times 10^3 = 6.5 \times 10^4 \text{ Å}^3$$

$$\frac{4 \text{ Pt atoms}}{60.42 \text{ Å}^3} \times 65.4498 \times 10^3 \text{ Å}^3 = 4333 = 4.3 \times 10^3 \text{ Pt atoms in a 5.0- nm sphere}$$

The surface area of the 5.0-nm sphere is

$$4\pi r^2 = 4\pi (2.5 \text{ nm})^2 \times \frac{10^2 \text{ Å}^2}{1^2 \text{ nm}^2} = 7853.98 = 7.9 \times 10^3 \text{ Å}^2$$

$$\frac{1 \text{ Pt atoms}}{6.1575 \text{ Å}^2} \times 7.854 \times 10^3 \text{ Å}^2 = 1275.5 = 1.3 \times 10^3 \text{ surface Pt atoms on a 5.0- nm sphere}$$

$$\frac{1276 \text{ surface Pt atoms}}{4333 \text{ total Pt atoms}} \times 100 = 29\% \text{ Pt atoms on the surface}$$

The calculations in parts (b) and (d) overestimate the number of Pt atoms on the surface of the sphere, because they do not account for empty space between atoms. For the purpose of comparison, it is most important that we use the same method for both spheres.

[Alternatively, use one face of a face-entered cubic unit cell as a model for the surface area that Pt atoms will occupy. On a face, there is the cross-section of 1 Pt atoms in the center and ¼ Pt atom at each corner. This amounts to the cross-sections of two Pt atoms in $(3.924 \text{ Å})^2 = 15.398 = 15.4 \text{ Å}^2$.]

$$\frac{2 \text{ Pt atoms}}{15.40 \text{ Å}^3} \times 1.2566 \times 10^3 \text{ Å}^2 = 163.2 = 1.6 \times 10^2 \text{ surface Pt atoms on a 2.0- nm sphere.}$$

$$\frac{163 \text{ surface Pt atoms}}{277 \text{ total Pt atoms}} \times 100 = 59\% \text{ Pt atoms on the surface}$$

[Similarly, in a 5.0-nm sphere, there are 1.0×10^3 Pt atoms on the surface, 24% of the total Pt atoms.]

(e)   Both surface models predict that the 2.0-nm sphere will be more catalytically active, because it has a much greater percentage of its atoms on the surface, where they can participate in the chemical reaction.

14.114   *Enzyme*: carbonic anhydrase; *substrate*: carbonic acid (H_2CO_3); *turnover number*: 1×10^7 molecules/s.

14.116   (a)   The rate law for the slow step is Rate = k_2[ES], where ES is an intermediate. Use relationships from the fast equilibrium step to substitute for [ES].

rate of the forward reaction = k_1[E][S]

rate of the reverse reaction = k_{-1}[ES]

For an equilibrium, rate forward = rate reverse, k_1[E][S] = k_{-1}[ES];

$$[ES] = \frac{k_1}{k_{-1}}[E][S]; \quad Rate = k_2[ES] = \frac{k_2 k_1}{k_{-1}}[E][S] = k[E][S]$$

(b)   E + I $\rightleftharpoons$ EI

Integrative Exercises

14.117   *Analyze/Plan.* $2 N_2O_5 \rightarrow 4 NO_2 + O_2$ Rate = $k[N_2O_5] = 1.0 \times 10^{-5}\ s^{-1}\ [N_2O_5]$

Use the integrated rate law for a first-order reaction, Equation 14.13, to calculate $k[N_2O_5]$ at 20.0 h. Build a stoichiometry table to determine mol O_2 produced in 20.0 h. Assuming that $O_2(g)$ is insoluble in chloroform, calculate the pressure of O_2 in the 10.0 L container.   *Solve.*

$$20.0\ h \times \frac{60\ min}{1\ h} \times \frac{60\ s}{1\ min} = 7.20 \times 10^4\ s;\ [N_2O_5]_0 = 0.600\ M$$

$$\ln[A]_t - \ln[A]_0 = -kt; \quad \ln[N_2O_5]_t = -kt + \ln[N_2O_5]_0$$

$$\ln[N_2O_5]_t = -1.0 \times 10^{-5}\ s^{-1}\ (7.20 \times 10^4\ s) + \ln(0.600) = -0.720 - 0.511 = -1.231$$

$$[N_2O_5]_t = e^{-1.231} = 0.292\ M$$

N_2O_5 was present initially as 1.00 L of 0.600 M solution.

mol $N_2O_5 = M \times L = 0.600$ mol N_2O_5 initial, 0.292 mol N_2O_5 at 20.0 h

	$2N_2O_5$	$\rightarrow$	$4NO_2$	+	O_2
t = 0	0.600 mol		0		0
change	−0.308 mol		0.616 mol		0.154 mol
t = 20 h	0.292 mol		0.616 mol		0.154 mol

[Note that the reaction stoichiometry is applied to the "change" line.]

PV = nRT; P = nRT/V; V = 10.0 L, T = 45 °C = 318 K, n = 0.154 mol

$$P = 0.154\ mol \times \frac{318\ K}{10.0\ L} \times \frac{0.08206\ L\text{-}atm}{mol\text{-}K} = 0.402\ atm$$

14.118   (a) $\ln k = -E_a/RT + \ln A$; $E_a = 86.8$ kJ/mol $= 8.68 \times 10^4$ J/mol; T = 35 °C + 273 = 308 K; $A = 2.10 \times 10^{11}\ M^{-1}\ s^{-1}$

$$\ln k = \frac{-8.68 \times 10^4\ J/mol}{308\ K} \times \frac{mol\text{-}K}{8.314\ J} + \ln(2.10 \times 10^{11}\ M^{-1}\ s^{-1})$$

$\ln k = -33.8968 + 26.0704 = -7.8264; \quad k = 3.99 \times 10^{-4}\ M^{-1}\ s^{-1}$

(b) $\dfrac{0.335 \text{ g KOH}}{0.250 \text{ L soln}} \times \dfrac{1 \text{ mol KOH}}{56.1 \text{ g KOH}} = 0.02389 = 0.0239 \ M \text{ KOH}$

$\dfrac{1.453 \text{ g } C_2H_5I}{0.250 \text{ L soln}} \times \dfrac{1 \text{ mol } C_2H_5I}{156.0 \text{ g } C_2H_5I} = 0.03726 = 0.0373 \ M \ C_2H_5I$

If equal volumes of the two solutions are mixed, the initial concentrations in the reaction mixture are 0.01194 M KOH and 0.01863 M C_2H_5I. Assuming the reaction is first order in each reactant:

Rate = $k[C_2H_5I][OH^-] = 3.99 \times 10^{-4} \ M^{-1} s^{-1}$ (0.01194 M)(0.01863 M) = $8.88 \times 10^{-8} \ M/s$

(c)   Because C_2H_5I and OH^- react in a 1 : 1 mole ratio and equal volumes of the solutions are mixed, the reactant with the smaller concentration, KOH, is the limiting reactant.

(d)   T = 50 °C + 273 = 323 K

$\ln k = \dfrac{-8.68 \times 10^4 \text{ J/mol}}{323 \text{ K}} \times \dfrac{\text{mol - K}}{8.314 \text{ J}} + \ln (2.10 \times 10^{11} \ M^{-1} s^{-1})$

$\ln k = -32.3227 + 26.0704 = -6.2523; \quad k = 1.93 \times 10^{-3} \ M^{-1} s^{-1}$

14.119   Obtaining data for an "Arrhenius" plot like this requires running the reaction several times, each at a different temperature. Different rates and rate constants (k) are obtained at each temperature. We expect a straight line, according to the relationship: $\ln k = -E_a/RT + \ln A$. The slope of the graph is $-E_a$ and the y-intercept is the orientation factor, A. The graph in the exercise demonstrates these characteristics, times two! The graph indicates that reaction requires two different activation energies, depending on temperature.

Assuming that reactants and products are the same at all temperatures, the reaction proceeds through different pathways, depending on temperature. This could mean two totally different reaction mechanisms, or a multi-step mechanism where different steps are rate-determining at different temperatures.

14.121   (a) $\Delta H_{rxn}^{\circ} = 2\Delta H_f^{\circ} \ H_2O(g) + 2\Delta H_f^{\circ} \ Br_2(g) - 4\Delta H_f^{\circ} \ HBr(g) - \Delta H_f^{\circ} \ O_2(g)$

$\Delta H_{rxn}^{\circ} = 2(-241.82) + 2(30.71) - 4(-36.23) - (0) = -277.30 \text{ kJ}$

(b)   Because the rate of the uncatalyzed reaction is very slow at room temperature, the magnitude of the activation energy for the rate-determining first step must be quite large. At room temperature, the reactant molecules have a distribution of kinetic energies (Chapter 10), but very few molecules even at the high end of the distribution have sufficient energy to form an activated complex. E_a for this step must be much greater than 3/2 RT, the average kinetic energy of the sample.

(c)   20 e^-, 10 e^- pr

$H - \overset{..}{\underset{..}{O}} - \overset{..}{\underset{..}{O}} - \overset{..}{\underset{..}{Br}} :$

The intermediate resembles hydrogen peroxide, H_2O_2.

14.122   (a)   D(Cl–Cl) = 242 kJ/mol Cl_2

$\dfrac{242 \text{ kJ}}{\text{mol } Cl_2} \times \dfrac{1000 \text{ J}}{\text{kJ}} \times \dfrac{1 \text{ mol}}{6.022 \times 10^{23} \text{ molecules}} = 4.019 \times 10^{-19} = 4.02 \times 10^{-19} \text{ J}$

$$\lambda = hc/E = \frac{6.626 \times 10^{-34} \text{ J-s} \times 2.998 \times 10^8 \text{ m/s}}{4.019 \times 10^{-19} \text{ J}} = 4.94 \times 10^{-7} \text{ m}$$

This wavelength, 494 nm, is in the visible portion of the spectrum.

(b)

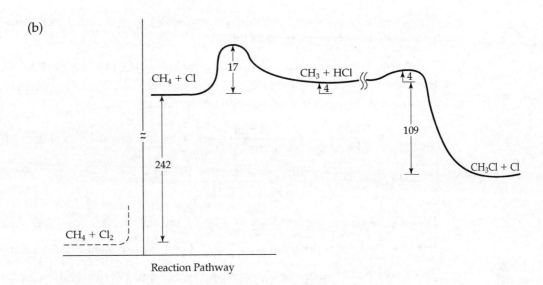

Reaction Pathway

(c)   Because D(Cl–Cl) is 242 kJ/mol, $CH_4(g) + Cl_2(g)$ should be about 242 kJ below the starting point on the diagram. For the reaction

$CH_4(g) + Cl_2(g) \rightarrow CH_3(g) + HCl(g) + Cl(g)$, E_a is 242 + 17 = 259 kJ.

(From bond dissociation enthalpies, ΔH for the overall reaction

$CH_4(g) + Cl_2(g) \rightarrow CH_3Cl(g) + Cl(g)$ is –104 kJ,

so the graph above is simply a sketch of the relative energies of some of the steps in the process.)

(d) CH_3, 7 valence e^-, odd electron species   H—C̈—H
                                                                                            |
                                                                                            H

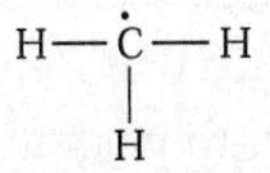

(e)   This sequence is called a chain reaction because Cl• radicals are regenerated in Reaction 4, perpetuating the reaction. Absence of Cl• terminates the reaction, so Cl• + Cl• → Cl_2 is a termination step.

14.124   (a)

Molecule	NO	NO_2	N_2
Valence e^-	11	17	10
e^- pairs	5.5	8.5	5
Lewis structure	:N̈=Ö:	:Ö—N̈=Ö:	:N≡N:
Bond order	2	1.5	3
Bond energy	607 kJ/mol	404 kJ/mol	941 kJ/mol

(b) The bond energies in the table above are from Table 8.4 of the text. The NO_2 molecule has two resonance forms and the bond order is 1.5. To obtain an approximate bond energy, average the energies for N=O and N–O:

(607 kJ + 201 kJ)/2 = –404 kJ/bond in NO_2

We know that resonance stabilizes a molecule, so the actual bond energy is probably somewhat greater than this value.

Use Avogadro's number to calculate energy in J/bond, then $\lambda = hc/E$ to calculate wavelength and region of the electromagnetic spectrum. These are the maximum wavelengths that would cause complete bond dissociation.

$$\frac{941\ kJ}{mol\ N_2} \times \frac{1000\ J}{kJ} \times \frac{1\ mol}{6.022\ \times\ 10^{23}\ molecules} = 1.5626 \times 10^{-18} = 1.56 \times 10^{-18}\ J$$

$$\lambda\ =\ hc/E\ =\ \frac{6.626\ \times\ 10^{-34}\ J\text{-}s \times 2.998\ \times\ 10^8\ m/s}{1.5626\ \times\ 10^{-18}\ J}\ =\ 1.27 \times 10^{-7}\ m$$

For NO, the energy per bond is 1.01×10^{-18} J and the wavelength is 1.97×10^{-7} m.

For NO_2, the energy per bond is 6.71×10^{-19} J and the wavelength is 2.96×10^{-7} m.

The three "bond dissociation" wavelengths, 127 nm, 197 nm, and 296 nm, are all in the ultraviolet region, near but not in the visible range of 400-700 nm. We expect longer wavelength electronic excitations for the three gases to be in the visible and near UV, in the same relative order as the bond dissociation wavelengths.

(c) The experiment requires a UV-VIS spectrometer and a gas flow cell that can be attached to the exhaust stream. According to Beer's law, as the concentration of absorbing species decreases, so does the absorbance. By monitoring a different wavelength of maximum absorption (longer than 297 nm) for each gas, we can measure the concentration of each gas at some point in time. We would monitor the stream before the catalytic converter to establish starting concentrations, then after the converter to observe changes. If the catalyst is working, we expect the two longer wavelength peaks for NO and NO_2 to decrease in size, and the shorter wavelength peak for N_2 to increase.

14.125 (a) No. Ethane, C_2H_6, has only sigma bonds. The electron density is all localized along the C–H and C–C bonds. The molecule has no pi electrons to interact with the metal surface.

(b) Yes. Ammonia, NH_3, has a nonbonding electron pair on nitrogen. This electron domain extends away from the sigma framework of the N–H bonds and is available to interact with the metal surface of the catalyst.

15 Chemical Equilibrium

Visualizing Concepts

15.2   Yes, the system is in equilibrium in boxes 4 and 5. The first box is pure reactant A. As the reaction proceeds, some A changes to B. In the fourth and fifth boxes, the relative amounts (concentrations) of A and B are constant. Although the reaction is ongoing the rates of A → B and B → A are equal, and the relative amounts of A and B are constant.

15.4   *Analyze/Plan.* Given that element A = red and element B = blue, evaluate the species in the reactant and product boxes, and write the reaction. Answer the remaining questions based on the balanced equation.   *Solve.*

(a)   reactants: $4A_2$ + 4B; products: $4A_2B$

balanced equation: A_2 + B → A_2B

(b) $K_c = \dfrac{[A_2B]}{[A_2][B]}$

(c)   Evaluate concentrations and the value of K for the box in equilibrium, the one on the right of the diagram. $[A_2B] = 0.4$ M; $[A_2] = 0.1$ M; $[B] = 0.1$ M

$K_c = \dfrac{[A_2B]}{[A_2][B]} = \dfrac{[0.4]}{[0.1][0.1]} = 40$

(d) $\Delta n = \Sigma n(\text{prod}) - \Sigma n(\text{react}) = 1 - 2 = -1$.

(e) $K_p = K_c(RT)^{\Delta n}$, Equation 15.14. Assume a temperature of $25\,^{\circ}C$ = 298 K.

$K_p = 40[(0.0821)(298)]^{-1} = \dfrac{40}{(0.0821)(298)} = 1.635 = 2$

15.8   *Analyze.* Given box diagrams, reaction type, and value of K_c, determine whether each reaction mixture is at equilibrium.

Plan. Analyze the contents of each box, express them as concentrations (see Solution 15.3). Write the equilibrium expression, calculate Q for each mixture, and compare it to K_c. If Q = K, the mixture is at equilibrium. If Q < K, the reaction shifts right (more product). If Q > K, the reaction shifts left (more reactant).

Solve. $K_c = \dfrac{[AB]^2}{[A_2][B_2]}$.

For this reaction, $\Delta n = 0$, so the volume terms cancel in the equilibrium expression. In this case, the number of each kind of particle can be used as a representation of moles (see Solution 5.3) and molarity.

(a)   Mixture (i): $1A_2$, $1B_2$, $6AB$; $Q = \dfrac{6^2}{(1)(1)} = 36$

Q > K$_c$, the mixture is not at equilibrium.

Mixture (ii): $3A_2$, $2B_2$, $3AB$; $Q = \dfrac{3^2}{(3)(2)} = 1.5$

Q = K$_c$, the mixture is at equilibrium.

Mixture (iii): $3A_2$, $3B_2$, $2AB$; $Q = \dfrac{2^2}{(3)(3)} = 0.44$

Q < K$_c$, the mixture is not at equilibrium.

(b)   Mixture (i) proceeds toward reactants.

Mixture (iii) proceeds toward products.

15.10   *Analyze.* Given the diagram and reaction type, calculate the equilibrium constant K$_c$.

Plan. Analyze the contents of the cylinder. Express them as concentrations, using number of particles as a measure of moles, and V = 2 L. Write the equilibrium expression in terms of concentration and calculate K$_c$.   *Solve.*

(a)   The mixture contains $2A_2$, 2B, 4AB. $[A_2] = 2/2 = 1$, $[B] = 2/2 = 1$, $[AB] = 4/2 = 2$.

$$K_c = \frac{[AB]^2}{[A_2][B]^2} = \frac{(2)^2}{(1)(1)^2} = 4$$

(b)   A decrease in volume favors the reaction with fewer moles of gas. This reaction has two moles of gas in products and three in reactants, so a decrease in volume favors products. The number of AB (product) molecules will increase.

Note that a change in volume does not change the value of K$_c$. If V decreases, the number of AB molecules must increase to maintain the equilibrium value of K$_c$.

15.12   (a)   Exothermic. In both reaction mixtures (orange and blue), [AB] decreases as T increases.

(b)   In the reaction, there are fewer moles of gas in products than reactants, so greater pressure favors production of products. At any single temperature, [AB] is greater at P = y than at P = x. Because the concentration of the product, AB, is greater at P = y, P = y is the greater pressure.

Equilibrium; The Equilibrium Constant (Sections 15.1–15.4)

15.14   (a)   Products will predominate at equilibrium. If K$_c$ is large, the numerator of the K$_c$ expression is much greater than the denominator.

(b)   The forward reaction has the greater rate constant. $K_c = k_f/k_r$; if K$_c$ is large, k_f is larger than k_r and the forward reaction has the greater rate constant.

15.16 (a) $K_c = \dfrac{[O_2]^3}{[O_3]^2}$ (b) $K_c = \dfrac{1}{[Cl_2]^2}$

       (c) $K_c = \dfrac{[C_2H_6]^2[O_2]}{[C_2H_4]^2[H_2O]^2}$ (d) $K_c = \dfrac{[CH_4]}{[H_2]^2}$

       (e) $K_c = \dfrac{[Cl_2]^2}{[HCl]^4[O_2]}$ (f) $K_c = \dfrac{[CO_2]^{16}[H_2O]^{18}}{[O_2]^{25}}$

       (g) $K_c = \dfrac{[CO_2]^{16}}{[O_2]^{25}}$

       homogeneous: (a), (c); heterogeneous: (b), (d), (e), (f), (g)

15.18 (a) equilibrium lies to right, favoring products ($K_p \gg 1$)

       (b) equilibrium lies to left, favoring reactants ($K_c \ll 1$)

15.20 (a) False. The value of Δn is not zero.

       (b) False. Equilibrium constants are not expressed with units.

       (c) True. For a gas phase equilibrium, increasing pressure (by decreasing volume) favors the reaction that produces fewer moles of gas. In this equilibrium, the forward reaction has fewer moles of gas. When the forward reaction is favored, the value of K increases.

15.22 $SO_2(g) + Cl_2(g) \rightleftharpoons SO_2Cl_2(g)$, $K_p = 34.5$. $\Delta n = 1 - 2 = -1$

       $K_p = K_c(RT)^{\Delta n}$; $34.5 = K_c(RT)^{-1} = K_c/RT$;

       $K_c = 34.5\ RT = 34.5(0.08206)(303) = 857.81 = 858$

15.24 $2\,H_2(g) + S_2(g) \rightleftharpoons 2\,H_2S(g)$, $K_c = 1.08 \times 10^7$ at 700 °C. $\Delta n = 2 - 3 = -1$.

       (a) $K_p = K_c(RT)^{\Delta n}$; $T = 700\ °C + 273 = 973\ K$.

            $K_p = 1.08 \times 10^7\ (RT)^{-1} = \dfrac{1.08 \times 10^7}{(0.08206)(973)} = 1.35 \times 10^5$

       (b) Mostly H_2S. Both K_p and K_c are much greater than one, so the product, H_2S, is favored at equilibrium.

       (c) $H_2(g) + \tfrac{1}{2}\,S_2(g) \rightleftharpoons H_2S(g)$; $K_{c2} = \dfrac{[H_2S]}{[H_2][S_2]^{1/2}}$

            $K_{c2} = (K_c)^{1/2} = (1.08 \times 10^7)^{1/2} = 3.29 \times 10^3$

            $K_{p2} = (K_p)^{1/2} = (1.35 \times 10^5)^{1/2} = 368$

15.26 $K_p = \dfrac{P_{HCl}^4 \times P_{O_2}}{P_{Cl_2}^2 \times P_{H_2O}^2} = 0.0752$

       (a) $K_p = \dfrac{P_{Cl_2}^2 \times P_{H_2O}^2}{P_{HCl}^4 \times P_{O_2}} = \dfrac{1}{0.0752} = 13.298 = 13.3$

(b) $K_p = \dfrac{P_{HCl}^2 \times P_{O_2}^{1/2}}{P_{Cl_2} \times P_{H_2O}} = (0.0752)^{1/2} = 0.2742 = 0.274$

(c) $K_p = K_c(RT)^{\Delta n}$; $\Delta n = 2.5 - 2 = 0.5$; $T = 480°C + 273 = 753 \text{ K}$

$K_p = K_c(RT)^{1/2}$, $K_c = K_p/(RT)^{1/2} = 0.2742/[0.08206 \times 753]^{1/2} = 0.03488 = 0.0349$

15.28 $2NO(g) + Br_2(g) \rightleftharpoons 2NOBr(g)$ $K_1 = 2.0$

$N_2(g) + O_2(g) \rightleftharpoons 2NO(g)$ $K_2 = \dfrac{1}{2.1 \times 10^{30}}$

$2NO(g) + Br_2(g) + N_2(g) + O_2(g) \rightleftharpoons 2NOBr(g) + 2NO(g)$

$N_2(g) + O_2(g) + Br_2(g) \rightleftharpoons 2NOBr(g)$

$K_c = K_1 \times K_2 = 2.0 \times \dfrac{1}{2.1 \times 10^{30}} = 9.524 \times 10^{-31} = 9.5 \times 10^{-31}$

15.30   (a) $K_p = 1/P_{SO_2}$

(b) $K_c = \dfrac{[Na_2SO_3]}{[Na_2O][SO_2]}$

Calculating Equilibrium Constants (Section 15.5)

15.32 $K_c = \dfrac{[H_2][I_2]}{[HI]^2} = \dfrac{(4.79 \times 10^{-4})(4.79 \times 10^{-4})}{(3.53 \times 10^{-3})^2} = 0.018413 = 0.0184$

15.34   (a) $K_p = \dfrac{P_{PCl_5}}{P_{PCl_3} \times P_{Cl_2}} = \dfrac{1.30 \text{ atm}}{0.124 \text{ atm} \times 0.157 \text{ atm}} = 66.8$

(b)   Because $K_p > 1$, products (the numerator of the K_p expression) are favored over reactants (the denominator of the K_p expression).

(c) $K_p = K_c(RT)^{\Delta n}$; $\Delta n = 1 - 2 = -1$; $K_p = K_c(RT)^{-1} = K_c/(RT)$

$K_c = K_p(RT) = 66.8(0.08206 \times 450) = 2.5 \times 10^3$

15.36   (a)   Calculate the initial concentrations of $H_2(g)$ and $Br_2(g)$ and the equilibrium concentration of $H_2(g)$. $M = \text{mol/L}$.

$[H_2]_{init} = 1.374 \text{ g H}_2 \times \dfrac{1 \text{ mol H}_2}{2.0159 \text{ g H}_2} \times \dfrac{1}{2.00 \text{ L}} = 0.34079 = 0.341\,M$

$[Br_2] = 70.31 \text{ g Br}_2 \times \dfrac{1 \text{ mol Br}_2}{159.81 \text{ g Br}_2} \times \dfrac{1}{2.00 \text{ L}} = 0.21998 = 0.220\,M$

$[H_2]_{equil} = 0.566 \text{ g H}_2 \times \dfrac{1 \text{ mol H}_2}{2.0159 \text{ g H}_2} \times \dfrac{1}{2.00 \text{ L}} = 0.14038 = 0.140\,M$

	$H_2(g)$	$+$	$Br_2(g)$	$\rightleftharpoons$	$2HBr(g)$
initial	0.34079 M		0.21998 M		0
change	−0.20041 M		−0.20041 M		+2(0.20041) M
equil.	0.14038 M		0.01957 M		0.40082 M

The change in H_2 is $(0.34079 - 0.14038 = 0.20041 = 0.200)$. The changes in $[Br_2]$ and $[HBr]$ are set by stoichiometry, resulting in the equilibrium concentrations shown in the table.

(b)　$K_c = \dfrac{[HBr]^2}{[H_2][Br_2]} = \dfrac{(0.40082)^2}{(0.14038)(0.01957)} = \dfrac{(0.401)^2}{(0.140)(0.020)} = 58.48 = 58$

The equilibrium concentration of Br_2 has 3 decimal places and 2 sig figs, so the value of K_c has 2 sig figs.

15.38　(a)

	$N_2O_4(g)$	$\rightleftharpoons$	$2NO_2(g)$
initial	1.500 atm		1.000 atm
change	+0.244 atm		–0.488 atm
equil.	1.744 atm		0.512 atm

The change in P_{NO_2} is $(1.000 - 0.512) = -0.488$ atm, so the change in $P_{N_2O_4}$ is $+(0.488/2) = +0.244$ atm.

(b)　$K_p = \dfrac{P_{NO_2}^2}{P_{N_2O_4}} = \dfrac{(0.512)^2}{(1.744)} = 0.1503 = 0.150$

(c)　$K_p = K_c(RT)^{\Delta n}$; $\Delta n = 2 - 1 = 1$; $K_p = K_c(RT)^1 = K_c(RT)$

$K_c = K_p/(RT) = 0.1503/(0.08206 \times 298) = 6.15 \times 10^{-3}$

15.40　The initial concentrations of drug candidate and protein are the same in the two experiments, and the two reactions have the same stoichiometry. At equilibrium, the concentration of B-protein complex is greater than the concentration of A-protein complex, so drug B is the better choice for further research. Calculation of equilibrium constants for the two reactions confirms this conclusion.

	$A(aq)$	+	$protein(aq)$	$\rightleftharpoons$	$A\text{-}protein(aq)$
initial	2.00×10^{-6} mM		1.50×10^{-6} mM		0
change	-1.00×10^{-6} mM		-1.00×10^{-6} mM		$+1.00 \times 10^{-6}$ mM
equil.	1.00×10^{-6} mM		0.50×10^{-6} mM		1.00×10^{-6} mM

$K_c = \dfrac{[A-\text{protein}]}{[A][\text{protein}]} = \dfrac{(1.00 \times 10^{-6})}{(1.00 \times 10^{-6})(0.50 \times 10^{-6})} = 2.0 \times 10^6$

	$B(aq)$	+	$protein(aq)$	$\rightleftharpoons$	$B\text{-}protein(aq)$
initial	2.00×10^{-6} mM		1.50×10^{-6} mM		0
change	-1.40×10^{-6} mM		-1.40×10^{-6} mM		$+1.40 \times 10^{-6}$ mM
equil.	0.60×10^{-6} mM		0.10×10^{-6} mM		1.40×10^{-6} mM

$K_c = \dfrac{[B-\text{protein}]}{[B][\text{protein}]} = \dfrac{(1.40 \times 10^{-6})}{(0.60 \times 10^{-6})(0.10 \times 10^{-6})} = 2.3 \times 10^7$

Applications of Equilibrium Constants (Section 15.6)

15.42   (a)   If $Q_c > K_c$, the reaction will proceed in the direction of more reactants, to the left.

       (b) $Q_c = 0$ if the concentration of any product is zero.

15.44   Calculate the reaction quotient in each case, compare with

$$K_p = \frac{P_{NH_3}^2}{P_{N_2} \times P_{H_2}^3} = 4.51 \times 10^{-5}$$

       (a) $Q = \dfrac{(98)^2}{(45)(55)^3} = 1.3 \times 10^{-3}$

            Because $Q > K_p$, the reaction will shift toward reactants to achieve equilibrium.

       (b) $Q = \dfrac{(57)^2}{(143)(0)^3} = \infty$

            Because $Q > K_p$, reaction must shift toward reactants to achieve equilibrium. There must be **some** H_2 present to achieve equilibrium. In this example, the only source of H_2 is the decomposition of NH_3.

       (c) $Q = \dfrac{(13)^2}{(27)(82)^3} = 1.1 \times 10^{-5}$

            Q is only slightly less than K_p, so the reaction will shift slightly toward products to achieve equilibrium.

15.46 $K_p = \dfrac{P_{SO_3}^2}{P_{SO_2}^2 \times P_{O_2}}$; $P_{SO_3} = \left(K_p \times P_{SO_2}^2 \times P_{O_2}\right)^{1/2} = [(0.345)(0.135)^2(0.455)]^{1/2} = 0.0535$ atm

15.48 $K_c = \dfrac{[HI]^2}{[H_2][I_2]} = 55.3$; $[HI] = (K_c[H_2][I_2])^{1/2}$

$$[H_2] = \frac{0.056\,\text{g}\,H_2}{2.00\,\text{L}} \times \frac{1\,\text{mol}\,H_2}{2.016\,\text{g}\,H_2} = 0.01389 = 0.014\,M$$

$$[I_2] = \frac{4.36\,\text{g}\,I_2}{2.00\,\text{L}} \times \frac{1\,\text{mol}\,I_2}{253.8\,\text{g}\,I_2} = 0.008589 = 0.00859\,M$$

$[HI] = [(55.3)(0.01389)(0.008589)]^{1/2} = 0.08122 = 0.081\,M$

$0.08122\,M\,HI \times 2.00\,\text{L} \times \dfrac{127.9\,\text{g}\,HI}{\text{mol}\,HI} = 20.78 = 21\,\text{g}\,HI$

Check. $K_c = \dfrac{(0.08122)^2}{(0.01389)(0.008589)} = 55.3$

15.50 $PV = nRT$; $P = \dfrac{gRT}{MM\,V}$

$$P_{SO_3} = \frac{1.17\,\text{g}\,SO_3}{80.06\,\text{g/mol}} \times \frac{0.08206\,\text{L-atm}}{\text{K-mol}} \times \frac{700\,\text{K}}{2.00\,\text{L}} = 0.4197 = 0.420\,\text{atm}$$

$$P_{O_2} = \frac{0.105\,g\,O_2}{32.00\,g/\,mol} \times \frac{0.08206\,L\text{-}atm}{K\text{-}mol} \times \frac{700\,K}{2.00\,L} = 0.09424 = 0.0942\,atm$$

$$K_p = 3.0 \times 10^4 = \frac{P_{SO_3}^2}{P_{SO_2}^2 \times P_{O_2}};\; P_{SO_2} = \left[P_{SO_3}^2/(K_p)(P_{O_2})\right]^{1/2}$$

$$P_{SO_2} = [(0.4197)^2/(3.0\times10^4)(0.09424)]^{1/2} = 7.894 \times 10^{-3} = 7.9 \times 10^{-3}\,atm$$

$$g\,SO_2 = \frac{MM\,PV}{RT} = \frac{64.06\,g\,SO_2}{mol\,SO_2} \times \frac{K\text{-}mol}{0.08206\,L\text{-}atm} \times \frac{7.894 \times 10^{-3}\,atm \times 2.00\,L}{700\,K}$$

$$= 0.01761 = 0.018\,g\;SO_2$$

Check. $K_p = [(0.4197)^2/(7.894 \times 10^{-3})^2(0.09424)] = 3.0 \times 10^4$

15.52 $[Br_2] = 0.25\,mol/3.0\,L = 0.08333 = 0.083\,M$; $[Cl_2] = 0.55\,mol/3.0\,L = 0.1833 = 0.18\,M$

	$Br_2(g)$	+	$Cl_2(g)$	$\rightleftharpoons$	$2BrCl(g)$	$K_c = \dfrac{[BrCl]^2}{[Br_2][Cl_2]} = 7.0$
initial	$0.083\,M$		$0.18\,M$		0	
change	$-x$		$-x$		$+2x$	
equil.	$(0.083 - x)$		$(0.18 - x)$		$+2x$	

$$7.0 = \frac{(2x)^2}{(0.08333-x)(0.1833-x)};\; 4x^2 = 7.0(0.0153 - 0.2666x + x^2);\; 0 = 0.1069 - 1.8662x + 3x^2$$

$$x = \frac{1.8662 \pm \sqrt{(-1.8662)^2 - 4(3)(0.1069)}}{2(3)} = 0.06387 = 0.064\,M$$

(The 0.56 *M* quadratic solution is not chemically meaningful.)

$[BrCl] = 2x = 0.1277 = 0.13\,M$; $[Br_2] = 0.08333 - 0.06387 = 0.01946 = 0.019\,M$

$[Cl_2] = 0.1833 - 0.06387 = 0.1195 = 0.12\,M$

Check. $K_c = (0.1277)^2/(0.01946)(0.1195) = 7.0125 = 7.0$

15.54 $K_c = [NH_3][H_2S] = 1.2 \times 10^{-4}$. Because of the stoichiometry, equilibrium concentrations of H_2S and NH_3 will be equal; call this quantity y. Then, $y^2 = 1.2 \times 10^{-4}$, $y = 0.010954 = 0.011\,M$.

15.56   (a)   *Analyze/Plan.* If only $PH_3BCl_3(s)$ is present initially, the equation requires that the equilibrium concentrations of $PH_3(g)$ and $BCl_3(g)$ are equal. Write the K_c expression and solve for $x = [PH_3] = [BCl_3]$. *Solve.*

         $K_c = [PH_3][BCl_3]$; $1.87 \times 10^{-3} = x^2$; $x = 0.043243 = 0.0432\,M\,PH_3$ and BCl_3

     (b)   Because the mole ratios are 1:1:1, mol $PH_3BCl_3(s)$ required = mol PH_3 or BCl_3 produced.

$$\frac{0.043243\,mol\,PH_3}{L} \times 0.250\,L = 0.01081 = 0.0108\,mol\,PH_3 = 0.0108\,mol\,PH_3BCl_3$$

$$0.01081\,mol\,PH_3BCl_3 \times \frac{151.2\,g\,PH_3BCl_3}{1\,mol\,PH_3BCl_3} = 1.6346 = 1.63\,g\,PH_3BCl_3$$

In fact, some $PH_3BCl_3(s)$ must remain for the system to be in equilibrium, so a bit more than 1.63 g PH_3BCl_3 is needed.

15.58 $CaCrO_4(s) \rightleftharpoons Ca^{2+}(aq) + CrO_4^{2-}(aq)$ $K_c = [Ca^{2+}][CrO_4^{2-}] = 7.1 \times 10^{-4}$

At equilibrium, $[Ca^{2+}] = [CrO_4^{2-}] = x$

$K_c = 7.1 \times 10^{-4} = x^2, x = 0.0266 = 0.027\ M\ Ca^{2+}$ and CrO_4^{2-}

15.60

	$CH_3COOH(solv)$ -	$CH_3CH_2OH(solv)$	$\rightleftharpoons$	$CH_3COOCH_2CH_3(g)$ +	$H_2O(solv)$
initial	0.275 M	3.85 M		0 M	0 M
change	–x M	–x M		+x M	+x M
equil.	0.275–x M	3.85–x M		+x M	+x M

$$K_c = 6.68 = \frac{x^2}{(0.275-x)(3.85-x)}; x^2 = 6.68(1.059 - 4.125\,x + x^2)$$

$$0 = 5.68\,x^2 - 27.56\,x + 7.072;\ x = \frac{27.56 \pm \sqrt{(-27.56)^2 - 4(5.68)(7.072)}}{2(5.68)} = 0.27185 = 0.272\ M$$

(The 4.58 M quadratic solution is not chemically meaningful.)

$$\frac{0.27185\ \text{mol ethyl acetate}}{L} \times 15.0\ L \times \frac{88.10\ \text{g ethyl acetate}}{mol} = 359.25 = 359\ \text{g ethyl acetate}$$

LeChâtelier's Principle (Section 15.7)

15.62 $4NH_3(g) + 5O_2(g) \rightleftharpoons 4NO(g) + 6\,H_2O(g)$

(a)   increase $[NH_3]$, increase yield NO

(b)   increase $[H_2O]$, decrease yield NO

(c)   decrease $[O_2]$, decrease yield NO

(d)   decrease container volume, decrease yield NO (fewer moles gas in reactants)

(e)   add catalyst, no change

(f)   increase temperature, decrease yield NO (reaction is exothermic)

15.64   (a)   The reaction must be endothermic ($+\Delta H$) if heating increases the fraction of products.

(b)   There must be more moles of gas in the products if increasing the volume of the vessel increases the fraction of products.

15.66   (a) $\Delta H° = \Delta H_f° \ CH_3OH(g) - \Delta H_f° \ CO(g) - 2\Delta H_f° \ H_2(g)$

= –201.2 kJ – (–110.5 kJ) – 0 kJ

= –90.7 kJ

(b)   The reaction is exothermic; an increase in temperature would decrease the value of K and decrease the yield. A low temperature is needed to maximize yield.

(c)   Increasing total pressure would increase the partial pressure of each gas, shifting the equilibrium toward products. The extent of conversion to CH_3OH increases as the total pressure increases.

15.68   (a)   Low temperature. For an exothermic reaction such as this, decreasing temperature increases the value of K and the amount of products at equilibrium.

(b)   No. Because there are equal numbers of moles of gas in the products and reactants, the equilibrium yield of products cannot be changed by changing pressure.

Additional Exercises

15.69   (a)   Because both the forward and reverse processes are elementary steps, we can write the rate laws directly from the chemical equation.

$$rate_f = k_f [CO][Cl_2] = rate_r = k_r [COCl][Cl]$$

$$\frac{k_f}{k_r} = \frac{[COCl][Cl]}{[CO][Cl_2]} = K$$

$$K_c = \frac{k_f}{k_r} = \frac{1.4 \times 10^{-28} M^{-1} s^{-1}}{9.3 \times 10^{10} M^{-1} s^{-1}} = 1.5 \times 10^{-39}$$

For a homogeneous equilibrium in the gas phase, we usually write K in terms of partial pressures. In this exercise, concentrations are more convenient because the rate constants are expressed in terms of molarity. For this reaction, the value of K is the same regardless of how it is expressed, because there is no change in the moles of gas in going from reactants to products.

(b)   Because the K is quite small, reactants are much more plentiful than products at equilibrium.

15.70 $2A(g) \rightleftharpoons B(g)$, $K_c = 1$

$\frac{[B]}{[A]^2} = 1$, $[B] = [A]^2$ and $[A] = [B]^{1/2}$

15.72 $[SO_2Cl_2] = \dfrac{2.00 \text{ mol}}{2.00 \text{ L}} = 1.00 \, M$

The change in $[SO_2Cl_2] = 0.56(1.00 \, M) = 0.56 \, M$

	$SO_2Cl_2(g)$	$\rightleftharpoons$	$SO_2(g)$	+	$Cl_2(g)$	$K_c = \dfrac{[SO_2][Cl_2]}{[SO_2Cl_2]}$
initial	1.00 M		0		0	
change	–0.56 M		+0.56 M		+0.56 M	
equil.	0.44 M		+0.56 M		+0.56 M	

(a) $K_c = \dfrac{(0.56)^2}{0.44} = 0.7127 = 0.71$

(b) $K_p = K_c(RT)^{\Delta n}$; $\Delta n = 2 - 1 = 1$; $K_p = (0.7127)(0.08206)(303) = 17.7214 = 18$

(c)   Increase. There are more moles of gas in the products, so increasing the container volume will shift equilibrium toward products.

(d) $[SO_2Cl_2] = \dfrac{2.00 \text{ mol}}{15.00 \text{ L}} = 0.13333 = 0.133 \, M$. Let x equal the change in $[SO_2Cl_2]$.

The equilibrium concentrations are: $[SO_2Cl_2] = (0.13333 - x)$; $[SO_2] = [Cl_2] = x$

$K_c = \dfrac{(x)^2}{(0.13333 - x)} = 0.7127$; solving the quadratic, x = 0.1148 = 0.11 M

(We expect the decomposition to be greater than 56%, so we must use the quadratic formula to solve for x.)

% decomposition = $(0.1148/0.1333) \times 100 = 86\%$; the increase in volume does shift the equilibrium toward products.

15.73   (a) $H_2(g) + S(s) \rightleftharpoons H_2S(g)$ $K_c = [H_2S]/[H_2]$

(b)   Calculate the molarities of H_2S and H_2.

$[H_2S] = \dfrac{0.46 \text{ g}}{34.1 \text{ g/mol}} \times \dfrac{1}{1.0 \text{ L}} = 0.01349 = 0.013 \, M$

$[H_2] = \dfrac{0.40 \text{ g}}{2.02 \text{ g/mol}} \times \dfrac{1}{1.0 \text{ L}} = 0.1980 = 0.20 \, M$

$K_c = 0.01349/0.1980 = 0.06812 = 0.068$

15.75   (a)

	A(g)	$\rightleftharpoons$	2B(g)
initial	0.75 atm		0
change	−0.39 atm		+0.78 atm
equil.	0.36 atm		0.78 atm

$P_t = P_A + P_B = 0.36 \text{ atm} + 0.78 \text{ atm} = 1.14 \text{ atm}$

(b) $K_P = \dfrac{(P_B)^2}{P_A} = \dfrac{(0.78)^2}{0.36} = 1.690 = 1.7$

(c)   Increasing the volume of the flask favors the reaction with more moles of gas. Doing the reaction in a larger flask maximizes the yield of B.

15.76   (a) $K_P = \dfrac{P_{NH_3}^2}{P_{N_2} \times P_{H_2}^3} = 4.34 \times 10^{-3}$; $T = 300 \, °C + 273 = 573 \, K$

$P_{NH_3} = \dfrac{gRT}{MM \times V} = \dfrac{1.05 \text{ g}}{17.03 \text{ g/mol}} \times \dfrac{0.08206 \text{ L-atm}}{\text{mol-K}} \times \dfrac{573 \text{ K}}{1.00 \text{ L}} = 2.899 = 2.90 \text{ atm}$

	$N_2(g) +$	$3H_2(g)$	$\rightleftharpoons$	$2NH_3(g)$
initial	0 atm	0 atm		?
change	x	3x		−2x
equil.	x atm	3x atm		2.899 atm

(Remember, only the change line reflects the stoichiometry of the reaction.)

$$K_p = \frac{(2.899)^2}{(x)(3x)^3} = 4.34 \times 10^{-3}; \quad 27 \, x^4 = \frac{(2.899)^2}{4.34 \times 10^{-3}}; \quad x^4 = 71.725$$

$$x = 2.910 = 2.91 \, \text{atm} = P_{N_2}; \quad P_{H_2} = 3x = 8.730 = 8.73 \, \text{atm}$$

$$g_{N_2} = \frac{MM \times PV}{RT} = \frac{28.02 \, \text{g N}_2}{\text{mol N}_2} \times \frac{\text{mol-K}}{0.08206 \, \text{L-atm}} \times \frac{2.910 \, \text{atm} \times 1.00 \, \text{L}}{573 \, \text{K}} = 1.73 \, \text{g N}_2$$

$$g_{H_2} = \frac{2.016 \, \text{g H}_2}{\text{mol H}_2} \times \frac{\text{mol-K}}{0.08206 \, \text{L-atm}} \times \frac{8.730 \, \text{atm} \times 1.00 \, \text{L}}{573 \, \text{K}} = 0.374 \, \text{g H}_2$$

(b)   The initial $P_{NH_3} = 2.899 \, \text{atm} + 2(2.910 \, \text{atm}) = 8.719 = 8.72 \, \text{atm}$

$$g_{NH_3} = \frac{17.03 \, \text{g NH}_3}{\text{mol NH}_3} \times \frac{\text{mol-K}}{0.08206 \, \text{L-atm}} \times \frac{8.719 \, \text{atm} \times 1.00 \, \text{L}}{573 \, \text{K}} = 3.16 \, \text{g NH}_3$$

(c) $P_t = P_{N_2} + P_{H_2} + P_{NH_3} = 2.910 \, \text{atm} + 8.730 \, \text{atm} + 2.899 \, \text{atm} = 14.54 \, \text{atm}$

15.78   (a) $K_p = 0.052; \quad K_p = K_c(RT)^{\Delta n}; \quad \Delta n = 2 - 0 = 2; \quad K_c = K_p/(RT)^2$

$K_c = 0.052/[0.08206)(333)]^2 = 6.964 \times 10^{-5} = 7.0 \times 10^{-5}$

(b) PH_3BCl_3 is a solid and its concentration is taken as a constant, C.

$$[BCl_3] = \frac{0.0500 \, \text{g BCl}_3}{1.500 \, \text{L}} \times \frac{\text{mol BCl}_3}{117.17 \, \text{g BCl}_3} = 2.8449 \times 10^{-4} = 2.84 \times 10^{-4} \, M \, BCl_3$$

	PH_3BCl_3	$\rightleftharpoons$	PH_3	+	BCl_3
initial	C		$0 \, M$		$2.84 \times 10^{-4} \, M$
change			$+x \, M$		$+x \, M$
equil.	C		$+x \, M$		$(2.84 \times 10^{-4} + x) \, M$

$K_c = [PH_3][BCl_3]; \quad 6.964 \times 10^{-5} = x(2.84 \times 10^{-4} + x); \quad x^2 + 2.84 \times 10^{-4} \, x - 6.964 \times 10^{-5} = 0$

$$x = \frac{-2.84 \times 10^{-4} \pm [(2.84 \times 10^{-4})^2 - 4(1)(-6.964 \times 10^{-5})]^{1/2}}{2(1)} = 0.008204 = 8.2 \times 10^{-3} \, M \, PH_3$$

Check. $K_c = (8.2 \times 10^{-3})(2.84 \times 10^{-4} + 8.2 \times 10^{-3}) = 7.0 \times 10^{-5}$.

15.79 $K_p = P_{NH_3} \times P_{H_2S}; \quad P_t = 0.614 \, \text{atm}$

If the equilibrium amounts of NH_3 and H_2S are solely due to the decomposition of $NH_4HS(s)$, the equilibrium pressures of the two gases are equal, and each is 1/2 of the total pressure.

$P_{NH_3} = P_{H_2S} = 0.614 \, \text{atm}/2 = 0.307 \, \text{atm}$

$K_p = (0.307)^2 = 0.0943$

15.81   In general, the reaction quotient is of the form $Q = \dfrac{P_{NOCl}^2}{P_{NO}^2 \times P_{Cl_2}}$.

(a) $Q = \dfrac{(0.11)^2}{(0.15)^2 (0.31)} = 1.7$

$Q > K_p$. The mixture is not at equilibrium. It will shift to the left and produce more reactants as it moves toward equilibrium.

(b) $Q = \dfrac{(0.050)^2}{(0.12)^2\,(0.10)} = 1.7$

$Q > K_p$. The mixture is not at equilibrium. It will shift to the left and produce more reactants as it moves toward equilibrium.

(c) $Q = \dfrac{(5.10 \times 10^{-3})^2}{(0.15)^2\,(0.20)} = 5.8 \times 10^{-3}$

$Q < K_p$. The mixture is not at equilibrium. It will shift to the right and produce more products as it moves toward equilibrium.

15.82 $K_c = [CO_2] = 0.0108;\ [CO_2] = \dfrac{g\ CO_2}{44.01\ g/mol} \times \dfrac{1}{10.0\ L}$

In each case, calculate $[CO_2]$ and determine the position of the equilibrium.

(a) $[CO_2] = \dfrac{4.25\ g}{44.01\ g/mol} \times \dfrac{1}{10.0\ L} = 9.657 \times 10^{-3} = 9.66 \times 10^{-3}\ M$

$Q = 9.66 \times 10^{-3} < K_c$. The reaction proceeds to the right to achieve equilibrium and the amount of $CaCO_3(s)$ decreases.

(b) $[CO_2] = \dfrac{5.66\ g\ CO_2}{44.01\ g/mol} \times \dfrac{1}{10.0\ L} = 0.0129\ M$

$Q = 0.0129 > K_c$. The reaction proceeds to the left to achieve equilibrium and the amount of $CaCO_3(s)$ increases.

(c)   6.48 g CO_2 means $[CO_2] > 0.0129\ M$; $Q > 0.0129 > K_c$, the amount of $CaCO_3$ increases.

15.84   (a) $[CO_2] = \dfrac{25.0\ g\ CO_2}{44.01\ g/mol} \times \dfrac{1}{3.00\ L} = 0.18935 = 0.189\ M$

$$C(s) \quad + \quad CO_2(g) \ \rightleftharpoons\ 2CO(g)$$

initial	excess	0.189	0
change	$-x$	$-x$	$+2x$
equil.		$0.189 - x$	$+2x$

$K_c = 1.9 = \dfrac{[CO]^2}{[CO_2]} = \dfrac{(2x)^2}{0.189 - x};\ \ 4x^2 = 1.9(0.189 - x);\ \ 4x^2 + 1.9x - 0.36 = 0.$

Solve the quadratic for x.

$x = \dfrac{-1.9 \pm \sqrt{(1.9)^2 - 4(4)(-0.36)}}{2(4)} = 0.14505 = 0.15\ M$

$[CO] = 2x = 2(0.14505) = 0.2901 = 0.29\ M$

$\dfrac{0.2901\ mol\ CO}{L} \times 3.00\ L \times \dfrac{28.01\ g\ CO}{mol} = 24.38 = 24\ g\ CO$

(b)   The amount of C(s) consumed is related to x. Change M to mol to g C.

$\dfrac{0.14505\ mol}{L} \times 3.00\ L \times 12.01\ g = 5.226 = 5.2\ g\ C$ consumed

(c)   A smaller vessel at the same temperature increases the total pressure of the mixture. The equilibrium shifts to form fewer total moles of gas, which favors reactants. The yield of CO product will be smaller in a smaller vessel.

(d)   The two K_c values are 0.133 at 298 K and 1.9 at 1000 K. The reaction is endothermic, because K is larger at higher temperature.

15.85 $K_p = \dfrac{P_{CO_2}}{P_{CO}} = 6.0 \times 10^2$

If P_{co} is 150 torr, P_{CO_2} can never exceed $760 - 150 = 610$ torr. Then $Q = 610/150 = 4.1$. Because this is far less than K, the reaction will shift in the direction of more product. Reduction will therefore occur.

15.86   The anecdote tells us that increasing the volume of the reaction container, the furnace, had no effect on the amount of unreacted CO(g), the amount of CO(g) expelled. This is true for reactions that have the same number of moles of gaseous products and reactants, as this one does. It also means that $K_p = K_c$.

15.87   (a)

$$CCl_4(g) \;\rightleftharpoons\; C(s) + 2Cl_2(g)$$

initial	2.00 atm	0 atm
change	–x atm	+2x atm
equil.	(2.00–x) atm	2x atm

$$K_p = 0.76 = \frac{P_{Cl_2}^2}{P_{CCl_4}} = \frac{(2x)^2}{(2.00 - x)}$$

$1.52 - 0.76x = 4x^2; \quad 4x^2 + 0.76x - 1.52 = 0$

Using the quadratic formula, $a = 4, b = 0.76, c = -1.52$

$$x = \frac{-0.76 \pm \sqrt{(0.76)^2 - 4(4)(-1.52)}}{2(4)} = \frac{-0.76 + 4.99}{8} = 0.5287 = 0.53 \text{ atm}$$

$$\text{Fraction } CCl_4 \text{ reacted} = \frac{x \text{ atm}}{2.00 \text{ atm}} = \frac{0.5287}{2.00} = 0.264 = 26\%$$

(b) $P_{Cl_2} = 2x = 2(0.5287) = 1.06 \text{ atm}$

        $P_{CCl_4} = 2.00 - x = 2.00 - 0.5287 = 1.47 \text{ atm}$

15.88   (a) $Q = \dfrac{P_{PCl_5}}{P_{PCl_3} \times P_{Cl_2}} = \dfrac{(0.20)}{(0.50)(0.50)} = 0.80$

        0.80 (Q) > 0.0870 (K), the reaction proceeds to the left.

(b)

$$PCl_3(g) \;+\; Cl_2(g) \;\rightleftharpoons\; PCl_5(g)$$

initial	0.50 atm	0.50 atm	0.20 atm
change	+x atm	+x atm	–x atm
equil.	(0.50 + x) atm	(0.50 + x) atm	(0.20 – x) atm

(Because the reaction proceeds to the left, P_{PCl_5} must decrease and P_{PCl_3} and P_{Cl_2} must increase.)

$$K_p = 0.0870 = \frac{(0.20-x)}{(0.50+x)(0.50+x)}; \quad 0.0870 = \frac{(0.20-x)}{(0.250+1.00\,x+x^2)}$$

$$0.0870(0.250 + 1.00x + x^2) = 0.20 - x; \; -0.17825 + 1.0870x + 0.0870\,x^2 = 0$$

$$x = \frac{-1.0870 \pm \sqrt{(1.0870)^2 - 4(0.0870)(-0.17825)}}{2(0.0870)} = \frac{-1.0870 + 1.1152}{0.174} = 0.162$$

$$P_{PCl_3} = (0.50+0.162)\, atm = 0.662 \qquad P_{Cl_2} = (0.50+0.162)\, atm = 0.662\, atm$$

$$P_{PCl_5} = (0.20-0.162)\, atm = 0.038\, atm$$

To two decimal places, the pressures are 0.66, 0.66, and 0.04 atm, respectively. When substituting into the K_p expression, pressures to three decimal places yield a result much closer to 0.0870.

(c)   Increasing the volume of the container favors the process where more moles of gas are produced, so the reverse reaction is favored and the equilibrium shifts to the left; the mole fraction of Cl_2 increases.

(d)   For an exothermic reaction, increasing the temperature decreases the value of K; more reactants and fewer products are present at equilibrium and the mole fraction of Cl_2 increases.

15.89   *Analyze/Plan.* Calculate the equilibrium pressures of H_2, I_2, and HI; use them to calculate K_p. Set up a new equilibrium table and calculate new equilibrium pressures. *Solve.*

$$\frac{P}{n} = \frac{RT}{V} = \frac{0.08206\ L\text{-}atm}{mol\text{-}K} \times \frac{731\ K}{5.00} = 11.997\ \frac{atm}{mol}$$

$$P_{H_2} = P_{I_2} = 0.112\ mol \times 11.997\ \frac{atm}{mol} = 1.344 = 1.34\ atm$$

$$P_{HI} = 0.775\ mol \times 11.997\ \frac{atm}{mol} = 9.298 = 9.30\ atm$$

$$H_2(g) + I_2(g) \rightleftharpoons 2HI(g); \quad K_p = \frac{P_{HI}^2}{P_{H_2} \times P_{I_2}} = \frac{(9.298)^2}{(1.344)^2} = 47.861 = 47.9$$

$$P_{HI}\ (added) = 0.200\ mol \times \frac{11.997\ atm}{mol} = 2.3994 = 2.40\ atm$$

	$H_2(g)$	+	$I_2(g)$	$\rightleftharpoons$	$2HI(g)$
initial	1.34 atm		1.34 atm		9.30 atm + 2.40 atm
change	+x atm		+x atm		−2x atm
equil.	(1.34+x) atm		(1.34+x) atm		(11.70−2x) atm

$$K_p = 47.86 = \frac{(11.70-2x)^2}{(1.34+x)^2}. \quad \text{Take the square root of both sides:}$$

$$6.918 = \frac{11.70 - 2x}{1.34 + x}; \; 9.270 + 6.918\,x = 11.70 - 2x; \; 8.918\,x = 2.430; \; x = 0.27248 = 0.272$$

$P_{H_2} = P_{I_2} = 1.34 + 0.272 = 1.612 = 1.61\,\text{atm}; P_{HI} = 11.70 - 2(0.272) = 11.156 = 11.16\,\text{atm}$

Check. $\dfrac{(11.156)^2}{(1.612)^2} = 47.89 = 47.9$

15.90   (a)   Because the volume of the vessel = 1.00 L, mol = M. The reaction will proceed to the left to establish equilibrium.

$$A(g) + 2B(g) \rightleftharpoons 2C(g)$$

	A(g) +	2B(g) $\rightleftharpoons$	2C(g)
initial	0 M	0 M	1.00 M
change	+x M	+2x M	–2x M
equil.	x M	2x M	(1.00 – 2x) M

At equilibrium, [C] = (1.00 – 2x) M, [B] = 2x M.

(b)   x must be less than 0.50 M (so that [C], 1.00 –2x, is not less than zero).

(c) $K_c = \dfrac{[C]^2}{[A][B]^2}$; $\dfrac{(1.00 - 2x)^2}{(x)(2x)^2} = 0.25$

$1.00 - 4x + 4x^2 = 0.25(4x)^3$; $x^3 - 4x^2 + 4x - 1 = 0$

(d)

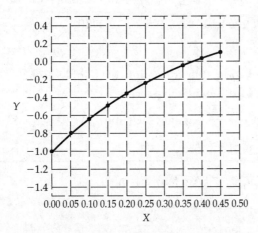

X	Y
0.0	−1.000
0.05	−0.810
0.10	−0.639
0.15	−0.487
0.20	−0.352
0.25	−0.234
0.35	−0.047
0.40	+0.024
0.45	+0.081
~0.383	0.00

(e)   From the plot, x ≈ 0.383 M

[A] = x = 0.383 M; [B] = 2x = 0.766 M

[C] = 1.00 – 2x = 0.234 M

Using the K_c expression as a check:

$K_c = 0.25$; $\dfrac{(0.234)^2}{(0.383)(0.766)^2} = 0.24$;   the estimated values are reasonable.

15.92   The patent claim is false. A catalyst does not alter the position of equilibrium in a system, only the rate of approach to the equilibrium condition.

Integrative Exercises

15.94   (a) $AgCl(s) \rightleftharpoons Ag^+(aq) + Cl^-(aq)$

      (b) $K_c = [Ag^+][Cl^-]$

      (c)   Using thermodynamic data from Appendix C, calculate ΔH for the reaction in part (a).

$$\Delta H° = \Delta H_f° \ Ag^+(aq) + \Delta H_f° \ Cl^-(aq) - \Delta H_f° \ AgCl(s)$$

$$\Delta H° = 105.90 \ kJ - 167.2 \ kJ - (-127.0 \ kJ) = 65.7 \ kJ$$

The reaction is endothermic (heat is a reactant), so the solubility of AgCl(s) in $H_2O(l)$ will increase with increasing temperature.

      (d)   Calculate the solubility of AgCl(s) in pure water as $[Ag^+]$, in 0.100 M NaCl(aq) as $[Ag^+]$, and in 0.100 M NaCl(aq) as $[AgCl_2]^-$.

In pure water, let $[Ag^+] = [Cl^-] = x$; $K_c = [Ag^+][Cl^-] = x^2 = 1.6 \times 10^{-10}$

$x = (1.6 \times 10^{-10})^{1/2}$; $[Ag^+] = 1.3 \times 10^{-5} \ M \ [Ag^+]$

In 0.100 M NaCl(aq), assuming no formation of $[AgCl_2^-]$,

$[Ag^+] = x$, $[Cl^-] = 0.100 + x$; $K_c = [Ag^+][Cl^-] = x(0.100 + x) = 1.6 \times 10^{-10}$

Assuming x is small relative to 0.100, $0.100 \ x = 1.6 \times 10^{-10}$,

$x = 1.6 \times 10^{-9} \ M \ [Ag^+]$

To account for complexation, the formation of soluble $AgCl_2^-$, sum the two reactions and calculate K for the overall process.

$AgCl(s) \rightleftharpoons Ag^+(aq) + Cl^-(aq)$	$K_1 = 1.6 \times 10^{-10}$
$Ag^+(aq) + 2Cl^-(aq) \rightleftharpoons AgCl_2^-(aq)$	$K_2 = 1.8 \times 10^5$

$AgCl(s) + Ag^+(aq) + 2Cl^-(aq) \rightleftharpoons Ag^+(aq) + Cl^-(aq) + AgCl_2^-(aq)$ $K_c = K_1 \times K_2$

$AgCl(s) + Cl^-(aq) \rightleftharpoons AgCl_2^-(aq)$ $K_c = 2.9 \times 10^{-5}$

$K_c = [AgCl_2^-]/[Cl^-]$; $[AgCl_2^-] = x$, $[Cl^-] = 0.100 - x$

Assuming x is small relative to 0.100, $2.9 \times 10^{-5} = x/(0.100)$;

$x = 2.9 \times 10^{-6} \ M \ AgCl_2^-$

These calculations show that the "solubility" of AgCl, the concentration of any soluble silver species, is actually greatest in pure water, $1.3 \times 10^{-5} \ M \ [Ag^+]$. In 0.100 M NaCl, complexation causes the solubility of silver, $2.9 \times 10^{-6} \ M \ [AgCl_2^-]$, to be greater than that predicted by the common ion effect alone, $1.6 \times 10^{-9} \ M \ [Ag^+]$, but not as great as its solubility in pure water.

15.95   Consider the energy profile for an exothermic reaction.

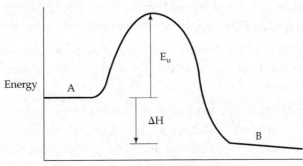

Reaction Pathway

The activation energy in the forward direction, E_{af}, equals E_u and the activation energy in the reverse reaction, E_{ar}, equals $E_u - \Delta H$. (The same is true for an endothermic reaction because the sign of ΔH is the positive and $E_{ar} < E_{af}$.) For the reaction in question,

$$K = \frac{k_f}{k_r} = \frac{A_f e^{-E_{af}/RT}}{A_r e^{-E_{ar}/RT}}$$

Because the ln form of the Arrhenius equation is easier to manipulate, we will consider ln K.

$$\ln K = \ln\left(\frac{k_f}{k_r}\right) = \ln k_f - \ln k_r = \frac{-E_{af}}{RT} + \ln A_f - \left[\frac{-E_{ar}}{RT} + \ln A_r\right]$$

Substituting E_u for E_{af} and $(E_u - \Delta H)$ for E_{ar}

$$\ln K = \frac{-E_u}{RT} + \ln A_f - \left[\frac{-(E_u - \Delta H)}{RT} + \ln A_r\right]; \quad \ln K = \frac{-E_u + (E_u - \Delta H)}{RT} + \ln A_f - \ln A_r$$

$$\ln K = \frac{-\Delta H}{RT} + \ln \frac{A_f}{A_r}$$

For the catalyzed reaction, $E_{cat} < E_u$ and $E_{af} = E_{cat}$, $E_{ar} = E_{cat} - \Delta H$. The catalyst does not change the value of ΔH.

$$\ln K_{cat} = \frac{-E_{cat} + (E_{cat} - \Delta H)}{RT} + \ln A_f - \ln A_r$$

$$\ln K_{cat} = \frac{-\Delta H}{RT} + \frac{\ln A_f}{A_r}$$

Thus, assuming A_f and A_r are not changed by the catalyst, $\ln K = \ln K_{cat}$ and $K = K_{cat}$.

15.96   (a)   For the reaction $SO_2(l) \rightleftharpoons SO_2(g)$, $K_p = P_{SO_2}$. From the phase diagram, as T increases P_{SO_2} and K_p increase. For an endothermic reaction, K increases as T increases. The phase diagram tells us that the vaporization of $SO_2(l)$ is an endothermic process.

         (b)   Read P_{SO_2} from the liquid-gas line on the phase diagram at $0\,°C$ and $100\,°C$. Note that the pressure axis of the phase diagram is logarithmic with respect to pressure, but linear with respect to logP. In terms of logP, the axis labels would be -1, 0, 1 and 2. The logP values at $0\,°C$ and $100\,°C$ are approximately 0.25 and 1.4. The values of P_{SO_2} and K_p are $10^{0.25}$ and $10^{1.4}$, 1.8, and 25, respectively.

(c)　It is not possible to calculate an equilibrium constant between the gas and liquid phases in the supercritical region, because they do not exist separately in this region. That is, the gas and liquid phases are indistinguishable in the supercritical region.

(d)　Gases are most ideal at high temperature and low pressure. The red dot at slightly greater than 240 $^{\circ}$C is the point where $SO_2(g)$ most closely approaches ideal behavior.

(e)　The point near 15 $^{\circ}$C is the one at the lowest temperature, but it is also at low pressure. In general, the closer the pressure and temperature conditions are to the point of phase transition, the less ideal the behavior of the gas (because it is nearly a liquid). This describes the point near 115 $^{\circ}$C and 20 atm, which is at relatively high pressure and near the liquid-gas line.

15.97　Mole % = pressure %. Because the total pressure is 1 atm, mol %/100 = mol fraction = partial pressure. $K_p = P_{CO}^2 / P_{CO_2}$.

Temp (K)	P_{CO_2} (atm)	P_{CO} (atm)	K_p
1123	0.0623	0.9377	14.1
1223	0.0132	0.9868	73.8
1323	0.0037	0.9963	2.7×10^2
1473	0.0006	0.9994	1.7×10^3 (2×10^3)

Because K grows larger with increasing temperature, the reaction must be endothermic in the forward direction.

15.98　(a)　$H_2O(l) \rightleftharpoons H_2O(g)$;　$K_p = P_{H_2O}$

(b)　At 30°C, the vapor pressure of $H_2O(l)$ is 31.82 torr. $K_p = P_{H_2O} = 31.82$ torr

$K_p = 31.82 \text{ torr} \times 1 \text{ atm}/760 \text{ torr} = 0.041868 = 0.04187$ atm

(c)　From part (b), the value of K_p is the vapor pressure of the liquid at that temperature. By definition, vapor pressure = atmospheric pressure = 1 atm at the normal boiling point. $K_p = 1$ atm

15.99　(a)　VSEPR indicates that each O atom has four electron domains about it and thus adopts tetrahedral geometry. One O atom has two covalent bonds to H and two hydrogen bonds to H atoms on the second water molecule. The O atom on the second water molecule has two nonbonding electron pairs. The water dimer is not symmetrical.

(b)　Hydrogen-bonding is the intermolecular interaction involved in water dimer formation.

(c)　Water dimer formation is exothermic, because the value of K decreases as temperature increases.

15.100   The O_2-binding reaction occurs in aqueous solution, so we will write a K_c expression. The amount of $O_2(g)$ will appear as a pressure. By convention, reactions which involve gaseous and aqueous substances have mixed equilibrium expressions written in terms of both pressures and molar concentrations.

$$K_c = \frac{[Hb\text{-}(O_2)_4]}{P_{O_2}^4 \times [Hb]}$$

The *P50 value* is the partial pressure at which 50% of the hemoglobin is saturated with $O_2(g)$. At this partial pressure, the concentrations of O_2-bound hemoglobin and free hemoglobin are equal, $[Hb - (O_2)_4] = [Hb]$. Substitute the two P50 values into the K_c expression and compare the values for fetal and adult hemoglobin.

$$\text{at P50, } K_{cF} = \frac{1}{P_{O_2}^4} = \frac{1}{19^4} = 7.7 \times 10^{-6}; \quad K_{cA} = \frac{1}{P_{O_2}^4} = \frac{1}{26.8^4} = 1.94 \times 10^{-6}$$

Comparing the two values, $K_{cF}/K_{cA} = 7.7 \times 10^{-6}/1.94 \times 10^{-6} \approx 4$. The equilibrium constant for O_2-binding by fetal hemoglobin is approximately four times that by adult hemoglobin. p

16 Acid–Base Equilibria

Visualizing Concepts

16.2   *Plan.* The stronger the acid, the greater the extent of ionization. The stronger the acid, the weaker its conjugate base. In an acid–base reaction, equilibrium will favor the side with the weaker acid and base.   *Solve.*

(a)   HY is stronger than HX. Starting with six HY molecules, four are dissociated; of six HX molecules, only two are dissociated. Because it is dissociated to a greater extent, HY is the stronger acid.

(b)   If HY is the stronger acid, Y^- is the weaker base and X^- is the stronger base.

(c)   HX and Y^-, the reactants, are the weaker acid and base. Equilibrium lies to the left, and $K_c < 1$.

16.3   (a)   True. Solution A is the color of methyl orange in an acidic solution.

(b)   False. Methyl orange turns yellow at a pH slightly greater than 4, so solution B could be at any pH greater than 4.

(c)   True. The basic color of any indicator occurs at higher pH than the acidic color does.

16.4   *Analyze/Plan.* The pH reading on the meter will identify the solution. Use the definition of pH or pOH to calculate the concentration of the solution.   *Solve.*

(a)   KOH(aq). The pH meter reads 12.08, indicating that the solution is strongly basic. KOH(aq) is the only choice that is a base.

(b)   pOH = 14.00 – pH; pOH = 14.00 – 12.08 = 1.92

$pOH = -\log [OH^-]$; $[OH^-] = 10^{-pOH}$; $[OH^-] = 10^{-1.92} = 0.01202 = 0.012\ M$

(c)   The pH scale for aqueous solutions is based on the value of K_w for water. K_w is an equilibrium constant whose value depends on temperature. If temperature is significantly different from 25 °C, the value of K_w is different from 1×10^{-14} and relationships like {pH + pOH = 14} do not hold true.

16.6   *Analyze/Plan.* We are shown a plot of concentration solution versus $[H^+]$. Use the definition of acids and bases along with relationships between total solute concentration and $[H^+]$ for strong or weak acids and bases to answer the questions. *Solve.*

(a)   The substance is a strong acid. It is an acid because $[H^+]$ increases as solute concentration increases. It is a strong acid because $[H^+]$ is directly proportional to solute concentration. This is not true for weak acids.

(b)   Yes. A strong acid is completely ionized in aqueous solution, so $[H^+]$ = solute concentration = 0.18 M. $pH = -\log[H^+] = -\log(0.18) = 0.74$.

(c)   No. In pure water, there is a finite concentration of H^+(aq) and $[OH^-]$, owing to the autoionization of water.

16.8   *Analyze/Plan.* Write the formula of each molecule and compare them to the entries in Tables 16.2 and 16.4. Select the molecule that fits the definition of an acid and the one that fits the definition of a base.   *Solve.*

(a)   Molecule A is hydroxyl amine, NH_2OH. It is an entry in Table 16.4. Molecule A is an H^+ acceptor because of the nonbonded electron pair on the N atom of the amine ($-NH_2$) group, not because it contains an $-OH$ group. The presence of an $-OH$ group in an organic molecule does not mean that the molecule is a base.

(b)   Molecule B is formic acid, $HCOOH$. It is similar to CH_3COOH, an entry in Table 16.2. The H atom bonded to O is ionizable and $HCOOH$ is an H^+ donor. In general, organic molecules that contain a carboxyl ($-COOH$) group are acids.

(c)   Molecule C is methanol, CH_3OH. In organic molecules, the $-OH$ functional group is an alcohol. The H atom bonded to O is not ionizable, and the $-OH$ group does not dissociate in aqueous solution. An alcohol is neither an acid nor a base.

16.9   (a)   Basic. Because of the amine group (N, with a lone electron pair, bound to two C atoms and a H atom), we expect phenylephrine solution to be basic. [Note that $-OH$ bound to a benzene ring is very weakly acidic (K_a for phenol is 1.3×10^{-10}), but probably not acidic enough to dominate the acid/base properties of the molecule.]

(b)   Phenylephrine is a neutral molecule, whereas phenylephrine hydrochloride is a salt. It is the product when HCl reacts with phenylephrine. The cation in the salt has an additional H atom bound to the N of phenylephrine. The anion is chloride.

(c)   Acidic. The cation of phenylephrine hydrochloride is the conjugate acid of phenylephrine, so a solution of the salt is acidic. (Chloride anion is a negligible base.)

16.10   Diagram C best represents an aqueous solution of NaF; it contains mostly Na^+ and F^-, along with a few HF molecules and OH^- ions. The HF and OH^- are present because F^- is a weak Brønsted–Lowry base; it accepts H^+ from a water molecule, producing HF and OH^-. The solution is basic because it contains OH^-.

16.12   (a)   *Plan.* Count valence electrons and draw the correct Lewis structures. Consider the definition of Lewis acids and bases.   *Solve.*

$$PCl_4^+ \qquad + \qquad Cl^- \qquad \longrightarrow \qquad PCl_5$$
$$32\,e^-, 16\,e^-\,pr \qquad 8\,e^-, 4\,e^-\,pr \qquad\qquad 40\,e^-, 20\,e^-\,pr$$

PCl_4^+ accepts an electron pair from Cl^-; PCl_4^+ is the Lewis acid and Cl^- is the Lewis base.

(b)   The hydrated cation is an oxyacid: the ionizable H is attached to O, which is bound to the central cation. As the charge on the cation increases, it attracts more electron density from the O–H bond, which becomes weaker and more polar. The degree of ionization and the equilibrium constant (K_a) increase.

Arrhenius and Brønsted–Lowry Acids and Bases (Sections 16.1 and 16.2)

16.14   (a)   According to the Arrhenius definition, a *base* when dissolved in water increases $[OH^-]$. According to the Brønsted–Lowry definition, a *base* is an H^+ acceptor regardless of physical state. A Brønsted–Lowry base is not limited to aqueous solution and need not contain OH^- or produce it in aqueous solution.

(b)   Yes, a substance can behave as an Arrhenius base even if it does not contain an OH group. One example is ammonia.

$NH_3(g) + H_2O(l) \rightleftharpoons NH_4^+(aq) + OH^-(aq)$. When NH_3 dissolves in water, it accepts H^+ from H_2O. In doing so, OH^- is produced and $[OH^-]$ in the aqueous solution increases. The OH^- produced was originally part of the H_2O molecule, not part of the NH_3 molecule. Other organic amines behave in a similar manner.

16.16   A conjugate base has one less H^+ than its conjugate acid. A conjugate acid has one more H^+ than its conjugate base.

(a)   (i) $HCOO^-$ (ii) PO_4^{3-}

(b)   (i) HSO_4^- (ii) $CH_3NH_3^+$

16.18

B-L Acid	+	**B-L Base**	$\rightleftharpoons$	**Conjugate Acid**	+	**Conjugate Base**
(a) $HBrO(aq)$		$H_2O(l)$		$H_3O^+(aq)$		$BrO^-(aq)$
(b) $HSO_4^-(aq)$		$HCO_3^-(aq)$		$H_2CO_3(aq)$		$SO_4^{2-}(aq)$
(c) $H_3O^+(aq)$		$HSO_3^-(aq)$		$H_2SO_3(aq)$		$H_2O(l)$

16.20   (a) $H_2C_6H_7O_5^-(aq) + H_2O(l) \rightleftharpoons H_3C_6H_7O_5(aq) + OH^-(aq)$

(b) $H_2C_6H_7O_5^-(aq) + H_2O(l) \rightleftharpoons HC_6H_7O_5^{2-}(aq) + H_3O^+(aq)$

(c) $H_3C_6H_7O_5$ is the conjugate acid of $H_2C_6H_7O_5^-$.

$HC_6H_7O_5^{2-}$ is the conjugate base of $H_2C_6H_7O_5^-$.

16.22   *Analyze/Plan.* Based on the chemical formula, decide whether the acid is strong, weak, or negligible. Is it one of the known seven strong acids (Section 16.5)? Also check Figure 16.4. To write the formula of the conjugate base, remove a single H and decrease the particle charge by one.   *Solve.*

(a) $HCOOH$, weak acid; $HCOO^-$, weak base

(b) H_2, negligible acid; H^-, strong base

(c) CH_4, negligible acid; CH_3^-, strong base

(d)   HF, weak acid; F^-, weak base

(e) NH_4^+, weak acid; NH_3, weak base

16.24 (a) $HClO_3$. It is one of the seven strong acids (Section 16.5). Also, in a series of oxyacids with the same central atom (Cl), the acid with more O atoms is stronger (Section 16.10).

(b) HS^-. H_2SO_4 is a stronger acid than H_2S, so HS^- is the stronger conjugate base. In fact, because H_2SO_4 is one of the seven strong acids, HSO_4^- is a negligible base.

16.26

Base	+	**Acid**	⇌	**Conjugate Acid**	+	**Conjugate Base**

(a) $OH^-(aq)$ + $NH_4^+(aq)$ ⇌ $H_2O(l)$ + $NH_3(aq)$

OH^- is a stronger base than NH_3 (Figure 16.4), so the equilibrium lies to the right.

(b) $CH_3COO^-(aq)$ + $H_3O^+(aq)$ ⇌ $CH_3COOH(aq)$ + $H_2O(l)$

H_3O^+ is a stronger acid than CH_3COOH (Figure 16.4), so the equilibrium lies to the right.

(c) $F^-(aq)$ + $HCO_3^-(aq)$ ⇌ $CO_3^{2-}(aq)$ + $HF(aq)$

CO_3^{2-} is a stronger base than F^-, so the equilibrium lies to the left.

Autoionization of Water (Section 16.3)

16.28 (a) $H_2O(l) \rightleftharpoons H^+(aq) + OH^-(aq)$

(b) $K_w = [H^+][OH^-]$

(c)   Statement (iii) is true. If a solution is basic, it contains more OH^- than H^+.

16.30 In pure water at 25 °C, $[H^+] = [OH^-] = 1 \times 10^{-7}\ M$. If $[OH^-] > 1 \times 10^{-7}\ M$, the solution is basic; if $[OH^-] < 1 \times 10^{-7}\ M$, the solution is acidic.

(a) $[OH^-] = \dfrac{K_w}{[H^+]} = \dfrac{1.0 \times 10^{-14}}{0.0505\,M} = 1.98 \times 10^{-13}\ M < 1 \times 10^{-7}\ M$; acidic

(b) $[OH^-] = \dfrac{K_w}{[H^+]} = \dfrac{1.0 \times 10^{-14}}{2.5 \times 10^{-10}\,M} = 4.0 \times 10^{-5}\ M > 1 \times 10^{-7}\ M$; basic

(c) $[H^+] = 1000[OH^-];\ K_w = 1000[OH^-][OH^-] = 1000[OH^-]^2$

$[OH^-] = (K_w / 1000)^{1/2} = 3.2 \times 10^{-9}\ M < 1 \times 10^{-7}\ M$; acidic

16.32 $K_w = [D^+][OD^-]$; for pure D_2O, $[D^+] = [OD^-]$; $8.9 \times 10^{-16} = [D^+]^2$;

$[D^+] = [OD^-] = 3.0 \times 10^{-8}\ M$

The pH Scale (Section 16.4)

16.34 $[H^+]_A = 250\ [H^+]_B$. From Solution 16.33, $\Delta pH = -\log \dfrac{[H^+]_B}{[H^+]_A}$

$$\Delta pH = -\log \frac{[H^+]_B}{250\,[H^+]_B} = -\log\left(\frac{1}{250}\right) = 2.40$$

The pH of solution A is 2.40 pH units lower than the pH of solution B, because $[H^+]_A$ is 250 times greater than $[H^+]_B$. The greater the $[H^+]$, the lower the pH of the solution.

16.36

pH	pOH	[H$^+$]	[OH$^-$]	acidic or basic
5.25	8.75	$5.6 \times 10^{-6}\,M$	$1.8 \times 10^{-9}\,M$	acidic
11.98	2.02	$1.1 \times 10^{-12}\,M$	$9.6 \times 10^{-3}\,M$	basic
9.36	4.64	$4.4 \times 10^{-10}\,M$	$2.3 \times 10^{-5}\,M$	basic
12.93	1.07	$1.2 \times 10^{-13}\,M$	$8.5 \times 10^{-2}\,M$	basic

16.38   The pH ranges from 5.2 to 5.6; pOH ranges from (14.0–5.2 =) 8.8 to (14.0–5.6 =) 8.4.

$[H^+] = 10^{-pH}$, $[OH^-] = 10^{-pOH}$

$[H^+] = 10^{-5.2} = 6.31 \times 10^{-6} = 6 \times 10^{-6}\,M$; $[H^+] = 10^{-5.6} = 2.51 \times 10^{-6} = 3 \times 10^{-6}\,M$

The range of $[H^+]$ is $6 \times 10^{-6}\,M$ to $3 \times 10^{-6}\,M$.

$[OH^-] = 10^{-8.8} = 1.58 \times 10^{-9} = 2 \times 10^{-9}\,M$; $[OH^-] = 10^{-8.4} = 3.98 \times 10^{-9} = 4 \times 10^{-9}\,M$

The range of $[OH^-]$ is $2 \times 10^{-9}\,M$ to $4 \times 10^{-9}\,M$.

(The pH has one decimal place, so concentrations are reported to 1 sig fig.)

16.40   (a)   Acidic. Bromthymol (or bromothymol) blue, which changes color at a lower pH than phenolphthalein, is yellow below pH 6, so the solution is acidic.

         (b)   (ii), a maximum pH. The solution is the lower-pH color for both indicators, so we know only that the maximum pH is 6.

         (c)   From Figure 16.8, methyl violet, thymol blue, methyl orange, and methyl red would help determine the pH of the solution more precisely. These indicators change colors at pH values from approximately 1 to 5. [One strategy would be to start at the low end of the pH range with methyl violet and work up.]

Strong Acids and Bases (Section 16.5)

16.42   (a)   True

         (b)   True

         (c)   False. Base strength should not be confused with solubility. Base strength describes the tendency of a dissolved molecule [formula unit for ionic compounds such as $Mg(OH)_2$] to dissociate into cations and hydroxide ions. $Mg(OH)_2$ is a strong base because each $Mg(OH)_2$ unit that dissolves also dissociates into $Mg^{2+}(aq)$ and $OH^-(aq)$. $Mg(OH)_2$ is not very soluble, so relatively few $Mg(OH)_2$ units dissolve when the solid compound is added to water.

16.44 For a strong acid, which is completely ionized, $[H^+]$ = the initial acid concentration.

(a) $0.0167\ M\ HNO_3 = 0.0167\ M\ H^+$; $pH = -\log(0.0167) = 1.777$

(b) $\dfrac{0.225\ g\ HClO_3}{2.00\ L\ soln} \times \dfrac{1\ mol\ HClO_3}{84.46\ g\ HClO_3} = 1.332 \times 10^{-3} = 1.33 \times 10^{-3}\ M\ HClO_3$

$[H^+] = 1.33 \times 10^{-3}\ M$; $pH = -\log(1.332 \times 10^{-3}) = 2.875$

(c) $M_c \times V_c = M_d \times V_d$; $0.500\ L = 500\ mL$

$1.00\ M\ HCl \times 15.00\ mL\ HCl = M_d\ HCl \times 500\ mL\ HCl$

$M_d\ HCl = \dfrac{1.00\ M \times 15.00\ mL}{500\ mL} = 3.00 \times 10^{-2}\ M\ HCl = 3.00 \times 10^{-2}\ M\ H^+$

$pH = -\log(3.00 \times 10^{-2}) = 1.523$

(d) $[H^+]_{total} = \dfrac{mol\ H^+\ from\ HCl + mol\ H^+\ from\ HI}{total\ L\ solution}$; $mol = M \times L$

$[H^+]_{total} = \dfrac{(0.020\ M\ HCl \times 0.0500\ L) + (0.010\ M\ HI \times 0.125\ L)}{0.175\ L}$

$[H^+]_{total} = \dfrac{1.0 \times 10^{-3}\ mol\ H^+ + 1.25 \times 10^{-3}\ mol\ H^+}{0.175\ L} = 0.01286 = 0.013\ M$

$pH = -\log(0.01286) = 1.89$

16.46 For a strong base, which is completely dissociated, $[OH^-]$ = the initial base concentration. Then, $pOH = -\log[OH^-]$ and $pH = 14 - pOH$.

(a) $0.182\ M\ KOH = 0.182\ M\ OH^-$; $pOH = -\log(0.182) = 0.740$; $pH = 14 - 740 = 13.260$

(b) $\dfrac{3.165\ g\ KOH}{0.5000\ L} \times \dfrac{1\ mol\ KOH}{56.106\ g\ KOH} = 0.112822 = 0.1128\ M = [OH^-]$

$pOH = -\log(0.112822) = 0.9476$; $pH = 14 - pOH = 13.0524$

(c) $M_c \times V_c = M_d \times V_d$

$0.0105\ M\ Ca(OH)_2 \times 10.0\ mL = M_d\ Ca(OH)_2 \times 500\ mL$

$M_d\ Ca(OH)_2 = \dfrac{0.0105\ M\ Ca(OH)_2 \times 10.0\ mL}{500.0\ mL} = 2.10 \times 10^{-4}\ M\ Ca(OH)_2$

$Ca(OH)_2(aq) \rightarrow Ca^{2+}(aq) + 2OH^-(aq)$

$[OH^-] = 2[Ca(OH)_2] = 2(2.10 \times 10^{-4}\ M) = 4.20 \times 10^{-4}\ M$

$pOH = -\log(4.20 \times 10^{-4}) = 3.377$; $pH = 14 - pOH = 10.623$

(d) $[OH^-]_{total} = \dfrac{mol\ OH^-\ from\ NaOH + mol\ OH^-\ from\ Ba(OH)_2}{total\ L\ solution}$

$\dfrac{(8.2 \times 10^{-3}\ M \times 0.0400\ L) + 2(0.015\ M \times 0.0200\ L)}{0.0600\ L}$

$[OH^-]_{total} = \dfrac{3.28 \times 10^{-4}\ mol\ OH^- + 6.0 \times 10^{-4}\ mol\ OH^-}{0.0600\ L} = 0.01547 = 0.015\ M\ OH^-$

$pOH = -\log(0.01547) = 1.81$; $pH = 14 - 1.81 = 12.19$

16.48 $pOH = 14 - pH = 14.00 - 10.05 = 3.95$

       $pOH = 3.95 = -\log[OH^-]; [OH^-] = 10^{-3.95} = 1.122 \times 10^{-4}\, M = 1.1 \times 10^{-4}\, M$

       $[OH^-] = 2[Ca(OH)_2]; [Ca(OH)_2] = [OH^-]\, /\, 2 = 1.122 \times 10^{-4}\, M\, /2 = 5.6 \times 10^{-5}\, M$

Weak Acids (Section 16.6)

16.50   (a) $C_6H_5COOH(aq) \rightleftharpoons H^+(aq) + C_6H_5COO^-(aq); \quad K_a = \dfrac{[H^+][C_6H_5COO^-]}{[HC_6H_5COOH]}$

            $C_6H_5COOH(aq) + H_2O(l) \rightleftharpoons H_3O^+(aq) + C_6H_5COO^-(aq);$

                                             $K_a = \dfrac{[H_3O^+][C_6H_5COO^-]}{[HC_6H_5COOH]}$

       (b) $HCO_3^-(aq) \rightleftharpoons H^+(aq) + CO_3^{2-}(aq); \quad K_a = \dfrac{[H^+][CO_3^{2-}]}{[HCO_3^-]}$

            $HCO_3^-(aq) + H_2O(l) \rightleftharpoons H_3O^+(aq) + CO_3^{2-}(aq); \quad K_a = \dfrac{[H_3O^+][CO_3^{2-}]}{[HCO_3^-]}$

16.52 $C_6H_5CH_2COOH(aq) \rightleftharpoons H^+(aq) + C_6H_5CH_2COO^-(aq); \quad K_a = \dfrac{[H^+][C_6H_5CH_2COO^-]}{[C_6H_5CH_2COOH]}$

       $[H^+] = [C_6H_5CH_2COO^-] = 10^{-2.68} = 2.09 \times 10^{-3} = 2.1 \times 10^{-3}\, M$

       $[C_6H_5CH_2COOH] = 0.085 - 2.09 \times 10^{-3} = 0.0829 = 0.083\, M$

       $K_a = \dfrac{(2.09 \times 10^{-3})^2}{0.0829} = 5.3 \times 10^{-5}$

16.54 $[H^+] = 0.132 \times [BrCH_2COOH]_{initial} = 0.0132\, M$

	$BrCH_2COOH(aq)$	$\rightleftharpoons$	$H^+(aq)$	$+$	$BrCH_2COO^-(aq)$
initial	0.100 M		0		0
equil.	0.087		0.0132 M		0.0132 M

       $K_a = \dfrac{[H^+][BrCH_2COO^-]}{[BrCH_2COOH]} = \dfrac{(0.0132)^2}{0.087} = 2.0 \times 10^{-3}$

16.56 $[H^+] = 10^{-pH} = 10^{-3.65} = 2.239 \times 10^{-4} = 2.2 \times 10^{-4}\, M$

       $K_a = 6.8 \times 10^{-4} = \dfrac{[H^+][F^-]}{[HF]} = \dfrac{(2.239 \times 10^{-4})^2}{x - 2.239 \times 10^{-4}}$

       $6.8 \times 10^{-4}(x - 2.239 \times 10^{-4}) = (2.239 \times 10^{-4})^2;$

       $6.8 \times 10^{-4}\, x = 1.522 \times 10^{-7} + 0.501 \times 10^{-7} = 2.024 \times 10^{-7}$

       $x = 2.976 \times 10^{-4} = 3.0 \times 10^{-4}\, M\ HF$

16.58

	$HClO_2(aq)$	$\rightleftharpoons$	$H^+(aq)$	$+$	$ClO_2^-(aq)$
initial	0.0125 M		0		0
equil.	(0.0125 − x) M		x M		x M

$$K_a = \frac{[H^+][ClO_2^-]}{[HClO_2]} = \frac{x^2}{(0.0125-x)} \approx \frac{x^2}{0.0125} = 1.1 \times 10^{-2}$$

Assuming x is small relative to 0.0125, $x^2 = 0.0125(0.011)$; $x = 1.2 \times 10^{-2} M$.

Clearly, x is not small relative to 0.0125, so we must solve the quadratic formula for $[H^+]$.

$x^2 = 0.011\,(0.0125 - x);\ x^2 + 0.011x - 1.38 \times 10^{-4} = 0$

$$x = \frac{-0.011 \pm \sqrt{(0.011)^2 - 4(1)(-1.38 \times 10^{-4})}}{2(1)} = 0.007452 = 0.0075 M\ ;$$

$[H^+] = [H_3O^+] = [ClO_2^-] = 0.0075\ M;\ [HClO_2] = 0.0125 - 0.0075 = 0.005045 = 5.0 \times 10^{-3}\ M$

Check. $K_a = \dfrac{(7.5 \times 10^{-3})^2}{5.0 \times 10^{-3}} = 0.011$; our results agree

16.60   (a)

	HOCl(aq)	$\rightleftharpoons$	H^+(aq)	+	OCl^- (aq)
initial	0.095 M		0		0
equil.	(0.095 – x) M		x M		x M

$$K_a = \frac{[H^+][OCl^-]}{[HOCl]} = \frac{x^2}{(0.095-x)} \approx \frac{x^2}{0.095} = 3.0 \times 10^{-8}$$

$x^2 = 0.095\,(3.0 \times 10^{-8});\ x = [H^+] = 5.3 \times 10^{-5}\ M,\ pH = 4.27$

Check. $\dfrac{5.3 \times 10^{-5}\ M\ H^+}{0.095\ M\ HOCl} \times 100 = 0.056\%$ ionization

The approximation is nearly valid. To 2 sig figs, the quadratic formula gives the same $[H^+]$.

(b)

	H_2NNH_2(aq) + H_2O(l)	$\rightleftharpoons$	$H_2NNH_3^+$(aq)	+	OH^- (aq)
initial	0.0085 M		0		0
equil.	(0.0085 – x) M		x M		x M

$$K_b = \frac{[H_2NNH_3^+][OH^-]}{[H_2NNH_2]} = \frac{x^2}{(0.0085-x)} \approx \frac{x^2}{0.0085} = 1.3 \times 10^{-6}$$

$x^2 = 0.0085\,(1.3 \times 10^{-6});\ x = [OH^-] = 1.051 \times 10^{-4} = 1.1 \times 10^{-4}\ M$

Clearly, $1.1 \times 10^{-4}\ M\ OH^-$ is not small compared to $8.5 \times 10^{-3}\ M\ H_2NNH_2$, and we must solve the quadratic.

$x^2 = 1.3 \times 10^{-6}(0.0085 - x);\ x^2 + 1.3 \times 10^{-6}\,x - 1.105 \times 10^{-8} = 0$

$$x = \frac{-1.3 \times 10^{-6} \pm \sqrt{(1.3 \times 10^{-6})^2 - 4(1)(-1.105 \times 10^{-8})}}{2(1)} = 1.0447 \times 10^{-4}$$

$$= 1.0 \times 10^{-4}\ M\ OH^-$$

$pOH = 3.981 = 3.98;\ pH = 14 - pOH = 14 - 3.981 = 10.019 = 10.02$

Check. Although this solution has more than 12% ionization, the difference in $[OH^-]$ between the estimate and the quadratic is not great.

(c)

$$HONH_2(aq) + H_2O(l) \rightleftharpoons HONH_3^+(aq) + OH^-(aq)$$

initial	0.165 M	0	0
equil.	(0.165 – x) M	x M	x M

$$K_b = \frac{[HONH_3^+][OH^-]}{[HONH_2]} = \frac{x^2}{(0.165-x)} \approx \frac{x^2}{0.165} = 1.1 \times 10^{-8}$$

$$x^2 = 0.165\,(1.1\times10^{-8}); \; x = [OH^-] = 4.3\times10^{-5}\,M, \text{pH} = 9.63$$

Check. $\dfrac{4.3 \times 10^{-5}\,M\,OH^-}{0.165\,M\,HONH_2} \times 100 = 0.026\%$ ionization; the approximation is valid

16.62   Calculate the initial concentration of $HC_9H_7O_4$.

$$2\ \text{tablets} \times \frac{500\ \text{mg}}{\text{tablet}} \times \frac{1\ \text{g}}{1000\ \text{mg}} \times \frac{1\ \text{mol}\ HC_9H_7O_4}{180.2\ \text{g}\ HC_9H_7O_4} = 0.005549 = 0.00555\ \text{mol}\ HC_9H_7O_4$$

$$\frac{0.005549\ \text{mol}\ HC_9H_7O_4}{0.250\ \text{L}} = 0.02220 = 0.0222\ M\ HC_9H_7O_4$$

	$HC_9H_7O_4\,(aq)$ $\rightleftharpoons$	$C_9H_7O_4^-$ +	$H^+(aq)$
initial	0.0222 M	0 M	0 M
equil.	(0.0222 – x)	x M	x M

$$K_a = 3.3 \times 10^{-4} = \frac{[H^+][C_9H_7O_4^-]}{[HC_9H_7O_4]} = \frac{x^2}{(0.0222 - x)}$$

Assuming x is small compared to 0.0222,

$$x^2 = 0.0222\,(3.3 \times 10^{-4}); \; x = [H^+] = 2.7 \times 10^{-3}\,M$$

$$\frac{2.7 \times 10^{-3}\,M\,H^+}{0.0222\,M\,HC_9H_7O_4} \times 100 = 12\%\ \text{ionization; the approximation is not valid}$$

Using the quadratic formula, $x^2 + 3.3 \times 10^{-4}\,x - 7.325 \times 10^{-6} = 0$

$$x = \frac{-3.3 \times 10^{-4} \pm \sqrt{(3.3 \times 10^{-4})^2 - 4(1)(-7.325 \times 10^{-6})}}{2(1)} = \frac{-3.3 \times 10^{-4} \pm \sqrt{2.941 \times 10^{-5}}}{2}$$

$$x = 2.547 \times 10^{-3} = 2.5 \times 10^{-3}\,M\,H^+;\ \text{pH} = -\log(2.547 \times 10^{-3}) = 2.594 = 2.59$$

16.64   (a) $C_2H_5COOH(aq) \rightleftharpoons H^+(aq) + C_2H_5COO^-(aq)$

$$K_a = 1.3 \times 10^{-5} = \frac{[H^+][C_2H_5COO^-]}{[C_2H_5COOH]} = \frac{x^2}{0.250 - x}$$

$$x^2 \approx 0.250\,(1.3 \times 10^{-5}); \; x = 1.803 \times 10^{-3} = 1.8 \times 10^{-3}\,M\,H^+$$

$$\%\ \text{ionization} = \frac{1.803 \times 10^{-3}\,M\,H^+}{0.250\,M\,C_2H_5COOH} \times 100 = 0.721\%$$

(b) $\dfrac{x^2}{0.0800} \approx 1.3 \times 10^{-5}; \; x = 1.020 \times 10^{-3} = 1.0 \times 10^{-3}\,M\,H^+$

$$\%\ \text{ionization} = \frac{1.020 \times 10^{-3}\,M\,H^+}{0.0800\,M\,C_2H_5COOH} \times 100 = 1.27\%$$

(c) $\dfrac{x^2}{0.0200} \approx 1.3 \times 10^{-5}; x = 5.099 \times 10^{-4} = 5.1 \times 10^{-4}\ M\,H^+$

$$\%\ ionization = \dfrac{5.099 \times 10^{-4}\ M\,H^+}{0.0200\ M\,C_2H_5COOH} \times 100 = 2.55\%$$

16.66 $H_2C_4H_4O_6(aq) \;\rightleftharpoons\; H^+(aq) \;+\; HC_4H_4O_6^-(aq) \qquad K_{a1} = 1.0 \times 10^{-3}$

$HC_4H_4O_6^-(aq) \;\rightleftharpoons\; H^+(aq) \;+\; C_4H_4O_6^{2-}(aq) \qquad K_{a2} = 4.6 \times 10^{-5}$

Begin by calculating the $[H^+]$ from the first ionization. The equilibrium concentrations are $[H^+] = [HC_4H_4O_6^-] = x$, $[H_2C_4H_4O_6] = 0.25 - x$.

$$K_{a1} = \dfrac{[H^+][HC_4H_4O_6^-]}{[H_2C_4H_4O_6]} = \dfrac{x^2}{0.25 - x}; \; x^2 + 1.0 \times 10^{-3}\,x - 2.5 \times 10^{-4} = 0$$

Using the quadratic formula, $x = 1.532 \times 10^{-2} = 0.015\ M\,H^+$ from the first ionization. Next, calculate the H^+ contribution from the second ionization.

	$HC_4H_4O_6^-(aq)$	$\rightleftharpoons$	$H^+(aq)$	$+$	$C_4H_4O_6^{2-}(aq)$
initial	0.015		0.015		0
equil.	$(0.015 - y)$		$(0.015 + y)$		y

$$K_{a2} = \dfrac{(0.015 + y)\,(y)}{(0.015 - y)} = 4.6 \times 10^{-5}; \text{ assuming } y \text{ is small compared to } 0.015,$$

$y = 4.6 \times 10^{-5}\ M\,C_4H_4O_6^{2-}(aq)$

This approximation is reasonable, because 4.6×10^{-5} is only 0.3% of 0.015.

$[H^+] = 0.015\ M$ (first ionization) $+ 4.6 \times 10^{-5}$ (second ionization)

Because 4.6×10^{-5} is 0.3% of 0.015 M, it can be safely ignored when calculating total $[H^+]$.

$pH = -\log(0.01532) = 1.18148 = 1.181$

Assumptions:

(1)   The ionization can be treated as a series of steps (valid by Hess's law).

(2)   The extent of ionization in the second step (y) is small relative to that from the first step (valid for this acid and initial concentration). This assumption was used twice, to calculate the value of y from K_{a2} and to calculate total $[H^+]$ and pH.

Weak Bases (Section 16.7)

16.68   (a)   *Analyze/Plan.* To determine relative strength, compare the K_b values of the two bases. *Solve.*

$$K_b \text{ for } OCl^- = \dfrac{K_w}{K_a \text{ for } HClO} = \dfrac{1.0 \times 10^{-14}}{3.0 \times 10^{-8}} = 3.3 \times 10^{-7}$$

K_b for hydroxylamine is 1.1×10^{-8}. OCl^- is a stronger base than hydroxylamine.

(b)   When OCl^- acts as a base, the O atom is the proton acceptor.

(c) $14\ e^-,\ 7\ e^-$ pairs $\quad \left[:\ddot{\underset{..}{Cl}}-\ddot{\underset{..}{O}}:\right]^-$

In OCl^-, the -1 formal charge is on O. H^+ attaches to the atom with the negative formal charge.

16.70   (a) $C_3H_7NH_2(aq) + H_2O(l) \rightleftharpoons C_3H_7NH_3^+(aq) + OH^-(aq);\ K_b = \dfrac{[C_3H_7NH_3^+][OH^-]}{[C_3H_7NH_2]}$

     (b) $HPO_4^{2-}(aq) + H_2O(l) \rightleftharpoons H_2PO_4^-(aq) + OH^-(aq);\ K_b = \dfrac{[H_2PO_4^-][OH^-]}{[HPO_4^{2-}]}$

     (c) $C_6H_5CO_2^-(aq) + H_2O(l) \rightleftharpoons C_6H_5CO_2H(aq) + OH^-(aq);\ K_b = \dfrac{[C_6H_5CO_2H][OH^-]}{[C_6H_5CO_2^-]}$

16.72

	$BrO^-(aq)$	$+\ H_2O(l)$	$\rightleftharpoons$	$HOBr(aq)$	$+$	$OH^-(aq)$
initial	$0.724\ M$			0		0
equil.	$(0.724-x)\ M$			$x\ M$		$x\ M$

$$K_b = \frac{[HOBr][OH^-]}{[BrO^-]} = \frac{x^2}{0.724-x} \approx \frac{x^2}{0.724} = 4.0 \times 10^{-6}$$

$$x^2 = 0.724\,(4.0 \times 10^{-6});\ x = [OH^-] = 1.70 \times 10^{-3} = 1.7 \times 10^{-3}\ M;\ pH = 11.23$$

Check. $\dfrac{1.7 \times 10^{-3}\ M\ OH^-}{0.724\ M\ BrO^-} \times 100 = 0.24\%$ hydrolysis; the approximation is valid

16.74   (a) $pOH = 14.00 - 9.95 = 4.05;\ [OH^-] = 10^{-4.05} = 8.91 \times 10^{-5} = 8.9 \times 10^{-5}\ M$

	$C_{18}H_{21}NO_3(aq)$	$+\ H_2O(l)$	$\rightleftharpoons$	$C_{18}H_{21}NO_3H^+(aq)$	$+$	$OH^-(aq)$
initial	$0.0050\ M$			0		0
equil.	$(0.0050 - 8.9 \times 10^{-5})$			$8.9 \times 10^{-5}\ M$		$8.9 \times 10^{-5}\ M$

$$K_b = \frac{[C_{18}H_{21}NO_3H^+][OH^-]}{[C_{18}H_{21}NO_3]} = \frac{(8.91 \times 10^{-5})^2}{(0.0050 - 8.91 \times 10^{-5})} = 1.62 \times 10^{-6} = 1.6 \times 10^{-6}$$

     (b) $pK_b = -\log(K_b) = -\log(1.62 \times 10^{-6}) = 5.79$

The K_a – K_b Relationship; Acid–Base Properties of Salts
(Sections 16.8 and 16.9)

16.76   The stronger a base, the weaker its conjugate acid. From the K_a values in Table 16.3, place the conjugate acids of these oxyanions in order of increasing K_a value, increasing acid strength, and decreasing conjugate base strength. Use K_{a2} for H_2SO_4, H_2CO_3, and H_2SO_3 and K_{a3} for H_3PO_4.

In order of increasing K_a value and acid strength: $HPO_4^{2-} < HCO_3^- < HSO_3^- < HSO_4^-$

In order of decreasing base strength: $PO_4^{3-} > CO_3^{2-} > SO_3^{2-} > SO_4^{2-}$

16.78 (a) Ammonia is the stronger base because it has the larger K_b value.

(b) Hydroxylammonium is the stronger acid because the weaker base, hydroxylamine, has the stronger conjugate acid.

(c) K_a for $NH_4^+ = K_w/K_b$ for $NH_3 = 1.0 \times 10^{-14}/1.8 \times 10^{-5} = 5.6 \times 10^{-10}$

K_a for $HONH_3^+ = K_w/K_b$ for $HONH_2 = 1.0 \times 10^{-14}/1.1 \times 10^{-8} = 9.1 \times 10^{-7}$

Note that K_a for $HONH_3^+$ is larger than K_a for NH_4^+.

16.80 (a) Proceeding as in Solution 16.79(a):

$$F^-(aq) + H_2O(l) \rightleftharpoons HF(aq) + OH^-(aq)$$

initial   0.105 M 0 M 0 M

equil.   (0.105 – x) M x M x M

$$K_b \text{ for } F^- = \frac{[HF][OH^-]}{[F^-]} = \frac{K_w}{K_a \text{ for } HF} = \frac{1.0 \times 10^{-14}}{6.8 \times 10^{-4}} = 1.47 \times 10^{-11} = 1.5 \times 10^{-11}$$

$$1.5 \times 10^{-11} = \frac{(x)(x)}{(0.105 - x)} ; \text{ assume the amount of } F^- \text{ that hydrolyzes is small}$$

$$x^2 = 0.105(1.47 \times 10^{-11}); x = [OH^-] = 1.243 \times 10^{-6} = 1.2 \times 10^{-6} M$$

pOH = 5.91; pH = 14 – 5.91 = 8.09

(b) $Na_2S(aq) \rightarrow S^{2-}(aq) + 2Na^+(aq)$

$S^{2-}(aq) + H_2O(l) \rightleftharpoons HS^-(aq) + OH^-(aq)$

As in part (a), $[OH^-] = [HS^-] = x; [S^{2-}] = 0.035 M$

$$K_b = \frac{[HS^-][OH^-]}{[S^{2-}]} = \frac{K_w}{K_a \text{ for } HS^-} = \frac{1.0 \times 10^{-14}}{1 \times 10^{-19}} = 1 \times 10^5$$

Because $K_b \gg 1$, this equilibrium lies far to the right and $[OH^-] = [HS^-] = 0.035 M$. K_b for $HS^- = 1.05 \times 10^{-7}$; $[OH^-]$ produced by further hydrolysis of HS^- amounts to $6.1 \times 10^{-5} M$. The second hydrolysis step does not make a significant contribution to the total $[OH^-]$ and pH.

$[OH^-] = 0.035 M$; pOH = 1.46, pH = 12.54

(c) As in Solution 16.79(c), calculate $[CH_3COO^-]$.

$[CH_3COO^-]_t = [CH_3COO^-]$ from $NaCH_3COO + [CH_3COO^-]$ from $Ba(CH_3COO)_2$

$[CH_3COO^-]_t = 0.045 M + 2(0.055 M) = 0.155 M$

The hydrolysis equilibrium is:

$CH_3COO^-(aq) + H_2O(l) \rightleftharpoons CH_3COOH(aq) + OH^-(aq)$

$$K_b = \frac{[CH_3COOH][OH^-]}{[CH_3COO^-]} = \frac{K_w}{K_a \text{ for } CH_3COOH} = \frac{1.0 \times 10^{-14}}{1.8 \times 10^{-5}} = 5.56 \times 10^{-10}$$

$$= 5.6 \times 10^{-10}$$

$[OH^-] = [CH_3COOH] = x; [CH_3COO^-] = 0.155 - x$

$$K_b = 5.56 \times 10^{-10} = \frac{x^2}{(0.155-x)}; \text{ assume x is small compared to } 0.155\ M$$

$$x^2 = 0.155\,(5.56 \times 10^{-10});\ x = [OH^-] = 9.280 \times 10^{-6} = 9.3 \times 10^{-6}$$

$$pH = 14 + \log(9.280 \times 10^{-6}) = 8.97$$

16.82   The hydrolysis equilibrium is: $C_5H_5NH^+(aq) \rightleftharpoons C_5H_5N(aq) + H^+(aq)$

$$K_a = \frac{[C_5H_5N][H^+]}{[C_5H_5NH^+]} = \frac{K_w}{K_b \text{ for } C_5H_5N} = \frac{1.0 \times 10^{-14}}{1.7 \times 10^{-9}} = 5.882 \times 10^{-6} = 5.9 \times 10^{-6}$$

$$[C_5H_5NH^+] = x;\ [C_5H_5N] = [H^+] = 10^{-pH} = 10^{-2.95} = 1.122 \times 10^{-3} = 1.1 \times 10^{-3}\ M$$

$$K_a = 5.882 \times 10^{-6} = \frac{(1.122 \times 10^{-3})^2}{x};\ x = 0.2140 = 0.21\ M\ C_5H_5NHBr$$

Note that no assumption was required in this calculation.

16.84   (a)   acidic; Al^{3+} is a highly charged metal cation and a Lewis acid; Cl^- is negligible.

       (b)   neutral; both Na^+ and Br^- are negligible.

       (c)   basic; ClO^- is the conjugate base of $HClO$; Na^+ is negligible.

       (d)   acidic; $CH_3NH_3^+$ is the conjugate acid of CH_3NH_2; NO_3^- is negligible.

       (e)   basic; SO_3^{2-} is the conjugate base of H_2SO_3; Na^+ is negligible.

16.86   *Plan.* Estimate pH of salt solution by evaluating the ions in the salts. Calculate to confirm if necessary. *Solve.*

     KBr: salt of strong acid and strong base, neutral solution. The unknown is probably KBr. Check the others to be sure.

     NH_4Cl: salt of a weak base and a strong acid, acidic solution

     KCN: salt of a strong base and a weak acid, basic solution

     K_2CO_3: salt of a strong base and a weak acid (HCO_3^-), basic solution

     Only KBr fits the acid–base properties of the unknown.

Acid–Base Character and Chemical Structure (Section 16.10)

16.88   (a)   For binary hydrides, acid strength increases going across a row, so HCl is a stronger acid than H_2S.

       (b)   For oxyacids, the more electronegative the central atom, the stronger the acid, so H_3PO_4 is a stronger acid than H_3AsO_4.

       (c) $HBrO_3$ has one more nonprotonated oxygen and a higher oxidation number on Br, so it is a stronger acid than $HBrO_2$.

       (d)   The first dissociation of a polyprotic acid is always stronger because H^+ is more tightly held by an anion, so $H_2C_2O_4$ is a stronger acid than $HC_2O_4^-$.

(e)   The conjugate base of benzoic acid, $C_6H_5COO^-$, is stabilized by resonance, whereas the conjugate base of phenol, $C_6H_5O^-$, is not. C_6H_5COOH has greater tendency to form its conjugate base and is the stronger acid.

16.90   (a) NO_2^- (HNO_3 is the stronger acid because it has more nonprotonated O atoms, so NO_2^- is the stronger base.)

         (b) PO_4^{3-} (K_a for $HAsO_4^{2-}$ is greater than K_a for HPO_4^{2-}, so K_b for PO_4^{3-} is greater and PO_4^{3-} is the stronger base. Note that P is more electronegative than As and H_3PO_4 is a stronger acid than H_3AsO_4, which could lead to the conclusion that AsO_4^{3-} is the stronger base. As in all cases, the measurement of base strength, K_b, supercedes the prediction. Chemistry is an experimental science.

         (c) CO_3^{2-} (The more negative the anion, the stronger the attraction for H^+.)

16.92   (a)   True.

         (b)   False. For oxyacids with the same structure but different central atom, the acid strength *increases* as the electronegativity of the central atom increases.

         (c)   False. HF is a weak acid, weaker than the other hydrogen halides, primarily because the H–F bond energy is exceptionally high.

Lewis Acids and Bases (Section 16.11)

16.94   No. If a substance is a Lewis acid, it is not necessarily a Brønsted–Lowry or an Arrhenius acid. The Lewis definition of an acid, an electron pair acceptor, is most general. A Lewis acid does not necessarily fit the more narrow description of a Brønsted–Lowry or Arrhenius acid. An electron pair acceptor is not necessarily an H^+ donor, nor must it produce H^+ in aqueous solution. An example is Al^{3+}, which is a Lewis acid, but has no ionizable hydrogen.

16.96

	Lewis Acid	**Lewis Base**
(a)	HNO_2 (or H^+)	OH^-
(b)	$FeBr_3$ (Fe^{3+})	Br^-
(c)	Zn^{2+}	NH_3
(d)	SO_2	H_2O

16.98   (a) $ZnBr_2$, smaller cation radius, same charge

         (b) $Cu(NO_3)_2$, higher cation charge

         (c) $NiBr_2$, smaller cation radius, same charge

Additional Exercises

16.99   (a)   Correct.

         (b)   Incorrect. A Brønsted–Lowry acid must have ionizable hydrogen. Lewis acids are electron pair acceptors, but need not have ionizable hydrogen.

(c)   Correct.

(d)   Incorrect. K^+ is a negligible Lewis acid because it is the conjugate of strong base KOH. Its relatively large ionic radius and low positive charge render it a poor attractor of electron pairs.

(e)   Correct.

16.100   Calculate moles OH^-, calculate moles H^+, determine which is in excess after neutralization, calculate pH

$$0.300 \text{ g Ca(OH)}_2 \times \frac{1 \text{ mol Ca(OH)}_2}{74.093 \text{ g Ca(OH)}_2} \times \frac{2 \text{ mol OH}^-}{1 \text{ mol Ca(OH)}_2} = 8.0979 \times 10^{-3} = 8.10 \times 10^{-3} \, M \, OH^-$$

$1.40 \, M \, HNO_3 \times 0.0500 \text{ L} = 0.0700 \text{ mol } H^+$; H^+ is in excess

$0.0700 \text{ mol } H^+ - 0.00810 \text{ mol } OH^- = 0.0619 \text{ mol } H^+$ remain

$0.0619 \text{ mol } H^+ / 0.0750 \text{ L} = 0.82536 = 0.825 \, M \, H^+$; pH $= -\log (0.82536) = 0.03356 = 0.0834$

16.102 $HX(aq) \rightleftharpoons H^+(aq) + X^-(aq)$; $K_a = \dfrac{[H^+][X^-]}{[HX]}$

$[H^+] = [X^-]$; assume the % ionization is small; $K_a = \dfrac{[H^+]^2}{[HX]}$; $[H^+] = K_a^{1/2} [HX]^{1/2}$

$pH = -\log K_a^{1/2} [HX]^{1/2} = -\log K_a^{1/2} - \log [HX]^{1/2}$; $pH = -1/2 \log K_a - 1/2 \log [HX]$

This is the equation of a straight line, where the intercept is $-1/2 \log K_a$, the slope is $-1/2$, and the independent variable is $\log [HX]$.

16.103   (a)   A higher O_2 concentration displaces protons from Hb, producing a more acidic solution, with lower pH in the lungs than in the tissues.

(b) $[H^+] = $ antilog $(-7.4) = 4.0 \times 10^{-8} \, M$. At body temperature, 37 °C, $K_w = 2.4 \times 10^{-14}$ (see Solution 16.37). At this temperature, a "neutral" solution has $[H^+] = 1.5 \times 10^{-7}$ and pH 6.81. Even though the frame of reference is a bit different at this temperature, blood at pH 7.4 is slightly basic.

(c)   The equilibrium indicates that a high $[H^+]$ shifts the equilibrium toward the proton-bound form HbH^+, which means a lower concentration of HbO_2 in the blood. Thus, the ability of hemoglobin to transport oxygen is impeded.

16.104   Upon dissolving, Li_2O dissociates to form Li^+ and O^{2-}. According to Equation 16.22, O^{2-} is completely protonated in aqueous solution.

$Li_2O(s) + H_2O(l) \rightarrow 2Li^+(aq) + 2OH^-(aq)$

Thus, initial $[Li_2O] = [O_2^-]$; $[OH^-] = 2[O^{2-}] = 2[Li_2O]$

$$[Li_2O] = \frac{\text{mol Li}_2O}{\text{L solution}} = 2.50 \text{ g Li}_2O \times \frac{1 \text{ mol Li}_2O}{29.88 \text{ g Li}_2O} \times \frac{1}{1.500 \text{ L}} = 0.0558 = 0.0558 \, M$$

$[OH^-] = 0.11156 = 0.112 \, M$; pOH $= 0.9525 = 0.953$; pH $= 14.00 - \text{pOH} = 13.0475 = 13.048$

16.105   (a)   The conjugate base of benzoic acid is benzoate anion, $C_6H_5COO^-$. The conjugate acid of aniline is anilinium cation, $C_6H_5NH_3^+$.

(b) To compare relative acidity, compare the K_a values for benzoic acid and anilinium ion.

$$K_a \text{ for } C_6H_5NH_3^+ = \frac{K_w}{K_b \text{ for } C_6H_5NH_2} = \frac{1.0 \times 10^{-14}}{4.3 \times 10^{-10}} = 2.3256 \times 10^{-5} = 2.3 \times 10^{-5}$$

K_a for C_6H_5COOH, 6.3×10^{-5}, is greater than K_a for $C_6H_5NH_3^+$, 2.3×10^{-5}. The 0.10 M solution of benzoic acid will be somewhat more acidic.

(c)
$$C_6H_5COOH \rightleftharpoons H^+ + C_6H_5COO^- \qquad K_a = 6.3 \times 10^{-5}$$
$$C_6H_5NH_2 + H_2O \rightleftharpoons C_6H_5NH_3^+ + OH^- \qquad K_b = 4.3 \times 10^{-10}$$
$$H^+ + OH^- \rightleftharpoons H_2O \qquad 1/K_w = 1/1.0 \times 10^{-14}$$

$$C_6H_5COOH + C_6H_5NH_2 + H_2O + H^+ + OH^- \rightleftharpoons C_6H_5COO^- + C_6H_5NH_3^+ + H^+ + OH^- + H_2O$$
$$C_6H_5COOH + C_6H_5NH_2 \rightleftharpoons C_6H_5COO^- + C_6H_5NH_3^+$$

$$K = \frac{K_a \times K_b}{K_w} = \frac{(6.3 \times 10^{-5})(4.3 \times 10^{-10})}{1.0 \times 10^{-14}} = 2.7090 = 2.7$$

16.107 (a) False. $H_2C_2O_4$ has no capacity to accept H^+.

(b) True.

(c) True. $HC_2O_4^-$ can act like either a Brønsted–Lowry acid or base. It is a stronger acid than base ($K_{a2} > K_{b1}$), so a solution of the salt will be acidic.

16.108 $H_2Suc(aq) \rightleftharpoons H^+(aq) + HSuc^-(aq) \qquad K_{a1} = 6.9 \times 10^{-5}$

          $HSuc^-(aq) \rightleftharpoons H^+(aq) + Suc^{2-}(aq) \qquad K_{a2} = 2.5 \times 10^{-6}$

(a) Calculating the $[H^+]$ from the first ionization. The equilibrium concentrations are $[H^+] = [HSuc^-] = x$; $[H_2Suc] = 0.32 - x$.

$$K_{a1} = \frac{[H^+][HSuc^-]}{[H_2Suc]} = \frac{x^2}{0.32 - x}; \text{ assume x is small relative to } 0.32$$

$$x^2 = (0.32)(6.9 \times 10^{-5}); \; x = 0.0046989 = 0.0047 \; M \; H^+; \; pH = 2.328 = 2.33$$

(b) To calculate $[Suc^{2-}]$, consider the second ionization equilibrium. Initial $[HSuc^-]$ and $[H^+]$ are 0.0047 M, from the first ionization. Then,

$[Suc^{2-}] = y$; $[H^+] = 0.0047 + y$; $[HSuc^-] = 0.0047 - y$.

$$K_{a2} = \frac{(0.0047 + y)(y)}{(0.0047 - y)} = 2.5 \times 10^{-6}; \text{assuming y is small compared to } 0.0047,$$

$y = 2.5 \times 10^{-6} \; M \; Suc^{2-}$. This assumption is reasonable, because 2.5×10^{-6} is only 0.05% of 0.0047.

(c) The assumption in part (a) is reasonable. From part (b), $[Suc^{2-}]$ and $[H^+]$ from the second ionization are 2.5×10^{-6} M. This is only 0.05% of 0.0047 M, $[H^+]$ from the first ionization. Only the first dissociation is relevant for calculating pH.

(d) Acidic. $HSuc^-$ can act like either a Brønsted–Lowry acid or base. It is a stronger acid than base ($K_{a2} > K_{b1}$), so a solution of the salt will be acidic.

16.110   *Analyze/Plan.* Evaluate the acid–base properties of the cation and anion to determine whether a solution of the salt will be acidic, basic, or neutral.   *Solve.*

(i) NH_4NO_3: NH_4^+, weak conjugate acid of NH_3; NO_3^-, negligible conjugate base of HNO_3; acidic solution

(ii) $NaNO_3$: Na^+, negligible conjugate acid of $NaOH$; NO_3^-, negligible conjugate base of HNO_3; neutral solution

(iii) CH_3COONH_4: NH_4^+, weak conjugate acid of NH_3, $K_a = K_w/1.8 \times 10^{-5} = 5.6 \times 10^{-10}$; CH_3COO^-, weak conjugate base of CH_3COOH, $K_b = K_w/1.8 \times 10^{-5} = 5.6 \times 10^{-10}$; neutral solution ($K_a$ for the cation and K_b for the anion are accidentally equal, producing a neutral solution)

(iv) NaF: Na^+, negligible conjugate acid of $NaOH$; F^-, weak conjugate base of HF, $K_b = K_w/6.8 \times 10^{-4} = 1.5 \times 10^{-11}$; basic solution

(v) CH_3COONa: Na^+, negligible; CH_3COO^-, weak base, $K_b = 5.6 \times 10^{-10}$; basic solution

In order of increasing acidity and decreasing pH:
0.1 *M* CH_3COONa > 0.1 *M* NaF > 0.1 *M* CH_3COONH_4 = 0.1 *M* $NaNO_3$ > 0.1 *M* NH_4NO_3; (v) > (iv) > (iii) ~ (ii) > (i)

(iv) and (v) are both bases, and (v) has the greater K_b value and higher pH. (ii) and (iii) are both neutral and (i) is acidic.

16.111   Calculate K_b for A^- and then K_a for HA.

$A^- = [NaA] = [A^-] = 0.25$ *M*; $[OH^-] = 10^{-pOH}$; $pOH = 14.00 - pH = 14.00 - 9.29 = 4.71$

$[OH^-] = 10^{-4.71} = 1.950 \times 10^{-5} = 2.0 \times 10^{-5}$ *M* $= [HX]$

$K_b = \dfrac{[OH^-][HA]}{[A^-]} = \dfrac{(1.950 \times 10^{-5})^2}{(0.25 - 1.950 \times 10^{-5})} = \dfrac{(1.950 \times 10^{-5})^2}{0.25} = 1.521 \times 10^{-9} = 1.5 \times 10^{-9}$

K_a for HA $= K_w/K_b$ for $A^- = 1.0 \times 10^{-14}/1.521 \times 10^{-9} = 6.576 \times 10^{-6} = 6.6 \times 10^{-6}$

16.112   The value of pK_{a2} can only be choice (iii).

Calculate K_{a1}, assuming that only the first ionization determines pH.

$[H_2A] = 0.10$ M; $[H^+] = [HA^-] = 10^{-pH} = 10^{-3.30} = 5.0119 \times 10^{-4} = 5.0 \times 10^{-4}$

$K_{a1} = \dfrac{[H^+][A^-]}{[HA]} = \dfrac{(5.0119 \times 10^{-4})^2}{(0.10 - 5.0119 \times 10^{-4})} = \dfrac{(5.0119 \times 10^{-4})^2}{0.0995} = 2.524 \times 10^{-6} = 2.5 \times 10^{-6}$

$pK_{a1} = 5.60$. pK_{a2} must be greater than pK_{a1}, which eliminates choices (i) and (ii).

If a solution of the salt NaHA is acidic, then $K_{a2} > K_{b1}$ and $pK_{a2} < pK_{b1}$.

$pK_{b1} = 14.00 - pK_{a1} = 14.00 - 5.60 = 8.84$. This eliminates choice (iv).

16.114   (a)   Consider the formation of the zwitterion as a series of steps (Hess's law).

$$NH_2-CH_2-COOH+H_2O \rightleftharpoons NH_2-CH_2-COO^-+H_3O^+ \qquad K_a$$
$$NH_2-CH_2-COOH+H_2O \rightleftharpoons {}^+NH_3-CH_2-COOH+OH^- \qquad K_b$$
$$H_3O^++OH^- \rightleftharpoons 2H_2O \qquad 1/K_w$$

$$NH_2-CH_2-COOH \rightleftharpoons {}^+NH_3-CH_2-COO^- \qquad \frac{K_a \times K_b}{K_w}$$

$$K=\frac{K_a \times K_b}{K_w}=\frac{(4.3 \times 10^{-3})(6.0 \times 10^{-5})}{1.0 \times 10^{-14}}=2.6 \times 10^7$$

(b)   As glycine exists as the zwitterion in aqueous solution, the pH is determined by the following equilibrium.

$${}^+NH_3-CH_2-COO^-+H_2O \rightleftharpoons NH_2-CH_2-COO^-+H_3O^+$$

$$K_a=\frac{[NH_2-CH_2-COO^-][H_3O^+]}{[{}^+NH_3-CH_2-COO^-]}=\frac{K_w}{K_b}=\frac{1.0 \times 10^{-14}}{6.0 \times 10^{-5}}=1.67 \times 10^{-10}=1.7 \times 10^{-10}$$

$$x=[H_3O^+]=[NH_2-CH_2-COO^-]; \ K_a=1.67 \times 10^{-10}=\frac{(x)(x)}{(0.050-x)} \approx \frac{x^2}{0.050}$$

$$x=[H_3O^+]=2.89 \times 10^{-6}=2.9 \times 10^{-6} \ M; \ pH=5.54$$

(c)   In strongly basic solution (pH 13), the $-NH_3^+$ group would be deprotonated, so glycine would be in the form $H_2NCH_2CO_2^-$. In a strongly acidic (pH 1) solution, the $-CO_2^-$ function would be protonated, so glycine would exist as ${}^+H_3NCH_2COOH$.

16.115   The general Lewis structures for these acids and their conjugate bases are shown below. X = H or Cl.

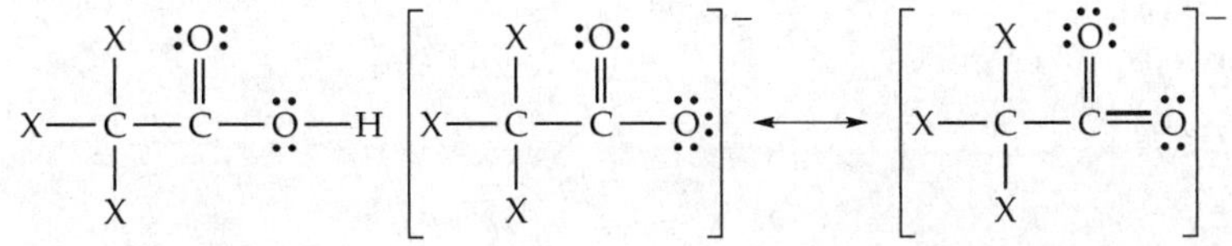

Replacement of H on the acid by the more electronegative chlorine atoms causes the central carbon to become more positively charged, thus withdrawing more electrons from the attached COOH group, in turn causing the O–H bond to be more polar, so that H^+ is more readily ionized.

For the conjugate base (two resonance structures), the electronegative X atoms delocalize negative charge and stabilize these forms relative to the unsubstituted anions. This favors products in the ionization equilibrium and increases the value of K_a.

To calculate pH, proceed as usual, except that the full quadratic formula must be used for all but acetic acid.

Acid	pH
acetic	3.37
chloroacetic	2.51
dichloroacetic	2.09
trichloroacetic	2.0

Integrative Exercises

16.117   *Analyze.* Based on the mass % and density of concentrated HCl, calculate volume of concentrated solution required to produce 10.0 L of HCl with pH = 2.05. *Plan.* Calculate molarity of concentrated solution from density and mass %. Calculate molarity of dilute solution from pH. Use the dilution formula to calculate volume (mL) of concentrated solution required.   *Solve.*

$$\frac{1.18\,\text{g conc. soln.}}{\text{mL conc. soln.}} \times \frac{36.0\,\text{g HCl}}{100\,\text{g conc. soln.}} \times \frac{1000\,\text{mL}}{1\,\text{L}} \times \frac{1\,\text{mol HCl}}{36.46\,\text{g HCl}} = 11.651\,\text{mol HCl/L}$$

$$= 11.7\,M\,\text{HCl/L}$$

For the dilute HCl solution, $[H^+] = 10^{-pH} = 10^{-2.05} = 8.913 \times 10^{-3} = 8.9 \times 10^{-3}\,M\,\text{HCl}$

$M_c \times L_c = M_d \times M_d;\; 11.651 \times L_c = 8.913 \times 10^{-3}\,M \times 10.0\,\text{L};$

$$L_c = 7.650 \times 10^{-3};\; 7.650 \times 10^{-3}\,\text{L} \times \frac{1000\,\text{mL}}{1\,\text{L}} = 7.65 = 7.7\,\text{mL conc. HCl}$$

16.118 $[H^+] = 10^{-pH} = 10^{-2} = 1 \times 10^{-2}\,M\,H^+;\; 1 \times 10^{-2}\,M \times 0.400\,\text{L} = 4.0 \times 10^{-3} = 4 \times 10^{-3}\,\text{mol}\,H^+$

$HCl(aq) + HCO_3^-(aq) \rightarrow Cl^-(aq) + H_2O(l) + CO_2(g)$

$$4 \times 10^{-3}\,\text{mol}\,H^+ = 4 \times 10^{-3}\,\text{mol}\,HCO_3^- \times \frac{84.01\,\text{g NaHCO}_3}{1\,\text{mol HCO}_3^-} = 0.336 = 0.3\,\text{g NaHCO}_3$$

16.120   (a) $K_w = [H^+][OH^-] = 5.48 \times 10^{-14}$. In pure water, $[H^+] = [OH^-]$.

            $[H^+]^2 = 5.48 \times 10^{-14}\;[H^+] = 2.34 \times 10^{-7}\,M;\;\; pH = 6.63$

     (b)   The value of K_w increases with increasing temperature, so the sign of ΔH is positive. The autoionization of water is endothermic.

16.121   (a)   24 valence e^-, 12 e^- pairs

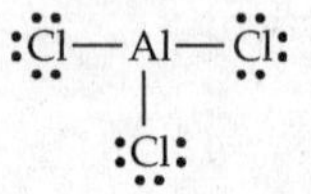

            :C̈l—Al—C̈l:
                  |
                :C̈l:

            The formal charges on all atoms are zero. Structures with multiple bonds lead to nonzero formal charges. There are three electron domains about Al. The electron-domain geometry and molecular structure are trigonal planar.

     (b)   The Al atom in $AlCl_3$ has an incomplete octet and is electron deficient. It "needs" to accept another electron pair, to act like a Lewis acid.

     (c)

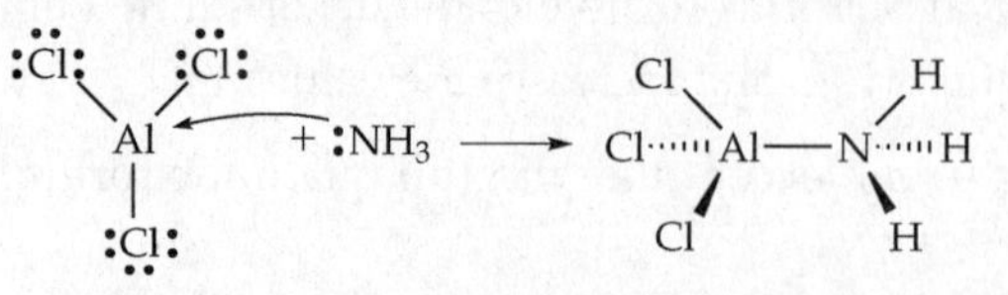

            Both the Al and N atoms in the product have tetrahedral geometry.

     (d)   The Lewis theory is most appropriate. H^+ and $AlCl_3$ are both electron pair acceptors, Lewis acids.

16.122   *Plan.* Use acid ionization equilibrium to calculate the total moles of particles in solution. Use density to calculate kg solvent. From the molality (m) of the solution, calculate ΔT_b and T_b. *Solve.*

	$HSO_4^-(aq)$	$\rightleftharpoons$	$H^+(aq)$	$+$	$SO_4^{2-}(aq)$
initial	0.10 M		0		0
equil.	$0.10 - x\ M$		$x\ M$		$x\ M$
	0.071 M		0.029 M		0.029 M

$$K_a = 1.2 \times 10^{-2} = \frac{[H^+][SO_4^{2-}]}{[HSO_4^-]} = \frac{x^2}{0.10 - x}; K_a \text{ is relatively large, so use the quadratic.}$$

$$x^2 + 0.012\,x - 0.0012 = 0;\ x = \frac{-0.012 \pm \sqrt{(0.012)^2 - 4(1)(-0.0012)}}{2}; x = 0.029\ M\ H^+, SO_4^{2-}$$

Total ion concentration = $0.10\ M\ Na^+ + 0.071\ M\ HSO_4^- + 0.029\ M\ H^+ + 0.029\ M\ SO_4^{2-}$

$$= 0.229 = 0.23\ M$$

Assume 100.0 mL of solution. 1.002 g/mL × 100.0 mL = 100.2 g solution.

$$0.10\ M\ NaHSO_4 \times 0.1000\ L = 0.010\ mol\ NaHSO_4 \times \frac{120.1\ g\ NaHSO_4}{mol\ NaHSO_4}$$

$$= 1.201 = 1.2\ g\ NaHSO_4$$

100.2 g soln – 1.201 g $NaHSO_4$ = 99.0 g = 0.099 kg H_2O

$$m = \frac{mol\ ions}{kg\ H_2O} = \frac{0.229\ M \times 0.1000\ L}{0.0990\ kg} = 0.231 = 0.23\ m\ \text{ions}$$

$\Delta T_b = K_b(m) = 0.52\ °C/m \times (0.23\ m) = +0.12\ °C;\ T_b = 100.0 + 0.12 = 100.1\ °C$

16.124   Calculate M of the solution from osmotic pressure and K_b using the equilibrium expression for the hydrolysis of cocaine. Let Coc = cocaine and $CocH^+$ be the conjugate acid of cocaine.

$$\Pi = MRT;\ M = \Pi/RT = \frac{52.7\ torr}{288\ K} \times \frac{1\ atm}{760\ torr} \times \frac{mol\text{-}K}{0.08206\ L\text{-}atm}$$

$$= 0.002934 = 2.93 \times 10^{-3}\ M\ Coc$$

pH = 8.53; pOH = 14 – pH = 5.47; $[OH^-] = 10^{-5.47} = 3.39 \times 10^{-6} = 3.4 \times 10^{-6}\ M$

	$Coc(aq) + H_2O(l)$	$\rightleftharpoons$	$CocH^+(aq)$	$+$	$OH^-(aq)$
initial	$2.93 \times 10^{-3}\ M$		0		0
equil.	$(2.93 \times 10^{-3} - 3.4 \times 10^{-6})\ M$		$3.4 \times 10^{-6}\ M$		$3.4 \times 10^{-6}\ M$

$$K_b = \frac{[CocH^+][OH^-]}{[Coc]} = \frac{(3.39 \times 10^{-6})^2}{(2.934 \times 10^{-3} - 3.39 \times 10^{-6})} = 3.9 \times 10^{-9}$$

Note that % hydrolysis is small in this solution, so $3.39 \times 10^{-6}\ M$ is small compared to $2.93 \times 10^{-3}\ M$ and can be ignored in the denominator of the calculation.

16.125   (a)   rate $= k[IO_3^-][SO_3^{2-}][H^+]$

         (b) $\Delta pH = pH_2 - pH_1 = 3.50 - 5.00 = -1.50$

             $\Delta pH = -\log[H^+]_2 - (-\log[H^+]_1); -\Delta pH = \log[H^+]_2 - \log[H^+]_1$

             $-\Delta pH = \log[H^+]_2 / [H^+]_1; [H^+]_2 / [H^+]_1 = 10^{-\Delta pH}$

             $[H^+]_2/[H^+]_1 = 10^{1.50} = 31.6 = 32$. The rate will increase by a factor of 32 if $[H^+]$ increases by a factor of 32. The reaction goes faster at lower pH.

         (c)   As H^+ does not appear in the overall reaction, it is either a catalyst or an intermediate. An intermediate is produced and then consumed during a reaction, so its contribution to the rate law can usually be written in terms of concentrations of other reactants (Sample Exercise 14.15). A catalyst is present at the beginning and end of a reaction and can appear in the rate law if it participates in the rate-determining step (Solution 14.78). This reaction is pH dependent because H^+ is a homogeneous catalyst that participates in the rate-determining step.

17 Additional Aspects of Aqueous Equilibria

Visualizing Concepts

17.2   (a)   The yellow solution has the higher pH. According to Figure 16.8, methyl orange is yellow above pH 4.5 and red (really pink) below pH 3.5. The beaker on the left has a pH greater than 4.5, and the one on the right has pH less than 3.5. (By calculation, pH of left beaker = 4.7, pH of right beaker = 2.9.) The right beaker, with lower pH and greater [H$^+$], is pure acetic acid. The left beaker contains equal amounts of the weak acid and its conjugate base, acetic acid and acetate ion. Adding the "common-ion" acetate (in the form of sodium acetate) shifts the acid ionization equilibrium to the left, decreases [H$^+$], and raises pH.

   (b)   When small amounts of NaOH are added, the left beaker is better able to maintain its pH. For solutions of the same weak acid, pH depends on the *ratio* of conjugate base to conjugate acid. Small additions of base (or acid) have the least effect when this ratio is close to one. The left beaker is a buffer because it contains a weak conjugate acid/conjugate base pair and resists rapid pH change upon addition of small amounts of strong base or acid.

17.3   Statement (b) is correct, [HA] > [A$^-$]. Buffers prepared from weak acids (HA) and their conjugate bases (A$^-$, usually in the form of a salt) have pH values in a range of approximately 2 pH units, centered around pK$_a$ for the weak acid. If concentration of the weak acid is greater than concentration of the conjugate base, pH < pK$_a$. If concentration of the conjugate base is greater than concentration of the weak acid, pH > pK$_a$. This is generally true for buffers containing a weak conjugate acid/conjugate base (CA/CB) pair.

   [CA] > [CB], pH of buffer < pK$_a$ of CA

   [CA] < [CB], pH of buffer > pK$_a$ of CA

17.4   *Analyze/Plan.* When strong acid is added to a buffer, it reacts with conjugate base (CB) to produce conjugate acid (CA). [CA] increases and [CB] decreases. The opposite happens when strong base is added to a buffer, [CB] increases and [CA] decreases. Match these situations to the drawings.   *Solve.*

   The buffer begins with equal concentrations of HX and X$^-$.

   (a)   After addition of strong acid, [HX] will increase and [X$^-$] will decrease. Drawing (3) fits this description.

   (b)   Adding of strong base causes [HX] to decrease and [X$^-$] to increase. Drawing (1) matches the description.

   (c)   Drawing (2) shows both [HX] and [X$^-$] to be smaller than the initial concentrations shown on the left. This situation cannot be achieved by adding strong acid or strong base to the original buffer.

17 Additional Aspects of Aqueous Equilibria

Solutions to Exercises

17.5 *Analyze/Plan.* Consider the reaction $HA + OH^- \rightarrow A^- + H_2O$. What are the major species present in solution at the listed stages of the titration? Which diagram represents these species? *Solve.*

(a) *Before addition of NaOH*, the solution is mostly HA. The only A^- is produced by the ionization equilibrium of HA and is too small to appear in the diagram. This situation is shown in diagram (iii), which contains only HA.

(b) *After addition of NaOH but before the equivalence point*, some, but not all, HA has been converted to A^-. The solution contains a mixture of HA and A^-; this is shown in diagram (i).

(c) *At the equivalence point*, all HA has been converted to A^-, with no excess HA or OH^- present. This is shown in diagram (iv).

(d) *After the equivalence point*, the same amount of A^- as at the equivalence point is present, plus some excess OH^-. This is diagram (ii).

17.6 *Analyze/Plan.* In each case, the first substance is in the buret, and the second is in the flask. If acid is in the flask, the initial pH is low; with base in the flask, the pH starts high. Strong acids have lower pH than weak acids; strong bases have higher pH than weak bases. Polyprotic acids and bases have more than one "jump" in pH. *Solve.*

(a) Strong base in flask, pH starts high, ends low as acid is added. Only diagram (ii) fits this description.

(b) Weak acid in flask, pH starts low, but not extremely low. Diagrams (i), (iii), and (iv) all start at low pH and get higher. Diagram (i) has very low initial pH, and likely has strong acid in the flask. Diagram (iv) has two pH jumps, so it has a polyprotic acid in the flask. Diagram (iii) best fits the profile of adding a strong base to a weak acid.

(c) Strong acid in the flask, pH starts very low, diagram (i).

(d) Polyprotic acid, more than one pH jump, diagram (iv).

17.8 *Analyze/Plan.* The beaker of saturated $Cd(OH)_2(aq)$ contains undissolved $Cd(OH)_2(s)$, $Cd^{2+}(aq)$, and $OH^-(aq)$. Decide how amounts of each of these three components change when HCl(aq) is added. *Solve.*

When HCl(aq) is added, it reacts with $OH^-(aq)$ to form $H_2O(l)$ and $Cl^-(aq)$. (Both have been omitted from the figure.) When $OH^-(aq)$ is removed from solution, more $Cd(OH)_2(s)$ dissolves to replace it; $[Cd^{2+}(aq)]$ increases, $[OH^-(aq)]$ decreases and the amount of undissolved $Cd(OH)_2(s)$ decreases. In the resulting solution, $[Cd^{2+}(aq)]$ is greater than $[OH^-(aq)]$ and there is less undissolved solid on the bottom of the beaker. Beaker A accurately represents the solution after equilibrium is reestablished.

17.10 *Analyze/Plan.* Calculate the molarity of the solution assuming all $Ca(OH)_2(s)$ dissolves. Use this concentration along with the K_{sp} expression for $Ca(OH)_2$ to answer the questions. *Solve.*

(a) $[Ca(OH)_2] = \dfrac{0.370 \text{ g } Ca(OH)_2}{0.500 \text{ L soln}} \times \dfrac{1 \text{ mol } Ca(OH)_2}{74.093 \text{ g } Ca(OH)_2} = 0.00998745 = 0.00999 \, M$

$[Ca^{2+}] = 0.00999 \, M; \, [OH^-] = 2(0.00998745) = 0.0199749 = 0.0200 \, M;$

$K_{sp} = [Ca^{2+}][OH^-]^2$. Calculate the reaction quotient using the calculated molarities. If it is equal to or greater than K_{sp}, the resulting solution is saturated. $Q = (0.0098745)(0.0199749)^2 = 3.99 \times 10^{-6}$. $Q < K_{sp}$ (6.5×10^{-6}) and the solution is not saturated.

(b)   Consider the beakers individually.

     (i)   The 50 mL of 1.0 M HCl is more than enough to neutralize 50 mL of 0.0200 M OH^-(aq). No precipitate forms.

     (ii)   NaCl does not react with $Ca(OH)_2$ and the two compounds contain no common ions. No precipitate forms.

     (iii) $CaCl_2$ does contain a common ion. Calculate Q for the resulting solution to see if $Ca(OH)_2$ precipitates. $[OH^-]$ in the new solution is 0.00999 M, because it is diluted by a factor of 2. $[Ca^{2+}] = (1.0 + 0.00999)/2 = 0.5050 \, M$. $Q = (0.5050)(0.00999)^2 = 5.04 \times 10^{-5}$. $Q > K_{sp}$ (6.5×10^{-6}) and $Ca(OH)_2$ precipitates.

     (iv)   A common ion with a different concentration; $[Ca^{2+}] = (0.10 + 0.00999)/2 = 0.0550 = 0.055 \, M$. $Q = (0.0550)(0.00999)^2 = 5.49 \times 10^{-6}$. $Q \approx K_{sp}$ (6.5×10^{-6}); the solution is very nearly saturated, but no precipitate forms.

17.11   Statement (c) explains the shape of the graph. Solubility is high initially, at low pH, and then decreases to a minimum as pH increases. This depicts formation of an insoluble hydroxide as $[H^+]$ decreases and $[OH^-]$ increases. Additional base then reacts with the insoluble hydroxide to dissolve it. This curve depicts the behavior of an amphoteric insoluble hydroxide; it dissolves upon addition of either acid or base.

17.12   According to Figure 17.23, the two precipitating agents are 6 M HCl (first) and H_2S in 0.2 M HCl (second).

Cation A = Ag^+ (precipitates as AgCl)

Cation B = Cu^+ (precipitates as CuS, acid insoluble)

Cation C = Ni^{2+} (remains in acidic solution)

The Common-Ion Effect (Section 17.1)

17.14   The added salt is soluble and increases $[HB^+]$ in the solution. For a generic weak base B,

$K_b = \dfrac{[HB^+][OH^-]}{[B]}$.

(a)   Stay the same. Addition of a common ion such as HB^+ does not change the equilibrium constant.

(b)   Increase. To maintain the value of the equilibrium constant, addition of HB^+ requires that [B] also increases.

(c)   Decrease. Additional HB^+ reacts with OH^-, lowering the pH of the solution.

17.16   *Analyze/Plan.* Follow the logic in Sample Exercise 17.1.   *Solve.*

(a)   HCOOH is a weak acid, and HCOONa contains the common ion $HCOO^-$, the conjugate base of HCOOH. Solve the common-ion equilibrium problem.

$$HCOOH(aq) \rightleftharpoons H^+(aq) + HCOO^-(aq)$$

i	0.100 M		0.250 M
c	$-x$	$+x$	$+x$
e	$(0.100 - x)\ M$	$+x\ M$	$(0.250 + x)\ M$

$$K_a = 1.8 \times 10^{-4} = \frac{[H^+][HCOO^-]}{[HCOOH]} = \frac{(x)(0.250 + x)}{(0.100 - x)} \approx \frac{0.250\,x}{0.100}$$

$$x = 7.20 \times 10^{-5} = 7.2 \times 10^{-5}\ M = [H^+],\ pH = 4.14$$

Check. Because the extent of ionization of a weak acid or base is suppressed by the presence of a conjugate salt, the 5% rule usually holds true in buffer solutions.

(b) C_5H_5N is a weak base, and C_5H_5NHCl contains the common ion $C_5H_5NH^+$, which is the conjugate acid of C_5H_5N. Solve the common-ion equilibrium problem.

$$C_5H_5N(aq) + H_2O(l) \rightleftharpoons C_5H_5NH^+(aq) + OH^-((aq)$$

i	0.510 M	0.450 M	
c	$-x$	$+x$	$+x$
e	$(0.510 - x)\ M$	$(0.450 + x)\ M$	$+x\ M$

$$K_b = 1.7 \times 10^{-9} = \frac{[C_5H_5NH^+][OH^-]}{[C_5H_5N]} = \frac{(0.450 + x)(x)}{(0.510 - x)} \approx \frac{0.450\,x}{0.510}$$

$$x = 1.927 \times 10^{-9} = 1.9 \times 10^{-9}\ M = [OH^-],\ pOH = 8.715,\ pH = 14.00 - 8.715 = 5.29$$

Check. In a buffer, if [conj. acid] > [conj. base], pH < pK_a of the conj. acid. If [conj. acid] < [conj. base], pH > pK_a of the conj. acid. In this buffer, pK_a of $C_5H_5NH^+$ is 5.23. $[C_5H_5NH^+] < [C_5H_5N]$ and pH = 5.29, greater than 5.23.

(c)   mol = $M \times L$; mol HF = 0.050 $M \times$ 0.055 L = 2.75×10^{-3} = 2.8×10^{-3} mol;
mol F^- = 0.10 $M \times$ 0.125 L = 0.0125 = 0.013 mol

$$HF(aq) \rightleftharpoons H^+(aq) + F^-(aq)$$

i	2.75×10^{-3} mol	0	0.0125 mol
c	$-x$	$+x$	$+x$
e	$(2.75 \times 10^{-3} - x)$ mol	$+x$	$(0.0125 + x)$ mol

$[HF] = (2.75 \times 10^{-3} + x)/0.180\ L;\ [F^-] = (0.0125 + x)/0.180\ L$

Note that the volumes will cancel when substituted into the K_a expression.

$$K_a = 6.8 \times 10^{-4} = \frac{[H^+][F^-]}{[HF]} = \frac{x(0.0125 + x)/0.180}{(2.75 \times 10^{-3} - x)/0.180} \approx \frac{x(0.0125)}{0.00275}$$

$$x = 1.50 \times 10^{-4} = 1.5 \times 10^{-4} \, M \, H^+; \, pH = 3.83$$

Check. K_a for HF = 3.17. [HF] < [F$^-$], pH of buffer = 3.83, greater than 3.17.

17.18 $\qquad CH_3CH(OH)COOH \rightleftharpoons H^+(aq) + CH_3CH(OH)COO^-$

equil (a) $\quad 0.125 - x \, M \qquad\qquad x \, M \qquad\qquad\quad x \, M$

equil (b) $\quad 0.125 - x \, M \qquad\qquad x \, M \qquad\qquad 0.0075 + x \, M$

$$K_a = \frac{[H^+][CH_3CH(OH)COO^-]}{[CH_3CH(OH)COOH]} = 1.4 \times 10^{-4}$$

(a) $\quad K_a = 1.4 \times 10^{-4} = \frac{x^2}{0.125 - x} \approx \frac{x^2}{0.125}; x = [H^+] = 4.18 \times 10^{-3} \, M = 4.2 \times 10^{-3} \, M \, H^+$

$$\% \text{ ionization} = \frac{4.2 \times 10^{-3} \, M \, H^+}{0.125 \, M \, CH_3CH(OH)COOH} \times 100 = 3.4\% \text{ ionization}$$

(b) $\quad K_a = 1.4 \times 10^{-4} = \frac{(x)(0.0075 + x)}{0.125 - x} \approx \frac{0.0075 \, x}{0.125}; x = 2.3 \times 10^{-3} \, M \, H^+$

$$\% \text{ ionization} = \frac{2.3 \times 10^{-3} \, M \, H^+}{0.125 \, M \, CH_3CH(OH)COOH} \times 100 = 1.9\% \text{ ionization}$$

Buffered Solutions (Section 17.2)

17.20 Only solution (a) is a buffer. NaOH is a strong base and will react with CH$_3$COOH to form CH$_3$COONa. As long as CH$_3$COOH is present in excess, the resulting solution will contain both the conjugate acid CH$_3$COOH(aq) and the conjugate base CH$_3$COO$^-$(aq), the requirements for a buffer.

Solution (b) contains a large excess and is essentially a strong base. Solution (c) is essentially a strong acid; HCl determines [H$^+$] and pH of the solution. Solution (d) contains two salts; one is the conjugate base of a weak acid. It can react with added acid, but no added base.

17.22 Assume that % ionization is small in these buffers (Solutions 17.17 and 17.18).

(a) The conjugate acid in this buffer is HCO$_3^-$, so use K_{a2} for H$_2$CO$_3$, 5.6×10^{-11}

$$K_a = \frac{[H^+][CO_3^{2-}]}{[HCO_3^-]}; [H^+] = \frac{K_a[HCO_3^-]}{[CO_3^{2-}]} = \frac{5.6 \times 10^{-11}(0.105)}{(0.125)}$$

$[H^+] = 4.70 \times 10^{-11} = 4.7 \times 10^{-11} \, M; \, pH = 10.33$

(b) mol = $M \times$ L; total volume = 140 mL = 0.140 L

$$[H^+] = \frac{K_a(0.20 \, M \times 0.065 \, L)/0.140 \, L}{(0.15 \, M \times 0.075 \, L)/0.140 \, L} = \frac{5.6 \times 10^{-11}(0.20 \times 0.065)}{(0.15 \times 0.075)}$$

$[H^+] = 6.47 \times 10^{-11} = 6.5 \times 10^{-11} \, M; \, pH = 10.19$

17.24 NH_4^+/NH_3 is a basic buffer. Either the hydrolysis of NH_3 or the dissociation of NH_4^+ can be used to determine the pH of the buffer. Using the dissociation of NH_4^+ leads directly to $[H^+]$ and facilitates use of the Henderson–Hasselbach relationship.

(a) $NH_4^+(aq) \rightleftharpoons H^+(aq) + NH_3(aq)$

$$K_a = \frac{K_w}{K_b} = \frac{1.0 \times 10^{-14}}{1.8 \times 10^{-5}} = 5.56 \times 10^{-10} = 5.6 \times 10^{-10}$$

$[NH_3] = 1.00\ M\ NH_3$

$$[NH_4^+] = \frac{10.0\ g\ NH_4Cl}{0.250\ L} \times \frac{1\ mol\ NH_4Cl}{53.50\ g\ NH_4Cl} = 0.74766 = 0.748\ M\ NH_4^+$$

$$K_a = \frac{[H^+][NH_3]}{[NH_4^+]}; [H^+] = \frac{K_a[NH_4^+]}{[NH_3]} = \frac{5.56\times10^{-10}\ (0.74766 - x)}{(1.00 + x)} \approx \frac{5.56\times10^{-10}\ (0.74766)}{(1.00)}$$

$[H^+] = 4.1537 \times 10^{-10} = 4.15 \times 10^{-10}\ M$, pH = 9.382

(b) $NH_3(aq) + H^+(aq) + NO_3^-(aq) \rightarrow NH_4^+(aq) + NO_3^-(aq)$

(c) $NH_4^+(aq) + Cl^-(aq) + K^+(aq) + OH^-(aq) \rightarrow NH_3(aq) + H_2O(l) + Cl^-(aq) + K^+(aq)$

17.26   (a) $C_6H_5COOH(aq) \rightleftharpoons H^+(aq) + C_6H_5COO^-(aq)$

$$K_a = 6.3\times10^{-5} = \frac{[H^+][C_6H_5COO^-]}{[C_6H_5COOH]}; [C_6H_5COOH] = 0.0200\ M;$$

$[H^+] = [C_6H_5COO^-] = x$

$$K_a = 6.3\times10^{-5} \approx \frac{x^2}{0.02000}; x = [H^+] = 1.123\times10^{-3} = 1.1\times10^{-3}\ M;\ pH = 2.95$$

Note that C_6H_5COOH is 5.6% ionized. Solving the quadratic for $[H^+]$ yields $(1.0914 \times 10^{-3} =)\ 1.1 \times 10^{-3}\ M\ H^+$, pH = 2.96; this is not a significant difference.

(b) $[H^+] = \dfrac{K_a[C_6H_5COOH]}{[C_6H_5COO^-]};\ \ [H^+] = 10^{-4.00} = 1.0 \times 10^{-4}\ M$

$[C_6H_5COOH] = 0.0200\ M$; calculate $[C_6H_5COO^-]$. Because the common ion $C_6H_5COO^-$ reduces % ionization, we assume the 5% approximation is valid.

$$[C_6H_5COO^-] = \frac{K_a[C_6H_5COOH]}{[H^+]} = \frac{6.3 \times 10^{-5}\ (0.0200)}{1.0 \times 10^{-4}} = 0.01260 = 0.013\ M$$

$$\frac{0.0126\ mol\ C_6H_5COONa}{L} \times 1.50\ L \times \frac{144.11\ g\ C_6H_5COONa}{1\ mol\ C_6H_5COONa}$$

$$= 2.724 = 2.7\ g\ C_6H_5COONa$$

17.28   (a) $K_a = \dfrac{[H^+][C_2H_5COO^-]}{[C_2H_5COOH]}; [H^+] = \dfrac{K_a[C_2H_5COOH]}{[C_2H_5COO^-]}$

Because this expression contains a ratio of concentrations, we can ignore total volume and work directly with moles.

$$[H^+] = \frac{1.3\times10^{-5}\ (0.15 - x)}{(0.10 + x)} \approx \frac{1.3\times10^{-5}\ (0.15)}{0.10} = 1.950\times10^{-5} = 2.0\times10^{-5}\ M,\ pH = 4.71$$

(b)

$C_2H_5COOH(aq)$	+	$OH^-(aq)$	$\rightarrow$	$C_2H_5COO^-(aq) + H_2O(l)$
0.15 mol		0.01 mol		0.10 mol
−0.01 mol		−0.01 mol		+0.01 mol
0.14 mol		0 mol		0.11 mol

$$[H^+] \approx \frac{1.3 \times 10^{-5}\,(0.14)}{(0.11)} = 1.6545 \times 10^{-5} = 1.7 \times 10^{-5}\,M;\ pH = 4.78$$

(c)

$C_2H_5COO^-(aq)$	+	$HI(aq)$	$\rightarrow$	$C_2H_5COOH(aq) + I^-(aq)$
0.10 mol		0.01 mol		0.15 mol
−0.01 mol		−0.01 mol		+0.01 mol
0.09 mol		0 mol		0.16 mol

$$[H^+] \approx \frac{1.3 \times 10^{-5}\,(0.16)}{(0.09)} = 2.3111 \times 10^{-5} = 2 \times 10^{-5}\,M;\ pH = 4.6$$

17.30

$$\frac{6.5\ g\ NaH_2PO_4}{0.355\ L\ soln} \times \frac{1\ mol\ NaH_2PO_4}{120\ g\ NaH_2PO_4} = 0.153 = 0.15\ M$$

$$\frac{8.0\ g\ Na_2HPO_4}{0.355\ L\ soln} \times \frac{1\ mol\ Na_2HPO_4}{142\ g\ Na_2HPO_4} = 0.159 = 0.16\ M$$

Use Equation 17.9 to find the pH of the buffer. K_a for $H_2PO_4^-$ is K_{a2} for H_3PO_4, 6.2×10^{-8}

$$pH = -\log(6.2 \times 10^{-8}) + \log\frac{0.159}{0.153} = 7.2076 + 0.0167 = 7.22$$

17.32 The solutes listed contain three possible conjugate acid/conjugate base (CA/CB) pairs. These are:

HCOOH/HCOONa, $pK_a = 3.74$

CH_3COOH/CH_3COONa, $pK_a = 4.74$

$HCN/NaCN$, $pK_a = 9.31$

For maximum buffer capacity, pK_a should be within 1 pH unit of the buffer. The acetic acid/acetate pair is most appropriate for a buffer with pH 5.00.

$$pH = pK_a + \log\frac{[CB]}{[CA]};\ 5.00 = 4.745 + \log\frac{[CH_3COONa]}{[CH_3COOH]}$$

$$\log\frac{[CH_3COONa]}{[CH_3COOH]} = 0.2553;\ \frac{[CH_3COONa]}{[CH_3COOH]} = 1.800 = 1.8$$

Because we are making a total of 1 L of buffer,
let y = vol CH_3COONa and (1 − y) = vol CH_3COOH.

$$1.800 = \frac{[CH_3COONa]}{[CH_3COOH]} = \frac{(0.10\ M \times y)/1.0\ L}{[0.10\ M \times (1-y)]/1.0\ L} = \frac{0.10\ y}{0.10 - 0.10\ y}$$

$1.800(0.10 - 0.10 \, y) = 0.10 \, y; \; 0.1800 = 0.2800 \, y; \; y = 0.6429 = 0.64 \, L$

640 mL of 0.10 M CH$_3$COONa, 360 mL of CH$_3$COOH

Check. pH (buffer) > pK$_a$ (CA) and the calculated amount of CB in the buffer is greater than the amount of CA.

Acid–Base Titrations (Section 17.3)

17.34   (a)   False. The quantity of base required to reach the equivalence point is the same in the two titrations, assuming both acids have the same initial concentrations.

         (b)   False. The pH is higher initially in the titration of a weak acid.

         (c)   False. The pH is higher at the equivalence point in the titration of a weak acid.

17.36   (a)   False. The pH at the beginning of the titration of the weaker acid, CH$_3$COOH, will be higher.

         (b)   True. Past the equivalence point, the titration curves are very similar (but not identical.

         (c)   False. According to Figures 17.12 and 17.13, methyl red is suitable for the titration of the strong acid HNO$_3$, but not for the titration of the weak acid CH$_3$COOH.

17.38   (a)   HCOOH(aq) + NaOH(aq) $\rightarrow$ HCOONa(aq) + H$_2$O(l)

            At the equivalence point, the major species are Na$^+$ and HCOO$^-$. Na$^+$ is negligible and HCOO$^-$ is the CB of HCOOH. The solution is basic, above pH 7.

         (b)   Ca(OH)$_2$(aq) + 2HClO$_4$(aq) $\rightarrow$ Ca(ClO$_4$)$_2$(aq) + 2H$_2$O(l)

            At the equivalence point, the major species are Ca^{2+} and ClO$_4^-$; both are negligible. The solution is at pH 7.

         (c)   C$_5$H$_5$N(aq) + HNO$_3$(aq) $\rightarrow$ C$_5$H$_5$NH$^+$NO$_3^-$(aq)

            At the equivalence point, the major species are C$_5$H$_5$NH$^+$ and NO$_3^-$. NO$_3^-$ is negligible and C$_5$H$_5$NH$^+$ is the CA of C$_5$H$_5$N. The solution is acidic, below pH 7.

17.40   (a)   At the equivalence point, moles HA added = moles B initially present = 0.10 M $\times$ 0.0300 L = 0.0030 moles HA added.

         (b)   BH$^+$(aq).

         (c)   Less than 7. The predominant form at equivalence, BH$^+$, is a weak acid.

         (d)   Because the pH at the equivalence point will be less than 7, methyl red would be more appropriate.

17.42   (a) $45.0 \text{ mL NaOH} \times \dfrac{0.0950 \text{ mol NaOH}}{1000 \text{ mL soln}} \times \dfrac{1 \text{ mol HCl}}{1 \text{ mol NaOH}} \times \dfrac{1000 \text{ mL soln}}{0.105 \text{ mol HCl}}$

                                                    = 40.7 mL HCl soln

(b) $22.5 \text{ mL NH}_3 \times \dfrac{0.118 \text{ mol NH}_3}{1000 \text{ mL soln}} \times \dfrac{1 \text{ mol HCl}}{1 \text{ mol NH}_3} \times \dfrac{1000 \text{ mL soln}}{0.105 \text{ mol HCl}}$

$= 25.3 \text{ mL HCl soln}$

(c) $125.0 \text{ mL} \times \dfrac{1.35 \text{ g NaOH}}{1000 \text{ mL}} \times \dfrac{1 \text{ mol NaOH}}{40.00 \text{ g NaOH}} \times \dfrac{1 \text{ mol HCl}}{1 \text{ mol NaOH}} \times \dfrac{1000 \text{ mL soln}}{0.105 \text{ mol HCl}}$

$= 40.2 \text{ mL HCl soln}$

17.44   moles $OH^- = M_{KOH} \times L_{KOH} = 0.150 \ M \times 0.0200 \text{ L} = 3.00 \times 10^{-3} \text{ mol}$

moles $H^+ = M_{HClO_4} \times L_{HClO_4} = 0.125 \ M \times L_{HClO_4}$

	mL_{KOH}	mL_{HClO_4}	Total Volume	Moles OH^-	Moles H^+	Molarity Excess Ion	pH
(a)	20.0	20.0	40.0	3.00×10^{-3}	2.50×10^{-3}	$0.013(OH^-)$	12.10
(b)	20.0	23.0	43.0	3.00×10^{-3}	2.88×10^{-3}	$2.9 \times 10^{-3}(OH^-)$	11.46
(c)	20.0	24.0	44.0	3.00×10^{-3}	3.00×10^{-3}	$1.0 \times 10^{-7}(OH^-)$	7.00
(d)	20.0	25.0	45.0	3.00×10^{-3}	3.13×10^{-3}	$2.8 \times 10^{-3}(H^+)$	2.56
(e)	20.0	30.0	50.0	3.00×10^{-3}	3.75×10^{-3}	$0.015(H^+)$	1.82

molarity of excess ion $= \dfrac{\text{moles ion}}{\text{total vol in L}}$

(a) $\dfrac{3.00 \times 10^{-3} \text{ mol OH}^- - 2.50 \times 10^{-3} \text{ mol H}^+}{0.0400 \text{ L}} = 0.0125 = 0.013 \ M \ OH^-$

(b) $\dfrac{3.00 \times 10^{-3} \text{ mol OH}^- - 2.875 \times 10^{-3} \text{ mol H}^+}{0.0430 \text{ L}} = 2.91 \times 10^{-3} = 2.9 \times 10^{-3} \ M \ OH^-$

(c) equivalence point, mol $H^+ = $ mol OH^-

$KClO_4$ does not hydrolyze, so $[H^+] = [OH^-] = 1 \times 10^{-7} \ M$

(d) $\dfrac{3.125 \times 10^{-3} \text{ mol H}^+ - 3.00 \times 10^{-3} \text{ mol OH}^-}{0.0450 \text{ L}} = 2.78 \times 10^{-3} = 2.8 \times 10^{-3} \ M \ H^+$

(e) $\dfrac{3.75 \times 10^{-3} \text{ mol H}^+ - 3.00 \times 10^{-3} \text{ mol OH}^-}{0.0500 \text{ L}} = 0.0150 = 0.015 \ M \ H^+$

17.46   (a)   Weak base problem: $K_b = 1.8 \times 10^{-5} = \dfrac{[NH_4^+][OH^-]}{[NH_3]}$

At equilibrium, $[OH^-] = x$, $[NH_3] = (0.030 - x)$; $[NH_4^+] = x$

$1.8 \times 10^{-5} = \dfrac{x^2}{(0.050 - x)} \approx \dfrac{x^2}{0.050} ; x = [OH^-] = 9.487 \times 10^{-4} = 9.5 \times 10^{-4} \ M$

pH $= 14.00 - 3.02 = 10.98$

(b–f) Calculate mol NH_3 and mol NH_4^+ after the acid–base reaction takes place. $0.050\ M\ NH_3 \times 0.0300\ L = 1.5 \times 10^{-3}$ mol NH_3 present initially.

$$NH_3(aq) \quad + \quad HCl(aq) \quad \rightarrow \quad NH_4^+(aq) + Cl^-(aq)$$

$$(0.025\ M \times 0.0200\ L) =$$

		NH_3	HCl	NH_4^+
(b)	before rx	1.5×10^{-3} mol	0.50×10^{-3} mol	0 mol
	after rx	$\mathbf{1.0 \times 10^{-3}\ mol}$	**0 mol**	$\mathbf{5.0 \times 10^{-4}\ mol}$

$$(0.025\ M \times 0.0590\ L) =$$

		NH_3	HCl	NH_4^+
(c)	before rx	1.5×10^{-3} mol	1.475×10^{-3} mol	0 mol
	after rx	$\mathbf{2.5 \times 10^{-5}\ mol}$	**0 mol**	$\mathbf{1.475 \times 10^{-3}\ mol}$

$$(0.025\ M \times 0.0600\ L) =$$

		NH_3	HCl	NH_4^+
(d)	before rx	1.5×10^{-3} mol	1.5×10^{-3} mol	0 mol
	after rx	**0 mol**	**0 mol**	$\mathbf{1.5 \times 10^{-3}\ mol}$

$$(0.025\ M \times 0.0610\ L) =$$

		NH_3	HCl	NH_4^+
(e)	before rx	1.5×10^{-3} mol	1.525×10^{-3} mol	0 mol
	after rx	**0 mol**	$\mathbf{2.5 \times 10^{-5}\ mol}$	$\mathbf{1.5 \times 10^{-3}\ mol}$

$$(0.025\ M \times 0.0650\ L) =$$

		NH_3	HCl	NH_4^+
(f)	before rx	1.5×10^{-3} mol	1.625×10^{-3} mol	0 mol
	after rx	**0 mol**	$\mathbf{1.25 \times 10^{-4}\ mol}$	$\mathbf{1.5 \times 10^{-3}\ mol}$

(b)   Using the acid dissociation equilibrium for NH_4^+ (so that we calculate $[H^+]$ directly), $NH_4^+(aq) \rightleftharpoons H^+(aq) + NH_3(aq)$

$$K_a = \frac{[H^+][NH_3]}{[NH_4^+]} = \frac{K_w}{K_b\ for\ NH_3} = \frac{1.0 \times 10^{-14}}{1.8 \times 10^{-5}} = 5.56 \times 10^{-10} = 5.6 \times 10^{-10}$$

$$[NH_3] = \frac{1.0 \times 10^{-3}\ mol}{0.0500\ L} = 0.020\ M; [NH_4^+] = \frac{5.0 \times 10^{-4}\ mol}{0.0500\ L} = 0.010\ M$$

$$[H^+] = \frac{5.56 \times 10^{-10}\ [NH_4^+]}{[NH_3]} \approx \frac{5.56 \times 10^{-10}\ (0.010)}{(0.020)} = 2.78 \times 10^{-10}; pH = 9.56$$

(We will assume $[H^+]$ is small compared to $[NH_3]$ and $[NH_4^+]$.)

(c) $[NH_3] = \dfrac{2.5 \times 10^{-5}\ mol}{0.0890\ L} = 2.8 \times 10^{-4}\ M; [NH_4^+] = \dfrac{1.475 \times 10^{-3}\ mol}{0.0890\ L} = 0.017\ M$

$$[H^+] = \frac{5.56 \times 10^{-10}\ (0.017)}{(2.8 \times 10^{-4})} = 3.38 \times 10^{-8} = 3.4 \times 10^{-8}\ M; pH = 7.47$$

(d)   At the equivalence point, $[H^+] = [NH_3] = x$

$$[NH_4^+] = \frac{1.5 \times 10^{-3} \; M}{0.0900 \; L} = 0.01667 = 0.017 \; M$$

$$5.56 \times 10^{-10} = \frac{x^2}{0.01667}; x = [H^+] = 3.043 \times 10^{-6} = 3.0 \times 10^{-6} \; M; \; pH = 5.52$$

(e)   Past the equivalence point, $[H^+]$ from the excess HCl determines the pH.

$$[H^+] = \frac{2.5 \times 10^{-5} \; mol}{0.0910 \; L} = 2.747 \times 10^{-4} = 2.7 \times 10^{-4} \; M; \; pH = 3.56$$

(f)   Past the equivalence point, $[H^+]$ from the excess HCl determines the pH.

$$[H^+] = \frac{1.25 \times 10^{-4} \; mol}{0.0950 \; L} = 1.316 \times 10^{-3} = 1.3 \times 10^{-3} \; M; \; pH = 2.88$$

17.48   The volume of NaOH solution required in all cases is

$$V_{base} = \frac{V_{acid} \times M_{acid}}{M_{base}} = \frac{(0.100) \, V_{acid}}{(0.080)} = 1.25 \, V_{acid}$$

The total volume at the equivalence point is $V_{base} + V_{acid} = 2.25 \, V_{acid}$

The concentration of the salt at the equivalence point is $\dfrac{M_{acid} \, V_{acid}}{2.25 \, V_{acid}} = \dfrac{0.100}{2.25} = 0.0444 \; M$

(a) $0.0444 \; M$ NaBr, pH = 7.00

(b) $0.0444 \; M$ NaClO$_2$; $ClO_2^-(aq) + H_2O(l) \rightleftharpoons HClO_2(aq) + OH^-(aq)$

$$K_b = \frac{[HClO_2][OH^-]}{[ClO_2^-]} = \frac{K_w}{K_a} = \frac{1.0 \times 10^{-14}}{1.1 \times 10^{-2}} = 9.09 \times 10^{-13} = 9.1 \times 10^{-13}$$

$[HClO_2] = [OH^-]; [ClO_2^-] \approx 0.0444 \; M$

$[OH^-]^2 \approx 0.0444(9.09 \times 10^{-13}); [OH^-] = 2.01 \times 10^{-7} = 2.0 \times 10^{-7} \; M, \; pOH = 6.70;$

$$pH = 7.30$$

Note that $HClO_2$ is a relatively strong acid (large K_a value), so the pH at the equivalence point is not much greater than 7.0. Because $[OH^-]$ from the hydrolysis of ClO_2^- is very small, the autoionization equilibrium should be considered for a more accurate value of the equivalence point pH.

Let $[H^+] = x$, $[OH^-] = (2.0 \times 10^{-7} \; M + x)$; $1.0 \times 10^{-14} = (x)(2.0 \times 10^{-7} \; M + x)$

Solving the quadratic equation gives a pH of 7.38.

(c) $C_6H_5COO^-(aq) + H_2O(l) \rightleftharpoons C_6H_5COOH(aq) + OH^-(aq)$

$$K_b = \frac{[C_6H_5COO^-][OH^-]}{[C_6H_5COOH]} = \frac{K_w}{K_a} = \frac{1.0 \times 10^{-14}}{6.3 \times 10^{-5}} = 1.59 \times 10^{-10} = 1.6 \times 10^{-10}$$

$[OH^-]^2 \approx 0.0444(1.59 \times 10^{-8}); [OH^-] = 2.655 \times 10^{-6} = 2.7 \times 10^{-6}, \; pH = 8.42$

Solubility Equilibria and Factors Affecting Solubility
(Sections 17.4 and 17.5)

17.50   (a) MZ_2 has the larger numerical value for the solubility product constant, $4s^2$ versus s^2.

(b)   The $[M^{2+}]$ is the same in the two saturated solutions, because the molar solubilities are the same and there is one mole of $[M^{2+}]$ in one mole of either salt.

(c) $[M^{2+}] = 4 \times 10^{-4}$ M. After the addition, mol M^{2+} increase, but so does the total volume. The $[M^{2+}]$ remains the same.

17.52   (a)   False. Solubility is the amount (grams, moles) of solute that will dissolve in a certain volume of solution. Solubility-product constant is an equilibrium constant, the product of the molar concentrations of all the dissolved ions in solution.

(b) $K_{sp} = [Mn^{2+}][CO_3^{2-}]$; $K_{sp} = [Hg^{2+}][OH^-]^2$; $K_{sp} = [Cu^{2+}]^3[PO_4^{3-}]^2$

17.54   (a) $PbBr_2(s) \rightleftharpoons Pb^{2+}(aq) + 2Br^-(aq)$

$K_{sp} = [Pb^{2+}][Br^-]^2$; $[Pb^{2+}] = 1.0 \times 10^{-2}$ M, $[Br^-] = 2.0 \times 10^{-2}$ M

$K_{sp} = (1.0 \times 10^{-2}\,M)(2.0 \times 10^{-2}\,M)^2 = 4.0 \times 10^{-6}$

(b) $AgIO_3(s) \rightleftharpoons Ag^+(aq) + IO_3^-(aq)$; $K_{sp} = [Ag^+][IO_3^-]$

$$[Ag^+] = [IO_3^-] = \frac{0.0490 \text{ g AgIO}_3}{1.00\,L \text{ soln}} \times \frac{1 \text{ mol AgIO}_3}{282.8 \text{ g AgIO}_3} = 1.733 \times 10^{-4} = 1.73 \times 10^{-4}\ M$$

$K_{sp} = (1.733 \times 10^{-4}\,M)\,(1.733 \times 10^{-4}\,M) = 3.00 \times 10^{-8}$

(c) $Ca(OH)_2(s) \rightleftharpoons Ca^{2+}(aq) + 2OH^-(aq)$; $K_{sp} = [Ca^{2+}][OH^-]^2$

$[Ca^{2+}] = x$, $[OH^-] = 2x$; $K_{sp} = 6.5 \times 10^{-6} = (x)(2x)^2$

$6.5 \times 10^{-6} = 4x^3$; $x = [Ca^{2+}] = 0.01176 = 0.012\ M$; $[OH^-] = 0.02351 = 0.024\ M$

$pH = 14 - pOH = 14 - 1.629 = 12.37$

17.56 $PbI_2(s) \rightleftharpoons Pb^{2+}(aq) + 2I^-(aq)$; $K_{sp} = [Pb^{2+}][I^-]^2$

$$[Pb^{2+}] = \frac{0.54 \text{ g PbI}_2}{1.00 \text{ L soln}} \times \frac{1 \text{ mol PbI}_2}{461.0 \text{ g PbI}_2} = 1.17 \times 10^{-3} = 1.2 \times 10^{-3}\ M$$

$[I^-] = 2[Pb^{2+}]$; $K_{sp} = [Pb^{2+}](2[Pb^{2+}])^2 = 4[Pb^{2+}]^3 = 4(1.17 \times 10^{-3})^3 = 6.4 \times 10^{-9}$

17.58 $LaF_3(s) \rightleftharpoons La^{3+}(aq) + 3F^-(aq)$; $K_{sp} = [La^{3+}][F^-]^3$

(a)   molar solubility $= x = [La^{3+}]$; $[F^-] = 3x$

$K_{sp} = 2 \times 10^{-19} = (x)(3x)^3$; $2 \times 10^{-19} = 27\,x^4$; $x = (7.41 \times 10^{-21})^{1/4}$, $x = 9.28 \times 10^{-6}$

$= 9 \times 10^{-6}\ M\ La^{3+}$

$$\frac{9.28 \times 10^{-6} \text{ mol LaF}_3}{1\,L} \times \frac{195.9 \text{ g LaF}_3}{1 \text{ mol}} = 1.82 \times 10^{-3} = 2 \times 10^{-3} \text{ g LaF}_3/L$$

(b) molar solubility = x = [La^{3+}]

There are two sources of F$^-$: KF(0.010 M) and LaF$_3$ (3x M)

K_{sp} = (x)(0.010 + 3x)^3; assume x is small compared to 0.010 M.

2×10^{-19} = (0.010)^3 x; x = 2×10^{-19}/1.0×10^{-6} = 2×10^{-13} M La^{3+}

$\dfrac{2 \times 10^{-13} \text{ mol LaF}_3}{1 \text{ L}} \times \dfrac{195.9 \text{ g LaF}_3}{1 \text{ mol}}$ = 3.92 $\times 10^{-11}$ = 4 $\times 10^{-11}$ g LaF$_3$/L

(c) molar solubility = x, [F$^-$] = 3x, [La^{3+}] = 0.050 M + x

K_{sp} = (0.050 + x)(3x)^3; assume x is small compared to 0.050 M.

2×10^{-19} = (0.050)(27 x^3) = 1.35 x^3; x = (1.48 $\times 10^{-19}$)$^{1/3}$ = 5.29 $\times 10^{-7}$ = 5 $\times 10^{-7}$ M

$\dfrac{5.29 \times 10^{-7} \text{ mol LaF}_3}{1 \text{ L}} \times \dfrac{195.9 \text{ g LaF}_3}{1 \text{ mol}}$ = 1.04 $\times 10^{-4}$ = 1 $\times 10^{-4}$ g LaF$_3$/L

17.60   As KI is added, [I$^-$] increases, K_{sp} is exceeded, and additional PbI$_2$ precipitates until equilibrium is reestablished. At the new equilibrium position:

(a) Increase. The additional I$^-$ from KI decreases the solubility of PbI$_2$.

(b) Decrease. After KI is added, additional PbI$_2$(s) precipitates.

(c) Increase. A small amount of the additional I$^-$ precipitates as PbI$_2$(s), but most of it increases the concentration of I$^-$ ions in solution.

17.62   Ni(OH)$_2$(s) $\rightleftharpoons$ Ni^{2+}(aq) + 2OH$^-$(aq); K_{sp} = 6.0 $\times 10^{-16}$

Because the [OH$^-$] is set by the pH of the solution, the solubility of Ni(OH)$_2$ is just [Ni^{2+}].

(a) pH = 8.0, pOH = 14 – pH = 6.0, [OH$^-$] = 10^{-pOH} = 1 $\times 10^{-6}$ M

K_{sp} = 6.0×10^{-16} = [Ni^{2+}](1.0×10^{-6})^2; [Ni^{2+}] = $\dfrac{6.0 \times 10^{-16}}{1.0 \times 10^{-12}}$ = 6.0×10^{-4} = 6×10^{-4} M

(b) pH = 10.0, pOH = 4.0, [OH$^-$] = 1.0 $\times 10^{-4}$ = 1 $\times 10^{-4}$ M

K_{sp} = 6.0×10^{-16} = [Ni^{2+}][1.0×10^{-4}]^2; [Ni^{2+}] = $\dfrac{6.0 \times 10^{-16}}{1.0 \times 10^{-8}}$ = 6.0×10^{-8} = 6×10^{-8} M

(c) pH = 12.0, pOH = 2.0, [OH$^-$] = 1.0 $\times 10^{-2}$ = 1 $\times 10^{-2}$ M

K_{sp} = 6.0×10^{-16} = [Ni^{2+}][1.0×10^{-2}]^2; [Ni^{2+}] = $\dfrac{6.0 \times 10^{-16}}{1.0 \times 10^{-4}}$ = 6.0×10^{-12} = 6×10^{-12} M

17.64   If the anion in the slightly soluble salt is the conjugate base of a strong acid, there will be no reaction.

(a) $MnS(s) + 2H^+(aq) \rightarrow H_2S(aq) + Mn^{2+}(aq)$

(b) $PbF_2(s) + 2H^+(aq) \rightarrow 2HF(aq) + Pb^{2+}(aq)$

(c) $AuCl_3(s) + H^+(aq) \rightarrow$ no reaction

(d) $Hg_2C_2O_4(s) + 2H^+(aq) \rightarrow H_2C_2O_4(aq) + Hg_2^{2+}(aq)$

(e) $CuBr(s) + H^+(aq) \rightarrow$ no reaction

17.66 $NiC_2O_4(s) \rightleftharpoons Ni^{2+}(aq) + C_2O_4^{2-}(aq); \quad K_{sp} = [Ni^{2+}][C_2O_4^{2-}] = 4 \times 10^{-10}$

When the salt has just dissolved, $[C_2O_4^{2-}]$ will be 0.020 M. Thus, $[Ni^{2+}]$ must be less than $4 \times 10^{-10} / 0.020 = 2 \times 10^{-8}$ M. To achieve this low $[Ni^{2+}]$, we must complex the Ni^{2+} ion with NH_3: $Ni^{2+}(aq) + 6NH_3(aq) \rightleftharpoons Ni(NH_3)_6^{2+}(aq)$. Essentially all Ni(II) is in the form of the complex, so $[Ni(NH_3)_6^{2+}] = 0.020$. Find K_f for $Ni(NH_3)_6^{2+}$ in Table 17.1.

$$K_f = \frac{[Ni(NH_3)_6^{2+}]}{[Ni^{2+}][NH_3]^6} = \frac{(0.020)}{(2 \times 10^{-8})[NH_3]^6} = 1.2 \times 10^9; [NH_3]^6 = 8.33 \times 10^{-4};$$

$$[NH_3] = 0.307 = 0.3 \ M$$

17.68   According Appendix D.3, K_{sp} for $Ag_2S(s)$ is of the type

$$Ag_2S(s) + H_2O(l) \rightleftharpoons 2Ag^+(aq) + HS^-(aq) + OH^-(aq) \qquad\qquad K_{sp}$$

$$HS^-(aq) + H^+(aq) \rightleftharpoons H_2S(aq) \qquad\qquad 1/K_{a1}$$

$$2[Ag^+(aq) + 2Cl^-(aq) \rightleftharpoons AgCl_2^-(aq)] \qquad\qquad K_f^2$$

$$Ag_2S(s) + H_2O(l) + H^+(aq) + 4Cl^-(aq) \rightleftharpoons 2AgCl_2^-(aq) + H_2S(aq)$$

Add $H^+(aq)$ to each side to obtain the overall reacation

$$Ag_2S(s) + 2H^+(aq) + 4Cl^-(aq) \rightleftharpoons 2AgCl_2^-(aq) + H_2S(aq)$$

$$K = \frac{K_{sp} \times K_f^2}{K_{a1}} = \frac{(6 \times 10^{-51})(1.1 \times 10^5)^2}{(9.5 \times 10^{-8})} = 7.64 \times 10^{-34} = 8 \times 10^{-34}$$

Precipitation and Separation of Ions (Section 17.6)

17.70   (a) $Co(OH)_2(s) \rightleftharpoons Co^{2+}(aq) + 2OH^-(aq); K_{sp} = [Co^{2+}][OH^-]^2 = 1.3 \times 10^{-15}$

pH = 8.5; pOH = 14 − 8.5 = 5.5; $[OH^-] = 10^{-5.5} = 3.16 \times 10^{-6} = 3 \times 10^{-6} \ M$

$Q = (0.020)(3.16 \times 10^{-6})^2 = 2 \times 10^{-13}; Q > K_{sp}, Co(OH)_2$ will precipitate.

(b) $AgIO_3(s) \rightleftharpoons Ag^+(aq) + IO_3^-(aq);\ K_{sp} = [Ag+][IO_3^-] = 3.1 \times 10^{-8}$

$$[Ag^+] = \frac{0.010\ M\ Ag^+ \times 0.020\ L}{0.030\ L} = 6.667 \times 10^{-3} = 6.7 \times 10^{-3}\ M$$

$$[IO_3^-] = \frac{0.015\ M\ IO_3^- \times 0.010\ L}{0.030\ L} = 5.000 \times 10^{-3} = 5.0 \times 10^{-3}\ M$$

$$Q = (6.667 \times 10^{-3})(5.00 \times 10^{-3}) = 3.3 \times 10^{-5};\ Q > K_{sp},\ AgIO_3\ \text{will precipitate.}$$

17.72 $PbI_2(s) \rightleftharpoons Pb^{2+}(aq) + 2I^-(aq);\ K_{sp} = [Pb^{2+}][I^-]^2 = 8.49 \times 10^{-9}$

(This K_{sp} value is taken from *CRC Handbook of Chemistry and Physics*, 74th edition.)

$$[Pb^{2+}] = \frac{0.10\ M \times 0.2\ mL}{10.2\ mL} = 1.96 \times 10^{-3} = 2 \times 10^{-3}\ M;$$

$$[I^-] = \left(\frac{8.49 \times 10^{-9}}{1.96 \times 10^{-3}\ M}\right)^{1/2} = 2.08 \times 10^{-3} = 2 \times 10^{-3}\ M$$

$$\frac{2.08 \times 10^{-3}\ mol\ I^-}{1\ L} \times \frac{126.90\ g\ I^-}{1\ mol\ I^-} \times 0.0102\ L = 2.69 \times 10^{-3}\ g\ I^- = 3 \times 10^{-3}\ g\ I^-$$

17.74 (a) Precipitation will begin when $Q = K_{sp}$.

$BaSO_4: K_{sp} = [Ba^{2+}][SO_4^{2-}] = 1.1 \times 10^{-10}$

$1.1 \times 10^{-10} = (0.010)[SO_4^{2-}];\ [SO_4^{2-}] = 1.1 \times 10^{-8}\ M$

$SrSO_4: K_{sp} = [Sr^{2+}][SO_4^{2-}] = 3.2 \times 10^{-7}$

$3.2 \times 10^{-7} = (0.010)[SO_4^{2-}];\ [SO_4^{2-}] = 3.2 \times 10^{-5}\ M$

The $[SO_4^{2-}]$ necessary to begin precipitation is the smaller of the two values, $1.1 \times 10^{-8}\ M\ SO_4^{2-}$.

(b) Ba^{2+} precipitates first, because it requires the smaller $[SO_4^{2-}]$.

(c) Sr^{2+} will begin to precipitate when $[SO_4^{2-}]$ in solution (not bound in $BaSO_4$) reaches $3.2 \times 10^{-5}\ M$.

17.76 It is not appropriate to compare K_{sp} values directly, because the stoichiometries of the the two precipitates are different.

(a) Precipitation will begin when $Q = K_{sp}$.

$CaSO_4: K_{sp} = [Ca^{2+}][SO_4^{2-}] = 2.4 \times 10^{-5}$

$2.4 \times 10^{-5} = (0.20)[SO_4^{2-}];\ [SO_4^{2-}] = 1.2 \times 10^{-4}\ M$

$Ag_2SO_4: K_{sp} = [Ag^+]^2[SO_4^{2-}] = 1.5 \times 10^{-5}$

$1.5 \times 10^{-5} = (0.30)^2[SO_4^{2-}];\ [SO_4^{2-}] = 1.7 \times 10^{-4}\ M$

$CaSO_4$ requires the smaller $[SO_4^{2-}]$ for precipitation and it will precipitate first.

(b)   The $[SO_4^{2-}]$ necessary to begin precipitation is the smaller of the two values, $1.2 \times 10^{-4}\ M\ SO_4^{2-}$.

$$1.2 \times 10^{-4}\ M = \frac{1.0\ M\ SO_4^{2-} \times x\ L}{(0.010 + x\ L)}; x = (0.010)1.2 \times 10^{-4} = 1.2 \times 10^{-6}\ L.$$

We assume x is small compared to 0.010 L. The required volume is then 1.2×10^{-6} L or 0.0012 mL or 1.2 μL. If one drop is approximately 0.2 mL, precipitation will begin as the first drop of 1.0 M Na_2SO_4 solution is added.

Qualitative Analysis for Metallic Elements (Section 17.7)

17.78   Initial solubility in water rules out CdS and HgO. Formation of a precipitate on addition of HCl indicates the presence of $Pb(NO_3)_2$ (formation of $PbCl_2$). Formation of a precipitate on addition of H_2S at pH 1 probably indicates $Cd(NO_3)_2$ (formation of CdS). (This test can be misleading because enough Pb^{2+} can remain in solution after filtering $PbCl_2$ to lead to visible precipitation of PbS.) Absence of a precipitate on addition of H_2S at pH 8 indicates that $ZnSO_4$ is not present. The yellow flame test indicates presence of Na^+. In summary, $Pb(NO_3)_2$ and Na_2SO_4 are definitely present, $Cd(NO_3)_2$ is probably present, and CdS, HgO, and $ZnSO_4$ are definitely absent.

17.80   (a)   Make the solution slightly acidic and saturate with H_2S; CdS will precipitate, but Na^+ remains in solution.

(b)   Make the solution acidic and saturate with H_2S; CuS will precipitate, but Mg^{2+} remains in solution.

(c)   Add HCl; $PbCl_2$ precipitates. (It is best to carry out the reaction in an ice-water bath to reduce the solubility of $PbCl_2$.)

(d)   Add dilute HCl; AgCl precipitates, but Hg^{2+} remains in solution.

17.82   The addition of $(NH_4)_2HPO_4$ could result in precipitation of salts from metal ions of the other groups. The $(NH_4)_2HPO_4$ will render the solution basic, so metal hydroxides as well as insoluble phosphates could form. It is essential to separate the metal ions of a group from other metal ions before carrying out the specific tests for that group.

Additional Exercises

17.84 $H_2CO_3(aq) \quad \rightleftharpoons \quad H^+(aq) \quad + \quad HCO_3^-(aq) \qquad K_{a1} = 4.3 \times 10^{-7} \qquad pK_{a1} = 6.37$

$HCO_3^-(aq \quad \rightleftharpoons \quad H^+(aq) \quad + \quad CO_3^{2-}(aq) \qquad K_{a2} = 5.6 \times 10^{-11} \qquad pK_{a2} = 10.25$

Use the two equilibrium constant expressions and the total carbonate concentration to solve for the three concentrations.

$[H^+] = 10^{-5.60} = 2.512 \times 10^{-6} = 2.5 \times 10^{-6}\ M.$

$$K_{a1} = \frac{[H^+][HCO_3^-]}{[H_2CO_3]}, \quad [H_2CO_3] = \frac{[H^+][HCO_3^-]}{K_{a1}}; \quad K_{a2} = \frac{[H^+][CO_3^{2-}]}{[HCO_3^-]}, \quad [CO_3^{2-}] = \frac{K_{a2}[HCO_3^-]}{[H^+]}$$

$[H_2CO_3] + [HCO_3^-] + [CO_3^{2-}] = 1.0 \times 10^{-5}$

$$\frac{[H^+][HCO_3^-]}{K_{a1}} + [HCO_3^-] + \frac{K_{a2}[HCO_3^-]}{[H^+]} = 1.0 \times 10^{-5}$$

Multiply by $K_{a1}[H^+]$.

$$[H^+]^2[HCO_3^-] + K_{a1}[H^+][HCO_3^-] + K_{a1}K_{a2}[HCO_3^-] = (1.0 \times 10^{-5})K_{a1}[H^+]$$

$$[HCO_3^-]([H^+]^2 + K_{a1}[H^+] + K_{a1}K_{a2}) = (1.0 \times 10^{-5})K_{a1}[H^+]$$

$$[HCO_3^-] = \frac{(1.0 \times 10^{-5})K_{a1}[H^+]}{[H^+]^2 + K_{a1}[H^+] + K_{a1}K_{a2}} = \frac{(1.0 \times 10^{-5})(4.3 \times 10^{-7})(2.5 \times 10^{-6})}{(2.5 \times 10^{-6})^2 + (4.3 \times 10^{-7})(2.5 \times 10^{-6}) + (4.3 \times 10^{-7})(5.6 \times 10^{-11})}$$

$$[HCO_3^-] = 1.468 \times 10^{-6} = 1.5 \times 10^{-6}\ M$$

$$[H_2CO_3] = \frac{[H^+][HCO_3^-]}{K_{a1}} = \frac{2.5 \times 10^{-6}(1.5 \times 10^{-6})}{4.3 \times 10^{-7}} = 8.7 \times 10^{-6}\ M$$

$$[CO_3^{2-}] = \frac{K_{a2}[HCO_3^-]}{[H^+]} = \frac{5.6 \times 10^{-11}(1.5 \times 10^{-6})}{2.5 \times 10^{-6}} = 3.4 \times 10^{-11}\ M$$

Check. First, pH of the raindrop (5.6) is less than pK_{a1} (6.37). We expect $[H_2CO_3]$ to be greater than $[HCO_3^-]$, and it is. Second, the calculated total carbon species concentration is

$$[H_2CO_3] + [HCO_3^-] + [CO_3^{2-}] = 8.7 \times 10^{-6} + 1.5 \times 10^{-6} + 3.4 \times 10^{-11} = 1.0 \times 10^{-5}\ M.$$

The calculated results are self-consistent.

17.85   The equilibrium of interest is

$$HC_5H_3O_3(aq) \rightleftharpoons H^+(aq) + C_5H_3O_3^-(aq);\ K_a = 6.76 \times 10^{-4} = \frac{[H^+][C_5H_3O_3^-]}{[HC_5H_3O_3]}$$

Begin by calculating $[HC_5H_3O_3]$ and $[C_5H_3O_3^-]$ for each case.

(a) $$\frac{25.0\ g\ HC_5H_3O_3}{0.250\ L\ soln} \times \frac{1\ mol\ HC_5H_3O_3}{112.1\ g\ HC_5H_3O_3} = 0.8921 = 0.892\ M\ HC_5H_3O_3$$

$$\frac{30.0\ g\ NaC_5H_3O_3}{0.250\ L\ soln} \times \frac{1\ mol\ NaC_5H_3O_3}{134.1\ g\ NaC_5H_3O_3} = 0.8949 = 0.895\ M\ C_5H_3O_3^-$$

$$[H^+] = \frac{K_a[HC_5H_3O_3]}{[C_5H_3O_3^-]} = \frac{6.76 \times 10^{-4}(0.8921 - x)}{(0.8949 + x)} \approx \frac{6.76 \times 10^{-4}(0.8921)}{(0.8949)}$$

$$[H^+] = 6.74 \times 10^{-4}\ M,\ pH = 3.171$$

(b)   For dilution, $M_1V_1 = M_2V_2$

$$[HC_5H_3O_3] = \frac{0.250\ M \times 30.0\ mL}{125\ mL} = 0.0600\ M$$

$$[C_5H_3O_3^-] = \frac{0.220\ M \times 20.0\ mL}{125\ mL} = 0.0352\ M$$

$$[H^+] \approx \frac{6.76 \times 10^{-4}(0.0600)}{0.0352} = 1.15 \times 10^{-3}\ M,\ pH = 2.938$$

(yes, $[H^+]$ is < 5% of 0.0352 M)

(c) $0.0850\ M \times 0.500\ L = 0.0425\ mol\ HC_5H_3O_3$

$1.65\ M \times 0.0500\ L = 0.0825\ mol\ NaOH$

	$HC_5H_3O_3(aq)$	$+$	$NaOH(aq)$	$\rightarrow$	$NaC_5H_3O_3(aq) + H_2O(l)$
initial	0.0425 mol		0.0825 mol		
reaction	−0.0425 mol		−0.0425 mol		+0.0425 mol
after	0 mol		0.0400 mol		0.0425 mol

The strong base NaOH dominates the pH; the contribution of $C_5H_3O_3^-$ is negligible. This combination would be "after the equivalence point" of a titration. The total volume is 0.550 L.

$$[OH^-] = \frac{0.0400\ mol}{0.550\ L} = 0.0727\ M;\ pOH = 1.138,\ pH = 12.862$$

17.86 $K_a = \dfrac{[H^+][In^-]}{[HIn]}$; at pH = 4.68, $[HIn] = [In^-]$; $[H^+] = K_a$; pH = pK_a = 4.68

17.87 (a) $HA(aq) + B(aq) \rightleftharpoons HB^+(aq) + A^-(aq)$ $K_{eq} = \dfrac{[HB^+][A^-]}{[HA][B]}$

(b) Note that the solution is slightly basic because B is a stronger base than HA is an acid. (Or, equivalently, that A^- is a stronger base than HB^+ is an acid.) Thus, a little of the A^- is used up in reaction: $A^-(aq) + H_2O(l) \rightleftharpoons HA(aq) + OH^-(aq)$. Because pH is not very far from neutral, it is reasonable to assume that the reaction in part (a) has gone far to the right, and that $[A^-] \approx [HB^+]$ and $[HA] \approx [B]$. Then,

$$K_a = \frac{[A^-][H^+]}{[HA]} = 8.0 \times 10^{-5};\ when\ pH = 9.2,\ [H^+] = 6.31 \times 10^{-10} = 6 \times 10^{-10}\ M$$

$$\frac{[A^-]}{[HA]} = 8.0 \times 10^{-5} / 6.31 \times 10^{-10} = 1.268 \times 10^5 = 1 \times 10^5$$

From the earlier assumptions, $\dfrac{[A^-]}{[HA]} = \dfrac{[HB^+]}{[B]}$, so $K_{eq} \approx \dfrac{[A^-]^2}{[HA]^2} = 1.608 \times 10^{10} = 2 \times 10^{10}$

(c) K_b for the reaction $B(aq) + H_2O(l) \rightleftharpoons BH^+(aq) + OH^-(aq)$ can be calculated by noting that the equilibrium constant for the reaction in part (a) can be written as $K = K_a\ (HA) \times K_b\ (B)\ /\ K_w$. (You should prove this to yourself.) Then,

$$K_b\ (B) = \frac{K \times K_w}{K_a\ (HA)} = \frac{(1.608 \times 10^{10})(1.0 \times 10^{-14})}{8.0 \times 10^{-5}} = 2.010 = 2$$

K_b (B) is larger than K_a (HA), as it must be if the solution is basic.

17.88 (a) $K_a = \dfrac{[H^+][HCOO^-]}{[HCOOH]}$; $[H^+] = \dfrac{K_a[HCOOH]}{[HCOO^-]}$

Buffer A : $[HCOOH] = [HCOO^-] = \dfrac{1.00\ mol}{1.00\ L} = 1.00\ M$

$$[H^+] = \frac{1.8 \times 10^{-4} \ (1.00 \ M)}{(1.00 \ M)} = 1.8 \times 10^{-4} \ M, \ pH = 3.74$$

Buffer B: $[HCOOH] = [HCOO^-] = \dfrac{0.010 \ mol}{1.00 \ L} = 0.010 \ M$

$$[H^+] = \frac{1.8 \times 10^{-4} \ (0.010 \ M)}{(0.010 \ M)} = 1.8 \times 10^{-4} \ M, \ pH = 3.74$$

The pH values of the two buffers are equal because they both contain HCOOH and HCOONa and the $[HCOOH] / [HCOO^-]$ *ratio* is the same in both solutions.

(b)   Buffer A has the greater capacity because it contains the greater absolute concentrations of HCOOH and $HCOO^-$.

(c)   Buffer A:

$HCOO^-$	+	HCl	$\rightarrow$	$HCOOH$	+ Cl^-
1.00 mol		0.001 mol		1.00 mol	
0.999 mol		0		1.001 mol	

$$[H^+] = \frac{1.8 \times 10^{-4} \ (1.001)}{(0.999)} = 1.8 \times 10^{-4} \ M, \ pH = 3.74$$

(In a buffer calculation, volumes cancel and we can substitute moles directly into the K_a expression.)

Buffer B:

$HCOO^-$	+	HCl	$\rightarrow$	$HCOOH$	+ Cl^-
0.010 mol		0.001 mol		0.010 mol	
0.009 mol		0		0.011 mol	

$$[H^+] = \frac{1.8 \times 10^{-4} \ (0.011)}{(0.009)} = 2.2 \times 10^{-4} \ M, \ pH = 3.66$$

(d)   Buffer A: $1.00 \ M \ HCl \times 0.010 \ L = 0.010 \ mol \ H^+$ added

mol $HCOOH = 1.00 + 0.010 = 1.01 \ mol$

mol $HCOO^- = 1.00 - 0.010 = 0.99 \ mol$

$$[H^+] = \frac{1.8 \times 10^{-4} \ (1.01)}{(0.99)} = 1.8 \times 10^{-4} \ M, \ pH = 3.74$$

Buffer B: mol $HCOOH = 0.010 + 0.010 = 0.020 \ mol = 0.020 \ M$

mol $HCOO^- = 0.010 - 0.010 = 0.000 \ mol$

The solution is no longer a buffer; the only source of $HCOO^-$ is the dissociation of HCOOH. Adding 10 mL of 1.00 M HCl exceeds the buffer capacity of this buffer.

$$K_a = \frac{[H^+][COO^-]}{[HCOOH]} = \frac{x^2}{(0.020 - x) \ M}$$

The extent of ionization is greater than 5%; from the quadratic formula, $x = [H^+] = 1.8 \times 10^{-3}, \ pH = 2.74$.

17.90   (a)   For a monoprotic acid (one H^+ per mole of acid), at the equivalence point moles OH^- added = moles H^+ originally present

$$M_B \times V_B = \text{g acid/molar mass}$$

$$MM = \frac{\text{g acid}}{M_B \times V_B} = \frac{0.2140 \text{ g}}{0.0950 \, M \times 0.0274 \text{ L}} = 82.21 = 82.2 \text{ g/mol}$$

(b)   initial mol HA $= \dfrac{0.2140 \text{g}}{82.21 \text{ g/mol}} = 2.603 \times 10^{-3} = 2.60 \times 10^{-3}$ mol HA

mol OH^- added to pH 6.50 $= 0.0950 \, M \times 0.0150$ L $= 1.425 \times 10^{-3}$

$$= 1.43 \times 10^{-3} \text{ mol } OH^-$$

	HA(aq)	+	NaOH(aq)	→	$NaA(aq) + H_2O$
before rx	2.603×10^{-3} mol		1.425×10^{-3} mol		0
change	-1.425×10^{-3} mol		-1.425×10^{-3} mol		$+1.425 \times 10^{-3}$ mol
after rx	1.178×10^{-3} mol		0		1.425×10^{-3} mol

$$[HA] = \frac{1.178 \times 10^{-3} \text{ mol}}{0.0400 \text{ L}} = 0.02945 = 0.0295 \, M$$

$$[A^-] = \frac{1.425 \times 10^{-3} \text{ mol}}{0.0400 \text{ L}} = 0.03563 = 0.0356 \, M; [H^+] = 10^{-6.50} = 3.162 \times 10^{-7}$$

$$= 3.2 \times 10^{-7} \, M$$

The mixture after reaction (a buffer) can be described by the acid dissociation equilibrium.

	HA(aq)	⇌	H^+(aq)	+	A^-(aq)
initial	0.0295 M		0		0.0356 M
equil.	$(0.0295 - 3.2 \times 10^{-7} \, M)$		$3.2 \times 10^{-7} \, M$		$(0.0356 + 3.2 \times 10^{-7}) \, M$

$$K_a = \frac{[H^+][A^-]}{[HA]} \approx \frac{(3.162 \times 10^{-7})\,(0.03563)}{(0.02945)} = 3.8 \times 10^{-7}$$

(Although we have carried 3 figures through the calculation to avoid rounding errors, the data dictate an answer with 2 sig figs.)

17.92   At the equivalence point of a titration, moles strong base added equals moles weak acid initially present. $M_B \times V_B$ = mol base added = mol acid initial.

At the half-way point, the volume of base is one-half of the volume required to reach the equivalence point, and the moles base delivered equals one-half of the moles acid initially present. This means that one-half of the weak acid HA is converted to the conjugate base A^-. If exactly half of the acid reacts, mol HA = mol A^- and [HA] = [A^-] at the half-way point.

From Equation 17.9, $pH = pK_a + \log \dfrac{[\text{conj. base}]}{[\text{conj. acid}]} = pK_a + \log \dfrac{[A^-]}{[HA^-]}$.

If $[A^-]/[HA] = 1$, $\log(1) = 0$ and $pH = pK_a$ of the weak acid being titrated.

17.93 If 50.0 mL base is required to reach the equivalence point, addition of 25.0 mL of base is half-way to the equivalence point. At this point, [A$^-$] = [HA] and [A$^-$]/[HA] = 1. The pK$_a$ of the weak acid being titrated is equal to the pH of the solution, 3.62.

17.94 Assume that H$_3$PO$_4$ will react with NaOH in a stepwise fashion: (This is not unreasonable, because the three K$_a$ values for H$_3$PO$_4$ are significantly different.)

$$H_3PO_4(aq) \quad + \quad NaOH(aq) \quad \rightarrow \quad H_2PO_4^-(aq) + Na^+(aq) + H_2O(l)$$

before   0.20 mol   0.30 mol   0 mol

after   0 mol   0.10 mol   0.20 mol

$$H_2PO_4^-(aq) \quad + \quad NaOH(aq) \quad \rightarrow \quad HPO_4^-(aq) + Na^+(aq) + H_2O(l)$$

before   0.20 mol   0.10 mol   0.25 mol

after   0.10 mol   0   0.35 mol

Thus, after all NaOH has reacted, the resulting 1.00 L solution is a buffer containing 0.10 mol H$_2$PO$_4^-$ and 0.35 mol HPO$_4^{2-}$. H$_2$PO$_4^-$(aq) $\rightleftharpoons$ H$^+$(aq) + HPO$_4^{2-}$(aq)

$$K_a = 6.2 \times 10^{-8} = \frac{[HPO_4^{2-}][H^+]}{[H_2PO_4^-]}; [H^+] = \frac{6.2 \times 10^{-8}\,(0.10\,M)}{0.35\,M} = 1.77 \times 10^{-8} = 1.8 \times 10^{-8}\,M;$$

$$pH = 7.75$$

17.95 The pH of a buffer system is centered around pK$_a$ for the conjugate acid component. For a diprotic acid, two conjugate acid/conjugate base pairs are possible.

$$H_2X(aq) \rightleftharpoons H^+(aq) + HX^-(aq); \quad K_{a1} = 2 \times 10^{-2}; \quad pK_{a1} = 1.70$$

$$HX^-(aq) \rightleftharpoons H^+(aq) + X^{2-}(aq); \quad K_{a2} = 5.0 \times 10^{-7}; \quad pK_{a2} = 6.30$$

Clearly, HX$^-$ / X^{2-} is the more appropriate combination for preparing a buffer with pH = 6.50. The [H$^+$] in this buffer = $10^{-6.50}$ = 3.16 × 10^{-7} = 3.2 × 10^{-7} M. Using the K$_{a2}$ expression to calculate the [X^{2-}] / [HX$^-$] ratio:

$$K_{a2} = \frac{[H^+][X^{2-}]}{[HX^-]}; \frac{K_{a2}}{[H^+]} = \frac{[X^{2-}]}{[HX^-]} = \frac{5.0 \times 10^{-7}}{3.16 \times 10^{-7}} = 1.58 = 1.6$$

Because X^{2-} and HX$^-$ are present in the same solution, the ratio of concentrations is also a ratio of moles.

$$\frac{[X^{2-}]}{[HX^-]} = \left(\frac{mol\ X^{2-}/L\ soln}{mol\ HX^-/L\ soln}\right) = \frac{mol\ X^{2-}}{mol\ HX^-} = 1.58;\ mol\ X^{2-} = (1.58)\ mol\ HX^-$$

In the 1.0 L of 1.0 M H$_2$X, there is 1.0 mol of material containing X^{2-}.

Thus, mol HX$^-$ + 1.58 (mol HX$^-$) = 1.0 mol. 2.58 (mol HX$^-$) = 1.0;

mol HX$^-$ = 1.0 / 2.58 = 0.39 mol HX$^-$; mol X^{2-} = 1.0 – 0.39 = 0.61 mol X^{2-}.

Thus, enough 1.0 M NaOH must be added to produce 0.39 mol HX$^-$ and 0.61 mol X^{2-}.

Considering the neutralization in a stepwise fashion (see discussion of titrations of polyprotic acids in Section 17.3).

$$H_2X(aq) \quad + \quad NaOH(aq) \quad \rightarrow \quad HX^-(aq) + H_2O(l)$$

before	1.0 mol	1 mol	0
after	0	0	1.0 mol

$$HX^-(aq) \quad + \quad NaOH(aq) \quad \rightarrow \quad X^{2-}(aq) + H_2O(l)$$

before	1.0		0.61
change	−0.61	−0.61	+0.61
after	0.39	0	0.61

Starting with 1.0 mol of H_2X, 1.0 mol of NaOH is added to completely convert it to 1.0 mol of HX^-. Of that 1.0 mol of HX^-, 0.61 mol must be converted to 0.61 mol X^{2-}. The total moles of NaOH added is $(1.00 + 0.61) = 1.61$ mol NaOH.

$$L\ NaOH = \frac{mol\ NaOH}{M\ NaOH} = \frac{1.61\ mol}{1.0\ M} = 1.6\ L\ of\ 1.0\ M\ NaOH .$$

17.96 $CH_3CH(OH)COO^-$ will be formed by reaction $CH_3CH(OH)COOH$ with NaOH.

$0.1000\ M \times 0.02500\ L = 2.500 \times 10^{-3}$ mol $CH_3CH(OH)COOH$; b = mol NaOH needed

$$CH_3CH(OH)COOH \quad + \quad NaOH \quad \rightarrow \quad CH_3CH(OH)COO^- + H_2O + Na^+$$

initial	2.500×10^{-3} mol	b mol	
rx	−b mol	−b mol	+b mol
after rx	$(2.500 \times 10^3 - b)$ mol	0	b mol

$$K_a = \frac{[H^+][CH_3CH(OH)COO^-]}{[CH_3CH(OH)COOH]}; K_a = 1.4 \times 10^{-4}; [H^+] = 10^{-pH} = 10^{-3.75} = 1.778 \times 10^{-4} = 1.8 \times 10^{-4}\ M$$

Because solution volume is the same for reaction $CH_3CH(OH)COOH$ and $CH_3CH(OH)COO^-$, we can use moles in the equation for $[H^+]$.

$$K_a = 1.4 \times 10^{-4} = \frac{1.778 \times 10^{-4}\ (b)}{(2.500 \times 10^{-3} - b)}; 0.7874\ (2.500 \times 10^{-3} - b) = b, 1.969 \times 10^{-3} = 1.7874\ b,$$

$b = 1.10 \times 10^{-3} = 1.1 \times 10^{-3}$ mol OH^-

(The precision of K_a dictates that the result has 2 sig figs.)

Substituting this result into the K_a expression gives $[H^+] = 1.8 \times 10^{-4}$. This checks and confirms our result. Calculate volume NaOH required from $M = mol/L$.

$$1.10 \times 10^{-3}\ mol\ OH^- \times \frac{1\ L}{1.000\ mol} \times \frac{1\ \mu L}{1 \times 10^{-6}\ L} = 1.1 \times 10^3\ \mu L\ (1.1\ mL)$$

17.98 (a) CdS: 8.0×10^{-28}; CuS: 6×10^{-37}. CdS has greater molar solubility.

(b) $PbCO_3$: 7.4×10^{-14}; $BaCrO_4$: 2.1×10^{-10}. $BaCrO_4$ has greater molar solubility.

(c) Because the stoichiometry of the two complexes is not the same, K_{sp} values can not be compared directly; molar solubilities must be calculated from K_{sp} values.

$Ni(OH)_2$: $K_{sp} = 6.0 \times 10^{-16} = [Ni^{2+}][OH^-]^2$; $[Ni^{2+}] = x$, $[OH^-] = 2x$

$6.0 \times 10^{-16} = (x)(2x)^2 = 4x^3$; $x = 5.3 \times 10^{-6}\ M\ Ni^{2+}$

Note that [OH⁻] from the autoionization of water is less than 1% of [OH⁻] from $Ni(OH)_2$ and can be neglected.

$NiCO_3$: $K_{sp} = 1.3 \times 10^{-7} = [Ni^{2+}][CO_3^{2-}]$; $[Ni^{2+}] = [CO_3^{2-}] = x$

$1.3 \times 10^{-7} = x^2$; $x = 3.6 \times 10^{-4} \, M \, Ni^{2+}$

$NiCO_3$ has greater molar solubility than $Ni(OH)_2$, but the values are much closer than expected from inspection of K_{sp} values alone.

(d) Again, molar solubilities must be calculated for comparison.

Ag_2SO_4: $K_{sp} = 1.5 \times 10^{-5} = [Ag^+]^2[SO_4^{2-}]$; $[SO_4^{2-}] = x$, $[Ag^+] = 2x$

$1.5 \times 10^{-5} = (2x)^2(x) = 4x^3$; $x = 1.6 \times 10^{-2} \, M \, SO_4^{2-}$

AgI: $K_{sp} = 8.3 \times 10^{-17} = [Ag^+][I^-]$; $[Ag^+] = [I^-] = x$

$8.3 \times 10^{-17} = x^2$; $x = 9.1 \times 10^{-9} \, M \, Ag^+$

Ag_2SO_4 has greater molar solubility than AgI.

17.99 (a) $K_{sp} = 4.5 \times 10^{-9} = [Ca^{2+}][CO_3^{2-}]$; $s = [Ca^{2+}] = [CO_3^{2-}]$

$s^2 = 4.5 \times 10^{-9}$, $s = 6.708 \times 10^{-5} = 6.7 \times 10^{-5}$

(b)
$$CaCO_3(s) \rightleftharpoons Ca^{2+}(aq) + CO_3^{2-}(aq) \qquad K_{sp}$$
$$CO_3^{2-}(aq) + H_2O(l) \rightleftharpoons HCO_3^-(aq) + OH^-(aq) \qquad K_b$$

$$CaCO_3(s) + H_2O(l) \rightleftharpoons Ca^{2+}(aq) + HCO_3^-(aq) + OH^-(aq) \qquad K$$

$K_b = K_w / K_a$ for HCO_3^-

$$K = K_{sp} \times K_b = \frac{K_{sp} \times K_w}{K_a \text{ for } HCO_3^-} = \frac{4.5 \times 10^{-9} \times 1 \times 10^{-14}}{5.6 \times 10^{-11}} = 8.036 \times 10^{-13} = 8.0 \times 10^{-13}$$

(c) $K = 8.036 \times 10^{-13} = [Ca^{2+}][HCO_3^-][OH^-] = s^3$; $s = 9.297 \times 10^{-5} = 9.3 \times 10^{-5} \, M$

(d) pH = 8.3, pOH = 14 − 8.3 = 5.7. $[OH^-] = 10^{-5.7} = 1.995 \times 10^{-6} = 2 \times 10^{-6} \, M$

$8.036 \times 10^{-13} = s^2(1.995 \times 10^{-6})$, $s = 6.346 \times 10^{-4} = 6 \times 10^{-4} \, M$

(e) pH = 7.5, pOH = 14 − 7.5 = 6.5. $[OH^-] = 10^{-6.5} = 3.162 \times 10^{-7} = 3 \times 10^{-7} \, M$

$8.036 \times 10^{-13} = s^2(3.162 \times 10^{-7})$, $s = 1.549 \times 10^{-3} = 2 \times 10^{-3} \, M$

The drop in pH from 8.3 to 7.5 significantly increases (from $6.7 \times 10^{-5} \, M$ to $1.5 \times 10^{-3} \, M$) the molar solubility of $CaCO_3(s)$.

17.100 (a) Hydroxyapatite: $K_{sp} = [Ca^{2+}]^5[PO_4^{3-}]^3[OH^-]$

Fluoroapatite: $K_{sp} = [Ca^{2+}]^5[PO_4^{3-}]^3[F^-]$

(b) For each mole of apatite dissolved, one mole of OH⁻ or F⁻ is formed. Express molar solubility, s, in terms of [OH⁻] and [F⁻].

Hydroxyapatite: $[OH^-] = s$, $[Ca^{2+}] = 5s$, $[PO_4^{3-}] = 3s$

$K_{sp} = 6.8 \times 10^{-27} = (5s)^5(3s)^3(s) = 84{,}375\ s^9$

$s^9 = 8.059 \times 10^{-32} = 8.1 \times 10^{-32}$.

Use logs to find s. $s = 3.509 \times 10^{-4} = 3.5 \times 10^{-4}\ M\ Ca_5(PO_4)_3OH$.

Fluoroapatite: $[F^-] = s,\ [Ca^{2+}] = 5s,\ [PO_4^{3-}] = 3s$

$K_{sp} = 1.0 \times 10^{-60} = (5s)^5(3s)^3(s) = 84{,}375\ s^9$

$s^9 = 1.185 \times 10^{-65} = 1.2 \times 10^{-65};\ s = 6.109 \times 10^{-8} = 6.1 \times 10^{-8}\ M\ Ca_5(PO_4)_3F$

17.101

$$Cr(OH)_3(s) \rightleftharpoons Cr^{3+}(aq) + 3OH^-(aq) \qquad K_{sp}$$

$$Cr^{3+}(aq) + 4OH^-(aq) \rightleftharpoons Cr(OH)_4^-(aq) \qquad K_f$$

$$Cr(OH)_3(s) + OH^-(aq) \rightleftharpoons Cr(OH)_4^-(aq) \qquad K$$

$K = K_{sp} \times K_f = (6.7 \times 10^{-31})(8 \times 10^{29}) = 0.536 = 0.5$

$K = \dfrac{[Cr(OH)_4^-]}{[OH^-]};\ pOH = 14 - pH = 14 - 10.0 = 4.0;\ [OH^-] = 1 \times 10^{-4}\ M$

$[Cr(OH)_4^-] = K \times [OH^-] = 0.536(1 \times 10^{-4}) = 5.36 \times 10^{-5} = 5 \times 10^{-5}\ M$

[The K_{sp} value for $Cr(OH)_3$ listed in Appendix D.3 is different from the value given in this exercise.]

17.103 $K_{sp} = [Ba^{2+}][MnO_4^-]^2 = 2.5 \times 10^{-10}$

$[MnO_4^-]^2 = 2.5 \times 10^{-10}/2.0 \times 10^{-8} = 0.0125;\quad [MnO_4^-] = \sqrt{0.0125} = 0.11\ M$

17.104 $[Ca^{2+}][CO_3^{2-}] = 4.5 \times 10^{-9};\ [Fe^{2+}][CO_3^{2-}] = 2.1 \times 10^{-11}$

Because $[CO_3^{2-}]$ is the same for both equilibria:

$[CO_3^{2-}] = \dfrac{4.5 \times 10^{-9}}{[Ca^{2+}]} = \dfrac{2.1 \times 10^{-11}}{[Fe^{2+}]};$ rearranging $\dfrac{[Ca^{2+}]}{[Fe^{2+}]} = \dfrac{4.5 \times 10^{-9}}{2.1 \times 10^{-11}} = 214 = 2.1 \times 10^2$

17.105 $PbSO_4(s) \rightleftharpoons Pb^{2+}(aq) + SO_4^{2-}(aq);\quad K_{sp} = 6.3 \times 10^{-7} = [Pb^{2+}][SO_4^{2-}]$

$SrSO_4(s) \rightleftharpoons Sr^{2+}(aq) + SO_4^{2-}(aq);\quad K_{sp} = 3.2 \times 10^{-7} = [Sr^{2+}][SO_4^{2-}]$

Let $x = [Pb^{2+}],\ y = [Sr^{2+}],\ x + y = [SO_4^{2-}]$

$\dfrac{x(x+y)}{y(x+y)} = \dfrac{6.3 \times 10^{-7}}{3.2 \times 10^{-7}};\ \dfrac{x}{y} = 1.9688 = 2.0;\ x = 1.969\ y = 2.0\ y$

$y(1.969\ y + y) = 3.2 \times 10^{-7};\ 2.969\ y^2 = 3.2 \times 10^{-7};\ y = 3.283 \times 10^{-4} = 3.3 \times 10^{-4}$

$x = 1.969\ y;\ x = 1.969(3.283 \times 10^{-4}) = 6.464 \times 10^{-4} = 6.5 \times 10^{-4}$

$[Pb^{2+}] = 6.5 \times 10^{-4}\ M,\ [Sr^{2+}] = 3.3 \times 10^{-4}\ M,\ [SO_4^{2-}] = (3.283 + 6.464) \times 10^{-4} = 9.7 \times 10^{-4}\ M$

17.106 $MgC_2O_4(s) \rightleftharpoons Mg^{2+}(aq) + C_2O_4^{2-}(aq)$

$K_{sp} = [Mg^{2+}][C_2O_4^{2-}] = 8.6 \times 10^{-5}$

If $[Mg^{2+}]$ is to be $3.0 \times 10^{-2} M$, $[C_2O_4^{2-}] = 8.6 \times 10^{-5}/3.0 \times 10^{-2} = 2.87 \times 10^{-3} = 2.9 \times 10^{-3} M$

The oxalate ion undergoes hydrolysis:

$$C_2O_4^{2-}(aq) + H_2O(l) \rightleftharpoons HC_2O_4^{-}(aq) + OH^{-}(aq)$$

$$K_b = \frac{[HC_2O_4^{-}][OH^{-}]}{[C_2O_4^{2-}]} = 1.0 \times 10^{-14}/6.4 \times 10^{-5} = 1.56 \times 10^{-10} = 1.6 \times 10^{-10}$$

$[Mg^{2+}] = 3.0 \times 10^{-2} M$, $[C_2O_4^{2-}] = 2.87 \times 10^{-3} = 2.9 \times 10^{-3} M$

$[HC_2O_4^{-}] = (3.0 \times 10^{-2} - 2.87 \times 10^{-3}) M = 2.71 \times 10^{-2} = 2.7 \times 10^{-2} M$

$$[OH^{-}] = 1.56 \times 10^{-10} \times \frac{[C_2O_4^{2-}]}{[HC_2O_4^{-}]} = 1.56 \times 10^{-10} \times \frac{(2.87 \times 10^{-3})}{(2.71 \times 10^{-2})} = 1.652 \times 10^{-11}$$

$[OH^{-}] = 1.7 \times 10^{-11} M$; pOH = 10.78, pH = 3.22

17.107 The student failed to account for the hydrolysis of the AsO_4^{3-} ion. If there were no hydrolysis, $[Mg^{2+}]$ would indeed be 1.5 times that of $[AsO_4^{3-}]$. However, as the reaction $AsO_4^{3-}(aq) + H_2O(l) \rightleftharpoons HAsO_4^{2-}(aq) + OH^{-}(aq)$ proceeds, the ion product $[Mg^{2+}]^3[AsO_4^{3-}]^2$ falls below the value for K_{sp}. More $Mg_3(AsO_4)_2$ dissolves, more hydrolysis occurs, and so on, until an equilibrium is reached. At this point, $[Mg^{2+}]$ in solution is much greater than 1.5 times free $[AsO_4^{3-}]$. However, it is exactly 1.5 times the total concentration of all arsenic-containing species. That is,

$$[Mg^{2+}] = 1.5 \, ([AsO_4^{3-}] + [HAsO_4^{2-}] + [H_2AsO_4^{-}] + [H_3AsO_4])$$

17.109 (a) $Cd(OH)_2(s) \rightleftharpoons Cd^{2+}(aq) + 2OH^{-}(aq)$; $K_{sp} = 2.5 \times 10^{-14} = [Cd^{2+}][OH^{-}]^2$.

$[Cd^{2+}] = s$; $[OH^{-}] = 2s$; $K_{sp} = 2.5 \times 10^{-14} = 4s^3$. $s = 1.8 \times 10^{-5} M$.

(b)
$$Cd(OH)_2(s) \rightleftharpoons Cd^{2+}(aq) + 2OH^{-}(aq) \qquad K_{sp} = 2.5 \times 10^{-14}$$
$$Cd^{2+}(aq) + 4Br^{-}(aq) \rightleftharpoons CdBr_4^{2-}(aq) \qquad K_f = 5 \times 10^{3}$$

$$Cd(OH)_2(s) + 4Br^{-}(aq) \rightleftharpoons CdBr_4^{2-}(aq) + 2OH^{-}(aq) \qquad K$$

$$K = K_{sp} \times K_f = (2.5 \times 10^{-14})(5 \times 10^{3}) = 1.25 \times 10^{-10} = 1 \times 10^{-10}$$

The desired molar solubility of $Cd(OH)_2$ is 1.0×10^{-3}. Assume all soluble Cd^{2+} is present as $CdBr_4^{2-}$. $[CdBr_4^{2-}] = 1.0 \times 10^{-3}$; $[OH^{-}] = 2(1.0 \times 10^{-3}) = 2.0 \times 10^{-3}$.

Let c = initial [NaBr] = initial $[Br^{-}]$; $[Br^{-}]$ at equilibrium =
$c - 4(1.0 \times 10^{-3}) = (c - 4.0 \times 10^{-3})$.

$$K = 1.25 \times 10^{-10} = \frac{[CdBr_4^{2-}][OH^{-}]^2}{[Br^{-}]^4} = \frac{(1.0 \times 10^{-3})(2.0 \times 10^{-3})^2}{(c - 4.0 \times 10^{-3})^4}$$

Assume c is large relative to 4.0×10^{-3}.

$(1.25 \times 10^{-10})c^4 = 4.0 \times 10^{-9}$; $c = (32)^{1/4} = 2.378 = 2 M$. The approximation is valid. 4.0×10^{-3} is about 0.2% of $2 M$. Check this result in the equilibrium expression.

$$K = \frac{(1.0 \times 10^{-3})(2.0 \times 10^{-3})^2}{(2.378 - 4.0 \times 10^{-3})^4} = 1.26 \times 10^{-10}. \text{ Our calculations are consistent.}$$

Integrative Exercises

17.110 (a) Complete ionic ($CHO_2^- = HCOO^-$)

$H^+(aq) + Cl^-(aq) + Na^+(aq) + HCOO^-(aq) \rightarrow HCOOH(aq) + Na^+(aq) + Cl^-(aq)$

Na^+ and Cl^- are spectator ions.

Net ionic: $H^+(aq) + HCOO^-(aq) \rightleftharpoons HCOOH(aq)$

(b) The net ionic equation in part (a) is the reverse of the dissociation of $HCOOH$.

$$K = \frac{1}{K_a} = \frac{1}{1.8 \times 10^{-4}} = 5.55 \times 10^3 = 5.6 \times 10^3$$

(c) For Na^+ and Cl^-, this is just a dilution problem.

$M_1 V_1 = M_2 V_2$; V_2 is 50.0 mL $+ 50.0$ mL $= 100.0$ mL

Cl^-: $\dfrac{0.15 \, M \times 50.0 \text{ mL}}{100.0 \text{ mL}} = 0.075 \, M$; Na^+: $\dfrac{0.15 \, M \times 50.0 \text{ mL}}{100.0 \text{ mL}} = 0.075 \, M$

H^+ and $HCOO^-$ react to form $HCOOH$. Because $K \gg 1$, the reaction essentially goes to completion.

0.15 M $\times 0.0500$ mL $= 7.5 \times 10^{-3}$ mol H^+

$\underline{0.15 \text{ M} \times 0.0500 \text{ mL} = 7.5 \times 10^{-3} \text{ mol } HCOO^-}$

$\qquad\qquad = 7.5 \times 10^{-3}$ mol $HCOOH$

Solve the weak acid problem to determine $[H^+]$, $[HCOO^-]$, and $[HCOOH]$ at equilibrium.

$K_a = \dfrac{[H^+][HCOO^-]}{[HCOOH]}$; $[H^+] = [HCOO^-] = x$ M; $[HCOOH] = \dfrac{(7.5 \times 10^{-3} - x) \text{ mol}}{0.100 \text{ L}}$

$\qquad\qquad\qquad\qquad\qquad\qquad\qquad\qquad\qquad\qquad = (0.075 - x)$ M

$1.8 \times 10^{-4} = \dfrac{x^2}{(0.075 - x)} \approx \dfrac{x^2}{0.075}$; $x = 3.7 \times 10^{-3}$ M H^+ and $HCOO^-$

$[HCOOH] = (0.075 - 0.0037) = 0.071$ M

$\dfrac{[H^+]}{[HCOOH]} \times 100 = \dfrac{3.7 \times 10^{-3}}{0.075} \times 100 = 4.9\%$ dissociation

In summary:

$[Na^+] = [Cl^-] = 0.075 \, M$, $[HCOOH] = 0.071 \, M$, $[H^+] = [HCOO^-] = 0.0037 \, M$

17.112 $n = \dfrac{PV}{RT} = 735 \text{ torr} \times \dfrac{1 \text{ atm}}{760 \text{ torr}} \times \dfrac{7.5 \text{ L}}{295 \text{ K}} \times \dfrac{K\text{-mol}}{0.08206 \text{ L-atm}} = 0.300 = 0.30$ mol NH_3

0.40 M $\times 0.50$ L $= 0.20$ mol HCl

	HCl(aq)	+	NH_3(g)	$\rightarrow$	NH_4^+(aq)	+	Cl^-(aq)
before	0.20 mol		0.30 mol				
after	0		0.10 mol		0.20 mol		0.20 mol

The solution will be a buffer because of the substantial concentrations of NH_3 and NH_4^+ present. Use K_a for NH_4^+ to describe the equilibrium.

$$NH_4^+(aq) \rightleftharpoons NH_3(aq) + H^+(aq)$$

equil. $\quad 0.20 - x \qquad 0.10 + x \qquad x$

$$K_a = \frac{1.0 \times 10^{-14}}{1.8 \times 10^{-5}} = 5.56 \times 10^{-10} = 5.6 \times 10^{-10}; \; K_a = \frac{[NH_3][H^+]}{[NH_4^+]}; \; [H^+] = \frac{K_a[NH_4^+]}{[NH_3]}$$

Because this expression contains a ratio of concentrations, volume will cancel and we can substitute moles directly. Assume x is small compared to 0.10 and 0.20.

$$[H^+] = \frac{5.56 \times 10^{-10} \, (0.20)}{(0.10)} = 1.111 \times 10^{-9} = 1.1 \times 10^{-9} \, M, \; pH = 8.95$$

17.114   According to Equation 13.4, $S_g = kP_g$

$$S_{CO_2} = 3.1 \times 10^{-2} \, \frac{mol}{L\text{-}atm} \times 1.10 \, atm = 0.0341 = \frac{0.034 \, mol}{L} = 0.034 \, M \, CO_2$$

$$CO_2(g) + H_2O(l) \rightarrow H_2CO_3(aq); \; 0.0341 \, M \, CO_2 = 0.0341 \, M \, H_2CO_3$$

Consider the stepwise dissociation of $H_2CO_3(aq)$.

$$H_2CO_3(aq) \rightleftharpoons H^+(aq) + HCO_3^-(aq)$$

initial $\quad 0.0341 \, M \qquad\qquad\quad 0 \qquad\qquad 0$

equil. $\quad (0.0341-x) \, M \qquad\qquad x \qquad\qquad x$

$$K_{a1} = \frac{[H^+][HCO_3^-]}{[H_2CO_3]} = \frac{x^2}{(0.0341-x)} \approx \frac{x^2}{0.0341} \approx 4.3 \times 10^{-7}$$

$$x^2 = 1.47 \times 10^{-8}; \; x = 1.2 \times 10^{-4} \, M \, H^+; \; pH = 3.92$$

$K_{a2} = 5.6 \times 10^{-11}$; assume the second ionization does not contribute significantly to $[H^+]$.

17.115 $Ca(OH)_2(aq) + 2HCl(aq) \rightarrow CaCl_2(aq) + 2H_2O$

mmol HCl $= M \times mL = 0.0983 \, M \times 11.23 \, mL = 1.1039 = 1.10$ mmol HCl

mmol $Ca(OH)_2$ = mmol HCl/2 = 1.1039/2 = 0.55195 = 0.552 mmol $Ca(OH)_2$

$$[Ca^{2+}] = \frac{0.55195 \, mmol}{50.00 \, mL} = 0.01104 = 0.0110 \, M$$

$$[OH^-] = 2[Ca^{2+}] = 0.02208 = 0.0221 \, M$$

$$K_{sp} = [Ca^{2+}][OH^-]^2 = (0.01104)(0.02208)^2 = 5.38 \times 10^{-6}$$

The value in Appendix D.3 is 6.5×10^{-6}, a difference of 17%. Because a change in temperature does change the value of an equilibrium constant, the solution may not have been kept at 25 °C. It is also possible that experimental errors led to the difference in K_{sp} values.

17.116 $\Pi = MRT, M = \dfrac{\Pi}{RT} = \dfrac{21 \text{ torr}}{298 \text{ K}} \times \dfrac{1 \text{ atm}}{760 \text{ torr}} \times \dfrac{K\text{-mol}}{0.08206 \text{ L-atm}} = 1.13 \times 10^{-3} = 1.1 \times 10^{-3} \ M$

$SrSO_4(s) \rightleftharpoons Sr^{2+}(aq) + SO_4^{2-}(aq); K_{sp} = [Sr^{2+}][SO_4^{2-}]$

The total particle concentration is $1.13 \times 10^{-3} \ M$. Each mole of $SrSO_4$ that dissolves produces 2 mol of ions, so $[Sr^{2+}] = [SO_4^{2-}] = 1.13 \times 10^{-3} \ M/2 = 5.65 \times 10^{-4} = 5.7 \times 10^{-4} \ M$.

$K_{sp} = (5.65 \times 10^{-4})^2 = 3.2 \times 10^{-7}$

17.118 To determine precipitation conditions, we must know K_{sp} for $CaF_2(s)$ and calculate Q under the specified conditions. $K_{sp} = 3.9 \times 10^{-11} = [Ca^{2+}][F^-]^2$

$[Ca^{2+}]$ and $[F^-]$. The term 1 ppb means 1 part per billion or 1 g solute per billion g solution. Assume that the density of this very dilute solution is the density of water.

$$1 \text{ ppb} = \frac{1 \text{ g solute}}{1 \times 10^9 \text{ g solution}} \times \frac{1 \text{ g solution}}{1 \text{ mL solution}} \times \frac{1 \times 10^3 \text{ mL}}{1 \text{ L}} = \frac{1 \times 10^{-6} \text{ g solute}}{1 \text{ L solution}}$$

$$\frac{1 \times 10^{-6} \text{ g solute}}{1 \text{ L solution}} \times \frac{1 \ \mu g}{1 \times 10^{-6} \text{ g}} = 1 \ \mu g/1 \text{ L}$$

$$8 \text{ ppb } Ca^{2+} \times \frac{1 \ \mu g}{1 \text{ L}} = \frac{8 \ \mu g \ Ca^{2+}}{1 \text{ L}} = \frac{8 \times 10^{-6} \text{ g } Ca^{2+}}{1 \text{ L}} \times \frac{1 \text{ mol } Ca^{2+}}{40 \text{ g}} = 2 \times 10^{-7} \ M \ Ca^{2+}$$

$$1 \text{ ppb } F^- \times \frac{1 \ \mu g}{1 \text{ L}} = \frac{1 \ \mu g \ F^-}{1 \text{ L}} = \frac{1 \times 10^{-6} \text{ g } F^-}{1 \text{ L}} \times \frac{1 \text{ mol } F^-}{19.0 \text{ g}} = 5 \times 10^{-8} \ M \ F^-$$

$Q = [Ca^{2+}][F^-]^2 = (2 \times 10^{-7})(5 \times 10^{-8})^2 = 5 \times 10^{-22}$

$5 \times 10^{-22} < 3.9 \times 10^{-11}$, $Q < K_{sp}$, no CaF_2 will precipitate

17.119 (a) $CH_3CH(OH)COOH(aq) + HCO_3^-(aq) \rightarrow H_2CO_3(aq) + CH_3CH(OH)COO^-(aq)$

$H_2CO_3(aq) \rightleftharpoons H_2O(l) + CO_2(g)$

(b) $\dfrac{2.16 \text{ g NaHCO}_3}{\text{mL}} \times \dfrac{236.6 \text{ mL}}{48 \text{ tsp}} \times \dfrac{1 \text{ mol NaHCO}_3}{84.01 \text{ g NaHCO}_3} \times 0.5 \text{ tsp} = 0.06337 = 0.0634 \text{ mol NaHCO}_3$

mol $NaHCO_3$ = mol HCO_3^-;
at neutralization, mol HCO_3^- = mol $CH_3CH(OH)COOH$

$\dfrac{0.06337 \text{ mol } CH_3CH(OH)COOH}{1 \text{ cup milk}} \times \dfrac{1 \text{ cup}}{236.6 \text{ mL}} \times \dfrac{1000 \text{ mL}}{\text{L}} = 0.2678 =$

$0.268 \ M \ CH_3CH(OH)COOH$

(c) $5/9(^\circ F - 32) = {}^\circ C; \ 5/9(350 - 32) = 176.67 = 177 \ ^\circ C; \ 177 \ ^\circ C + 273 = 450 \text{ K}$

mol CO_2 = mol HCO_3^-

$V = \dfrac{nRT}{P} = 0.06337 \text{ mol} \times \dfrac{450 \text{ K}}{1 \text{ atm}} \times \dfrac{0.08206 \text{ L-atm}}{\text{mol-K}} = 2.34 \text{ L } CO_2$

17.120 Statements (b) and (d) follow from this observation.

When HF behaves like this, it is acting like a base, rather than an acid. This agrees with statements (b) and (d) and renders (a) incorrect. As for (c), if HF were thermodynamically unstable, it would decompose rather than gain H^+.

18 Chemistry of the Environment

Visualizing Concepts

18.2   Molecules in the upper atmosphere tend to have multiple bonds because they have sufficiently high bond dissociation enthalpies (Table 8.4) to survive the incoming high-energy radiation from the Sun. According to Table 8.4, for the same two bonded atoms, multiple bonds have higher bond dissociation enthalpies than single bonds. Molecules with single bonds are likely to undergo photodissociation in the presence of the high-energy, short-wavelength solar radiation present in the upper atmosphere.

18.4   *Analyze.* Given granite, marble, bronze, and other solid materials, what observations and measurements indicate whether the material is appropriate for an outdoor sculpture? If the material changes (erodes) over time, what chemical processes are responsible?

Plan. An appropriate material resists chemical and physical changes when exposed to environmental conditions. An inappropriate material undergoes chemical reactions with substances in the troposphere, degrading the structural strength of the material and the sculpture.   *Solve.*

(a)   The appearance and mass of the material upon environmental exposure are indicators of both chemical and physical changes. If the appearance and mass of the material are unchanged after a period of time, the material is well suited for the sculpture because it is inert to chemical and physical changes. Changes in the color or texture of the material's surface indicate that a chemical reaction has occurred, because a different substance with different properties has formed. A decrease in mass indicates that some of the material has been lost, by either chemical reaction or physical change. An increase in mass indicates corrosion. If the mass of the material is unchanged, it is probably inert to chemical and physical environmental changes and suitable for sculpture.

(b)   The two main chemical processes that lead to erosion are reaction with acid rain and corrosion or air oxidation, which is encouraged by acid conditions (see Section 20.8).

Acid rain is primarily H_2SO_3 and/or H_2SO_4, which reacts directly with carbonate minerals such as marble and limestone. Acidic conditions created by acid rain encourage corrosion of metals such as iron, steel, and bronze. Corrosion produces metal oxides, which may or may not cling to the surface of the material. If the oxides are washed away, the material will lose mass after corrosion. Physical erosion due to the effects of wind and rain on soft materials such as sandstone also causes mass to decrease.

18.6   *Saltwater*, *freshwater*, and *groundwater* differ in salt content, location, and percentage of Earth's total water.

         Saltwater contains high concentrations of dissolved salts and solids and includes the world ocean (97.2% of all water) and brackish or salty water (0.1% of all water) in places such as the Great Salt Lake and the Chesapeake Bay. The world ocean averages about 35 g of dissolved salts per kg of water, or 35,000 ppm.

         Freshwater (0.6% of all water on Earth) refers to natural waters that have low concentrations (less than 500 ppm) of dissolved salts and solids. Freshwater includes the waters of lakes, rivers, ponds, and streams.

         Groundwater is freshwater that is under the soil. It resides in aquifers, porous rock that holds water, and composes 20% of the world's freshwater.

18.8   *Analyze/Plan.* Explain how an ion-exchange column "softens" water. See the Closer Look box on "Water Softening" in Section 18.4.

         Solve. The plastic beads in an ion-exchange column contain covalently bound anionic groups such as $R–COO^-$ and $R–SO_3^-$. These groups have Na^+ cations associated with them for charge balance. When "hard" water containing Ca^{2+} and other divalent cations passes over the beads, the 2+ cations are attracted to the anionic groups and Na^+ is displaced. The higher charge on the divalent cations leads to greater electrostatic attractions, which promote the cation exchange. The "soft" water that comes out of the column contains two Na^+ ions in place of each divalent cation, mostly Ca^{2+} and Mg^{2+}, that remains in the column associated with the plastic beads.

18.10   Some of the missing CO_2 is absorbed by "land plants" (vegetation other than trees) and incorporated into the soil. Soil is the largest land-based carbon reservoir. The amount of carbon-storing capacity of soil is affected by erosion, soil fertility, and other complex factors. For more details, search the Internet for "carbon budget."

Earth's Atmosphere (Section 18.1)

18.12   (a)   Boundaries between regions of the atmosphere are at maxima and minima (peaks and valleys) in the atmospheric temperature profile. For example, in the troposphere, temperature decreases with altitude, whereas in the stratosphere, it increases with altitude. The temperature minimum is the tropopause boundary.

         (b)   From Figure 18.1, atmospheric pressure in the troposphere ranges from 760 torr to 200 torr, whereas pressure in the stratosphere ranges from 200 torr to 20 torr. Gas density (g/L) is directly proportional to pressure. The much lower density of the stratosphere means it has the smaller mass, despite having a larger volume than the troposphere.

18.14 $P_{Ar} = X_{Ar} \times P_{atm}; P_{Ar} = 0.00934\,(1.05\,bar) = 0.009807 = 9.81 \times 10^{-3}\,bar$

         $P_{Ar} = 0.009807\,bar \times \dfrac{10^5\,Pa}{bar} \times \dfrac{760\,torr}{101,325\,Pa} = 7.3559 = 7.36\,torr$

         $P_{CO_2} = X_{CO_2} \times P_{atm}; P_{CO_2} = 0.000400\,(1.05\,bar) = 0.0004200 = 4.20 \times 10^{-4}\,bar$

$$P_{CO_2} = 0.0004200 \text{ bar} \times \frac{10^5 \text{ Pa}}{\text{bar}} \times \frac{760 \text{ torr}}{101,325 \text{ Pa}} = 0.3150 = 0.315 \text{ torr}$$

18.16   (a)   ppm Ne = mol Ne/1×10^6 mol air; $X_{Ne} = 1.818 \times 10^{-5}$ mol Ne/mol air

$$\frac{1.818 \times 10^{-5} \text{ mol Ne}}{1 \text{ mol air}} = \frac{x \text{ mol Ne}}{1 \times 10^6 \text{ mol air}}; x = 18.18 \text{ ppm Ne}$$

       (b) $P_{Ne} = X_{Ne} \times P_{atm} = 1.818 \times 10^{-5} \times 730 \text{ torr} \times \dfrac{1 \text{ atm}}{760 \text{ torr}} = 1.7462 \times 10^{-5} = 1.75 \times 10^{-5} \text{ atm}$

            T = 296 K

$$\frac{n_{Ne}}{V} = \frac{P_{Ne}}{RT} = \frac{1.7462 \times 10^{-5} \text{ atm}}{296 \text{ K}} \times \frac{\text{mol-K}}{0.08206 \text{ L-atm}} = 7.1892 \times 10^{-7} = 7.19 \times 10^{-7} \text{ mol/L}$$

$$\frac{7.1892 \times 10^{-7} \text{ mol Ne}}{L} \times \frac{6.022 \times 10^{23} \text{ atoms}}{\text{mol}} = 4.3293 \times 10^{17}$$

$$= 4.33 \times 10^{17} \text{ Ne atoms/L}$$

18.18 $\dfrac{339 \times 10^3 \text{ J}}{1 \text{ mol}} \times \dfrac{1 \text{ mol}}{6.022 \times 10^{23} \text{ molecules}} = 5.6294 \times 10^{-19} = 5.63 \times 10^{-19} \text{ J/molecule}$

$$\lambda = \frac{hc}{E} = \frac{(6.626 \times 10^{-34} \text{ J-sec}) (3.00 \times 10^8 \text{ m/sec})}{5.6294 \times 10^{-19} \text{ J}} = 3.53 \times 10^{-7} \text{ m} = 353 \text{ nm}$$

$$\frac{293 \times 10^3 \text{ J}}{1 \text{ mol}} \times \frac{1 \text{ mol}}{6.022 \times 10^{23} \text{ molecules}} = 4.8655 \times 10^{-19} = 4.87 \times 10^{-19} \text{ J/molecule}$$

$$\lambda = \frac{(6.626 \times 10^{-34} \text{ J-sec}) (3.00 \times 10^8 \text{ m/sec})}{4.8655 \times 10^{-19} \text{ J}} = 4.09 \times 10^{-7} \text{ m} = 409 \text{ nm}$$

Photons of wavelengths longer than 409 nm cannot cause rupture of the C–Cl bond in either CF_3Cl or CCl_4. Photons with wavelengths between 409 and 353 nm can cause C–Cl bond rupture in CCl_4, but not in CF_3Cl.

18.20   Photodissociation of N_2 is relatively unimportant compared to photodissociation of O_2 for two reasons. The bond dissociation energy of N_2, 941 kJ/mol, is much higher than that of O_2, 495 kJ/mol. Photons with a wavelength short enough to photodissociate N_2 are not as abundant as the ultraviolet photons that lead to photodissociation of O_2. Also, N_2 does not absorb these photons as readily as O_2 so even if a short-wavelength photon is available, it may not be absorbed by an N_2 molecule.

Human Activities and Earth's Atmosphere (Section 18.2)

18.22   Reactions (a) and (b).

It is those star* reactions, Equations 18.3–18.5. For example, the reaction of $O_2(g) + O(g)$ is exothermic and produces a high-energy $O_3^*(g)$ molecule with 105 kJ of energy to disperse. This energy is transferred through collisions, primarily to $N_2(g)$ and $O_2(g)$ molecules. The overall kinetic energy (translational, vibrational, and rotational energy) of these molecules

(M*) increases and the temperature of the stratosphere is kept relatively high. (Recall that the temperature of a gas is directly proportional to its average kinetic energy.)

18.24   32 e⁻, 16 e⁻ pr

$$:\ddot{F}:$$
$$|$$
$$:\ddot{C}l-C-\ddot{C}l:$$
$$|$$
$$:\ddot{C}l:$$

CFC–11, $CFCl_3$, contains C–Cl bonds that can be cleaved by UV light in the stratosphere to produce Cl atoms. It is chlorine in atomic form that catalyzes the destruction of stratospheric ozone. CFC–11 is chemically inert and resists decomposition in the troposphere, so that it eventually reaches the stratosphere in molecular form.

18.26   Yes. Assuming $CFBr_3$ reaches the stratosphere intact, it contains C–Br bonds that are even more susceptible to cleavage by UV light than C–Cl bonds. According to Table 8.4, the average C–Br bond dissociation energy is 276 kJ/mol, compared to 328 kJ/mol for C–Cl bonds. Once in atomic form, Br atoms catalyze the destruction of ozone by a mechanism similar to that of Cl atoms.

18.28   Rainwater is naturally acidic because of the presence of $CO_2(g)$ in the atmosphere. All oxides of nonmetals produce acidic solutions when dissolved in water. Even in the absence of polluting gases such as SO_2, SO_3, NO, and NO_2, CO_2 causes rainwater to be acidic. The important equilibria are:

$$CO_2(g) + H_2O(l) \rightleftharpoons H_2CO_3(aq) \rightleftharpoons H^+(aq) + HCO_3^-(aq)$$

18.30   (a) $Fe(s) + O_2(g) + 4H_3O^+(aq) \rightarrow Fe^{2+}(aq) + 6H_2O(l)$

(b)   No. Silver is a "noble" metal. It is relatively resistant to oxidation, and much more resistant than iron. In Table 4.5, The Activity Series of Metals in Aqueous Solution, Ag is much, much lower than Fe and it is below hydrogen, whereas Fe is above hydrogen. This means that Fe is susceptible to oxidation by acid, but Ag is not.

18.32   (a)   Visible (Figure 6.4).

(b) $E_{photon} = hc/\lambda = \dfrac{6.626 \times 10^{-34} \text{ J-s} \times 3.00 \times 10^8 \text{ m/s}}{420 \times 10^{-9} \text{ m}} = 4.733 \times 10^{-19}$

$$= 4.73 \times 10^{-19} \text{ J/photon}$$

$$\dfrac{4.733 \times 10^{-19} \text{ J}}{1 \text{ photon}} \times \dfrac{6.022 \times 10^{23} \text{ photons}}{1 \text{ mol}} \times \dfrac{1 \text{ kJ}}{1000 \text{ J}} = 285 \text{ kJ/mol}$$

(c) $\ddot{O}=\dot{N}-\ddot{O}: + h\nu \longrightarrow \ddot{O}=\dot{N}\cdot + :\ddot{O}\cdot$

18.34   (a)   A *greenhouse gas* absorbs energy in the 10,000–30,000 nm or infrared region. It absorbs wavelengths of radiation emitted by Earth and returns it as heat. A non-greenhouse gas is transparent to radiation in this wavelength range.

(b) $CH_4(g)$ contains 4 C–H bonds, whereas N_2 has one strong triple bond. Infrared radiation has insufficient energy to cause electron transitions or bond cleavage, but it has an appropriate amount of energy to cause molecular deformations, bond stretching, and angle bending. The CH_4 molecule absorbs infrared radiation while undergoing these deformations, but symmetrical diatomic gases such as N_2 cannot "use" infrared radiation and are transparent to it.

Earth's Water (Section 18.3)

18.36   If the phosphorous is present as $H_2PO_4^-$, there is a 1:1 ratio between the molarity of phosphorus and molarity of phosphate. Thus, we can calculate the molarity based on the given mass of P.

$$\frac{0.07 \text{ g P}}{1\times10^6 \text{ g H}_2\text{O}} \times \frac{1 \text{ mol P}}{31 \text{ g P}} \times \frac{1 \text{ mol PO}_4^{3-}}{1 \text{ mol P}} \times \frac{1\times10^3 \text{ g H}_2\text{O}}{1 \text{ L H}_2\text{O}} = 2.26\times10^{-6} = 2\times10^{-6} \, M \text{ PO}_4^{3-}$$

18.38   (a) $$\frac{168 \text{ W}}{\text{m}^2} \times \frac{1 \text{ J/s}}{1 \text{ W}} = \frac{168 \text{ J}}{\text{m}^2\text{-s}}$$

$$\frac{168 \text{ J}}{\text{m}^2\text{-s}} \times 1.00 \text{ m}^2 \times 12 \text{ h} \times \frac{60 \text{ min}}{1 \text{ h}} \times \frac{60 \text{ s}}{1 \text{ min}} \times \frac{1 \text{ kJ}}{1000 \text{ J}} = 7257.6 = 7.26\times10^3 \text{ kJ}$$

$$7257.6 \text{ kJ} \times \frac{1 \text{ mol H}_2\text{O}}{6.01 \text{ kJ}} \times \frac{18.02 \text{ g H}_2\text{O}}{1 \text{ mol H}_2\text{O}} = 21,761 = 2.18\times10^4 \text{ g H}_2\text{O (ice)}$$

   (b) $$1.00 \text{ m}^2 \times 1.00 \text{ cm} \times \frac{(100)^2 \text{ cm}^2}{1 \text{ m}^2} \times \frac{0.99987 \text{ g}}{1 \text{ cm}^3} = 9998.7 = 1.00\times10^4 \text{ g H}_2\text{O (ice at 0 }^\circ\text{C)}$$

$$7257.6 \text{ kJ} \times \frac{1000 \text{ J}}{1 \text{ kJ}} \times \frac{1 \text{ g-}^\circ\text{C}}{2.032 \text{ J}} \times \frac{1}{9998.7 \text{ g H}_2\text{O}} = 357.21 = 357.2 \, ^\circ\text{C}$$

Assuming no phase changes, the final temperature is –5 $^\circ$C + 357 $^\circ$C = 352 $^\circ$C. Clearly the ice melts. This agrees with the result from part (a), which shows that sunlight striking 1.00 square meter of ice for 12 hours provides enough energy to melt 2.18 × 10^4 g ice, twice the mass in the first centimeter of a square meter of ice.

18.40   0.05 ppb Au = 0.05 g Au/1 × 10^9 g seawater

$$\$1,000,000 \times \frac{1 \text{ troy oz Au}}{\$1300} \times \frac{31.1035 \text{ g}}{\text{troy oz}} = 2.3926\times10^4 \text{ g} = 2.39\times10^4 \text{ g Au needed}$$

$$2.3926\times10^4 \text{ g Au} \times \frac{1\times10^9 \text{ g seawater}}{0.05 \text{ g Au}} \times \frac{1 \text{ mL seawater}}{1.03 \text{ g seawater}} \times \frac{1 \text{ L}}{1000 \text{ mL}} = 4.6458\times10^{11}$$

$$= 5\times10^{11} \text{ L seawater}$$

5 × 10^{11} L seawater is needed if the process is 100% efficient; because it is only 50% efficient, twice as much seawater is needed.

4.6458 × 10^{11} × 2 = 9.2916 × 10^{12} = 9 × 10^{12} L seawater

Note that the 1 sig fig in 0.05 ppb Au limits the precision of the calculation.

18.42   (a)   We need to replace the 18 billion gal of water per day used for irrigation. One billion is 1 × 10^9.

$$\frac{18 \times 10^9 \text{ gal}}{d} \times \frac{365 \text{ d}}{yr} \times \frac{3.7854 \text{ L}}{gal} \times \frac{1 \text{ dm}^3}{L} \times \frac{1 \text{ m}^3}{(10)^3 \text{ dm}^3} \times \frac{1 \text{ km}^3}{(1000)^3 \text{ m}^3} \times \frac{1}{6 \times 10^5 \text{ km}^2}$$

$$= 4.145 \times 10^{-5} = 4 \times 10^{-5} \text{ km/ yr}$$

$$\frac{4.145 \times 10^{-5} \text{ km}}{yr} \times \frac{1000 \text{ m}}{km} \times \frac{100 \text{ cm}}{m} \times \frac{1 \text{ in}}{2.54 \text{ cm}} = 1.632 = 2 \text{ in/ yr is needed}$$

However, only 2% of rainfall actually recharges to aquifer, so $(1.632/0.02) = 81.59$ = 80 in/year annual rainfall is required to replace water removed for irrigation. (Data limits the calculated result to 1 sig fig.)

(b)   The process of dissolving accounts for the presence of arsenic in well water. If rocks in the ground around the well contain minerals that are somewhat soluble, they can release arsenic-containing molecules and ions into the groundwater. Some of this groundwater eventually becomes arsenic-containing well water.

Human Activities and Water Quality (Section 18.4)

18.44   Calculate the total ion concentration of seawater by summing the molarities given in Table 18.5. Then use $\Pi = \Delta MRT$ to calculate pressure.

$$M_{total} = 0.55 + 0.47 + 0.028 + 0.054 + 0.010 + 0.010 + 2.3 \times 10^{-3} + 8.3 \times 10^{-4}$$

$$+ 4.3 \times 10^{-4} + 9.1 \times 10^{-5} + 7.0 \times 10^{-5} = 1.1257 = 1.13 \text{ } M$$

$$\Pi = \frac{(1.1257 - 0.02) \text{ mol}}{L} \times \frac{0.08206 \text{ L} \times \text{atm}}{\text{mol} \cdot \text{K}} \times 297 \text{ K} = 26.948 = 26.9 \text{ atm}$$

Check. The largest numbers in the molarity sum have 2 decimal places, so M_{total} has 2 decimal places and 3 sig figs. ΔM also has 2 decimal places and 3 sig figs so the calculated pressure has 3 sig figs. Units are correct.

18.46   (a)   Decomposition of organic matter by aerobic bacteria depletes dissolved O_2. A low dissolved oxygen concentration indicates the presence of organic pollutants.

(b)   In general, gas solubility decreases with increasing temperature. According to Figure 13.16, the solubility of $O_2(g)$ at 20 °C is approximately 1.4 mM, and at 30 °C is 1.2 mM. This is a 14.3% decrease in solubility over a typical atmospheric temperature range.

Colder natural water has a greater maximum possible $O_2(g)$ solubility. A general increase in global average temperature accompanied by an increase in water temperature decreases water quality by decreasing the amount of dissolved oxygen.

18.48   Water at 9 ppm O_2 is 50% depleted when the concentration drops by 4.5 ppm.

$$1,200,000 \text{ persons} \times \frac{59 \text{ g O}_2}{1 \text{ person}} \times \frac{1 \times 10^6 \text{ g H}_2\text{O}}{4.5 \text{ g O}_2} \times \frac{1 \text{ L H}_2\text{O}}{1 \times 10^3 \text{ g H}_2\text{O}} = 1.57 \times 10^{10} = 2 \times 10^{10} \text{ L H}_2\text{O}$$

18.50   (a) Ca^{2+}, Mg^{2+}, Fe^{2+}.

   (b)   Divalent cations (ions with 2+ charges) contribute to water hardness. These ions react with soap to form scum on surfaces or leave undesirable deposits on surfaces, particularly inside pipes, upon heating.

18.52 $Ca(OH)_2$ is added to remove Ca^{2+} and HCO_3^- as $CaCO_3(s)$, and Na_2CO_3 removes the remaining Ca^{2+}.

$$Ca^{+2}(aq) + 2HCO_3^-(aq) + [Ca^{2+}(aq) + 2OH^-(aq)] \rightarrow 2CaCO_3(s) + 2H_2O(l)$$

One mole $Ca(OH)_2$ is needed for each 2 moles of $HCO_3^-(aq)$ present.

$$5.0 \times 10^7 \text{ L H}_2\text{O} \times \frac{1.7 \times 10^{-3} \text{ mol HCO}_3^-}{1 \text{ L H}_2\text{O}} \times \frac{1 \text{ mol Ca(OH)}_2}{2 \text{ mol HCO}_3^-} \times \frac{74 \text{ g Ca(OH)}_2}{1 \text{ mol Ca(OH)}_2}$$

$$= 3.1 \times 10^6 \text{ g Ca(OH)}_2$$

Half of the native HCO_3^- precipitates the added Ca^{2+} so this operation reduces the Ca^{2+} concentration from 5.7×10^{-3} M to $(5.7 \times 10^{-3} - 8.5 \times 10^{-4})$ $M = 4.85 \times 10^{-3} =$ 4.9×10^{-3} M. Next, we must add sufficient Na_2CO_3 to further reduce $[Ca^{2+}]$ to 1.1×10^{-3} M (20% of the original $[Ca^{2+}]$). We thus need to reduce $[Ca^{2+}]$ by $(4.85 \times 10^{-3} - 1.1 \times 10^{-3})$ $M = 3.75 \times 10^{-3} = 3.8 \times 10^{-3}$ M.

$$Ca^{2+}(aq) + CO_3^{-2}(aq) \rightarrow CaCO_3(s)$$

$$5.0 \times 10^7 \text{ L H}_2\text{O} \times \frac{3.75 \times 10^{-3} \text{ mol Ca}^{2+}}{1 \text{ L H}_2\text{O}} \times \frac{1 \text{ mol Na}_2\text{CO}_3}{1 \text{ mol Ca}^{2+}} \times \frac{106 \text{ g Na}_2\text{CO}_3}{1 \text{ mol Na}_2\text{CO}_3}$$

$$= 2.0 \times 10^7 \text{ g Na}_2\text{CO}_3$$

18.54 $Al_2(SO_4)_3$ is a typical coagulant in municipal water purification. It reacts with OH^- in a slightly basic solution to form a gelatinous precipitate that occludes very small particles and bacteria. The precipitate settles slowly and is removed by sand filtration.

Properties of $Al_2(SO_4)_3$ and other useful coagulants are:

   •   They react with low concentrations of $OH^-(aq)$. That is, K_{sp} of the hydroxide precipitate is very small. The capacity to form a hydroxide precipitate means that no extra salts must be added to form the precipitate. Also, the [OH^-] can be easily adjusted by $Ca(OH)_2$ and other reagents that are part of the purification process.

   •   The hydroxide precipitate is composed of very small, evenly dispersed particles that do not settle quickly. This is required to remove very small bacteria and viruses from all parts of the liquid, not just the sites of solid formation.

18.56   (a)   The most likely origin of bromate ion, BrO_3^-, in municipal water supplies is oxidation of dissolved bromide ion, Br^-. Bromide can react with ozone in a two-step process to form bromate. The ozone might be produced photochemically, or be part of the water disinfection process.

   (b) BrO_3^- is an oxidizing agent. Hyponitrite ion, NO^-, has one less O atom than nitrite ion, NO_2^-.

$$BrO_3^-(aq) + 2NO^-(aq) \rightarrow BrO^-(aq) + 2NO_2^-(g) \text{ or}$$

$$BrO_3^-(aq) + NO^-(aq) \rightarrow BrO^-(aq) + NO_3^-(g)$$

Green Chemistry (Section 18.5)

18.58   Catalysts increase the rate of a reaction by lowering activation energy, E_a. For an uncatalyzed reaction that requires extreme temperatures and pressures to generate product at a viable rate, finding a suitable catalyst reduces the required temperature and/or pressure, which reduces the amount of energy used to run the process. A catalyst can also increase rate of production, which would reduce the net time and thus energy required to generate a certain amount of product.

18.60 $scCO_2$ achieves maximum conversion much faster than CH_2Cl_2 solvent. This reduces processing time, temperature, and energy requirements. It also results in fewer unwanted by-products to be separated and processed. Although use of $scCO_2$ can increase the amount of a greenhouse gas released to the environment, it eliminates use of CH_2Cl_2, which is implicated in stratospheric ozone depletion. Use of $scCO_2$ rather than CH_2Cl_2 is a good green trade-off. (If the CO_2 used to form $scCO_2$ can be captured from some other industrial process, the net release of CO_2 is the same, and the use of CH_2Cl_2 is avoided.)

(In either solvent, the reaction is catalyzed, which usually leads to decreased processing temperatures and times, and greater energy efficiency.)

18.62   (a)   The catalyzed reaction that can be run close to room temperature and for a shorter time is definitely greener, according to criteria (6) design for energy efficiency and (9) catalysis.

(b)   The reagent obtained from corn husks is greener, by criteria (7) use of renewable feedstocks.

(c)   Neither process is totally "ungreen," because recycling of unavoidable by-products is always desirable. However, by criterion (2) atom economy, the process that produces no by-products is greener.

Additional Exercises

18.63   (a)   *Acid rain* is rain with a larger $[H^+]$ and thus a lower pH than expected. The additional H^+ is produced by the dissolution of sulfur and nitrogen oxides such as $SO_3(g)$ and $NO_2(g)$ in rain droplets to form sulfuric and nitric acid, $H_2SO_4(aq)$ and $HNO_3(aq)$.

(b)   A *greenhouse gas* absorbs infrared or "heat" radiation emitted from Earth's surface and serves to maintain a relatively constant temperature on the surface. These include $H_2O(g)$, CH_4, and CO_2. A significant buildup of greenhouse gases in the atmosphere could cause a corresponding increase in the average surface temperature and stimulate global climate.

(c)   *Photochemical smog* is an unpleasant collection of atmospheric pollutants initiated by photochemical dissociation of NO_2 to form NO and O atoms. The major components are $NO(g)$, $NO_2(g)$, $CO(g)$, and unburned hydrocarbons, all produced by automobile engines, and $O_3(g)$, ozone.

(d)   *Ozone depletion* is the reduction of O_3 concentration in the stratosphere, most notably over Antarctica. It is caused by reactions between O_3 and Cl atoms originating from CFCs, CF_xCl_{4-x}. Depletion of the ozone layer allows damaging ultraviolet radiation disruptive to the plant and animal life in our ecosystem to reach Earth.

18.64 MM_{avg} at the surface = 83.8(0.17) + 16.0(0.38) + 32.0(0.45) = 34.73 = 35 g/mol
Next, calculate the percentage composition at 200 km. The fractions can be "normalized" by saying that the 0.45 fraction of O_2 is converted into *two* 0.45 fractions of O atoms, then dividing by the total fractions, 0.17 + 0.38 + 0.45 + 0.45 = 1.45:

$$MM_{avg} = \frac{83.8(0.17) + 16.0(0.38) + 16.0(0.90)}{1.45} = 23.95 = 24 \text{ g/mol}$$

18.65   Stratospheric ozone is formed and destroyed in a cycle of chemical reactions. The decomposition of O_3 to O_2 and O produces oxygen atoms, an essential ingredient for the production of ozone. Although single O_3 molecules exist for only a few seconds, new O_3 molecules are constantly reformed. This cyclic process ensures a finite concentration of O_3 in the stratosphere available to absorb ultraviolet radiation. (This explanation assumes that the cycle is not disrupted by outside agents such as CFCs.)

18.67   CFCs, primarily $CFCl_3$ and CF_2Cl_2, are chemically inert and water insoluble. These properties make them valuable as propellants, refrigerants, and foaming agents because they are virtually unreactive in the *troposphere* (lower atmosphere) and do not initiate or propagate undesirable reactions. Further, they are water insoluble and not removed from the atmosphere by rain; they do not end up in the fresh water supply.

These properties render CFCs a long-term problem in the *stratosphere*. Because CFCs are inert and water insoluble, they are not removed from the troposphere by reaction or dissolution and have very long lifetimes. Virtually the entire mass of released CFCs eventually diffuses into the stratosphere where conditions are right for photo-dissociation and the production of Cl atoms. Cl atoms catalyze the destruction of ozone, O_3.

18.68   (a)   The production of Cl atoms in the stratosphere is the result of the photodissociation of a C–Cl bond in the CFC molecule.

$$CF_2Cl_2(g) \xrightarrow{h\nu} CF_2Cl(g) + Cl(g)$$

According to Table 8.4, the bond dissociation energy of a C–Br bond is 276 kJ/mol, whereas the value for a C–Cl bond is 328 kJ/mol. Photodissociation of $CBrF_3$ to form Br atoms requires less energy than the production of Cl atoms and should occur readily in the stratosphere.

      (b) $CBrF_3(g) \xrightarrow{h\nu} CF_3(g) + Br(g)$

$Br(g) + O_3(g) \rightarrow BrO(g) + O_2(g)$

Also, under certain conditions

$BrO(g) + BrO(g) \rightarrow Br_2O_2(g)$

$Br_2O_2(g) + h\nu \rightarrow O_2(g) + 2Br(g)$

18.70   From Section 18.2:

$$N_2(g) + O_2(g) \rightleftharpoons 2NO(g) \quad \Delta H = +180.8 \text{ kJ} \quad [18.11]$$

$$2\,NO(g) + O_2(g) \rightleftharpoons 2NO_2(g) \quad \Delta H = -113.1 \text{ kJ} \quad [18.12]$$

In an endothermic reaction, heat is a reactant. As the temperature of the reaction increases, the addition of heat favors formation of products and the value of K increases. The reverse is true for exothermic reactions; as temperature increases, the value of K decreases. Thus, K for reaction [18.11], which is endothermic, increases with increasing temperature and K for reaction [18.12], which is exothermic, decreases with increasing temperature.

18.72   (a) $2\,SO_2(g) + O_2(g) \rightarrow 2\,SO_3(g)$

               $SO_2(g) + O_3(g) \rightarrow SO_3(g) + O_2(g)$

               $SO_2(g) + H_2O(l) \rightarrow H_2SO_3(l, \text{aerosol})$

               $SO_3(g) + H_2O(l) \rightarrow H_2SO_4(l, \text{aerosol})$

    (b)   The finely dispersed liquid droplets in the aerosol reflect sunlight into space. Less warming solar radiation reaches Earth's surface.

    (c)   In the stratosphere, aerosol particles act as a heterogeneous catalyst for ozone destruction by halogens. That is, they provide a platform to attract and orient the reactants in ozone depletion processes. The depletion reactions occur at a greater rate than in the absence of the aerosol catalyst.

18.73   (a)   According to Section 13.3, the solubility of gases in water decreases with increasing temperature. From the graph, this is also true for $CO_2(g)$. Comparing the graph in this exercise with Figure 13.16, the general shape of the solubility versus temperature curve for $CO_2(g)$ is similar to that for other gases. [Although the solubility units are different on the two graphs, it seems that $CO_2(g)$ is significantly more soluble than the gases in Figure 13.16.]

    (b)   If the solubility of $CO_2(g)$ in the ocean decreased because of climate change, more $CO_2(g)$ would be released into the atmosphere, perpetuating a cycle of increasing temperature and concomitant release of $CO_2(g)$ from the ocean.

18.74   Most of the 390 watts/m^2 radiated from Earth's surface is in the infrared region of the spectrum. Tropospheric gases, particularly $H_2O(g)$, $CH_4(g)$, and $CO_2(g)$, absorb much of this radiation and prevent it from escaping into space (Figures 18.11 and 18.12). The energy absorbed by these so-called greenhouse gases warms the atmosphere close to Earth's surface and makes the planet livable.

18.76   (a) $NO(g) + h\nu \rightarrow N(g) + O(g)$

    (b) $NO(g) + h\nu \rightarrow NO^+(g) + e^-$

    (c) $NO(g) + O_3(g) \rightarrow NO_2(g) + O_2(g)$

    (d) $3NO_2(g) + H_2O(l) \rightarrow 2HNO_3(aq) + NO(g)$

18.78   Because NO has an odd electron, such as Cl(g), it could act as a catalyst for decomposition of ozone in the stratosphere. The increased destruction of ozone by NO would result in less absorption of short-wavelength UV radiation now being screened out primarily by the ozone. Radiation in this wavelength range is known to be harmful to humans; it causes skin cancer. There is evidence that many plants do not tolerate it very well either, though more research is needed to test this idea.

In Chapter 22, the oxidation of NO to NO_2 by oxygen is described. On dissolving in water, NO_2 disproportionates into NO_3^- (aq) and NO(g). Thus, over time the NO in the troposphere will be converted into NO_3^-, which is in turn incorporated into soils.

18.79   *Plan.* Calculate the volume of air above Los Angeles and the volume of pure O_3 that would be present at the 84 ppb level. For gases at the same temperature and pressure, volume fractions equal mole fractions.   *Solve.*

$$V_{air} = 4000 \text{ mi}^2 \times \frac{(1.6093)^2 \text{ km}^2}{\text{mi}^2} \times \frac{(1000)^2 \text{ m}^2}{1 \text{ km}^2} \times 100 \text{ m} \times \frac{1 \text{ L}}{1 \times 10^{-3} \text{ m}^3} = 1.036 \times 10^{15}$$

$$= 1.0 \times 10^{15} \text{ L air}$$

$$84 \text{ ppb O}_3 = \frac{84 \text{ mol O}_3}{1 \times 10^9 \text{ mol air}} = 8.4 \times 10^{-8} = X_{O_3}$$

$$V \text{ (pure O}_3) = 8.4 \times 10^{-8} (1.036 \times 10^{15} \text{ L air}) = 8.702 \times 10^7 = 8.7 \times 10^7 \text{ L O}_3$$

Values for P and T are required to calculate mol O_3 from volume O_3, using the ideal-gas law. Because these are not specified in the exercise, we will make a reasonable assumption for a sunny April day in Los Angeles. The city is near sea level and temperatures are moderate throughout the year, so P = 1 atm and T = 25 °C (78 °F) are reasonable values. PV = nRT, n = PV/RT.

$$n = 1.000 \text{ atm} \times \frac{8.702 \times 10^7 \text{ L}}{298 \text{ K}} \times \frac{\text{mol-K}}{0.08206 \text{ L-atm}} = 3.558 \times 10^6 = 3.6 \times 10^6 \text{ mol O}_3$$

Check. Using known conditions to make reasonable estimates and assumptions is a valuable skill for problem solving. Knowing when assumptions are required is an important step in the learning process.

Integrative Exercises

18.80   (a) $0.016 \text{ ppm NO}_2 = \dfrac{0.016 \text{ mol NO}_2}{1 \times 10^6 \text{ mol air}} = 1.6 \times 10^{-8} = X_{NO_2}$

$$P_{NO_2} = X_{NO_2} \times P_{atm} = 1.6 \times 10^{-8} (755 \text{ torr}) = 1.208 \times 10^{-5} = 1.2 \times 10^{-5} \text{ torr}$$

(b) $n = \dfrac{PV}{RT}$; molecules $= n \times \dfrac{6.022 \times 10^{23} \text{ molecules}}{\text{mol}} = \dfrac{PV}{RT} \times \dfrac{6.022 \times 10^{23} \text{ molecules}}{\text{mol}}$

$$V = 15 \text{ ft} \times 14 \text{ ft} \times 8 \text{ ft} \times \frac{12^3 \text{ in}^3}{\text{ft}^3} \times \frac{2.54^3 \text{ cm}^3}{\text{in}^3} \times \frac{1 \text{ L}}{1000 \text{ cm}^3} = 4.757 \times 10^4 = 5 \times 10^4 \text{ L}$$

$$1.208 \times 10^{-5} \text{ torr} \times \frac{1 \text{ atm}}{760 \text{ torr}} \times \frac{4.757 \times 10^4 \text{ L}}{293 \text{ K}} \times \frac{\text{mol - K}}{0.08206 \text{ L - atm}}$$

$$\times \frac{6.022 \times 10^{23} \text{ molecules}}{\text{mol}} = 1.894 \times 10^{19} = 2 \times 10^{19} \text{ molecules}$$

18.82   *Coarse sand* is removed by coarse sand filtration. *Finely divided particles* and some *bacteria* are removed by precipitation with aluminum hydroxide. Remaining *harmful bacteria* are removed by ozonation. *Trihalomethanes* are removed by either aeration or activated carbon filtration; use of activated carbon might be preferred because it does not involve release of TCMs into the atmosphere. *Dissolved organic substances* are oxidized (and rendered less harmful, but not removed) by both aeration and ozonation. Dissolved *nitrates* and *phosphates* are not removed by any of these processes, but are rendered less harmful by adequate aeration.

18.83   (a) $\cdot \ddot{\text{O}} - \text{H}$

        (b) HNO_3 is a major component in acid rain.

        (c)   Although it removes CO, the reaction produces NO_2. The photodissociation of NO_2 to form O atoms is the first step in the formation of tropospheric ozone and photochemical smog.

        (d)   Again, NO_2 is the initiator of photochemical smog. Also, methoxyl radical, OCH_3, is a reactive species capable of initiating other undesirable reactions.

        (e)   Beer's law is A = εbc, where A is measured absorbance, b is path length, and c is concentration of analyte. Measured absorbance is directly proportional to both path length and concentration. Because concentration of hydroxyl radical in the troposphere is quite small, a long path length compensates and makes observation of absorbance possible.

        (f)   The first reaction in (d) shows the ability to remove CH_4, a greenhouse gas, from the atmosphere. The third reaction shows the ability to remove NO. (The products of these reactions may be worse pollutants than the reactants; the cure may be worse than the disease.)

18.84   Calculate the molar concentration of impurity that would have an absorbance of 0.0001. This is the minimum concentration of the impurity detectable by absorption spectroscopy.

        A = εbc; A = absorbance, ε = extinction coefficient, b = path length, c = molarity. The common path length is 1 cm.

        $$c = \frac{A}{\varepsilon b} = 0.0001 \times \frac{M \times cm}{3.45 \times 10^3} \times \frac{1}{1\,cm} = 2.8986 \times 10^{-8} = 3 \times 10^{-8}\ M$$

        Because we do not have the identity of the impurity, we cannot calculate the corresponding concentration in ppb. We can calculate a maximum molar mass for the impurity, such that a 3×10^{-3} M solution is 50 ppb. A concentration of 50 ppb corresponds to 50 g impurity per 10^9 L solution.

        $$\frac{50\,g\,impurity}{10^9\,g\,solution} \times \frac{1000\,g\,solution}{L\,solution} \times \frac{1\,L\,solution}{2.8986 \times 10^{-8}\,mol\,impurity} = 1725\,g\,impurity/mol$$

        In this calculation, molar mass is directly proportional to ppm concentration. This means that a 50 ppm solution or any impurity with a molar mass less than or equal to 1725 g/mol will be observable by absorption spectroscopy. Concentrations less than 50 ppm are probably observable, because 1725 is a large molar mass. (The calculated molar mass is more correctly represented with one sig fig as 2×10^{-3} g/mol.)

18.86   According to Equation 14.12, $\ln([A]_t / [A]_o) = -kt.$ $[A]_t = 0.10\,[A]_o.$

$\ln(0.10\,[A]_o / [A]_o) = \ln(0.10) = -(2 \times 10^{-6}\,s^{-1})\,t$

$t = -\ln(0.10) / 2 \times 10^{-6}\,s^{-1} = 1.151 \times 10^6\,s$

$1.151 \times 10^6\,s \times \dfrac{1\,min}{60\,s} \times \dfrac{1\,h}{60\,min} \times \dfrac{1\,day}{24\,h} = 13.3\,days\,(1 \times 10\,days)$

The value of the rate constant limits the result to 1 sig fig. This implies that there is minimum uncertainty of ±1 in the tens place of our answer. Realistically, the remediation could take anywhere from 1 to 20 days.

18.88   (a)   Assume the density of water at 20 °C is the same as at 25 °C.

$1.00\,gal \times \dfrac{4\,qt}{1\,gal} \times \dfrac{1\,L}{1.057\,qt} \times \dfrac{1000\,mL}{1\,L} \times \dfrac{0.99707\,g\,H_2O}{1\,mL} = 3773$

$= 3.77 \times 10^3\,g\,H_2O$

The $H_2O(l)$ must be heated from 20 °C to 100 °C and then vaporized at 100 °C.

$3.773 \times 10^3\,g\,H_2O \times \dfrac{4.184\,J}{g\,°C} \times 80\,°C \times \dfrac{1\,kJ}{1000\,J} = 1263 = 1.3 \times 10^3\,kJ$

$3.773 \times 10^3\,g\,H_2O \times \dfrac{1\,mol\,H_2O}{18.02\,g\,H_2O} \times \dfrac{40.67\,kJ}{mol\,H_2O} = 8516 = 8.52 \times 10^3\,kJ$

energy $= 1263\,kJ + 8516\,kJ = 9779 = 9.8 \times 10^3\,kJ/gal\,H_2O$

(b)   According to Solution 5.18, 1 kwh $= 3.6 \times 10^6\,J.$

$\dfrac{9779\,kJ}{gal\,H_2O} \times \dfrac{1000\,J}{kJ} \times \dfrac{1\,kwh}{3.6 \times 10^6\,J} \times \dfrac{\$0.085}{kwh} = \$0.23/gal$

(c) $\dfrac{\$0.23}{\$1.26} \times 100 = 18\%$ of the total cost is energy

18.89   (a)   A rate constant of $M^{-1}s^{-1}$ is indicative of a reaction that is second order overall. For the reaction given, the rate law is probably rate $= k[O][O_3]$. (Although rate $= k[O]^2$ or $k[O_3]^2$ are possibilities, it is difficult to envision a mechanism consistent with either one that would result in two molecules of O_2 being produced.)

(b)   Yes. Most atmospheric processes are initiated by collision. One could imagine an activated complex of four O atoms collapsing to form two O_2 molecules. Also, the rate constant is large, which is less likely for a multistep process. The reaction is analogous to the destruction of O_3 by Cl atoms (Equation 18.7), which is also second order with a large rate constant.

(c) $\Delta H° = 2\Delta H_f° O_2(g) - \Delta H_f° O(g) - \Delta H_f° O_3(g)$

$\Delta H° = 0 - 247.5\,kJ - 142.3\,kJ = -389.8\,kJ$

The reaction is exothermic, so energy is released; the reaction would raise the temperature of the stratosphere.

18.91   rate = $k[CF_3CH_2F][OH]$. $k = 1.6 \times 10^8\,M^{-1}\,s^{-1}$ at 4 °C.

$[CF_3CH_2F] = 6.3 \times 10^8$ molecules/cm^3, $[OH] = 8.1 \times 10^5$ molecules/cm^3

Change molecules/cm^3 to mol/L (M) and substitute into the rate law.

$$\frac{6.3 \times 10^8 \text{ molecules}}{\text{cm}^3} \times \frac{1 \text{ mol}}{6.022 \times 10^{23} \text{ molecules}} \times \frac{1000 \text{ cm}^3}{1 \text{ L}} =$$

$$1.0462 \times 10^{-12} = 1.0 \times 10^{-12}\,M\ CF_3CH_2F$$

$$\frac{8.1 \times 10^5 \text{ molecules}}{\text{cm}^3} \times \frac{1 \text{ mol}}{6.022 \times 10^{23} \text{ molecules}} \times \frac{1000 \text{ cm}^3}{1 \text{ L}} =$$

$$1.3451 \times 10^{-15} = 1.3 \times 10^{-15}\,M\ OH$$

rate $= \dfrac{1.6 \times 10^8}{M\text{-s}} \times 1.0462 \times 10^{-12}\,M \times 1.3451 \times 10^{-15}\,M = 2.2515 \times 10^{-19} = 2.3 \times 10^{-19}\,M/s$

18.92   (a)   According to Table 18.1, the mole fraction of CO_2 in air is 0.000375.

$P_{CO_2} = X_{CO_2} \times P_{atm} = 0.000400\ (1.00\ atm) = 4.00 \times 10^{-4}\ atm$

$C_{CO_2} = kP_{CO_2} = 3.1 \times 10^{-2}\,M/atm \times 4.00 \times 10^{-4}\ atm = 1.24 \times 10^{-5} = 1.2 \times 10^{-5}\,M$

(b) H_2CO_3 is a weak acid, so the $[H^+]$ is regulated by the equilibria:

$H_2CO_3(aq) \rightleftharpoons H^+(aq) + HCO_3^-(aq)$ $K_{a1} = 4.3 \times 10^{-7}$

$HCO_3^-(aq) \rightleftharpoons H^+(aq) + CO_3^{2-}(aq)$ $K_{a2} = 5.6 \times 10^{-11}$

Because the value of K_{a2} is small compared to K_{a1}, we will assume that most of the $H^+(aq)$ is produced by the first dissociation.

$K_{a1} = 4.3 \times 10^{-7} = \dfrac{[H^+][HCO_3^-]}{[H_2CO_3]}; [H^+] = [HCO_3^-] = x, [H_2CO_3] = 1.24 \times 10^{-5} - x$

Because K_{a1} and $[H_2CO_3]$ have similar values, we cannot assume x is small compared to 1.2×10^{-5}.

$4.3 \times 10^{-7} = \dfrac{x^2}{(1.24 \times 10^{-5} - x)}; 5.332 \times 10^{-12} - 4.3 \times 10^{-7}x = x^2$

$0 = x^2 + 4.3 \times 10^{-7}x - 5.332 \times 10^{-12}$

$x = \dfrac{-4.3 \times 10^{-7} \pm \sqrt{(4.3 \times 10^{-7})^2 - 4(1)(-5.332 \times 10^{-12})}}{2(1)}$

$x = \dfrac{-4.3 \times 10^{-7} \pm \sqrt{1.85 \times 10^{-13} + 2.133 \times 10^{-11}}}{2} = \dfrac{-4.3 \times 10^{-7} \pm 4.638 \times 10^{-6}}{2}$

The negative result is meaningless; $x = 2.104 \times 10^{-6} = 2.1 \times 10^{-6}\,M\,H^+; pH = 5.68$

Because this $[H^+]$ is quite small, the $[H^+]$ from the autoionization of water might be significant. Calculation shows that for $[H^+] = 2.1 \times 10^{-6}\,M$ from H_2CO_3, $[H^+]$ from $H_2O = 5.2 \times 10^{-9}\,M$, which we can ignore.

18.93   (a) $Al(OH)_3(s) \rightleftharpoons Al^{3+}(aq) + 3OH^-(aq)$ $K_{sp} = 1.3 \times 10^{-33} = [Al^{3+}][OH^-]^3$

This is a precipitation conditions problem. At what $[OH^-]$ (we can get pH from $[OH^-]$) will $Q = 1.3 \times 10^{-33}$, the requirement for the onset of precipitation?

$Q = 1.3 \times 10^{-33} = [Al^{3+}][OH^-]^3$. Find the molar concentration of $Al_2(SO_4)_3$ and thus $[Al^{3+}]$.

$$\frac{5.0 \text{ lb } Al_2(SO_4)_3}{2000 \text{ gal } H_2O} \times \frac{453.6 \text{ g}}{1 \text{ lb}} \times \frac{1 \text{ mol } Al_2(SO_4)_3}{342.2 \text{ g } Al_2(SO_4)_3} \times \frac{1 \text{ gal}}{4 \text{ qt}} \times \frac{1 \text{ qt}}{0.946 \text{ L}}$$

$$= 8.758 \times 10^{-4} M \ Al_2(SO_4)_3 = 1.752 \times 10^{-3} = 1.8 \times 10^{-3} M \ Al^{3+}$$

$Q = 1.3 \times 10^{-33} = (1.752 \times 10^{-3})[OH^-]^3; [OH^-]^3 = 7.42 \times 10^{-31}$

$[OH^-] = 9.054 \times 10^{-11} = 9.1 \times 10^{-11} M$; pOH = 10.04; pH = 14 − 10.04 = 3.96

(b) $CaO(s) + H_2O(l) \rightarrow Ca^{2+}(aq) + 2OH^-(aq)$; $[OH^-] = 9.054 \times 10^{-11}$ mol/L

$$\text{mol } OH^- = \frac{9.054 \times 10^{-11} \text{ mol}}{1 L} \times 2000 \text{ gal} \times \frac{4 \text{ qt}}{1 \text{ gal}} \times \frac{0.946 \text{ L}}{1 \text{ qt}} = 6.852 \times 10^{-7}$$

$$= 6.9 \times 10^{-7} \text{ mol } OH^-$$

$$6.852 \times 10^{-7} \text{ mol } OH^- \times \frac{1 \text{ mol } CaO}{2 \text{ mol } OH^-} \times \frac{56.1 \text{ g } CaO}{1 \text{ mol } CaO} \times \frac{1 \text{ lb}}{453.6 \text{ g}} = 4.2 \times 10^{-8} \text{ lb } CaO$$

This is a *very* small amount of CaO, about 20 μg.

18.95   (a)   The various forms of carbonate in water are related by the following equilibria:

$H_2CO_3(aq) \rightleftharpoons H^+(aq) + HCO_3^-(aq)$ $K_{a1} = 4.3 \times 10^{-7}$

$HCO_3^-(aq) \rightleftharpoons H^+(aq) + CO_3^{2-}(aq)$ $K_{a2} = 5.6 \times 10^{-11}$

$$K_{a1} = 4.3 \times 10^{-7} = \frac{[H^+][HCO_3^-]}{[H_2CO_3]}; \quad K_{a2} = 5.6 \times 10^{-11} = \frac{[H^+][CO_3^{2-}]}{[HCO_3^-]}$$

$[H^+] = 10^{-pH} = 10^{-5.6} = 2.5119 \times 10^{-6} = 3 \times 10^{-6} M$

Also, $[H_2CO_3] + [HCO_3^-] + [CO_3^{2-}] = 1.0 \times 10^{-5} M$

We now have 3 equations in 3 unknowns, so we can solve explicitly for one. Solve for $[HCO_3^-]$ (because it appears in both K_a expressions) and then substitute to find $[H_2CO_3]$ and $[CO_3^{2-}]$.

$$1.0 \times 10^{-5} = \frac{[H^+][HCO_3^-]}{K_{a1}} + [HCO_3^-] + \frac{K_{a2}[HCO_3^-]}{[H^+]}$$

$$1.0 \times 10^{-5} = \frac{2.5119 \times 10^{-6}[HCO_3^-]}{4.3 \times 10^{-7}} + [HCO_3^-] + \frac{5.6 \times 10^{-11}[HCO_3^-]}{2.5119 \times 10^{-6}}$$

$$1.0 \times 10^{-5} = 5.8416[HCO_3^-] + [HCO_3^-] + 2.2294 \times 10^{-5}[HCO_3^-]$$

$$[HCO_3^-] = \frac{1.0 \times 10^{-5}}{6.8416} = 1.4616 \times 10^{-6} = 1.5 \times 10^{-6}\, M$$

Note that $[CO_3^{2-}]$ is very small compared to $[H_2CO_3]$ and $[HCO_3^-]$.

$$[H_2CO_3] = \frac{(2.5119 \times 10^{-6})(1.4616 \times 10^{-6})}{4.3 \times 10^{-7}} = 8.5383 \times 10^{-6} = 8.5 \times 10^{-6}\, M$$

$$[CO_3^{2-}] = \frac{(5.6 \times 10^{-11})(1.4616 \times 10^{-6})}{2.5119 \times 10^{-6}} = 3.2586 \times 10^{-11} = 3.3 \times 10^{-11}\, M$$

Check. $1.5 \times 10^{-6}\, M + 8.5 \times 10^{-6}\, M + 3.3 \times 10^{-11}\, M = 1.0 \times 10^{-5}\, M$

(b)   To test for sulfur-containing species, we must first remove the various forms of carbonate. One method is to exploit the solubility differences between carbonate and sulfate salts. Most sulfates are soluble, whereas most carbonates are not. However, K_{sp} values for carbonates are relatively large, and $[CO_3^{2-}]$ in the raindrop is very small. Precipitating insoluble carbonates will shift the acid dissociation equilibria to the right, but precipitation may not be the best method for effectively removing carbonates.

A different method involves removing carbonates as $CO_2(g)$. Heating the rainwater will decrease the solubility of $CO_2(g)$, which will bubble off as a gas. Slightly acidifying the solution will encourage this process, by shifting the acid dissociation equilibria toward H_2CO_3 and $CO_2(g)$.

After removal of carbonates, sulfates are precipitated with $Ba^{2+}(aq)$. The amount of precipitate is small, but it does cause turbidity in the solution. Turbidity is detected by instrumental methods that measure light scattering by colloids.

19 Chemical Thermodynamics

Visualizing Concepts

19.2   (a)   Based on experience, the process is spontaneous. We know that 1,1-difluoro-ethane is a gas at atmospheric pressure, so the pressure inside the can must be much greater than atmospheric in order for the substance to be liquefied. When the nozzle is pressed and the system is open to the lower pressure of the atmosphere, the liquid vaporizes spontaneously. The 1,1-difluoroethane gas escapes the nozzle without external assistance.

           (b)   We expect q_{sys} to be positive. We know that ΔH is positive for the vaporization of a gas. Because the change does not occur at constant pressure, q_{sys} and ΔH are not equal, but the sign of q_{sys} is still positive.

           (c) ΔS is definitely positive for this process. Because the process is spontaneous and ΔH is positive, ΔS must be positive and large so that ΔG is negative. It is also true that the system, the 1,1-difluoroethane molecules, occupy a larger volume and have greater motional freedom after vaporization.

           (d)   The operation of the keyboard cleaner definitely depends more on entropy change than heat flow.

19.3   (a)   The process depicted is a change of state from a solid to a gas. ΔS is positive because of the greater motional freedom of the particles. ΔH is positive because both melting and boiling are endothermic processes.

           (b)   The sign of ΔS_{surr} is negative, and the magnitude is less than or equal to ΔS_{sys}. If the process is spontaneous, the second law states that $\Delta S_{univ} \geq 0$. Because ΔS_{sys} is positive, ΔS_{surr} must be negative. If the change occurs via a reversible pathway, $\Delta S_{univ} = 0$ and $\Delta S_{surr} = -\Delta S_{sys}$. If the pathway is irreversible, the magnitude of ΔS_{sys} is greater than the magnitude of ΔS_{surr}, but the sign of ΔS_{surr} is still negative.

19.5   *Analyze/Plan.* Consider the physical changes that occur when a substance is heated. How do these changes affect the entropy of the substance?   *Solve.*

           (a)   Both 1 and 2 represent changes in entropy at constant temperature; these are phase changes. Because 1 happens at a lower temperature, it represents melting (fusion), and 2 represents vaporization.

           (b)   The substance changes from solid to liquid in 1, from liquid to gas in 2. The larger volume and greater motional freedom of the gas phase causes ΔS for vaporization to (always) be larger than ΔS for fusion.

           (c)   For a perfect crystal at $T = 0$ K, the value of S is zero. This is the third law of thermodynamics.

19.6   (a)   We expect the enthalpy of combustion of the two isomers to be very similar. The molecular formulas of the two molecules are the same, so the balanced chemical equations for the two combustion reactions are identical. In the calculation of combustion enthalpy from standard enthalpies of formation of products and reactants, the only difference will be in the standard enthalpies of formation of the two isomers.

       (b)   We expect n-pentane to have the higher standard molar entropy. The rod-shaped n-pentane has more possible vibrational and rotational motions than the almost-spherical neopentane. That is, n-pentane has greater motional energy, which results in a higher standard molar entropy than that of neopentane.

19.8   (a)   At equilibrium, $\Delta G = 0$. On the diagram, $\Delta G = 0$ at 250 K. The system is at equilibrium at 250 K.

       (b)   A reaction is spontaneous when ΔG is negative. The reaction is spontaneous at temperatures greater than 250 K.

       (c) $\Delta G = \Delta H - T\Delta S$, in the form of $y = b + mx$. ΔH is the y intercept of the graph (where $T = 0$) and is positive.

       (d)   The slope of the graph is $-\Delta S$. The slope is negative, so ΔS is positive. [Also, ΔG decreases as T increases, so the $T\Delta S$ term must become more negative and ΔS is positive.]

19.9   (a)   *Analyze.* The boxes depict three different mixtures of reactants and products for the reaction $A_2 + B_2 \rightleftharpoons 2AB$.

            Plan. $K_c = 1 = \dfrac{[AB]^2}{[A][B]}$. Calculate Q for each box, using number of molecules as a measure of concentration. If $Q = 1$, the system is at equilibrium. *Solve.*

            Box 1: $K = \dfrac{(3)^2}{(3)(3)} = 1$

            Box 2: $Q = \dfrac{(1)^2}{(4)(4)} = \dfrac{1}{16} = 0.0625 = 0.06$

            Box 3: $Q = \dfrac{(7)^2}{(1)(1)} = \dfrac{49}{1} = 49$

            Box 1 is at equilibrium.

       (b)   Box 2.

       (c)   Qualitatively, Box 3 is farthest from equilibrium, so it has the largest magnitude of ΔG (driving force to reach equilibrium), then Box 2, and then Box 1, where $\Delta G = 0$.

            Box 1 < Box 2 < Box 3

            Quantitatively, $\Delta G = \Delta G° - RT\ln Q$. For Box 1, $\Delta G = 0$ and $K = 1$, so $\Delta G° = 0$.

            Box 2: $\Delta G = 0 - RT\ln(0.0625) = 2.77\ RT$

            Box 3: $\Delta G = 0 - RT\ln(49) = -3.89\ RT$

            Quantitative treatment confirms the order for magnitude of ΔG as Box 1 < Box 2 < Box 3.

Spontaneous Processes (Section 19.1)

19.12   (a)   Nonspontaneous; at 1 atm, ice does not melt spontaneously at temperatures below its normal melting point.

(b)   Nonspontaneous; a mixture cannot be separated without outside intervention.

(c)   Spontaneous.

(d)   Spontaneous. The reaction is spontaneous but slow unless encouraged by a catalyst or spark.

(e)   Spontaneous; the very polar HCl molecules readily dissolve in water to form concentrated HCl(aq).

19.14   (a)   The sign of $\Delta S°$ at room temperature is positive, because there are more moles of gas in the products than the reactants. (Also, because the process is spontaneous and $\Delta H°$ is positive at room temperature, $\Delta S°$ must be positive so that ΔG is negative.)

(b) $\Delta S°$ is defined as the entropy change for the reaction with all reactants and products in their standard states, calculated at 298 K. The value of $\Delta S°$ does not change with a change in reaction conditions. The value of ΔG will change, depending on the partial pressure of water vapor in the container.

19.16   (a)   Exothermic. If melting requires heat and is endothermic, freezing must be exothermic.

(b)   At 1 atm (indicated by the term *normal* freezing point), the freezing of *n*-octane is spontaneous at temperatures below –57 °C.

(c)   At 1 atm, the freezing of *n*-octane is nonspontaneous at temperatures above –57 °C.

(d)   At 1 atm and –57 °C, the normal freezing point of *n*-octane, the solid and liquid phases are in equilibrium. That is, at the freezing point, *n*-octane molecules escape to the liquid phase at the same rate as liquid *n*-octane solidifies, assuming no heat is exchanged between *n*-octane and the surroundings.

19.18   (a)   True.

(b)   True.

(c)   False. For a reversible process, the change in entropy of the surroundings is equal in magnitude and opposite in sign for the change in entropy of the system.

(d)   False. The reversible pathway for a process produces the maximum possible work.

19.20   (a)   Yes, because ΔE is a state function. $(1 \rightarrow 2) = -\Delta E \ (2 \rightarrow 1)$

(b)   No. We can say nothing about the values of q and w because we have no information about the paths.

(c)   The magnitudes of the work are equal, but the signs are opposite. If the changes of state are reversible, the two paths are the same and $w \ (1 \rightarrow 2) = -w \ (2 \rightarrow 1)$. This is the maximum realizable work from this system.

19.22   (a)   The detonation of an explosive is definitely spontaneous, once it is initiated.

(b)   The quantity q is related to ΔH. As the detonation is highly exothermic, q is large and negative.

If only P–V work is done and P is constant, $\Delta H = q$. Although these conditions probably do not apply to a detonation, we can still predict the sign of q, based on ΔH, if not its exact magnitude.

(c) The sign (and magnitude) of w depend on the path of the process, the exact details of how the detonation is carried out. It seems clear, however, that work will be done by the system on the surroundings in almost all circumstances (buildings collapse, earth and air are moved), so the sign of w is probably negative.

(d) $\Delta E = q + w$. If q and w are both negative, then the sign of ΔE is negative, regardless of the magnitudes of q and w.

Entropy and the Second Law of Thermodynamics (Section 19.2)

19.24 Both vaporizations are *isothermal*; they occur at constant temperature. For an isothermal process, $\Delta S = q_{rev}/T$.

(a) Assuming that q_{rev} is closely related to enthalpy of vaporization and is about the same at the two temperatures, ΔS is larger at 25 °C than at 100 °C.

(b) No. Because ΔS is a state function, it is independent of path. We can calculate ΔS for a reversible pathway, even if the change does not occur that way.

19.26 (a) Ga(l) → Ga(s), ΔS is negative, less motional freedom

(b) $\Delta H = 60.0 \text{ g Ga} \times \dfrac{1 \text{ mol Ga}}{69.723 \text{ g Ga}} \times \dfrac{-5.59 \text{ kJ}}{\text{mol Ga}} = -4.81046 = -4.81 \text{ kJ}$

$\Delta S = \dfrac{\Delta H}{T} = -4.81046 \text{ kJ} \times \dfrac{1000 \text{ J}}{1 \text{ kJ}} \times \dfrac{1}{(273.15 + 29.8)\text{K}} = -15.9 \text{ J/K}$

19.28 (a) Not necessarily. The only thing we know for sure is that the entropy of the universe increases for a spontaneous process.

(b) ΔS_{surr} is positive and greater than the magnitude of the decrease in ΔS_{sys}.

(c) $\Delta S_{sys} = 78 \text{ J/K}$.

19.30 (a) According to Boyle's law, pressure and volume are inversely proportional at constant amount and temperature. If the pressure of an ideal gas increases, volume decreases. We expect a decrease in entropy, or negative ΔS, for the isothermal compression of an ideal gas, owing to the smaller volume available for motion of the particles.

(b) According to Boyle's law, $P_1V_1 = P_2V_2$ at constant n and T.

0.750 atm × V_1 = 1.20 atm × V_2; V_2/V_1 = 0.750 atm/1.20 atm = 0.62500 = 0.625

ΔS_{sys} = nR ln (V_2/V_1) = 0.600 mol (8.314 J/mol-K)(ln 0.625) = –2.34 J/K

Check. An increase in pressure results in a decrease in volume at constant T, so we expect ΔS to be negative, and it is.

(c) The temperature at which the compression (increase in pressure, decrease in volume) occurs is not needed to calculate the entropy change, as long as the process is isothermal.

The Molecular Interpretation of Entropy and the Third Law of Thermodynamics (Section 19.3)

19.32   (a)   A thermodynamic *state* is a set of conditions, usually temperature and pressure, that defines the properties of a bulk material. A *microstate* is a single possibility for all the positions and kinetic energies of all the molecules in a sample; it is a snapshot of positions and speeds at a particular instant.

     (b)   According to Equation 19.5 (Boltzmann law), the more possible microstates for a macroscopic state, the greater the entropy of the state. If S decreases going from A to B, then A has more microstates than B. Or, if ΔS is negative, the number of microstates decreases.

     (c)   According to part (b), if the number of microstates available to a system decreases, ΔS_{sys} is negative. For a spontaneous process, ΔS_{univ} is positive, so ΔS_{surr} is positive (and the magnitude is greater than that of ΔS_{sys}).

19.34   (a) ΔH_{vap} for H_2O at 25 °C = 44.02 kJ/mol; at 100 °C = 40.67 kJ/mol

$$\Delta S = \frac{q_{rev}}{T} = \frac{44.02\,kJ}{mol} \times \frac{1000\,J}{kJ} \times \frac{1}{298\,K} = 148\,J/mol\text{-}K$$

$$\Delta S = \frac{q_{rev}}{T} = \frac{40.67\,kJ}{mol} \times \frac{1000\,J}{kJ} \times \frac{1}{373\,K} = 109\,J/mol\text{-}K$$

     (b)   At both temperatures, the liquid → gas phase transition is accompanied by an increase in entropy, as expected. That the magnitude of the increase is greater at the lower temperature requires some explanation.

         In the liquid state, there are significant hydrogen bonding interactions between H_2O molecules. This reduces the number of possible molecular positions and the number of microstates. Liquid water at 100° has sufficient kinetic energy to have broken many hydrogen bonds, so the number of microstates for $H_2O(l)$ at 100° is greater than the number of microstates for $H_2O(l)$ at 25 °C. The difference in the number of microstates upon vaporization at 100 °C is smaller, and the magnitude of ΔS is smaller.

19.36   (a)   Solids are much more ordered than gases, so ΔS is negative.

     (b)   The entropy of the system increases in Exercise 19.12 (a) and (e). There is more motional freedom for the system in both cases. In (b), (c), and (d), there is less motional freedom after the change and the entropy of the system decreases.

19.38   (a)   When temperature increases, the range of accessible molecular speeds and kinetic energies increases. This produces more microstates and an increase in entropy.

     (b)   When the volume of a gas increases (even at constant T), there are more possible positions for the particles, more microstates, and greater entropy.

     (c)   When equal volumes of two miscible liquids are mixed, the volume of the sample and, therefore, the number of possible arrangements increases. This produces more microstates and an increase in entropy.

19.40   (a)   True. (From the Boltzmann relationship, $S = k \ln W$.)

(b)   True. Because CO_2 has more than one atom, the thermal energy can be distributed as translational, vibrational, or rotational motion.

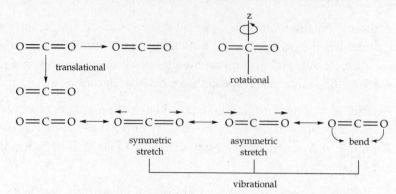

(c)   False. At a given temperature, $CO_2(g)$ has more microstates and, thus, greater entropy than $Ar(g)$. Because $CO_2(g)$ is a triatomic molecule, it has multiple rotational and vibrational microstates not available to monatomic $Ar(g)$.

19.42   (a)   1 mol of $As_4(g)$ at 300 °C, 0.01 atm (As_4 has more massive atoms in a comparable system at the same temperature.)

           (b)   1 mol $H_2O(g)$ at 100 °C, 1 atm [larger volume occupied by $H_2O(g)$]

           (c)   0.5 mol $CH_4(g)$ at 298 K, 20-L volume (more complex molecule, more rotational and vibrational degrees of freedom)

           (d)   100 g of $Na_2SO_4(aq)$ at 30 °C (more motional freedom in aqueous solution)

19.44   (a) $Au(l) \rightarrow Au(s)$; negative ΔS, less motional freedom in the solid

           (b) $Cl_2(g) \rightarrow 2Cl(g)$; positive ΔS, moles of gas increase

           (c) $CO(g) + 2 H_2(g) \rightarrow CH_3OH(l)$; negative ΔS, moles of gas decrease

           (d) $3 Ca(NO_3)_2(aq) + 2 (NH_4)_3PO_4(aq) \rightarrow Ca_3(PO_4)_2(s) + 6 NH_4NO_3(aq)$; ΔS is negative, less motional freedom, fewer moles of ions in aqueous solution

Entropy Changes in Chemical Reactions (Section 19.4)

19.46   Melting $= -126.5$ °C; boiling $= 97.4$ °C

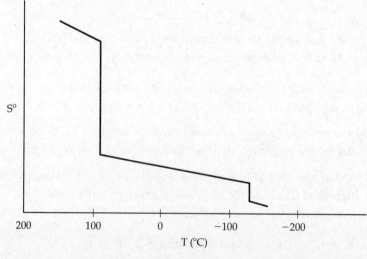

19.48   Propylene will have a higher $S°$ at 25 °C. At this temperature, both are gases, so there are no lattice effects. Because they have the same molecular formula, only the details of their structures are different. In propylene, there is free rotation around the C–C single bond, whereas in cyclopropane the 3-membered ring severely limits rotation. The greater motional freedom of the propylene molecule leads to a higher absolute entropy.

19.50   (a)   CuO(s), 42.59 J/mol-K; Cu_2O(s), 92.36 J/mol-K. Molecules in the solid state have only vibrational motion available to them. The more complex Cu_2O molecule has more vibrational degrees of freedom and a larger standard entropy.

       (b)   1 mol N_2O_4(g), 304.3 J/K; 2 mol NO_2(g), 2(240.45) = 480.90 J/K. More particles have a greater number of arrangements.

       (c) SiO_2(s), 41.84 J/mol-K; CO_2(g), 213.6 J/mol-K. Molecules in the gas phase have a larger volume and more motional freedom than molecules in the solid state. SiO_2 is a covalent network solid so its molecular motion is even more restrained than a typical molecular solid.

       (d)   CO(g), 197.9 J/mol-K; CO_2(g), 213.6 J/mol-K. The more complex CO_2 molecule has more vibrational degrees of freedom and a slightly higher entropy.

19.52   (a)   C(diamond), $S°$ = 2.43 J/mol-K; C(graphite), $S°$ = 5.69 J/mol-K. Diamond is a network covalent solid with each C atom tetrahedrally bound to four other C atoms. Graphite consists of sheets of fused planar 6-membered rings with each C atom bound in a trigonal planar arrangement to three other C atoms. The internal entropy in graphite is greater because there is translational freedom among the planar sheets of C atoms whereas there is very little translational or vibrational freedom within the covalent-network diamond lattice.

       (b) $S°$ for buckminsterfullerene will be ≥ 10 J/mol-K. $S°$ for graphite is twice $S°$ for diamond, and $S°$ for the fullerene should be higher than that of graphite. The 60-atom "bucky" balls have more flexibility than graphite sheets. Also, the balls have translational freedom in three dimensions, whereas graphite sheets have it in only two directions. Because of the ball structure, there is more empty space in the fullerene lattice than in graphite or diamond; essentially, 60 C-atoms in fullerene occupy a larger volume than 60 C-atoms in graphite or diamond. Thus, the fullerene has additional "molecular" complexity, more degrees of translational freedom, and occupies a larger volume, all features that point to a higher absolute entropy.

19.54   (a) $\Delta S° = S° \, NH_4NO_3(s) - S° \, HNO_3(g) - S° \, NH_3(g)$

           $= 151 - 266.4 - 192.5 = -307.9 = -308 \, J/K$

       $\Delta S°$ is large and negative because all reactants are gases (2 moles) and the product is a solid.

       (b) $\Delta S° = 4S° \, Fe(s) + 3S° \, O_2(g) - 2 \, S° \, Fe_2O_3(s)$

           $= 4(27.15) + 3(205.0) - 2(89.96) = 543.68 = 543.7 \, J/K$

       $\Delta S°$ is large and positive because the reaction produces 3 moles of gas and the reactant is a solid.

(c) $\Delta S° = S°\ CaCl_2(s)\ + S°\ CO_2(g) + S°\ H_2O(l) - S°\ CaCO_3(s) - 2S°\ HCl(g)$

$$= 104.6 + 213.6 + 69.91 - 92.88 - 2(186.69) = -78.15\ J/K$$

$\Delta S°$ is small and negative because the products contain one fewer mole of gas, but one more mole of liquid. Note the very small standard entropy for $H_2O(l)$, owing to its strength of hydrogen bonding. If the products included one mole of a different liquid, the magnitude of the entropy change would be even smaller.

(d) $\Delta S° = S°\ C_6H_6(l) + 6S°\ H_2(g) - 3S°\ C_2H_6(g)$

$$= 172.8 + 6(130.58) - 3(229.5) = 267.78 = 267.8\ J/K$$

$\Delta S°$ is positive because there are more moles of gas in the products.

Gibbs Free Energy (Sections 19.5 and 19.6)

19.56   (a)   No. $\Delta G = \Delta G° + RT\ lnQ$. The relative magnitudes of ΔG and $\Delta G°$ depend on the value of Q.

        (b)   For a process that occurs at constant temperature and pressure, the system is at equilibrium when $\Delta G = 0$.

        (c)   No. Activation energy is related to the rate constant, k. The sign and magnitude of ΔG give no information about rate.

19.58   (a) $\Delta H°$ is positive; the reaction is endothermic.

        (b) $\Delta S°$ is positive; the reaction leads to an increase in disorder.

        (c) $\Delta G° = \Delta H° - T\Delta S° = 23.7\ kJ - 298\ K\ (0.0524\ kJ/K) = 8.0848 = 8.08\ kJ$

        (d)   At 298 K, $\Delta G°$ is positive. If all reactants and products are present in their standard states, the reaction is spontaneous in the reverse direction at this temperature; it is nonspontaneous in the forward direction.

19.60   (a)   There is no thermodynamic data in Appendix C for CrO_3; we will substitute Cr_2O_3. $\ \ 4Cr(s) + 3O_2(g) \rightarrow 2Cr_2O_3(s)$

            $\Delta H° = 2(-1139.7) - 4(0) + 3(0) = -2279.4\ kJ$

            $\Delta S° = 2(81.2) - 4(23.6) - 3(205.0) = -547.0\ J/K$

            $\Delta G° = 2(-1058.1) - 4(0) - 3(0) = -2116.2\ kJ$

            $\Delta G° = -2279.4\ kJ - 298\ K(-0.5470\ kJ/K) = -2116.4\ kJ$

        (b) $\Delta H° = -553.5 - 393.5 - (-1216.3) = 269.3\ kJ$

            $\Delta S° = 70.42 + 213.6 - 112.1 = 171.92 = 171.9\ J/K$

            $\Delta G° = -525.1 - 394.4 - (-1137.6) = 218.1\ kJ$

            $\Delta G° = 269.3\ kJ - 298\ K\ (0.1719\ kJ/K) = 218.1\ kJ$

        (c)   Assume the reactant is P(g), not P(s).

            $\Delta H° = 2(-1594.4) + 5(0) - 2(316.4) - 10(-268.61) = -1135.5\ kJ$

            $\Delta S° = 2(300.8) + 5(130.58) - 2(163.2) - 10(173.51) = -807.0\ J/K$

$\Delta G° = 2(-1520.7) + 5(0) - 2(280.0) - 10(-270.70) = -894.4$ kJ

$\Delta G° = -1135.5$ kJ $- 298$ K$(-0.8070$ kJ/K$) = -895.014 = -895.0$ kJ

(The small discrepancy in $\Delta G°$ values is because of experimental uncertainties in tabulated thermodynamic data.)

(d) $\Delta H° = -284.5 - (0) - (0) = -284.5$ kJ

$\Delta S° = 122.5 - 64.67 - 205.0 = -147.2$ J/K

$\Delta G° = -240.6 - (0) - (0) = -240.6$ kJ

$\Delta G° = -284.5$ kJ $- 298$ K $(-0.1472$ kJ/K$) = -240.634 = -240.6$ kJ

19.62   (a) $\Delta G° = 2\Delta G°$ AgCl(s) $- [2\Delta G°$ Ag(s) $+ \Delta G°$ Cl$_2$(g)$]$

   $= 2(-109.7) - 2(0) - 0 = -219.4$ kJ, spontaneous

(b)   P$_4$O$_{10}$(s) $+ 16$H$_2$(g) $\rightarrow$ 4PH$_3$(g) $+ 10$H$_2$O(g)

$\Delta G° = 4\Delta G°$ PH$_3$(g) $+ 10\Delta G°$ H$_2$O(g) $- [\Delta G°$ P$_4$O$_{10}$(s) $+ 16\Delta G°$ H$_2$(g)$]$

   $= 4(13.4) + 10(-228.57) - [-2675.2] - 16(0) = 443.1$ kJ, nonspontaneous

(c) $\Delta G° = \Delta G°$ CF$_4$(g) $+ 4\Delta G°$ HF(g) $- [\Delta G°$ CH$_4$(g) $+ 4\Delta G°$ F$_2$(g)$]$

   $= -635.1 + 4(-270.70) - (-50.8) - 4(0) = -1667.1$ kJ, spontaneous

(d) $\Delta G° = 2\Delta G°$ H$_2$O(l) $+ \Delta G°$ O$_2$(g) $- 2\Delta G°$ H$_2$O$_2$(l)

   $= 2(-237.13) + 0 - 2(-120.4) = -233.5$ kJ, spontaneous

19.64   (a) $\Delta G°$ should be less negative than $\Delta H°$. Products contain fewer moles of gas, so $\Delta S°$ is negative. $\Delta G° = \Delta H° - T\Delta S°$; $-T\Delta S°$ is positive, so $\Delta G°$ is less negative than $\Delta H°$.

(b)   We can estimate $\Delta S°$ using a similar reaction and then use $\Delta G° = \Delta H° - T\Delta S°$ (estimate) to get a ballpark figure. There are no sulfite salts listed in Appendix C, so use a reaction such as CO$_2$(g) + CaO(s) $\rightarrow$ CaCO$_3$(s) or CO$_2$(g) + BaO(s) $\rightarrow$ BaCO$_3$(s). Or, calculate both $\Delta S°$ values and use the average as your estimate.

19.66 $\Delta G° = \Delta H° - T\Delta S°$

(a) $\Delta G° = -844$ kJ $- 298$ K$(-0.165$ kJ/K$) = -795$ kJ, spontaneous

(b) $\Delta G° = +572$ kJ $- 298$ K$(0.179$ kJ/K$) = +519$ kJ, nonspontaneous

To be spontaneous, ΔG must be negative ($\Delta G < 0$).

Thus, $\Delta H° - T\Delta S° < 0$; $\Delta H° < T\Delta S°$; $T > \Delta H°/\Delta S°$; $T > \dfrac{572 \text{ kJ}}{0.179 \text{ kJ/K}} = 3.20 \times 10^3$ K

19.68   At 45 °C or 318 K, $\Delta G > 0$. $\Delta G = \Delta H - T\Delta S > 0$

$\Delta H - 318$ K $(72$ J/K$) > 0$; $\Delta H > +2.3 \times 10^4$ J; $\Delta H > +23$ kJ

The reaction is nonspontaneous and has a positive ΔS, so it must be endothermic. Because it is "barely" nonspontaneous, the magnitude will not be much greater than 23 kJ.

19.70　　ΔG is negative when TΔS > ΔH or T > ΔH/ΔS.

ΔH° = ΔH° CH₃OH + ΔH° CO(g) – ΔH° CH₃COOH(l)

$\quad$ = –201.2 – 110.5 – (–487.0) = 175.3 kJ

ΔS° = S° CH₃OH + S° CO(g) - S° CH₃COOH(l) = 237.6 + 197.9 – 159.8 = 275.7 J/K

$$T > \frac{175.3 \text{ kJ}}{0.2757 \text{ kJ/K}} = 635.8 \text{ K}$$

The reaction is spontaneous above 635.8 K (363 °C).

19.72　　(a)　　$\Delta H° = \Delta H_f° \text{ CH}_3\text{OH(g)} - \Delta H_f° \text{ CH}_4\text{(g)} - 1/2\, \Delta H_f° \text{ O}_2\text{(g)}$

$\quad\quad$ = –201.2 – (–74.8) – (1/2)(0) = –126.4 kJ

$\Delta S° = S° \text{ CH}_3\text{OH(g)} - S° \text{ CH}_4\text{(g)} - 1/2\, S° \text{ O}_2\text{(g)}$

$\quad\quad$ = 237.6 – 186.3 – 1/2(205.0) = –51.2 J/K = –0.0512 kJ/K

(b)　　ΔG° = ΔH° – TΔS°. –TΔS° is positive, so ΔG° will increase (becomes more positive) as temperature increases.

(c)　　ΔG° = ΔH° – TΔS° = –126.4 kJ – 298 K(–0.0512 kJ/K) = –111.1 kJ

The reaction is spontaneous at 298 K because ΔG° is negative at this temperature. In this case, ΔG° could have been calculated from $\Delta G_f°$ values in Appendix C, because these values are tabulated at 298 K.

(d)　　No. The reaction is at equilibrium when ΔG° = 0.

ΔG° = ΔH° – TΔS° = 0. ΔH° = TΔS°, T = ΔH°/ΔS°

T = –126.4 kJ/–0.0512 kJ/K = 2469 = 2470 K

This temperature is so high that the reactants and products are likely to decompose. At standard conditions, equilibrium is functionally unattainable for this reaction.

19.74　　(a)　　As in Sample Exercise 19.10, $T_{sub} = \Delta H_{sub}° / \Delta S_{sub}°$.

Use data from Appendix C to calculate $\Delta H_{sub}°$ and $\Delta S_{sub}°$ for I_2(s).

I_2(s) → I_2(l) melting

I_2(l) → I_2(g) boiling

I_2(s) → I_2(g) sublimation

$\Delta H_{sub}° = \Delta H_f° I_2(g) - \Delta H_f° I_2(s) = 62.25 - 0 = 62.25 \text{ kJ}$

$\Delta S_{sub}° = S° I_2(g) - S° I_2(s) = 260.57 - 116.73 = 143.84 \text{ J/K} = 0.14384 \text{ kJ/K}$

$$T_{sub} = \frac{\Delta H_{sub}°}{\Delta S_{sub}} = \frac{62.25 \text{ kJ}}{0.14384 \text{ kJ/K}} = 432.8 \text{ K} = 159.6 \text{ °C}$$

In making this estimate, we assume that at equilibrium, both I_2(s) and I_2(g) are present in their standard states of pure solid and vapor at 1 atm and, consequently, $\Delta G_{sub} = \Delta G_{sub}° = 0$. We also assume that the values of $\Delta H_{sub}°$ and $\Delta S_{sub}°$ are the same at 298 K and at the sublimation temperature.

(b) T_m for $I_2(s) = 386.85$ K $= 113.7$ °C; $T_b = 457.4$ K $= 184.3$ °C
(from WebElements™ 2013)

(c)   The boiling point of I_2 is closer to the sublimation temperature. Both boiling and sublimation begin with molecules in a condensed phase (little space between molecules) and end in the gas phase (large intermolecular distances). Separation of the molecules is the main phenomenon that determines both ΔH and ΔS, so it is not surprising that the ratio of $\Delta H/\Delta S$ is similar for sublimation and boiling.

19.76   (a) $CH_4(g) + 2O_2(g) \rightarrow CO_2(g) + 2H_2O(l)$

$$\Delta H° = \Delta H_f° \ CO_2(g) + 2\Delta H_f° \ H_2O(l) - \Delta H_f° \ CH_4(g) - 2\Delta H_f° \ O_2(g)$$

$$= (-393.5) + 2(-285.83) - (-74.8) - 2(0) = -890.4 \text{ kJ/mol } CH_4 \text{ burned}$$

(b) $w_{max} = \Delta G° = \Delta G_f° \ CO_2(g) + 2\Delta G_f° \ H_2O(l) - \Delta G_f° \ CH_4(g) - 2\Delta G_f° \ O_2(g)$

$$= (-394.4) + 2(-237.13) - (-50.8) - 2(0) = -817.9 \text{ kJ}$$

The system can accomplish at most 817.86 kJ of work per mole of CH_4 on the surroundings.

Free Energy and Equilibrium (Section 19.6)

19.78   Consider the relationship $\Delta G = \Delta G° + RT \ln Q$, where Q is the reaction quotient.

(a) ΔG decreases. $H_2(g)$ appears in the denominator of Q for this reaction. An increase in pressure of H_2 decreases Q and ΔG becomes smaller or more negative. Increasing the concentration or partial pressure of a reactant increases the tendency for a reaction to occur.

(b) ΔG increases. $H_2(g)$ appears in the numerator of Q for this reaction. Increasing the pressure of H_2 increases Q and ΔG becomes more positive. Increasing the concentration or partial pressure of a product decreases the tendency for the reaction to occur.

(c) ΔG decreases. $H_2(g)$ appears in the denominator of Q for this reaction. An increase in pressure of H_2 decreases Q and ΔG becomes smaller or more negative. Increasing the concentration or partial pressure of a reactant increases the tendency for a reaction to occur.

19.80   (a) $\Delta G° = \Delta G° \ C_3H_8(g) + 2\Delta G° \ H_2(g) \rightarrow 3\Delta G° \ CH_4(g)$

$$= -23.47 + 2(0) - 3(-50.8) = 128.9 \text{ kJ}$$

(b) $\Delta G = \Delta G° + RT \ln[P_{C_3H_8} \times P_{H_2}^2 / P_{CH_4}^3]$

$$= 128.9 + \frac{8.314 \times 10^{-3} \text{ kJ}}{\text{mol-K}} \times 298 \text{ K} \times \ln[(0.0100) \times (0.0180)^2 /(40.0)^3]$$

$$= 128.9 - 58.735 = 70.165 = 70.2 \text{ kJ}$$

19.82 $\Delta G° = -RT \ln K$; $\ln K = -\Delta G° / RT$; at 298 K, RT = 2.4776 = 2.478 kJ

(a) $\Delta G° = \Delta G° \ NaOH(s) + \Delta G° \ CO_2(g) - \Delta G° \ NaHCO_3(s)$

$$= -379.5 + (-394.4) - (-851.8) = +77.9 \text{ kJ}$$

$$\ln K = \frac{-\Delta G^\circ}{RT} = \frac{-77.9 \text{ kJ}}{2.478 \text{ kJ}} = -31.442 = -31.4; \quad K = 2 \times 10^{-14}$$

$$K = P_{CO_2} = 2 \times 10^{-14}$$

(b) $\Delta G^\circ = 2\Delta G^\circ \text{ HCl(g)} + \Delta G^\circ \text{ Br}_2(g) - 2\Delta G^\circ \text{ HBr(g)} - \Delta G^\circ \text{ Cl}_2(g)$

           $= 2(-95.27) + 3.14 - 2(-53.22) - 0 = -80.96 \text{ kJ}$

$$\ln K = \frac{-(-80.96)}{2.4776} = +32.68; \quad K = 1.6 \times 10^{14}$$

$$K = \frac{P_{HCl}^2 \times P_{Br_2}}{P_{HBr}^2 \times P_{Cl_2}} = 1.6 \times 10^{14}$$

(c)   From Solution 19.61(a), ΔG° at 298 K = −140.0 kJ.

$$\ln K = \frac{-\Delta G^\circ}{RT} = \frac{-(-140.0)}{2.4776} = 56.51; \quad K = 3.5 \times 10^{24}$$

$$K = \frac{P_{SO_3}^2}{P_{SO_2}^2 \times P_{O_2}} = 3.5 \times 10^{24}$$

19.84 $K = P_{CO_2}$. Calculate ΔG° at the two temperatures using $\Delta G^\circ = \Delta H^\circ - T\Delta S^\circ$ and then calculate K and P_{CO_2}.

        $\Delta H^\circ = \Delta H^\circ \text{ PbO(s)} + \Delta H^\circ \text{ CO}_2(g) - \Delta H^\circ \text{ PbCO}_3(s)$

           $= -217.3 - 393.5 + 699.1 = 88.3 \text{ kJ}$

        $\Delta S^\circ = S^\circ \text{ PbO(s)} + S^\circ \text{ CO}_2(g) - S^\circ \text{ PbCO}_3(s)$

           $= 68.70 + 213.6 - 131.0 = 151.3 \text{ J/K or } 0.1513 \text{ kJ/K}$

(a) $\Delta G^\circ = \Delta H^\circ - T\Delta S^\circ$. At 673 K, $\Delta G^\circ = 88.3 \text{ kJ} - 673 \text{ K}(0.1513 \text{ kJ/K}) = -13.525$

                                         $= -13.5 \text{ kJ}$

$$\ln K = \frac{-\Delta G^\circ}{RT} = \frac{-(-13.525 \times 10^3) \text{J}}{8.314 \text{ J/K} \times 673 \text{ K}} = 2.4172 = 2.42$$

$$K = P_{CO_2} = 11.214 = 11 \text{ atm}$$

(b) $\Delta G^\circ = \Delta H^\circ - T\Delta S^\circ$. At 453 K, $\Delta G^\circ = 88.3 \text{ kJ} - 453 \text{ K } (0.1513 \text{ kJ}) = 19.7611$

                                         $= 19.8 \text{ kJ}$

$$\ln K = \frac{-(19.7611 \times 10^3 \text{ J})}{8.314 \text{ J/K} \times 453 \text{ K}} = -5.2469 = -5.25; \quad K = P_{CO_2} = 5.3 \times 10^{-3} \text{ atm}$$

19.86   (a) $CH_3NH_2(aq) + H_2O(l) \rightleftharpoons CH_3NH_3^+(aq) + OH^-(aq)$

      (b) $\Delta G^\circ = -RT \ln K_b = -(8.314 \times 10^{-3})(298) \ln (4.4 \times 10^{-4}) = 19.148 = 19.1 \text{ kJ}$

      (c) $\Delta G = 0$ at equilibrium

      (d) $\Delta G = \Delta G^\circ + RT \ln Q; \; [OH^-] = 1 \times 10^{-14}/6.7 \times 10^{-9} = 1.4925 \times 10^{-6} = 1.5 \times 10^{-6}$

$$= 19.148 + (8.314 \times 10^{-3})(298) \ln \frac{(2.4 \times 10^{-3})(1.4925 \times 10^{-6})}{0.098} = -23.28 = -23.3 \text{ kJ}$$

Additional Exercises

19.88   (a)   False. The essential question is whether the reaction proceeds far to the right before arriving at equilibrium. The position of equilibrium, which is the essential aspect, is dependent not only on ΔH but on the entropy change as well.

(b)   True.

(c)   True.

(d)   False. *Non*spontaneous processes in general require that work be done to force them to proceed. Spontaneous processes occur without application of work.

(e)   False. Such a process *might* be spontaneous, but would not necessarily be so. Spontaneous processes are those that are exothermic and/or that lead to increased disorder in the system.

19.89

Process	ΔH	ΔS
(a)	+	+
(b)	−	−
(c)	+	+
(d)	+	+
(e)	−	+

19.90   There is no mistake in the calculation. The second law states that in any spontaneous process there is an increase in the entropy of the universe. Although there may be a decrease in entropy of the system, as in the present case, this decrease is more than offset by an increase in entropy of the surroundings.

19.92   (a)   Microstates are possible arrangements for the system. For each die, there are six possibilities for the top face, resulting in (6)(6) = 36 possible arrangements or microstates. (The face that appears on top of one die is not related to or determined by the face on top of the other die.) The two arrangements of top faces shown in the exercise are two of the 36 possible microstates.

(b)   The left pair of dice belongs to state III; the right pair belongs to state VII.

(c)   There are eleven possible states (II through XII; I is not a possibility).

(d)   The state with the most microstates has the highest entropy. State VII has six microstates and the highest entropy. The microstates are (1+6), (2+5), (3+4), (4+3), (5+2), and (6+1). States VI and VIII, on either side of VII, have five microstates. Moving farther away from VII, the number of microstates decreases until we reach the two extremes, II and XII, which each have one microstate.

(e)   States II and XII, with one microstate each, have the lowest entropy.

(f)   The Boltzmann relationship is $S = k \ln W$, where S is the absolute entropy, k is the Botzmann constant, and W is the total number of microstates of the system.

$$S = 1.38 \times 10^{-23} \text{ J/K } (\ln 36) = 4.9453 \times 10^{-23} = 4.95 \times 10^{-23} \text{ J/K}$$

19.93 If $NH_4NO_3(s)$ dissolves spontaneously in water, $\Delta G = \Delta H - T\Delta S$. If ΔG is negative and ΔH is positive, the sign of ΔS must be positive. Furthermore, $T\Delta S > \Delta H$ at room temperature.

19.94 (a) The sign of q for expansion is (+). Vaporization is an endothermic process; the enthaphy of the system increases and q is positive. Our system is the refrigerant. Because the expansion does not occur at constant pressure, q is not exactly equal to ΔH, but its sign is positive.

(b) The sign of q for compression is (–). Compression is the reverse of expansion, and it has the opposite sign.

(c) The expansion chamber is inside the house and the compression chamber is outside. During expansion, q_{sys} increases and q_{sur} decreases. The air surrounding the expansion chamber is cooled and then distributed throughout the house to cool it. If expansion occurred outside, the cool air would be wasted. Compression releases heat to the surroundings; it occurs outside so that the released heat can be dissipated by the outside air.

(d) No. Heat can flow reversibly between a system and its surroundings only if the two have an infinitesimally small difference in temperature and the amount of heat transferred is infinitesimally small. There is no mechanism in our system to regulate the amount of heat transferred. When the liquid flows into the low pressure chamber, all the liquid vaporizes, not an infinitesimally small amount.

(e) A spontaneous process occurs without outside intervention. In an air conditioner, expansion (vaporization) of the refrigerant is spontaneous, but compression (condensation) to the liquid state is nonspontaneous. Cooling the house from 31 $^{\circ}$C to 24 $^{\circ}$C is nonspontaneous. [Note that all spontaneous processes are irreversible, but not all irreversible processes are spontaneous.]

19.95 At the normal boiling point of a liquid, $\Delta G = 0$ and $\Delta H_{vap} = T\Delta S_{vap}$; $T = \Delta H_{vap}/\Delta S_{vap}$. By Trouton's rule, $\Delta S_{vap} = 88$ J/mol-K. The process of vaporization is:

(a) $Br_2(l) \rightleftharpoons Br_2(g)$

$$\Delta H_{vap} = \Delta H_f^{\circ} \, Br_2(g) - \Delta H_f^{\circ} \, Br_2(l) = 30.71 \text{ kJ} - 0 = 30.71 \text{ kJ}$$

$$T_b = \frac{\Delta H_{vap}}{\Delta S_{vap}} = \frac{30.71 \text{ kJ}}{88 \text{ J/ mol-K}} \times \frac{1000 \text{ J}}{\text{kJ}} = 349 = 3.5 \times 10^2 \text{ K}$$

(b) According to WebElements^{TM} 2013, the normal boiling point of $Br_2(l)$ is 332 K. Trouton's rule provides a good "ballpark" estimate.

Trouton's rule assumes that the entropy of vaporization of most molecules is due to the much greater motional freedom of the gaseous state relative to the liquid state. This assumption is incorrect for liquids with strong intermolecular forces (usually hydrogen bonding) in either the liquid or the gaseous state.

There are also the usual experimental uncertainties in the measurement of ΔH_f° for $Br_2(g)$ and the normal boiling point of $Br_2(l)$.

19.97　(a)　(i)　Ti(s) + 2Cl$_2$(g) → TiCl$_4$(g)

ΔH° = ΔH° TiCl$_4$(g) − ΔH° Ti(s) − 2ΔH° Cl$_2$(g)

= −763.2 − 0 − 2(0) = −763.2 kJ

ΔS° = 354.9 − 30.76 − 2(222.96) = −121.78 = −121.8 J/K

ΔG° = −726.8 − 0 − 2(0) = −726.8 kJ

$\ln K = \dfrac{-\Delta G°}{RT} = \dfrac{-(-726.8)\ kJ}{2.4777\ kJ} = 293.337 = 293.3;\ K = 2 \times 10^{127}$

(ii)　C$_2$H$_6$(g) + 7Cl$_2$(g) → 2CCl$_4$(g) + 6HCl(g)

ΔH° = 2ΔH° CCl$_4$(g) + 6ΔH° HCl(g) − ΔH° C$_2$H$_6$(g) − 7ΔH° Cl$_2$(g)

= 2(−106.7) + 6(−92.30) − (−84.68) − 7(0) = −682.52 = −682.5 kJ

ΔS° = 2(309.4) + 6(186.69) − 229.5 − 7(222.96) = −51.28 = −51.4 J/K

ΔG° = 2(−64.0) + 6(−95.27) − (−32.89) − 7(0) = −666.73 = −666.7 kJ

$\ln K = \dfrac{-\Delta G°}{RT} = \dfrac{-(-666.73)\ kJ}{2.4777\ kJ} = 269.0923 = 269.1;\ K = 7 \times 10^{116}$

(iii)　BaO(s) + CO$_2$(g) → BaCO$_3$(s)

ΔH° = ΔH° BaCO$_3$(s) − ΔH° BaO(s) − ΔH° CO$_2$(g)

= −1216.3 − (−553.5) − (−393.5) = −269.3 kJ

ΔS° = 112.1 − 70.42 − 213.6 = −171.9 J/K

ΔG° = −1137.6 − (−525.1) − (−394.4) = −218.1 kJ

$\ln K = \dfrac{-\Delta G°}{RT} = \dfrac{-(-218.1)\ kJ}{2.4777\ kJ} = 88.0252 = 88.03;\ K = 1.7 \times 10^{38}$

(b)　(i), (ii), and (iii) all have negative ΔG° values and are spontaneous at standard conditions and 25 °C. Essentially, these reactions all go to completion; they have unimaginably large K values.

(c)　ΔG° = ΔH° − TΔS°. All three reactions have −ΔH° and −ΔS°. They all have −ΔG° at 25 °C, and ΔG° becomes more positive as T increases.

19.98　ΔG = ΔG° + RT ln Q; ln K = −ΔG°/RT

(a)　$Q = \dfrac{P_{NH_3}^2}{P_{N_2} \times P_{H_2}^3} = \dfrac{(1.2)^2}{(2.6)(5.9)^3} = 2.697 \times 10^{-3} = 2.7 \times 10^{-3}$

ΔG° = 2ΔG° NH$_3$(g) − ΔG° N$_2$(g) − 3ΔG° H$_2$(g)

= 2(−16.66) − 0 − 3(0) = −33.32 kJ

$\ln K = \dfrac{-\Delta G°}{RT} = \dfrac{-(-33.32)\ kJ}{2.4777\ kJ} = 13.448 = 13.45;\ K = 6.9 \times 10^5$

$\Delta G = -33.32\ kJ + \dfrac{8.314 \times 10^{-3}\ kJ}{mol\text{-}K} \times 298\ K \times \ln(2.69 \times 10^{-3})$

ΔG = −33.32 − 14.66 = −47.98 = −48.0 kJ

(b)　$Q = \dfrac{P_{N_2}^3 \times P_{H_2O}^4}{P_{N_2H_4}^2 \times P_{NO_2}^2} = \dfrac{(0.5)^3(0.3)^4}{(5.0 \times 10^{-2})^2(5.0 \times 10^{-2})^2} = 162 = 2 \times 10^2$

$\Delta G° = 3\Delta G°\, N_2(g) + 4\Delta G°\, H_2O(g) - 2\Delta G°\, N_2H_4(g) - 2\Delta G°\, NO_2(g)$

$= 3(0) + 4(-228.57) - 2(159.4) - 2(51.84) = -1336.8 \text{ kJ}$

$\ln K = \dfrac{-\Delta G°}{RT} = \dfrac{-(-1336.8)\text{ kJ}}{2.4777\text{ kJ}} = 539.533 = 539.53;\ K = 2.1 \times 10^{234}$

$\Delta G = -1336.8 \text{ kJ} + 2.478 \ln 162 = -1324.2 = -1.32 \times 10^3 \text{ kJ}$

(c)　$Q = \dfrac{P_{N_2} \times P_{H_2}^2}{P_{N_2H_4}} = \dfrac{(1.5)(2.5)^2}{0.5} = 18.75 = 2 \times 10^1$

$\Delta G° = \Delta G°\, N_2(g) + 2\Delta G°\, H_2(g) - \Delta G°\, N_2H_4(g)$

$= 0 + 2(0) - 159.4 = -159.4 \text{ kJ}$

$\ln K = \dfrac{-\Delta G°}{RT} = \dfrac{-(-159.4)\text{ kJ}}{2.4777\text{ kJ}} = 64.334 = 64.33;\ K = 8.7 \times 10^{27}$

$\Delta G = -159.4 \text{ kJ} + 2.478 \ln 18.75 = -152.1 = -152 \text{ kJ}$

19.99　**Reaction**　**(a) Sign of ΔH°**　**(a) Sign of ΔS°**　**(b) K > 1?**　**(c) Variation in K as Temp. Increases**

Reaction	Sign of ΔH°	Sign of ΔS°	K > 1?	Variation in K as Temp. Increases
(i)	−	−	yes	decrease
(ii)	+	+	no	increase
(iii)	+	+	no	increase
(iv)	+	+	no	increase

(a)　Note that at a particular temperature, positive ΔH° leads to a smaller value of K, whereas positive ΔS° increases the value of K.

19.100　(a)　$K = \dfrac{\chi_{CH_3COOH}}{\chi_{CH_3OH}P_{CO}}$

$\Delta G° = -RT \ln K;\ \ln K = -\Delta G/RT$

$\Delta G° = \Delta G°\, CH_3COOH(l) - \Delta G°\, CH_3OH(l) - \Delta G°\, CO(g)$

$= -392.4 - (-166.23) - (-137.2) = -89.0 \text{ kJ}$

$\ln K = \dfrac{-(-89.0\text{ kJ})}{(8.314 \times 10^{-3}\text{ kJ/K})(298\text{ K})} = 35.922 = 35.9;\ K = 4 \times 10^{15}$

(b)　$\Delta H° = \Delta H°\, CH_3COOH(l) - \Delta H°\, CH_3OH(l) - \Delta H°\, CO(g)$

$= -487.0 - (-238.6) - (-110.5) = -137.9 \text{ kJ}$

The reaction is exothermic, so the value of K will decrease with increasing temperature, and the mole fraction of CH_3COOH will also decrease. Elevated temperatures must be used to increase the speed of the reaction. Thermodynamics cannot predict the rate at which a reaction reaches equilibrium.

(c) $\Delta G° = -RT \ln K$; $K = 1$, $\ln K = 0$, $\Delta G° = 0$

$\Delta G° = \Delta H° - T\Delta S°$; when $\Delta G° = 0$, $\Delta H° = T\Delta S°$

$\Delta S° = S°\ CH_3COOH(l) - S°\ CH_3OH(l) - S°\ CO(g)$

      $= 159.8 - 126.8 - 197.9 = -164.9\ J/K = -0.1649\ kJ/K$

$-137.9\ kJ = T(-0.1649\ kJ/K)$, $T = 836.3\ K$

The equilibrium favors products up to 836 K or 563 °C, so the elevated temperatures to increase the rate of reaction can be safely employed.

19.102   (a) $\Delta G° = -RT \ln K$ (Equation 19.20); $\ln K = -\Delta G°/RT$

Use $\Delta G° = \Delta H° - T\Delta S°$ to get $\Delta G°$ at the two temperatures. Calculate $\Delta H°$ and $\Delta S°$ using data in Appendix C.

$2CH_4(g) \rightarrow C_2H_6(g) + H_2(g)$

$\Delta H° = \Delta H°\ C_2H_6(g) + \Delta H°\ H_2(g) - 2\Delta H°\ CH_4(g) = -84.68 + 0 - 2(-74.8) = 64.92$

                                                  $= 64.9\ kJ$

$\Delta S° = S°\ C_2H_6(g) + S°\ H_2(g) - 2S°\ CH_4(g) = 229.5 + 130.58 - 2(186.3) = -12.52$

                                                  $= -12.5\ J/K$

at 298 K, $\Delta G = 64.92\ kJ - 298\ K(-12.52 \times 10^{-3}\ kJ/K) = 68.65 = 68.7\ kJ$

$$\ln K = \frac{-68.65\ kJ}{(8.314 \times 10^{-3}\ kJ/K)(298\ K)} = -27.709 = -27.7, K = 9.25 \times 10^{-13} = 9 \times 10^{-13}$$

at 773 K, $\Delta G = 64.9\ kJ - 773\ K(-12.52 \times 10^{-3}\ J/K) = 74.598 = 74.6\ kJ$

$$\ln K = \frac{-74.598\ kJ}{(8.314 \times 10^{-3}\ kJ/K)(773\ K)} = -11.607 = -11.6,\ K = 9.1 \times 10^{-6}$$

Because the reaction is endothermic, the value of K increases with an increase in temperature.

$2CH_4(g) + 1/2\ O_2(g) \rightarrow C_2H_6(g) + H_2O(g)$

$\Delta H° = \Delta H°\ C_2H_6(g) + \Delta H°\ H_2O(g) - 2\Delta H°\ CH_4(g) - 1/2\ \Delta H°\ O_2(g)$

      $= -84.68 + (-241.82) - 2(-74.8) - 1/2\ (0) = -176.9\ kJ$

$\Delta S° = S°\ C_2H_6(g) + S°\ H_2O(g) - 2S°\ CH_4(g) - 1/2\ S°\ O_2(g)$

      $= 229.5 + 188.83 - 2(186.3) - 1/2\ (205.0) = -56.77 = -56.8\ J/K$

at 298 K, $\Delta G = -176.9\ kJ - 298\ K(-56.77 \times 10^{-3}\ kJ/K) = -159.98 = -160.0\ kJ$

$$\ln K = \frac{-(-159.98\ kJ)}{(8.314 \times 10^{-3}\ kJ/K)(298\ K)} = 64.571 = 64.57;\ K = 1.1 \times 10^{28}$$

at 773 K, $\Delta G = -176.9\ kJ - 773\ K\ (-56.77 \times 10^{-3}\ kJ/K) = -133.02 = -133.0\ kJ$

$$\ln K = \frac{-(-133.02\ kJ)}{(8.314 \times 10^{-3}\ kJ/K)(773\ K)} = 20.698 = 20.70;\ K = 9.750 \times 10^8 = 9.8 \times 10^8$$

Because this reaction is exothermic, the value of K decreases with increasing temperature.

(b)   The difference in $\Delta G°$ for the two reactions is primarily enthalpic; the first reaction is endothermic and the second exothermic. Both reactions have $-\Delta S°$, which inhibits spontaneity.

(c)   This is an example of coupling a useful but nonspontaneous reaction with a spontaneous one to spontaneously produce a desired product.

$$2CH_4(g) \rightarrow C_2H_6(g) + H_2(g) \qquad \Delta G_{298}^o = +68.7 \text{ kJ, nonspontaneous}$$

$$H_2(g) + 1/2\,O_2(g) \rightarrow H_2O(g) \qquad \Delta G_{298}^o = -228.57 \text{ kJ, spontaneous}$$

$$2CH_4(g) + 1/2\,O_2(g) \rightarrow C_2H_6(g) + H_2O(g) \qquad \Delta G_{298}^o = -159.9 \text{ kJ, spontaneous}$$

(d) $CH_4(g) + 2O_2(g) \rightarrow CO_2(g) + 2H_2O(g)$

19.103 $\Delta G°$ for the metabolism of glucose is:

$6\Delta G° \, CO_2(g) + 6\Delta G° \, H_2O(l) - \Delta G° \, C_6H_{12}O_6(s) - 6\Delta G° \, O_2(g)$

$\Delta G° = 6(-394.4) + 6(-237.13) - (-910.4) + 6(0) = -2878.8 \text{ kJ}$

mol ATP $= -2878.8 \text{ kJ} \times 1 \text{ mol ATP} / (-30.5 \text{ kJ}) = 94.4 \text{ mol ATP} / \text{mol glucose}$

Note that this calculation is done at standard conditions, not metabolic conditions. A more accurate answer would be obtained using ΔG values that reflect actual concentration, partial pressure, and pH in a cell.

19.104   (a)   The equilibrium of interest here can be written as:

K^+ (plasma) $\rightleftharpoons$ K^+ (muscle)

Because the fluid is aqueous on both sides of the cell membrane, assume that the equilibrium constant for this process is exactly 1; that is, $\Delta G° = 0$. However, ΔG is not zero because the concentrations are not the same on both sides of the membrane. Use Equation 19.19 to calculate ΔG:

$$\Delta G = \Delta G° + RT \ln \frac{[K^+ \text{ (muscle)}]}{[K^+ \text{ (plasma)}]}$$

$$= 0 + (8.314)(310) \ln \frac{(0.15)}{(5.0 \times 10^{-3})} = 8766 \text{ J} = 8.8 \text{ kJ}$$

(b)   Note that ΔG is positive. This means that work must be done on the system (blood plasma plus muscle cells) to move the K^+ ions "uphill," as it were. The minimum amount of work possible is given by the value for ΔG. This value represents the minimum amount of work (8.8 kJ) required to transfer one mole of K^+ ions from the blood plasma at 5×10^{-3} M to muscle cell fluids at 0.15 M, assuming constancy of concentrations. In practice, a larger than minimum amount of work is required.

19.105 $S = k \ln W$ (Equation 19.5), $k = R/N$, $W \propto V^m$

$\Delta S = S_2 - S_1; S_1 = k \ln W_1, S_2 = k \ln W_2$

$\Delta S = k \ln W_2 - k \ln W_1; W_2 = cV_2^m; W_1 = cV_1^m$

(The number of particles, m, is the same in both states.)

$$\Delta S = k \ln cV_2^{\,m} - k \ln cV_1^{\,m}; \quad \ln a^b = b \ln a$$

$$\Delta S = k\, m \ln cV_2 - k\, m \ln cV_1; \quad \ln a - \ln b = \ln (a/b)$$

$$\Delta S = k\, m \ln\left(\frac{cV_2}{cV_1}\right) = k\, m \ln\left(\frac{V_2}{V_1}\right) = \frac{R}{N}\, m \ln\left(\frac{V_2}{V_1}\right)$$

$$\frac{m}{N} = \frac{particles}{6.022\times10^{23}} = n(mol); \quad \Delta S = nR \ln\left(\frac{V_2}{V_1}\right)$$

19.106   (a) $(T_{high} - T_{low})/T_{high} = (700 - 288)/700 = 0.58857 = 58.9\%$ efficiency

           (b)   The cooler the exit temperature of the engine or generator, the more efficient the engine. If a body of water can be used to naturally reduce the exit temperature, the efficiency of energy production increases.

           (c)   The closer the exit temperature to 0 K, the more efficient the heat engine.

           (d)   Refer to Figure 5.10.

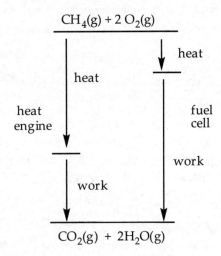

Integrative Exercises

19.108   The activated complex in Figure 14.15 is a single "particle" or entity that contains four atoms. It is formed from an atom A and a triatomic molecule, ABC, that must collide with exactly the correct energy and orientation to form the single entity. There are many fewer degrees of freedom for the activated complex than the separate reactant particles, so the *entropy of activation* is negative.

19.109   Calculate $\Delta G°$, $\Delta H°$, and $\Delta S°$ using data from Appendix C. Use $\Delta G° = \Delta H° - T\Delta S°$ to calculate the temperature at which $\Delta G°$ changes sign. Couple this temperature with the sign of $\Delta G°$ at 298 K to state a temperature range over which the reaction is spontaneous.

$$\Delta G° = 2\Delta G°\, CO_2(g) + 3\Delta G°\, Fe(s) - 2\Delta G°\, Fe_3O_4(s) - 2\,\Delta G°\, C(s, graphite)$$

$$= 2(-394.4) + 3(0) - (-1014.2) - 2(0) = 225.4 \text{ kJ}$$

$\Delta H° = 2\Delta H° \, CO_2(g) + 3\Delta H° \, Fe(s) - 2\Delta H° \, Fe_3O_4(s) - 2\,\Delta H° \, C(s, graphite)$

$\quad = 2(-393.5) + 3(0) - (-1117.1) - 2(0) = 330.1 \text{ kJ}$

$\Delta S° = 2\Delta S° \, CO_2(g) + 3\Delta S° \, Fe(s) - 2\Delta S° \, Fe_3O_4(s) - 2\,\Delta S° \, C(s, graphite)$

$\quad = 2(213.6) + 3(27.15) - 146.4 - 2(5.69) = 350.87 = 350.9 \text{ J/K} = 0.3509 \text{ kJ/K}$

$\Delta H° = T\Delta S°; \ T = \Delta H°/\Delta S° = 330.1 \text{ kJ}/0.3509 \text{ J/K} = 940.7 \text{ K} = 667.6 \text{ °C}$

The reaction will be spontaneous at temperatures above 940.7 K or 667.6 °C. Even though $\Delta S°$ is large and positive, a very high temperature is required to overcome the large positive enthalpy of reaction.

19.110   (a) $O_2(g) \xrightarrow{h\nu} 2O(g)$; S increases because there are more moles of gas in the products.

(b) $O_2(g) + O(g) \rightarrow O_3(g)$; S decreases because there are fewer moles of gas in the products.

(c)   S increases as the gas molecules diffuse into the larger volume of the stratosphere; there are more possible positions and, therefore, more motional freedom.

(d) $NaCl(aq) \rightarrow NaCl(s) + H_2O(l)$; ΔS decreases as the mixture (seawater, greater disorder) is separated into pure substances (fewer possible arrangements, more order).

19.111   The heat lost by the hot water (iv) in the cup equals the heat gained by the ice cube. The ice cube is heated from –20 °C to 0 °C (i), the ice melts at 0 °C (ii), the new liquid heats to the final temperature (iii). (iv) = (i) + (ii) + (iii). The 500 mL of hot water weighs 500 g; the ice cube weighs 20 g.

(i) $20 \text{ g} \times \dfrac{2.03 \text{ J}}{1 \text{ g-°C}} \times (20 \text{ °C}) = 812 = 8.1 \times 10^2 \text{ J}$

(ii) $20 \text{ g} \times \dfrac{1 \text{ mol}}{18.02 \text{ g}} \times \dfrac{6.01 \text{ kJ}}{\text{mol}} \times \dfrac{1000 \text{ J}}{\text{kJ}} = 6670.4 = 6.7 \times 10^3 \text{ J}$

(iii) $20 \text{ g} \times \dfrac{4.184 \text{ J}}{1 \text{ g-°C}} \times (T_f - 0 \text{ °C}) = 83.60 \,(T_f - 0 \text{ °C}) = 84 \,(T_f)$

(iv) $500 \text{ g} \times \dfrac{4.184 \text{ J}}{1 \text{ g-°C}} \times (83 \text{ °C} - T_f) = 2,092 \,(83 \text{ °C} - T_f) = (2.09 \times 10^3)(83 \text{ °C} - T_f)$

$2092(83 \text{ °C} - T_f) = 812 \text{ J} + 6670.4 \text{ J} + 83.60(T_f);$

$173,636 - [2092 \,(T_f)] = 7482.4 + 83.60(T_f)$

$166,153.6 = 2175.6(T_f); \ T_f = 76.37 = 76 \text{ °C}$

19.112   (a)   16 e⁻, 8 e⁻ pairs. The C–S bond order is approximately 2.

$\overset{..}{\underset{..}{S}} = C = \overset{..}{\underset{..}{S}}$

(b)   2 e⁻ domains around C, linear e⁻ domain geometry, linear molecular structure.

(c) $CS_2(l) + 3O_2(g) \rightarrow CO_2(g) + 2SO_2(g)$

(d) $\Delta H° = \Delta H° \, CO_2(g) + 2\Delta H° \, SO_2(g) - \Delta H° \, CS_2(l) - 3\Delta H° \, O_2(g)$

             $= -393.5 + 2(-296.9) - (89.7) - 3(0) = -1077.0 \text{ kJ}$

             $\Delta G° = \Delta G° \, CO_2(g) + 2\Delta G° \, SO_2(g) - \Delta G° \, CS_2(l) - 3 \, \Delta G° \, O_2(g)$

             $= -394.4 + 2(-300.4) - (65.3) - 3(0) = -1060.5 \text{ kJ}$

The reaction is exothermic $(-\Delta H°)$ and spontaneous $(-\Delta G°)$ at 298 K.

(e)   vaporization: $CS_2(l) \rightarrow CS_2(g)$

             $\Delta G°_{vap} = \Delta H°_{vap} - T\Delta S°_{vap}; \quad \Delta S°_{vap} = (\Delta H°_{vap} - \Delta G°_{vap})/T$

             $\Delta G°_{vap} = \Delta G° \, CS_2(g) - \Delta G° \, CS_2(l) = 67.2 - 65.3 = 1.9 \text{ kJ}$

             $\Delta H°_{vap} = \Delta H° \, CS_2(g) - \Delta H° \, CS_2(l) = 117.4 - 89.7 = 27.7 \text{ kJ}$

             $\Delta S°_{vap} = (27.7 - 1.9) \text{ kJ}/298 \text{ K} = 0.086577 = 0.0866 \text{ kJ/K} = 86.6 \text{ J/K}$

ΔS_{vap} is always positive, because the gas phase occupies a greater volume and has more motional freedom and a larger absolute entropy than the liquid.

(f)   At the boiling point, $\Delta G = 0$ and $\Delta H_{vap} = T_b \Delta S_{vap}$.

             $T_b = \Delta H_{vap}/\Delta S_{vap} = 27.7 \text{ kJ}/0.086577 \text{ kJ/K} = 319.9 = 320 \text{ K}$

             $T_b = 320 \text{ K} = 47 \text{ °C. } CS_2$ is a liquid at 298 K, 1 atm

19.113   (a) $Ag(s) + 1/2 \, N_2(g) + 3/2 \, O_2(g) \rightarrow AgNO_3(s)$; S decreases because there are fewer moles of gas in the product.

     (b) $\Delta G°_f = \Delta H°_f - T\Delta S°; \quad \Delta S° = (\Delta G°_f - \Delta H°_f)/(-T) = (\Delta H°_f - \Delta G°_f)/T$

             $\Delta S° = -124.4 \text{ kJ} - (-33.4 \text{ kJ}) / 298 \text{ K} = -0.305 \text{ kJ/K} = -305 \text{ J/K}$

             $\Delta S°$ is relatively large and negative, as anticipated from part (a).

     (c)   Dissolving of $AgNO_3$ can be expressed as

             $AgNO_3(s) \rightarrow AgNO_3 \, (aq, 1 \, M)$

             $\Delta H° = \Delta H° \, AgNO_3(aq) - \Delta H° \, AgNO_3(s) = -101.7 - (-124.4) = +22.7 \text{ kJ}$

             $\Delta H° = \Delta H° \, MgSO_4(aq) - \Delta H° \, MgSO_4(s) = -1374.8 - (-1283.7) = -91.1 \text{ kJ}$

             Dissolving $AgNO_3(s)$ is endothermic $(+\Delta H°)$, but dissolving $MgSO_4(s)$ is exothermic $(-\Delta H°)$.

     (d) $AgNO_3: \Delta G° = \Delta G°_f \, AgNO_3(aq) - \Delta G°_f \, AgNO_3(s) = -34.2 - (-33.4) = -0.8 \text{ kJ}$

             $\Delta S° = (\Delta H° - \Delta G°) / T = [22.7 \text{ kJ} - (-0.8 \text{ kJ})] / 298 \text{ K} = 0.0789 \text{ kJ/K} = 78.9 \text{ J/K}$

             $MgSO_4: \Delta G° = \Delta G°_f \, MgSO_4(aq) - \Delta G°_f \, MgSO_4(s) = -1198.4 - (-1169.6) = -28.8 \text{ kJ}$

             $\Delta S° = (\Delta H° - \Delta G°) / T = [-91.1 \text{ kJ} - (-28.8 \text{ kJ})] / 298 \text{ K} = -0.209 \text{ kJ/K} = -209 \text{ J/K}$

(e)   In general, we expect dissolving a crystalline solid to be accompanied by an increase in positional disorder and an increase in entropy; this is the case for $AgNO_3$ ($\Delta S° = +78.9$ J/K). However, for dissolving $MgSO_4(s)$, there is a substantial decrease in entropy ($\Delta S = -209$ J/K). According to Section 13.5, ion-pairing is a significant phenomenon in electrolyte solutions, particularly in concentrated solutions where the charges of the ions are greater than 1. According to Table 13.4, a 0.1 m $MgSO_4$ solution has a van't Hoff factor of 1.21. That is, for each mole of $MgSO_4$ that dissolves, there are only 1.21 moles of "particles" in solution instead of 2 moles of particles. For a 1 m solution, the factor is even smaller. Also, the exothermic enthalpy of mixing indicates substantial interactions between solute and solvent. Substantial ion-pairing coupled with ion–dipole interactions with H_2O molecules lead to a decrease in entropy for $MgSO_4(aq)$ relative to $MgSO_4(s)$.

19.115 (a) $\Delta G° = 3\Delta G_f°\ S(s) + 2\Delta G_f°\ H_2O(g) - \Delta G_f°\ SO_2(g) - 2\Delta G_f°\ H_2S(g)$

$= 3(0) + 2(-228.57) - (-300.4) - 2(-33.01) = -90.72 = -90.7$ kJ

$\ln K = \dfrac{-\Delta G°}{RT} = \dfrac{-(-90.72\ \text{kJ})}{(8.314 \times 10^{-3}\ \text{kJ/K})(298\ \text{K})} = 36.6165 = 36.6;\ K = 7.99 \times 10^{15}$

$= 8 \times 10^{15}$

(b)   The reaction is highly spontaneous at 298 K and feasible in principle. However, use of $H_2S(g)$ produces a severe safety hazard for workers and the surrounding community.

(c) $P_{H_2O} = \dfrac{25\ \text{torr}}{760\ \text{torr/atm}} = 0.033$ atm

$K = \dfrac{P_{H_2O}^2}{P_{SO_2} \times P_{H_2S}^2};\quad P_{SO_2} = P_{H_2S} = x$ atm

$K = 7.99 \times 10^{15} = \dfrac{(0.033)^2}{x(x)^2};\quad x^3 = \dfrac{(0.033)^2}{7.99 \times 10^{15}}$

$x = 5 \times 10^{-7}$ atm

(d) $\Delta H° = 3\Delta H_f°\ S(s) + 2\Delta H_f°\ H_2O(g) - \Delta H_f°\ SO_2(g) - 2\Delta H_f°\ H_2S(g)$

$= 3(0) + 2(-241.82) - (-296.9) - 2(-20.17) = -146.4$ kJ

$\Delta S° = 3S°\ S(s) + 2S°\ H_2O(g) - S°\ SO_2(g) - 2S°\ H_2S(g)$

$= 3(31.88) + 2(188.83) - 248.5 - 2(205.6) = -186.4$ J/K

The reaction is exothermic ($-\Delta H$), so the value of K_{eq} will decrease with increasing temperature. The negative $\Delta S°$ value means that the reaction will become nonspontaneous at some higher temperature. The process will be less effective at elevated temperatures.

19.116   (a)   When the rubber band is stretched, the molecules become more ordered, so the entropy of the system decreases, ΔS_{sys} is negative.

        (b) $\Delta S_{sys} = q_{rev}/T$. Because ΔS_{sys} is negative, q_{rev} is negative and heat is emitted by the system.

        (c)   The unstretched rubber band feels cooler. This confirms our answer to (b). If heat is emitted by the system when it is stretched, the surroundings feel warmer. Upon return to the initial state, heat is absorbed by the system (the rubber band) and the surroundings (your lip) feel cooler.

20 Electrochemistry

Visualizing Concepts

20.2   Unintended oxidation reactions in the body lead to unwanted health effects, just as unwanted oxidation of metals leads to corrosion. Antioxidants probably have modes of action similar to anti-corrosion agents. They can preferentially react with oxidizing agents (cathodic protection), create conditions that are unfavorable to the oxidation–reduction reaction, or physically coat or surround the molecule being oxidized to prevent the oxidant from attacking it. The first of these modes of action is likely to be safest in biological systems. Adjusting reaction conditions in our body can be dangerous, and physical protection is unlikely to provide lasting protection against oxidation. Anti-oxidants are likely to be reductants that preferentially react with oxidizing agents.

20.4   *Analyze/Plan.* Consider the voltaic cell pictured in Figure 20.5 as a model. The reaction in a voltaic cell is spontaneous. To generate a *standard* emf, substances must be present in their standard states.

   (a)   A concentration of 1 M is the standard state for ions in solution. Ions, but not solution, must be able to flow between compartments to complete the circuit, so that the cell can develop an emf. Add 1 M A^{2+}(aq) to the beaker with the A(s) electrode. Add 1 M B^{2+}(aq) to the beaker with the B(s) electrode. Add a salt bridge to enable the flow of ions from one compartment to the other.

   (b)   Reduction occurs at the cathode. For the reaction to occur spontaneously (and thus generate an emf), the half-reaction with the greater E°_{red} will be the reduction half-reaction. In this cell, it is the half-reaction involving A(s) and A^{2+}(aq). The A electrode functions as the cathode.

   (c)   According to Figure 20.5, electrons flow through the external circuit from the anode to the cathode. In this example, B is the anode and A is the cathode, so electrons flow from B to A through the external circuit.

   (d) $E^{\circ}_{cell} = E^{\circ}_{red}(\text{cathode}) - E^{\circ}_{red}(\text{anode})$

   $E^{\circ}_{cell} = -0.10\ \text{V} - (-1.10\ \text{V}) = 1.00\ \text{V}$

20.5 $A(aq) + B(aq) \rightarrow A^{-}(aq) + B^{+}(aq)$

   (a)   A gains electrons and is being reduced; it is the cathode. B loses electrons and is being oxidized; it is the anode.

   (b)   Reduction occurs at the cathode; oxidation occurs at the anode.

   $A(aq) + 1e^{-} \rightarrow A^{-}(aq)$ occurs at the cathode.

   $B(aq) \rightarrow B^{+}(aq) + 1e^{-}$ occurs at the anode.

(c)   In a voltaic cell, the anode is at higher potential energy than the cathode. The anode reaction, $B(aq) \rightarrow B^+(aq) + 1e^-$, is higher in potential energy.

(d) $\Delta G° = -nFE°$; the signs of $\Delta G°$ and $E°$ (or ΔG and E) are opposite. Because this is a spontaneous reaction, $\Delta G°$ is negative and $E°$ is positive.

20.6   *Analyze.* Given a series of reduction half-reactions and their standard electrode potentials (E_{red}^o), draw conclusions about their relative strengths as oxidizing and reducing agents. *Plan.* The reactant with the largest E_{red}^o is the easiest to reduce and the strongest oxidizing agent. The reduced form of this substance, the product of the reduction half-reaction, is the most difficult to oxidize and the weakest reducing agent. Conversely, the reactant with the smallest E_{red}^o is the hardest to reduce and the weakest oxidizing agent. The reduced form of this substance, the product of the reduction half-reaction, is the easiest to oxidize and the strongest reducing agent. *Solve.*

(a) $A^+(aq)$ is the strongest oxidizing agent, and D^{3+} is the weakest oxidizing agent.

(b) $D(s)$ is the strongest reducing agent, and $A(s)$ is the weakest reducing agent.

(c)   Reactants with more positive E_{red}^o than $C^{3+}(aq)$ will oxidize $C^{2+}(aq)$. Both $A^+(aq)$ and $B^{2+}(aq)$ will oxidize $C^{2+}(aq)$.

20.8   *Analyze.* Given the voltaic cell shown in the diagram, answer questions about the cell and the effect of solution concentration on cell potential, E.

Plan. Use the definition of a voltaic cell and standard emf, along with the Nernst equation, $E = \Delta E° - (0.0592 \text{ V}/n)\log Q$, to answer the questions. *Solve.*

(a)   A voltaic cell involves a spontaneous redox reaction, one with positive E_{cell}^o. To achieve a positive E_{cell}^o, the half-reaction with the more positive E_{red}^o occurs at the cathode. For this cell the two half reactions are

$$Ag^+(aq) + e^- \rightarrow Ag(s), \quad E_{red}^o = 0.799\text{V}$$

$$Fe^{2+}(aq) + 2e^- \rightarrow Fe(s), \quad E_{red}^o = -0.440 \text{ V}$$

The Ag(s) electrode is the cathode.

(b)   The standard emf is just E_{cell}^o.

$$E_{cell}^o = E_{red}^o(\text{cathode}) - E_{red}^o(\text{anode}) = 0.799\text{V} - (-0.440\text{V}) = 1.239\text{V}$$

The cell in the diagram is at standard conditions, with solid metal electrodes and 1 *M* aqueous solutions, so the potential on the meter in the circuit is the standard emf.

(c) $2Ag^+(aq) + Fe(s) \rightarrow 2Ag(s) + Fe^{2+}(aq);\ n = 2;\ E = E° - \dfrac{0.0592}{2}\log\dfrac{[Fe^{2+}]}{[Ag^+]^2}$

The solution in the cathode half-cell is $Ag^+(aq)$. If $[Ag^+(aq)]$ increases by a factor of 10, the change in cell voltage is $E - E° = -\dfrac{0.0592}{2}\log\dfrac{[1]}{[10]^2} = 0.0592 \text{ V}$.

(d)   The solution in the anode half-cell is $Fe^{2+}(aq)$. If $[Fe^{2+}(aq)]$ increases by a factor of 10, the change in cell voltage is $E - E° = -\dfrac{0.0592}{2}\log\dfrac{[10]}{[1]^2} = -0.0296 \text{ V}$.

20.9   *Analyze/Plan.* Consider the Nernst equation, which describes the variation of potential (emf) with respect to changes in concentration.

Solve.

$$E = E° - \frac{0.0592}{n} \log Q.$$

(a)   For this half-reaction, $E_{red}^o = 0.799$ V; $Q = 1/[Ag^+]$

$$E = 0.80 - \frac{0.0592}{1} \log \frac{1}{[Ag^+]}; \quad E = 0.80 + \frac{0.0592}{1} \log[Ag^+]$$

The y-intercept of the graph is $E°$. The slope of the line is $+0.0592$. So, as $[Ag^+]$ and $\log[Ag^+]$ increase, E increases. The line that describes this behavior is line 1.

(b)   When $\log[Ag^+] = 0$, $[Ag^+] = 1$ M; this is the standard state for $Ag^+(aq)$, so $E_{red} = E_{red}^o = 0.799$V.

20.11   Beaker B. The process of iron corrosion includes Fe being oxidized and O_2 being reduced. The reduction of O_2 requires H^+. Beaker B, which contains dilute HCl(aq), provides the H^+ required to encourage corrosion. Iron in contact with a solution above pH 9 does not corrode.

20.12   (a) $MgCl_2(l)$

         (b)   Oxidation occurs at the anode. Formation of $Cl_2(g)$ from $2Cl^-(l)$ is oxidation, so the carbon electrode is the anode.

               Reduction occurs at the cathode. Formation of Mg(l) from $Mg^{2+}(l)$ is reduction, so the steel electrode is the cathode.

               Check. Note that this electrolysis cell does not comply with the convention of drawing anode on the left and cathode on the right.

         (c) $MgCl_2(l) \rightarrow Mg(l) + Cl_2(g)$ overall

                     $2\,Cl^- \rightarrow Cl_2(g) + 2\,e^-$ anode(oxidation)

             $Mg^{2+} + 2\,e^- \rightarrow Mg(l)$ cathode (reduction)

         (d)   Magnesium is an active metal. It must be separated from the $Cl_2(g)$ that is also formed by electrolysis (see screen in diagram) or $MgCl_2$ will spontaneously reform. Also, the Mg(l) and Mg(g) should not come in contact with air (O_2) or moisture (H_2O).

Oxidation–Reduction Reactions (Section 20.1)

20.14   (a)   *Reduction* is the gain of electrons.

         (b)   The electrons appear on the reactants side (left side) of a reduction half-reaction.

         (c)   The *reductant* is the reactant that is oxidized; it provides the electrons that are gained by the substance being reduced.

         (d)   A *reducing agent* is the substance that promotes reduction. It donates the electrons gained by the substance that is reduced. It is the same as the reductant.

20.16   (a)   False. If something is reduced, it gains electrons.

         (b)   True.

         (c)   True. Oxidation can be thought of as a gain of oxygen atoms. Looking forward, this view will be useful for organic reactions, Chapter 24.

20.18 (a) Mn is reduced from +7 to +4; S is oxidized from –2 to 0.

   (b) Cl is reduced from +7 to +3; O is oxidized from –1 to 0.

   (c) Cl is reduced from +4 to +3; O is oxidized from –1 to 0.

20.20 (a) No oxidation–reduction.

   (b) Pb is reduced from +4 to +2; O is oxidized from –2 to 0.

   (c) S is reduced from +6 to +4; Br is oxidized from –1 to 0.

Balancing Oxidation–Reduction Reactions (Section 20.2)

20.22 (a) $2\,N_2H_4(g) + N_2O_4(g) \rightarrow 3\,N_2(g) + 4\,H_2O(g)$

   (b) $N_2H_4(g)$ is oxidized; $N_2O_4(g)$ is reduced.

   (c) $N_2H_4(g)$ serves as the reducing agent; it is itself oxidized. $N_2O_4(g)$ serves as the oxidizing agent; it is itself reduced.

20.24 (a) $Mo^{3+}(aq) + 3\,e^- \rightarrow Mo(s)$, reduction

   (b) $H_2SO_3(aq) + H_2O(l) \rightarrow SO_4^{2-}(aq) + 4H^+(aq) + 2\,e^-$, oxidation

   (c) $NO_3^-(aq) + 4\,H^+(aq) + 3\,e^- \rightarrow NO(g) + 2\,H_2O(l)$, reduction

   (d) $O_2(g) + 4\,H^+(aq) + 4\,e^- \rightarrow 2\,H_2O(l)$, reduction

   (e) $O_2(g) + 2\,H_2O(l) + 4\,e^- \rightarrow 4\,OH^-(aq)$, reduction
       (O_2 is reduced to OH^-, not H_2O, in basic solution)

   (f) $Mn^{2+}(aq) + 4\,OH^-(aq) \rightarrow MnO_2(s) + 2\,H_2O(l) + 2\,e^-$, oxidation

   (g) $Cr(OH)_3(s) + 5\,OH^-(aq) \rightarrow CrO_4^{2-}(aq) + 4\,H_2O(l) + 3\,e^-$, oxidation

20.26 (a)

$$3\,[NO_2^-(aq) + H_2O(l) \rightarrow NO_3^-(aq) + 2\,H^+(aq) + 2\,e^-]$$
$$Cr_2O_7^{2-}(aq) + 14\,H^+(aq) + 6\,e^- \rightarrow 2\,Cr^{3+}(aq) + 7\,H_2O(l)$$

Net:
$$3\,NO_2^-(aq) + Cr_2O_7^{2-}(aq) + 8\,H^+(aq) \rightarrow 3\,NO_3^-(aq) + 2\,Cr^{3+}(aq) + 4\,H_2O(l)$$

oxidizing agent, $Cr_2O_7^{2-}$; reducing agent, NO_2^-

   (b) The oxidation half-reaction involves S, and is listed in Appendix E. The reduction half-reaction involves N, and must be written and balanced, according to the procedure in Section 20.2.

$$HNO_3(aq) \rightarrow N_2O(g)$$
$$2\,HNO_3(aq) \rightarrow N_2O(g)$$
$$2\,HNO_3(aq) \rightarrow N_2O(g) + 5\,H_2O(l)$$
$$2\,HNO_3(aq) + 8\,H^+(aq) \rightarrow N_2O(g) + 5\,H_2O(l)$$
$$2\,HNO_3(aq) + 8\,H^+ + 8\,e^- \rightarrow N_2O(g) + 5\,H_2O(l)$$

$$2\,[S(s) + 3\,H_2O(l) \rightarrow H_2SO_3(aq) + 4\,H^+ + 4\,e^-]$$

$$2\,HNO_3(aq) + 8\,H^+ + 8\,e^- \rightarrow N_2O(g) + 5\,H_2O(l)$$

$$\overline{2\,HNO_3(aq) + 2\,S(s) + H_2O(l) \rightarrow 2\,H_2SO_3(aq) + N_2O(g)}$$

oxidizing agent, HNO_3; reducing agent, S.

(c)
$$2\,[Cr_2O_7^{2-}(aq) + 14\,H^+(aq) + 6\,e^- \rightarrow 2\,Cr^{3+}(aq) + 7\,H_2O(l)]$$

$$3\,[CH_3OH(aq) + H_2O(l) \rightarrow HCO_2H(aq) + 4\,H^+(aq) + 4\,e^-]$$

$$\overline{2\,Cr_2O_7^{2-}(aq) + 3\,CH_3OH(aq) + 16\,H^+(aq) \rightarrow 4\,Cr^{3+}(aq) + 3\,HCO_2H(aq) + 11\,H_2O(l)}$$

Net:

oxidizing agent, $Cr_2O_7^{2-}$; reducing agent, CH_3OH

(d) The half-reaction involving N_2H_4 is given in Appendix E in base. We add $4\,H^+(aq)$ to each side and reverse the reaction to obtain oxidation half-reaction shown below.

$$2\,[2BrO_3^-(aq) + 12\,H^+(aq) + 10\,e^- \rightarrow Br_2(l) + 6\,H_2O(l)]$$

$$5\,[N_2H_4(aq) \rightarrow N_2(g) + 4\,H^+(aq) + 4\,e^-]$$

Net: $$4\,BrO_3^-(aq) + 5\,N_2H_4(aq) + 4\,H^+(aq) \rightarrow 2\,Br_2(l) + 5\,N_2(g) + 12\,H_2O(l)$$

oxidizing agent, BrO_3^-; reducing agent, N_2H_4

(e) Write and balance each half-reaction, and then sum to get the overall reaction. Follow the procedure in Sample Exercise 20.3 for reactions in basic solution.

oxidation: $Al(s) \rightarrow AlO_2^-(aq)$

$$Al(s) + 2\,H_2O(l) \rightarrow AlO_2^-(aq) + 4\,H^+(aq) + 3\,e^-$$

$$Al(s) + 4\,OH^-(aq) \rightarrow AlO_2^-(aq) + 2\,H_2O(l) + 3\,e^-$$

reduction: $NO_2^-(aq) \rightarrow NH_4^+(aq)$

$$NO_2^-(aq) + 8\,H^+(aq) + 6\,e^- \rightarrow NH_4^+(aq)) + 2\,H_2O(l)$$

$$NO_2^-(aq) + 6\,H_2O(l) + 6\,e^- \rightarrow NH_4^+(aq) + 8\,OH^-(aq)$$

$$NO_2^-(aq) + 6\,H_2O(l) + 6\,e^- \rightarrow NH_4^+(aq) + 8\,OH^-(aq)$$

$$2[Al(s) + 4\,OH^-(aq) \rightarrow AlO_2^-(aq) + 2\,H_2O(l) + 3\,e^-]$$

$$\overline{NO_2^-(aq) + 6\,H_2O(l) + 6\,e^- \rightarrow NH_4^+(aq) + 8\,OH^-(aq)}$$

$$NO_2^-(aq) + 2\,Al(s) + 2\,H_2O(l) \rightarrow NH_4^+(aq) + 2\,AlO_2^-(aq)$$

oxidizing agent, NO_2^-; reducing agent, Al

(f) $$H_2O_2(aq) + 2\,e^- \rightarrow O_2(g) + 2\,H^+(aq)$$

Because the reaction is in base, the H^+ can be "neutralized" by adding $2\,OH^-$ to each side of the equation: $H_2O_2(aq) + 2\,OH^-(aq) + 2\,e^- \rightarrow O_2(g) + 2\,H_2O(l)$. The other half reaction is: $2\,[ClO_2(aq) + e^- \rightarrow ClO_2^-(aq)]$.

Net: $H_2O_2(aq) + 2\,ClO_2(aq) + 2\,OH^-(aq) \rightarrow O_2(g) + 2\,ClO_2^-(aq) + 2\,H_2O(l)$

oxidizing agent, ClO_2; reducing agent, H_2O_2

Voltaic Cells (Section 20.3)

20.28   (a)   The porous glass dish in Figure 20.4 provides a mechanism by which ions not directly involved in the redox reaction can migrate into the anode and cathode compartments to maintain charge neutrality of the solutions. Ionic conduction within the cell, through the glass disk, completes the cell circuit.

       (b)   In the anode compartment of Figure 20.5, Zn atoms are oxidized to Zn^{2+} cations, increasing the number of positively charged particles in the compartment. NO_3^- anions migrate into the compartment to maintain charge balance as Zn^{2+} ions are produced.

20.30   (a)   Al(s) is oxidized, Ni^{2+}(aq) is reduced.

       (b) $Al(s) \rightarrow Al^{3+}(aq) + 3e^-$; $Ni^{2+}(aq) + 2e^- \rightarrow Ni(s)$

       (c)   Al(s) is the anode; Ni(s) is the cathode.

       (d)   Al(s) is negative (–); Ni(s) is positive (+).

       (e)   Electrons flow from the Al(–) electrode toward the Ni(+) electrode.

       (f)   Cations migrate toward the Ni(s) cathode; anions migrate toward the Al(s) anode.

Cell Potentials under Standard Conditions (Section 20.4)

20.32   (a)   In a voltaic cell, the anode has the higher potential energy for electrons. To achieve a lower potential energy, electrons flow from the anode to the cathode.

       (b)   The units of electrical potential are volts. A potential of one volt imparts one joule of energy to one coulomb of charge.

20.34   (a)   For a reduction potential to be a *standard reduction potential,* the substances in the reaction or half-reaction must be at standard conditions, 1 *M* aqueous solutions and 1 atm gas pressures.

       (b) $E_{red}^o = 0\ V$ for a standard hydrogen electrode.

       (c)   It is not possible to measure the standard reduction potential of a single half-reaction because each voltaic cell consists of two half-reactions and only the potential of a complete cell can be measured.

20.36   (a) $PdCl_4^{2-}(aq) + 2\ e^- \rightarrow Pd(s) + 4\ Cl^-$ cathode $E_{red}^o = ?$

                     $Cd(s) \rightarrow Cd^{2+}(aq) + 2\ e^-$ anode $E_{red}^o = -0.403\ V$

       (b) $E_{cell}^o = E_{red}^o\ (cathode) - E_{red}^o\ (anode); 1.03\ V = E_{red}^o - (-0.403\ V);$

           $E_{red}^o = 1.03\ V - 0.403 = 0.63\ V$

(c)

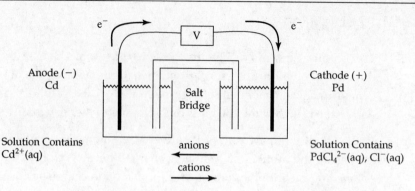

20.38 (a) $F_2(g) + 2 e^- \rightarrow 2 F^-(aq)$ $\qquad\qquad E^o_{red} = 2.87 \text{ V}$

$H_2(g) \rightarrow 2 H^+(aq) + 2 e^-$ $\qquad\qquad E^o_{red} = 0.00 \text{ V}$

$E^o = 2.87 \text{ V} - 0.00 \text{ V} = 2.87 \text{ V}$

(b) $Cu^{2+}(aq) + 2 e^- \rightarrow Cu(s)$ $\qquad\qquad E^o_{red} = 0.337 \text{ V}$

$Ca(s) \rightarrow Ca^{2+}(aq) + 2 e^-$ $\qquad\qquad E^o_{red} = -2.87 \text{ V}$

$E^o = 0.337 \text{ V} - (-2.87 \text{ V}) = 3.21 \text{ V}$

(c) $Fe^{2+}(aq) + 2 e^- \rightarrow Fe(s)$ $\qquad\qquad E^o_{red} = -0.440 \text{ V}$

$2[Fe^{2+}(aq) \rightarrow Fe^{3+}(aq) + 1 e^-]$ $\qquad\qquad E^o_{red} = 0.771 \text{ V}$

$E^o = -0.440 \text{ V} - 0.771 \text{ V} = -1.211 \text{ V}$

(d) $2 ClO_3^-(aq) + 12 H^+ + 10 e^- \rightarrow Cl_2(g) + 6 H_2O(l)$ $\quad E^o_{red} = 1.47 \text{ V}$

$5 [2 Br^-(aq) \rightarrow Br_2(l) + 2 e^-]$ $\qquad\qquad E^o_{red} = 1.065 \text{ V}$

$E^o = 1.47 \text{ V} - 1.065 \text{ V} = 0.405 = 0.41 \text{ V}$

20.40 (a)

$2 [Au(s) + 4 Br^-(aq) \rightarrow AuBr_4^-(aq) + 3 e^-]$ $\qquad E^o_{red} = -0.858 \text{ V}$

$3 [2e^- + IO^-(aq) + H_2O(l) \rightarrow I^-(aq) + 2 OH^-(aq)]$ $\qquad E^o_{red} = 0.49 \text{ V}$

$2 Au(s) + 8 Br^-(aq) + 3 IO^-(aq) + 3 H_2O(l) \rightarrow 2 AuBr_4^-(aq) + 3 I^-(aq) + 6 OH^-(aq)$

$E^o = 0.49 - (-0.858) = 1.35 \text{ V}$

(b)

$2 [Eu^{2+}(aq) \rightarrow Eu^{3+}(aq) + 1 e^-]$ $\qquad\qquad E^o_{red} = -0.43 \text{ V}$

$Sn^{2+}(aq) + 2 e^- \rightarrow Sn(s)$ $\qquad\qquad E^o_{red} = -0.14 \text{ V}$

$2 Eu^{2+}(aq) + Sn^{2+}(aq) \rightarrow 2 Eu^{3+}(aq) + Sn(s)$ $\qquad E^o = -0.14 - (-0.43) = 0.29 \text{ V}$

20.42 (a) The two half-reactions are:

$Cd^{2+}(aq) + 2 e^- \rightarrow Cd(s)$ $\qquad\qquad E^o = -0.403 \text{ V}$

$Cl_2(g) + 2 e^- \rightarrow 2 Cl^-(aq)$ $\qquad\qquad E^o = 1.359 \text{ V}$

Because E^o for the reduction of Cl_2 is greater, Cl_2 is reduced at the cathode, the Pt electrode. $Cd(s)$ is oxidized at the anode, the Cd electrode.

(b) The Cd anode loses mass as $Cd^{2+}(aq)$ is produced.

(c) $Cl_2(g) + Cd(s) \rightarrow Cd^{2+}(aq) + 2Cl^-(aq)$

(d) $E° = 1.359 \text{ V} - (-0.403 \text{ V}) = 1.762 \text{ V}$

Strengths of Oxidizing and Reducing Agents (Section 20.4)

20.44   Follow the logic in Sample Exercise 20.8. In each case, choose the half-reaction with the more positive reduction potential and with the given substance on the left.

       (a) $Cl_2(g)$ (1.359 V vs. 1.065 V)

       (b) $Cd^{2+}(aq)$ (−0.403 V vs. −0.763 V)

       (c) $ClO_3^-(aq)$ ($Cl^-(aq)$ is in its minimum oxidation state and cannot act as an oxidizing agent)

       (d) $O_3(g)$ (2.07 V vs. 1.776 V)

20.46   If the substance is on the left of a reduction half-reaction, it will be an oxidant; if it is on the right, it will be a reductant. The sign and magnitude of the E^o_{red} determine whether it is strong or weak.

       (a) $Ce^{3+}(aq)$: very weak reductant (on the right, $E^o_{red} = 1.61 \text{ V}$)

       (b) $Ca(s)$: strong reductant (on the right, $E^o_{red} = -2.87 \text{ V}$)

       (c) $ClO_3^-(aq)$: strong oxidant (on the left, $E^o_{red} = 1.47 \text{ V}$)

       (d) $N_2O_5(g)$: oxidant (N has maximum oxidation number, +5; can only be reduced and act as oxidant.)

20.48   (a)   The strongest oxidizing agent is the species most readily reduced, as evidenced by a large, positive reduction potential. That species is H_2O_2. The weakest oxidizing agent is the species that least readily accepts an electron. We expect that it will be very difficult to reduce $Zn(s)$; indeed, $Zn(s)$ acts as a comparatively strong **reducing** agent. No potential is listed for reduction of $Zn(s)$, but we can safely assume that it is less readily reduced than any of the other species present.

       (b)   The strongest reducing agent is the species most easily oxidized (the largest negative reduction potential). Zn, $E^o_{red} = -0.76 \text{ V}$, is the strongest reducing agent and F^-, $E^o_{red} = 2.87 \text{ V}$, is the weakest.

20.50   Any oxidized species from Appendix E with a reduction potential greater than 0.59 V will oxidize RuO_4^{2-} to RuO_4^-. From the list of possible oxidants in the exercise, $Br_2(l)$ and $BrO_3^-(aq)$ will definitely oxidize RuO_4^{2-} to RuO_4^-. $Sn^{2+}(aq)$ will not, and $O_2(g)$ depends on conditions. In base, it will not, but in strongly acidic solution, it will.

Free Energy and Redox Reactions (Section 20.5)

20.52   (a)

$$2\,I^-(aq) \rightarrow I_2(s) + 2\,e^- \qquad\qquad E^o_{red} = 0.536 \text{ V}$$

$$Hg_2^{2+}(aq) + 2\,e^- \rightarrow 2\,Hg(l) \qquad\qquad E^o_{red} = 0.789 \text{ V}$$

$$2\,I^-(aq) + Hg_2^{2+}(aq) \rightarrow I_2(s) + 2\,Hg(l) \qquad E^o = 0.789 - 0.536 = 0.253 \text{ V}$$

$$\Delta G^\circ = -nFE^\circ = -2\ \text{mol e}^- \times \frac{96.5\ \text{kJ}}{\text{V - mol e}^-} \times 0.253\ \text{V} = -48.829 = -48.8\ \text{kJ}$$

$$\ln K = \frac{-(-4.8829 \times 10^4\ \text{J})}{(8.314\ \text{J/mol - K})(298\ \text{K})} = 19.708 = 19.7;\quad K = e^{19.7} = 3.61 \times 10^8 = 3.6 \times 10^8$$

(b)
$$3\ [Cu^+(aq) \rightarrow Cu^{2+}(aq) + 1\ e^-] \qquad E^\circ_{red} = 0.153\ \text{V}$$
$$NO_3^-(aq) + 4\ H^+(aq) + 3\ e^- \rightarrow NO(g) + H_2O(l) \qquad E^\circ_{red} = 0.96\ \text{V}$$
___
$$3\ Cu^+(aq) + NO_3^-(aq) + 4\ H^+(aq) \rightarrow 3\ Cu^{2+}(aq) + NO(g) + 2\ H_2O(l)$$

$$E^\circ = 0.96 - 0.153 = 0.81\ \text{V};\ \Delta G^\circ = -3(96.5)(0.81) = -2.345 \times 10^2\ \text{kJ} = -2.3 \times 10^5\ \text{J}$$

$$\ln K = \frac{-(-2.345 \times 10^5\ \text{J})}{(8.314\ \text{J/mol - K})(298\ \text{K})} = 94.65 = 95;\quad K = e^{95} = 1.3 \times 10^{41} = 10^{41}$$

(c)
$$2\ [Cr(OH)_3(s) + 5\ OH^-(aq) \rightarrow CrO_4^{2-}(aq) + 4\ H_2O(l) + 3\ e^-]\ E^\circ_{red} = -0.13\ \text{V}$$
$$3\ [ClO^-(aq) + H_2O(l) + 2\ e^- \rightarrow Cl^-(aq) + 2\ OH^-(aq)] \qquad E^\circ_{red} = 0.89\ \text{V}$$
___
$$2\ Cr(OH)_3(s) + 3\ ClO^-(aq) + 4\ OH^-(aq) \rightarrow 2\ CrO_4^{2-}(aq) + 3\ Cl^-(aq) + 5\ H_2O(l)$$

$$E^\circ = 0.89 - (-0.13) = 1.02\ \text{V};\ \Delta G^\circ = -6(96.5)(1.02) = -590.58\ \text{kJ} = -5.91 \times 10^5\ \text{J}$$

$$\ln K = \frac{-(-5.9058 \times 10^5\ \text{J})}{(8.314\ \text{J/mol - K})(298\ \text{K})} = 238.37 = 238;\quad K = 3.3 \times 10^{103} = 10^{103}$$

This is an unimaginably large number.

20.54 $K = 8.7 \times 10^4;\ \Delta G^\circ = -RT \ln K;\ E^\circ = -\Delta G^\circ/nF;\ n = 1;\ T = 298\ \text{K}$

$$\Delta G^\circ = -8.314\ \text{J/mol-K} \times 298\ \text{K} \times \ln(8.7 \times 10^4) = -2.818 \times 10^4\ \text{J} = -28.2\ \text{kJ}$$

$$E^\circ = -\Delta G^\circ/nF = \frac{-(-28.18\ \text{kJ})}{1e^- \times 96.5\ \text{kJ/V - mol e}^-} = 0.292\ \text{V}$$

20.56   At 298 K, $E^\circ = \frac{0.0592\ \text{V}}{n} \log K;\ \log K = \frac{nE^\circ}{0.0592\ \text{V}}.$ See Solution 20.55 for a more complete explanation.

(a) $E^\circ = 0.799\ \text{V} - 0.337\ \text{V} = 0.462\ \text{V};\ n = 2\ (2Ag^+ + 2e^- \rightarrow 2Ag)$

$$\log K = \frac{2(0.462\ \text{V})}{0.0592\ \text{V}} = 15.6081 = 15.6;\ K = 4.056 \times 10^{15} = 4 \times 10^{15}$$

(b) $E^\circ = 1.61\ \text{V} - 0.32\ \text{V} = 1.29\ \text{V};\ n = 3\ (3Ce^{4+} + 3e^- \rightarrow 3Ce^{3+})$

$$\log K = \frac{3(1.29)}{0.0592} = 65.372 = 65.4;\ K = 2.35 \times 10^{65} = 2 \times 10^{65}$$

(c) $E^\circ = 0.36\ \text{V} - (-0.23\ \text{V}) = 0.59\ \text{V};\ n = 4\ (4Fe(CN)_6^{3-} + 4e^- \rightarrow 4Fe(CN)_6^{4-})$

$$\log K = \frac{4(0.59)}{0.0592} = 39.865 = 40;\ K = 7.3 \times 10^{39} = 10^{40}$$

20.58   At 298 K, $E^\circ = \dfrac{0.0592 \text{ V}}{n}\log K$; $n = \dfrac{0.0592 \text{ V}}{E^\circ}\log K$. See Solution 20.55 for a more complete development.

$$n = \frac{0.0592 \text{ V}}{0.17 \text{ V}}\log(5.5\times10^5); n = 2$$

20.60   For this cell at standard conditions, $E^\circ = 1.10$ V.

$$w_{max} = \Delta G^\circ = -nFE^\circ = -2(96.5)(1.10) = -212.3 = -212 \text{ kJ/mol Cu}$$

$$50.0 \text{ g Cu} \times \frac{1 \text{ mol Cu}}{63.55 \text{ g Cu}} \times \frac{-212.3 \text{ kJ}}{\text{mol Cu}} = -167 \text{ kJ} = -1.67\times10^5 \text{ J}$$

Cell EMF under Nonstandard Conditions (Section 20.6)

20.62   (a)   Decrease. As the spontaneous chemical reaction of the voltaic cell proceeds, the concentrations of products increase and the concentrations of reactants decrease.

(b)   If concentration of products increases, Q increases and E decreases.

20.64 $Al(s)+3Ag^+(aq) \rightarrow Al^{3+}(aq)+3Ag(s)$; $E = E^\circ - \dfrac{0.0592}{n}\log Q$; $Q = \dfrac{[Al^{3+}]}{[Ag^+]^3}$

Any change that causes the reaction to be less spontaneous (that causes Q to increase and ultimately shifts the equilibrium to the left) will result in a less positive value for E.

(a)   Increases E by decreasing $[Al^{3+}]$ on the right side of the equation, which decreases Q.

(b)   No effect; the "concentrations" of pure solids and liquids do not influence the value of K for a heterogeneous equilibrium.

(c)   No effect; the concentration of Ag^+ and the value of Q are unchanged.

(d)   Decreases E; forming AgCl(s) decreases the concentration of Ag^+, which increases Q.

20.66   (a)

$$3[Ce^{4+}(aq)+1 e^- \rightarrow Ce^{3+}(aq)] \qquad E^\circ_{red} = 1.61 \text{ V}$$
$$Cr(s) \rightarrow Cr^{3+}(aq)+3 e^- \qquad E^\circ_{red} = -0.74 \text{ V}$$

$$3 Ce^{4+}(aq)+Cr(s) \rightarrow 3 Ce^{3+}(aq)+Cr^{3+}(aq) \qquad E^\circ = 1.61-(-0.74)=2.35 \text{ V}$$

(b) $E = E^\circ - \dfrac{0.0592}{n}\log\dfrac{[Ce^{3+}]^3[Cr^{3+}]}{[Ce^{4+}]^3}$; $n = 3$

$$E = 2.35 - \frac{0.0592}{3}\log\frac{(0.10)^3(0.010)}{(3.0)^3} = 2.35 - \frac{0.0592}{3}\log(3.704\times10^{-7})$$

$$E = 2.35 - \frac{0.0592(-6.431)}{3} = 2.35+0.127 = 2.48 \text{ V}$$

(c)
$$E = 2.35 - \frac{0.0592}{3} \log \frac{(2.0)^3 (1.5)}{(0.010)^3} = 2.35 - 0.1397 = 2.21 \text{ V}$$

20.68 (a)

$$2 \, [Fe^{3+}(aq) + 1 \, e^- \rightarrow Fe^{2+}(aq)] \qquad\qquad E^o_{red} = 0.771 \text{ V}$$

$$H_2(g) \rightarrow 2H^+(aq) + 2 \, e^- \qquad\qquad E^o_{red} = 0.000 \text{ V}$$

$$\overline{2 \, Fe^{3+}(aq) + H_2(g) \rightarrow 2 \, Fe^{2+}(aq) + 2 \, H^+(aq) \qquad E^o = 0.771 - 0.000 = 0.771 \text{ V}}$$

(b)

$$E = E^o - \frac{0.0592}{n} \log \frac{[Fe^{2+}]^2 [H^+]^2}{[Fe^{3+}]^2 P_{H_2}}; \; [H^+] = 10^{-pH} = 1.0 \times 10^{-4}, \; n = 2$$

$$E = 0.771 - \frac{0.0592}{2} \log \frac{(0.0010)^2 (1.0 \times 10^{-4})^2}{(3.50)^2 (0.95)} = 0.771 - \frac{0.0592}{2} \log (8.6 \times 10^{-16})$$

$$E = 0.771 - \frac{0.0592(-15.066)}{2} = 0.771 + 0.446 = 1.217 \text{ V}$$

20.70 (a) The compartment with 0.0150 M Cl$^-$ (aq) is the cathode.

(b) $E^o = 0$ V

(c) $E = E^o - \dfrac{0.0592}{n} \log Q; \; Q = [Cl^-, \text{dilute}] / [Cl^-, \text{conc.}]$

$$E = 0 - \frac{0.0592}{1} \log \frac{(0.0150)}{(2.55)} = -0.13204 = -0.1320 \text{ V}$$

(d) In the anode compartment, [Cl$^-$] will decrease from 2.55 M. In the cathode, [Cl$^-$] will increase from 0.0150 M.

20.72 (a) $E^o = -0.136 \text{ V} - (-0.126 \text{ V}) = -0.010 \text{ V}; n = 2$

$$0.22 = -0.010 - \frac{0.0592}{2} \log \frac{[Pb^{2+}]}{[Sn^{2+}]} = -0.010 - \frac{0.0592}{2} \log \frac{[Pb^{2+}]}{1.00}$$

$$\log [Pb^{2+}] = \frac{-0.23 (2)}{0.0592} = -7.770 = -7.8; [Pb^{2+}] = 1.7 \times 10^{-8} = 2 \times 10^{-8} \text{ M}$$

(b) For PbSO$_4$(s), $K_{sp} = [Pb^{2+}] [SO_4^{2-}] = (1.0)(1.7 \times 10^{-8}) = 1.7 \times 10^{-8}$

Batteries and Fuel Cells (Section 20.7)

20.74 (a) The overall cell reaction is:

$$2 \, MnO_2(s) + Zn(s) + 2 \, H_2O(l) \rightarrow 2 \, MnO(OH)(s) + Zn(OH)_2(s)$$

$$4.50 \text{ g Zn} \times \frac{1 \text{ mol Zn}}{65.39 \text{ g Zn}} \times \frac{2 \text{ mol MnO}_2}{1 \text{ mol Zn}} \times \frac{86.94 \text{ g MnO}_2}{1 \text{ mol MnO}_2} = 12.0 \text{ g MnO}_2$$

(b) Two mol e$^-$ are transferred for every mol of Zn reacted. 96,485 C/mol e$^-$

$$4.50 \text{ g Zn} \times \frac{1 \text{ mol Zn}}{65.39 \text{ g Zn}} \times \frac{2 \text{ mol e}^-}{1 \text{ mol Zn}} \times \frac{96,485 \text{ C}}{1 \text{ mol e}^-} = 13,280 = 1.32 \times 10^4 \text{ C}$$

20 Electrochemistry

20.76 **(a)** $HgO(s) + Zn(s) \rightarrow Hg(l) + ZnO(s)$

(b) $E^o_{cell} = E^o_{red}$ (cathode) $- E^o_{red}$ (anode)

E^o_{red} (anode) $= E^o_{red} - E^o_{cell} = 0.098 - 1.35 = -1.25$ V

(c) E^o_{red} is different from Zn^{2+} (aq) $+ 2e^- \rightarrow Zn(s)$ (–0.76 V) because in the battery the process happens in the presence of base and Zn^{2+} is stabilized as ZnO(s). Stabilization of a reactant in a half-reaction decreases the driving force, so E^o_{red} is more negative.

20.78 **(a)** $NiO(OH)(s) + H_2O(l) + 1 e^- \rightarrow Ni(OH)_2(s) + OH^-(aq)$

(b) $Zn(s) + 2 OH^-(aq) \rightarrow Zn(OH)_2(s) + 2 e^-$

(c) A nickel-zinc battery will produce about 0.94 V. E^o_{red} for Cd (–0.40 V) is less negative than E^o_{red} for Zn (–0.76 V), so E_{cell} for the nickel-zinc battery will be less positive than that for the nickel-cadmium battery by 0.36 V. That is, the potential of the nickel-zinc battery will be approximately (1.30 V – 0.36 V) = 0.94 V.

(d) Higher. Specific energy density is the amount of energy stored per unit mass of the battery. The potential of the nickel-zinc battery is about 72 % of the nickel-cadmium battery (0.94 V/1.30 V); the molar mass of zinc is about 58 % of the molar mass of cadmium (65.38 g/112.4 g). The reduction in mass by exchanging zinc for cadmium more than compensates for the reduction in cell potential.

20.80 **(a)** $LiMn_2O_4$: FW = 6.941 + 2(54.938) + 4(15.9994) = 180.815 amu

% Li $= \dfrac{6.941 \text{ g}}{180.815 \text{ g}} \times 100 = 3.8387 = 3.839$ %

$LiCoO_2$: FW = 6.941 + 58.9332 + 2(15.9994) = 97.873 amu

% Li $= \dfrac{6.941 \text{ g}}{97.873 \text{ g}} \times 100 = 7.0918 = 7.092$ %

(b) $LiCoO_2$ has almost twice as great a mass percent Li as $LiMn_2O_4$. This definitely explains why $LiMn_2O_4$ cathodes deliver less power on discharging. A Li ion battery produces power by the migration of Li^+, but not all the Li^+ in the cathode material migrates during a full charge. The greater percentage of Li in the cathode material, the more Li^+ that migrates and the greater the power that can be produced upon discharge.

(c) Assume 100.000 g of cathode material. A $LiCoO_2$ cathode has 7.092 g of Li; if half of the Li migrates during charging, 3.546 g Li migrates. A $LiMn_2O_4$ cathode contains 3.839 g of Li. To deliver the same amount of Li to the graphite anode,

$\dfrac{3.546 \text{ g}}{3.839 \text{ g}} \times 100 = 92.368 = 92$ % of the Li in the $LiMn_2O_4$ cathode would have to migrate.

20.82 **(a)** Both batteries and fuel cells are electrochemical power sources. Both take advantage of spontaneous oxidation–reduction reactions to produce a certain voltage. The difference is that batteries are self-contained (all reactants and

373
Copyright © 2015 Pearson Education, Inc.

products are present inside the battery casing), whereas fuel cells require continuous supply of reactants and exhaust of products.

(b) No. The fuel in a fuel cell must be fluid, either gas or liquid. Because fuel must be continuously supplied to the fuel cell, it must be capable of flow; the fuel cannot be solid.

Corrosion (Section 20.8)

20.84 (a) Calculate E_{cell}^o for the given reactants at standard conditions.

$$O_2(g) + 4\,H^+(aq) + 4\,e^- \rightarrow 2\,H_2O(l) \qquad\qquad E_{red}^o = 1.23\ V$$

$$2\,[Cu(s) \rightarrow Cu^{2+}(aq) + 2\,e^-] \qquad\qquad E_{red}^o = 0.337\ V$$

$$\overline{2\,Cu(s) + O_2(g) + 4\,H^+(aq) \rightarrow 2\,Cu^{2+}(aq) + 2\,H_2O(l) \qquad E^\circ = 1.23 - 0.337 = 0.89\ V}$$

At standard conditions with $O_2(g)$ and $H^+(aq)$ present, the oxidation of Cu(s) has a positive E° value and is spontaneous. Cu(s) will oxidize (corrode) in air in the presence of acid.

(b) Fe^{2+} has a more negative reduction potential (–0.440 V) than Cu^{2+} (+0.337 V), so Fe(s) is more readily oxidized than Cu(s). If the two metals are in contact, Fe(s) would act as a sacrificial anode and oxidize (corrode) in preference to Cu(s); this would weaken the iron support skeleton of the statue. The teflon spacers prevent contact between the two metals and insure that the iron skeleton doesn't corrode when the Cu(s) skin comes in contact with atmospheric $O_2(g)$ and $H^+(aq)$.

20.86 No. To afford cathodic protection, a metal must be more difficult to reduce (have a more negative reduction potential) than Fe^{2+}. $E_{red}^o\ Co^{2+} = -0.28\ V$, $E_{red}^o\ Fe^{2+} = -0.44\ V$.

20.88 The principal metallic component of steel is Fe. E_{red}^o for Fe, –0.763 V, is more negative than that of Cu, 0.337 V. When the two are in contact, Fe acts as the sacrificial anode and corrodes (oxidizes) preferentially in the presence of $O_2(g)$.

$2\,Fe(s) + O_2(g) + 4\,H^+(aq) \rightarrow 2\,Fe^{2+}(aq) + 2\,H_2O(l)$

E° = 1.23 V – (–0.440 V) = 1.67 V

$2\,Cu(s) + O_2(g) + 4\,H^+(aq) \rightarrow 2\,Cu^{2+}(aq) + 2\,H_2O(l)$

E° = 1.23 V – (0.337 V) = 0.893 V

Both reactions are spontaneous, but the corrosion of Fe has the larger E° value and happens preferentially.

Electrolysis; Electrical Work (Section 20.9)

20.90 (a) An *electrolytic cell* is the vessel in which electrolysis occurs. It consists of a power source and two electrodes in a molten salt or aqueous solution.

(b) It is the cathode. In an electrolysis cell, as in a voltaic cell, electrons are consumed (via reduction) at the cathode. Electrons flow from the negative terminal of the voltage source and then to the cathode.

(c)   A small amount of $H_2SO_4(aq)$ present during the electrolysis of water acts as a change carrier, or supporting electrolyte. This facilitates transfer of electrons through the solution and at the electrodes, speeding up the reaction. [Considering $H^+(aq)$ as the substance reduced at the cathode changes the details of the half-reactions, but not the overall E° for the electrolysis. $SO_4^{2-}(aq)$ cannot be oxidized.]

(d)   If the active metal salt is present as an aqueous solution during electrolysis, water is reduced [to $H_2(g)$] rather than the metal ion being reduced to the metal. This is true for any active metal with an E_{red}^o value more negative than –0.83 V.

20.92   Coulombs = amps-s; because this is a 2e⁻ reduction, each mole of Mg(s) requires 2 Faradays.

(a) $4.55\ A \times 4.50\ d \times \dfrac{24\ h}{1\ d} \times \dfrac{60\ min}{1\ h} \times \dfrac{60\ s}{1\ min} \times \dfrac{1\ C}{1\ amp\text{-}s} \times \dfrac{1\ F}{96{,}485\ C}$

$$\times \dfrac{1\ mol\ Mg}{2\ F} \times \dfrac{24.31\ g\ Mg}{1\ mol\ Mg} = 223\ g\ Mg$$

(b) $25.00\ g\ Mg \times \dfrac{1\ mol\ Mg}{24.31\ g\ Mg} \times \dfrac{2\ F}{1\ mol\ Mg} \times \dfrac{96{,}485\ C}{F} \times \dfrac{1\ amp\text{-}s}{C} \times \dfrac{1\ min}{60\ s} \times \dfrac{1}{3.50\ A}$

$$= 945\ min$$

20.94   (a) $7.5 \times 10^3\ A \times 48\ h \times \dfrac{3600\ s}{1\ h} \times \dfrac{1\ C}{1\ amp\text{-}s} \times \dfrac{1\ F}{96{,}485\ C} \times \dfrac{1\ mol\ Ca}{2\ F} \times 0.68 \times \dfrac{40.0\ g\ Ca}{1\ mol\ Ca}$

$$= 1.830 \times 10^5 = 1.8 \times 10^5\ g\ Ca$$

(b) $E_{cell}^o = E_{red}^o(\text{cathode}) - E_{red}^o(\text{anode}) = -2.87\ V - (1.359\ V) = -4.229 = -4.23\ V$

The minimum voltage required to drive the reaction is the magnitude of E_{cell}^o, 4.23 V.

20.96   The standard reduction potential for Te^{4+}, 0.57 V, is more positive than that of Cu^{2+}, 0.34 V. This means the Te^{4+} is "easier" to reduce than Cu^{2+}, but Te is harder to oxidize and less active than Cu. During electrorefining, although Cu is oxidized from the crude anode, Te will not be oxidized. It is likely to accumulate along with other impurities less active than Cu, in the so-called anode sludge.

Additional Exercises

20.98   (a)

$Fe(s) \rightarrow Fe^{2+}(aq) + 2\ e^-$

$\underline{2Ag^+(aq) + 2\ e^- \rightarrow 2\ Ag(s)}$

$Fe(s) + 2\ Ag^+(aq) \rightarrow Fe^{2+}(aq) + 2\ Ag(s)$

(b)

$$Zn(s) \rightarrow Zn^{2+}(s) + 2\,e^-$$

$$\underline{2\,H^+(aq) + 2\,e^- \rightarrow H_2(g)}$$

$$Zn(s) + 2\,H^+(aq) \rightarrow Zn^{2+}(aq) + H_2(g)$$

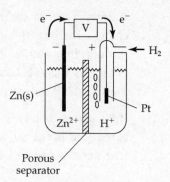

(c) $Cu \mid Cu^{2+} \parallel ClO_3^-, Cl^- \mid Pt$. Here, both the oxidized and reduced forms of the cathode solution are in the same phase, so we separate them by a comma and then indicate an inert electrode.

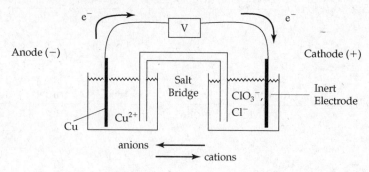

20.100   (a)   The reduction potential for $O_2(g)$ in the presence of acid is 1.23 V. $O_2(g)$ cannot oxidize Au(s) to $Au^+(aq)$ or $Au^{3+}(aq)$, even in the presence of acid.

         (b)   The possible oxidizing agents need a reduction potential greater than 1.50 V. These include $Co^{3+}(aq)$, $F_2(g)$, $H_2O_2(aq)$, and $O_3(g)$. Marginal oxidizing agents (those with reduction potential near 1.50 V) from Appendix E are $BrO_3^-(aq)$, $Ce^{4+}(aq)$, $HClO(aq)$, $MnO_4^-(aq)$, and $PbO_2(s)$.

         (c) $4\,Au(s) + 8\,NaCN(aq) + 2\,H_2O(l) + O_2(g) \rightarrow 4\,Na[Au(CN)_2](aq) + 4\,NaOH(aq)$

$$Au(s) + 2\,CN^-(aq) \rightarrow [Au(CN)_2]^- + 1\,e^-$$

$$O_2(g) + 2\,H_2O(l) + 4\,e^- \rightarrow 4\,OH^-(aq)$$

            Au(s) is being oxidized and $O_2(g)$ is being reduced.

         (d) $2\,[Na[Au(CN)_2](aq) + 1\,e^- \rightarrow Au(s) + 2\,CN^-(aq) + Na^+(aq)]$

$$\underline{Zn(s) \rightarrow Zn^{2+}(aq) + 2\,e^-}$$

$$2\,Na[Au(CN)_2](aq) + Zn(s) \rightarrow 2\,Au(s) + Zn^{2+}(aq) + 2\,Na^+(aq) + 4\,CN^-(aq)$$

            Zn(s) is being oxidized and $[Au(CN)_2]^-(aq)$ is being reduced. Although $OH^-(aq)$ is not included in this redox reaction, its presence in the reaction mixture probably causes $Zn(OH)_2(s)$ to form as the product. This increases the driving force (and E°) for the overall reaction.

20.101   (a) $2\,[Ag^+(aq)+1\,e^- \rightarrow Ag(s)]$ $E^o_{red} = 0.80$ V

$Ni(s) \rightarrow Ni^{2+}(aq)+2\,e^-$ $E^o_{red} = -0.28$ V

_____

$2\,Ag^+(aq) + Ni(s) \rightarrow 2\,Ag(s)+Ni^{2+}(aq)$ $E^o = 0.80-(-0.28)=1.08$ V

(b)   As the reaction proceeds, $Ni^{2+}(aq)$ is produced, so $[Ni^{2+}]$ increases as the cell operates.

(c) $E = E^o - \dfrac{0.0592}{n}\log Q; 1.12 = 1.08 - \dfrac{0.0592}{2}\log\dfrac{[Ni^{2+}]}{[Ag^+]^2}$

$-\dfrac{0.04(2)}{0.0592} = \log(0.0100) - \log[Ag^+]^2; \log[Ag^+]^2 = \log(0.0100) + \dfrac{0.04(2)}{0.0592}$

$\log[Ag^+]^2 = -2.000 + 1.351 = -0.649; [Ag^+]^2 = 0.255\ M; [Ag^+] = 0.474 = 0.5\ M$

[Strictly speaking, $[E - E^o]$ having only 1 sig fig leads (after several steps) to the answer having only 1 sig fig. This is not a very precise or useful result.]

20.102   (a) $I_2(s)+2\,e^- \rightarrow 2\,I^-(aq)$ $E^o_{red} = 0.536$ V

$2\,[Cu(s) \rightarrow Cu^+(aq)+1\,e^-]$ $E^o_{red} = 0.521$ V

_____

$I_2(s) + 2\,Cu(s) \rightarrow 2\,Cu^+(aq)+2\,I^-(aq)$ $E^o = 0.536-0.521 = 0.015$ V

$E = E^o - \dfrac{0.0592}{n}\log Q = 0.015 - \dfrac{0.0592}{2}\log\,[Cu^+]^2\,[I^-]^2$

$E = +0.015 - \dfrac{0.0592}{2}\log\,(0.25)^2(3.5)^2 = +0.015 + 0.0034 = 0.018$ V

(b)   Because the cell potential is positive at these concentration conditions, the reaction as written in part (a) is spontaneous in the forward direction. Cu is oxidized and Cu(s) is the anode.

(c)   Yes. E^o is positive, so Cu is oxidized and Cu(s) is the anode at standard conditions.

(d) $E = 0;\quad +0.015 = \dfrac{0.0592}{2}\log\,(0.15)^2\,[I^-]^2;\quad \dfrac{2(0.015)}{0.0592} = \log\,(0.15)^2 + 2\log\,[I^-];$

$\log[I^-] = 1.0773 = 1.08;\ [I^-] = 10^{1.08} = 11.95 = 12\ M\ I^-$

20.104   (a)   In discharge, $Cd(s) + 2NiO(OH)(s) + 2H_2O(l) \rightarrow Cd(OH)_2(s) + 2Ni(OH)_2(s)$.
In charging, the reverse reaction occurs.

(b) $E^o = 0.49$ V $-$ $(-0.76$ V$)$ $= 1.25$ V

(c)   The 1.25 V calculated in part (b) is the standard cell potential, E^o. The concentrations of reactants and products inside the battery are adjusted so that the cell output is greater than E^o. Note that most of the reactants and products are pure solids or liquids, which do not appear in the Q expression. It must be $[OH^-]$ that is other than 1.0 M, producing an emf of 1.30 rather than 1.25.

(d) $\quad E° = \dfrac{0.0592}{n} \log K; \ \log k = \dfrac{nE°}{0.0592}$

$\log K = \dfrac{2 \times 1.30}{0.0592} = 43.92 = 43.9; \ K = 8.3 \times 10^{43} = 8 \times 10^{43}$

20.105   (a)   The battery capacity expressed in units of mAh indicates the total amount of electrical charge that can be delivered by the battery.

      (b)   Quantity of electrical charge is measured in coulombs, C. $C = A \cdot s$

$2850 \ \text{mAh} \times \dfrac{1 \ \text{A}}{1000 \ \text{mA}} \times \dfrac{3600 \ \text{s}}{\text{h}} \times \dfrac{1 \ \text{C}}{1 \ \text{A-s}} = 10{,}260 \ \text{C}$

The battery can deliver 10,260 C. Work, electrical or otherwise, is measured in J. $J = V \times C$. If the battery voltage decreases linearly from 1.55 V to 0.80 V, assume an average voltage of 1.175 = 1.2 V.

10, 260 C × 1.175 V = 12,055.5 = 12 × 10³ J = 12 kJ

The total maximum electrical work of the battery is 12 kJ.

(This is 3.3×10^{-3} kWh or 3.3 Wh.)

20.107   (a)   Total volume of Cr $= 2.5 \times 10^{-4} \ \text{m} \times 0.32 \ \text{m}^2 = 8.0 \times 10^{-5} \ \text{m}^3$

$\text{mol Cr} = 8.0 \times 10^{-5} \ \text{m}^3 \ \text{Cr} \times \dfrac{100^3 \ \text{cm}^3}{1 \ \text{m}^3} \times \dfrac{7.20 \ \text{g Cr}}{1 \ \text{cm}^3} \times \dfrac{1 \ \text{mol Cr}}{52.0 \ \text{g Cr}} = 11.077$

$= 11 \ \text{mol Cr}$

The electrode reaction is:

$CrO_4^{2-}(aq) + 4 \ H_2O(l) + 6 \ e^- \rightarrow Cr(s) + 8 \ OH^-(aq)$

$\text{Coulombs required} = 11.077 \ \text{mol Cr} \times \dfrac{6 \ \text{F}}{1 \ \text{mol Cr}} \times \dfrac{96{,}485 \ \text{C}}{1 \ \text{F}} = 6.41 \times 10^6$

$= 6.4 \times 10^6 \ \text{C}$

      (b) $6.41 \times 10^6 \ \text{C} \times \dfrac{1 \ \text{amp-s}}{1 \ \text{C}} \times \dfrac{1}{10.0 \ \text{s}} = 6.4 \times 10^5 \ \text{amp}$

      (c)   If the cell is 65% efficient, $(6.41 \times 10^6/0.65) = 9.867 \times 10^6 = 9.9 \times 10^6$ C is required to plate the bumper.

$6.0 \ \text{V} \times 9.867 \times 10^6 \ \text{C} \times \dfrac{1 \ \text{J}}{1 \ \text{C-V}} \times \dfrac{1 \ \text{kWh}}{3.6 \times 10^6 \ \text{J}} = 16.445 = 16 \ \text{kWh}$

20.108   (a)   The standard reduction potential for $H_2O(l)$ is much greater than that of $Mg^{2+}(aq)(-0.83$ V vs. -2.37 V). In aqueous solution, $H_2O(l)$ would be preferentially reduced and no Mg(s) would be obtained.

      (b) $97{,}000 \ \text{A} \times 24 \ \text{h} \times \dfrac{3600 \ \text{s}}{1 \ \text{h}} \times \dfrac{1 \ \text{C}}{1 \ \text{A-s}} \times \dfrac{1 \ \text{F}}{96{,}485 \ \text{C}} \times \dfrac{1 \ \text{mol Mg}}{2 \ \text{F}} \times \dfrac{24.31 \ \text{g Mg}}{1 \ \text{mol Mg}} \times 0.96$

$= 1.0 \times 10^6 \ \text{g Mg} = 1.0 \times 10^3 \ \text{kg Mg}$

20.110 (a) $7 \times 10^8 \text{ mol } H_2 \times \dfrac{2\,F}{1\,\text{mol } H_2} \times \dfrac{96{,}485\,C}{1\,F} = 1.35 \times 10^{14} = 1 \times 10^{14}\,C$

(b)
$$2\,H_2O(l) \rightarrow O_2(g) + 4\,H^+(aq) + 4\,e^- \qquad E^\circ_{red} = 1.23\,V$$
$$2\,[2\,H^+(aq) + 2\,e^- \rightarrow H_2(g)] \qquad E^\circ_{red} = 0\,V$$
$$2\,H_2O(l) \rightarrow O_2(g) + 2\,H_2(g) \qquad E^\circ = 0.00 - 1.23 = -1.23\,V$$

$P_t = 300\,\text{atm} = P_{O_2} + P_{H_2}$. Because $H_2(g)$ and $O_2(g)$ are generated in a 2:1 mole ratio, $P_{H_2} = 200\,\text{atm}$ and $P_{O_2} = 100\,\text{atm}$.

$$E = E^\circ - \frac{0.0592}{4}\log(P_{O_2} \times P_{H_2}^2) = -1.23\,V - \frac{0.0592}{4}\log[100 \times (200)^2]$$

$E = -1.23\,V - 0.100\,V = -1.33\,V; \; E_{min} = 1.33\,V$

(c) $\text{Energy} = nFE = 2(7 \times 10^8\,\text{mol})(1.33\,V)\dfrac{96{,}485\,J}{V\text{-}mol} = 1.80 \times 10^{14} = 2 \times 10^{14}\,J$

(d) $1.80 \times 10^{14}\,J \times \dfrac{1\,kWh}{3.6 \times 10^6\,J} \times \dfrac{\$0.85}{kWh} = \$4.24 \times 10^7 = \$4 \times 10^7$

It would cost more than $40 million for the electricity alone.

Integrative Exercises

20.111 $N_2(g) + 3\,H_2(g) \rightarrow 2\,NH_3(g)$

(a) The oxidation number of $H_2(g)$ and $N_2(g)$ is 0. The oxidation number of N in NH_3 is −3, H in NH_3 is +1. H_2 is being oxidized and N_2 is being reduced.

(b) Calculate ΔG° from ΔG_f° values in Appendix C. Use $\Delta G^\circ = -RT \ln K$ to calculate K.

$\Delta G^\circ = 2\Delta G_f^\circ\,NH_3(g) - \Delta G_f^\circ\,N_2(g) - 3\Delta G_f^\circ\,H_2(g)$

$\Delta G^\circ = 2(-16.66\,kJ) - 0 - 3(0) = -33.32\,kJ$

$\Delta G^\circ = -RT \ln K, \ln K = \dfrac{-\Delta G^\circ}{RT} = \dfrac{-(-33.32 \times 10^3\,J)}{(8.314\,J/\text{mol-K})(298\,K)} = 13.4487 = 13.45$

$K = e^{13.4487} = 6.9 \times 10^5$

(c) $\Delta G^\circ = -nfE^\circ; \quad E^\circ = \dfrac{-\Delta G^\circ}{nF}; \quad n = ?$

2 N atoms change from 0 to −3 or 6 H atoms change from 0 to +1.
Either way, n = 6.

$E^\circ = \dfrac{-(-33.32\,kJ)}{6 \times 96.5\,kJ/V} = 0.05755\,V$

20.112 The redox reaction is: $2Ag^+(aq) + H_2(g) \rightarrow 2Ag(s) + 2H^+(aq)$. n = 2 for this reaction.

$E^\circ_{cell} = E^\circ_{red}\,\text{cathode} - E^\circ_{red}\,\text{anode} = 0.799\,V - 0\,V = 0.799$

$E = E^\circ - \dfrac{0.0592}{n}\log\dfrac{[H^+]^2}{[Ag^+]^2\,P_{H_2}}$

$[H^+]$ in the cell is held essentially constant by the benzoate buffer.

$$C_6H_5COOH(aq) \rightleftharpoons H^+(aq) + C_6H_5COO^-(aq) \quad K_a = ?$$

$$K_a = \frac{[H^+][C_6H_5COO^-]}{[C_6H_5COOH]}; [H^+] = \frac{K_a[C_6H_5COOH]}{[C_6H_5COO^-]} = \frac{0.10 \text{ M}}{0.050 \text{ M}} \times K_a = 2K_a$$

Solve the Nernst expression for $[H^+]$ and calculate K_a and pK_a as shown above.

$$1.030 \text{ V} = 0.799 \text{ V} - \frac{0.0592}{n} \log \frac{[H^+]^2}{(1.00)^2(1.00)}$$

$$0.231 \times \frac{2}{0.0592} = -\log [H^+]^2 = -2 \log[H^+]$$

$$\frac{0.231}{0.0592} = -\log [H^+] = pH; pH = 3.9020 = 3.90; [H^+] = 10^{-3.902} = 1.253 \times 10^{-4} = 1.3 \times 10^{-4};$$

$$[H^+] = 2K_a; K_a = [H^+]/2 = 6.265 \times 10^{-5} = 6.3 \times 10^{-5}; pK_a = 4.20$$

Check. According to Appendix D, K_a for benzoic acid is 6.3×10^{-5}.

20.113   (a)   The oxidation potential of A is equal in magnitude but opposite in sign to the reduction potential of A^+.

       (b)   Li(s) has the highest oxidation potential, Au(s) the lowest.

       (c)   The relationship is reasonable because both oxidation potential and ionization energy describe removing electrons from a substance. Ionization energy is a property of gas phase atoms or ions, whereas oxidation potential is a property of the bulk material.

20.115   (a) $\Delta H° = 2 \Delta H° \, H_2O(l) - 2 \Delta H° \, H_2(g) - \Delta H° \, O_2(g) = 2(-285.83) - 2(0) - 0 = -571.66 \text{ kJ}$

            $\Delta S° = 2 \, S° \, H_2O(l) - 2 \, S° \, H_2(g) - \Delta S° \, O_2(g)$

                 $= 2(69.91) - 2(130.58) - (205.0) = -326.34 \text{ J}$

       (b)   Because $\Delta S°$ is negative, $-T\Delta S$ is positive and the value of ΔG will become more positive as T increases. The reaction will become nonspontaneous at a fairly low temperature, because the magnitude of $\Delta S°$ is large.

       (c) $\Delta G = w_{max}$. The larger the negative value of ΔG, the more work the system is capable of doing on the surroundings. As the magnitude of ΔG decreases with increasing temperature, the usefulness of H_2 as a fuel decreases.

       (d)   The combustion method increases the temperature of the system, which quickly decreases the magnitude of the work that can be done by the system. Even if the effect of temperature on this reaction could be controlled, only about 40% of the energy from any combustion can be converted to electrical energy, so combustion is intrinsically less efficient than direct production of electrical energy via a fuel cell.

20.116   First, balance the equation:

$$4\,CyFe^{2+}(aq) + O_2(g) + 4\,H^+(aq) \rightarrow 4\,CyFe^{3+}(aq) + 2\,H_2O(l);\ E = +0.60\ V;\ n = 4$$

(a)   From Equation 20.11, we can calculate ΔG for the process under the conditions specified for the measured potential E:

$$\Delta G = -nFE = -(4\ mol\ e^-) \times \frac{96.485\ kJ}{1\ V\text{-}mol\ e^-}(0.60\ V) = -231.6 = -232\ kJ$$

(b)   The moles of ATP synthesized per mole of O_2 is given by:

$$\frac{231.6\ kJ}{O_2\ molecule} \times \frac{1\ mol\ ATP\ formed}{37.7\ kJ} = \text{approximately 6 mol ATP/mol } O_2$$

20.118   The reaction can be written as a sum of the steps:

$$Pb^{2+}(aq) + 2\,e^- \rightarrow Pb(s) \qquad\qquad E^o_{red} = -0.126\ V$$
$$PbS(s) \rightarrow Pb^{2+}(aq) + S^{2-}(aq) \qquad\qquad ``E^o" = ?$$
$$\overline{PbS(s) + 2\,e^- \rightarrow Pb(s) + S^{2-}(aq) \qquad\qquad E^o_{red} = ?}$$

"E^o" for the second step can be calculated from K_{sp}.

$$E^o = \frac{0.0592}{n}\log K_{sp} = \frac{0.0592}{2}\log(8.0 \times 10^{-28}) = \frac{0.0592}{2}(-27.10) = -0.802\ V$$

E^o for the half-reaction = $-0.126\ V + (-0.802\ V) = -0.928\ V$

Calculating an imaginary E^o for a nonredox process like step 2 may be a disturbing idea. Alternatively, one could calculate K for step 1 (5.4×10^{-5}), K for the reaction in question ($K = K_1 \times K_{sp} = 4.4 \times 10^{-32}$), and then E^o for the half-reaction. The result is the same.

20.119   The two half-reactions in the electrolysis of $H_2O(l)$ are:

$$2\,[2H_2O(l) + 2\,e^- \rightarrow H_2(g) + 2\,OH^-]$$
$$\underline{2\,H_2O(l) \rightarrow O_2(g) + 4\,H^+ + 4\,e^-}$$
$$2\,H_2O(l) \rightarrow 2\,H_2(g) + O_2(g)$$

4 mol e^-/2 mol $H_2(g)$ or 2 mol e^-/mol $H_2(g)$

Using partial pressures and the ideal-gas law, calculate the mol $H_2(g)$ produced, and the current required to do so.

$P_t = P_{H_2} + P_{H_2O}$. From Appendix B, P_{H_2O} at $25.5\,^\circ C$ is approximately 24.5 torr.

$P_{H_2} = 768\ torr - 24.5\ torr = 743.5 = 744\ torr$

$$n = PV/RT = \frac{(743.5/760)\ atm \times 0.0123\ L}{298.5\ K \times 0.08206\ L\text{-}atm/mol\text{-}K} = 4.912 \times 10^{-4} = 4.91 \times 10^{-4}\ mol\ H_2$$

$$4.912 \times 10^{-4}\ mol\ H_2 \times \frac{2\ mol\ e^-}{mol\ H_2} \times \frac{96,485\ C}{1\ mol\ e^-} \times \frac{1\ amp\text{-}s}{1\ C} \times \frac{1\ min}{60\ s} \times \frac{1}{2.00\ min} = 0.790\ amp$$

21 Nuclear Chemistry

Visualizing Concepts

21.2   *Analyze/Plan.* From the diagram, determine the atomic number (number of protons) and mass number (number of protons plus neutrons) of the two nuclides involved. Based on the relationship between the two nuclides, decide whether the reaction is alpha or beta decay, positron emission or electron capture. Complete the nuclear reaction, balancing atomic numbers and mass numbers.

Solve. The two nuclides in the diagram are $^{109}_{46}\text{Pd}$ and $^{109}_{47}\text{Ag}$, so the second product is a beta particle. The balanced reaction is:

$$^{109}_{46}\text{Pd} \rightarrow \,^{109}_{47}\text{Ag} + \,^{0}_{-1}\text{e}$$

Check. Atomic number and mass number balance.

21.3   *Analyze/Plan.* Determine the number of protons and neutrons present in the two heavy nuclides in the reaction. Draw a graph with appropriate limits and plot the two points. Draw an arrow from reactant to product.   *Solve.*

Bi: 83 p, 128 n; Tl: 81 p, 126 n.

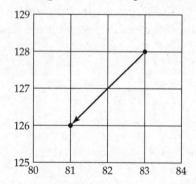

Check. An alpha particle has 2 p and 2 n. The diagram shows a decrease in 2 p and 2 n for the reaction.

21.4   *Analyze/Plan.* Count the protons and neutrons in each particle. Use a periodic table and Table 21.2 to identify the particles. Refer to Sample Exercise 21.4 when writing the reaction using condensed notation. Use Figure 21.2 to determine stability of the product nucleus.

(a)   From left to right, the particles are $^{10}_{5}\text{B}$, $^{1}_{0}\text{n}$, $^{7}_{3}\text{Li}$, and $^{4}_{2}\text{He}$

(b) $^{10}_{5}\text{B}(\text{n}, \alpha)\,^{7}_{3}\text{Li}$

(c)   The $^{7}_{3}\text{Li}$ product nucleus is probably stable. It appears to be in the belt of stability in Figure 21.2. The product nucleus has an odd number of protons but an even number of neutrons. There are 50 stable "odd, even" nuclei.

21 Nuclear Chemistry Solutions to Exercises

21.5   *Analyze/Plan.* Use the information in Table 21.3 to identify the type of radioactive decay for each step.

(a)   (i) Alpha decay. Mass number decreases by 4 and atomic number decreases by 2.

(ii) Beta decay. Mass number is the same, atomic number increases by 1.

(iii) Beta decay. Mass number is the same, atomic number increases by 1.

(b) $^{228}_{89}Ac$ has the highest activity; it has the shortest half-life.

(c) $^{232}_{90}Th$ has the lowest activity; it has the longest half-life.

(d) $^{224}_{88}Rn$. Mass number decreases by 4 and atomic number decreases by 2, relative to the the last isotope shown.

21.8   *Analyze/Plan.* Express the particles in the diagram as a nuclear reaction. Determine the mass number and atomic number of the unknown particle by balancing these quantities in the nuclear reaction.   *Solve.*

(a) $^{239}_{94}Pu + ^{1}_{0}n \rightarrow ^{95}_{40}Zr + ? + 2\,^{1}_{0}n$

The unidentified particle has an atomic number of (94–40) = 54; it is Xe. The mass number of the nuclide is [(239 + 1) – (95 + 2)] = 143. The unknown particle is $^{143}_{54}Xe$.

(b) ^{95}Zr: 40 p, 55 n is stable. ^{143}Xe: 54 p, 89 n is above the belt of stability and is not stable; it will probably undergo beta decay.

Radioactivity (Section 21.1)

21.10   p = protons, n = neutrons, e = electrons; number of protons = atomic number; number of neutrons = mass number – atomic number

(a) $^{129}_{53}I$: 53p, 76n   (b) ^{138}Ba: 56p, 82n   (c) ^{237}Np: 93p, 144n

21.12   (a) $^{1}_{1}p$ or $^{1}_{1}H$ (b) $^{0}_{-1}e$ or β^- (c) $^{0}_{+1}e$

21.14   (a) $^{213}_{83}Bi \rightarrow ^{209}_{81}Tl + ^{4}_{2}He$ (b) $^{13}_{7}N + ^{0}_{-1}e$ (orbital electron) $\rightarrow ^{13}_{6}C$

(c) $^{98}_{43}Tc + ^{0}_{-1}e$ (orbital electron) $\rightarrow ^{98}_{42}Mo$ (d) $^{188}_{79}Au \rightarrow ^{188}_{78}Pt + ^{0}_{+1}e$

21.16   (a) $^{24}_{11}Na \rightarrow ^{24}_{12}Mg + ^{0}_{-1}e$; a beta particle is produced

(b) $^{188}_{80}Hg \rightarrow ^{188}_{79}Au + ^{0}_{+1}e$; a positron is produced

(c) $^{122}_{53}I \rightarrow ^{122}_{54}Xe + ^{0}_{-1}e$; a beta particle is produced

(d) $^{242}_{94}Pu \rightarrow ^{238}_{92}U + ^{4}_{2}He$; an alpha particle is produced

21.18   This decay series represents a change of (232–208 =) 24 mass units. Because only alpha emissions change the nuclear mass, and each changes the mass by four, there must be a total of 6 alpha emissions. Each alpha emission causes a decrease of two in atomic number.

Therefore, the 6 alpha emissions, by themselves, would cause a decrease in atomic number of 12. The series as a whole involves a decrease of 8 in atomic number. Thus, there must be a total of 4 beta emissions, each of which increases atomic number by one. Overall, there are 6 alpha emissions and 4 beta emissions.

Patterns of Nuclear Stability (Section 21.2)

21.20   (a) $^{3}_{1}H$ - high neutron/proton ratio, beta emission

        (b) $^{89}_{38}Sr$ - (slightly) high neutron/proton ratio, beta emission

        (c) $^{120}_{53}I$ - low neutron/proton ratio, positron emission

        (d) $^{102}_{47}Ag$ - low neutron/proton ratio, positron emission

21.22   Use criteria listed in Table 21.4.

        (a) $^{40}_{20}Ca$; stable, magic numbers of protons and neutrons

             $^{45}_{20}Ca$, radioactive, high neutron/proton ratio.

        (b) $^{12}_{6}C$, stable, even proton, even neutron

             $^{14}_{6}C$, radioactive, high neutron/proton ratio

        (c) $^{206}_{82}Pb$, stable, magic number of protons, even proton, even neutron

             $^{230}_{90}Th$, radioactive, atomic number greater than 84

21.24   The high points on the graph all correspond to elements with even atomic numbers. Stable isotopes of elements with even numbers of protons are much more abundant than isotopes of elements with odd atomic numbers. If radioactive isotopes of elements with odd atomic numbers were once abundant, they would have decayed to more stable nuclides.

21.26   The criterion employed in judging whether the nucleus is likely to be radioactive is the position of the nucleus on the plot shown in Figure 21.2. If the neutron/proton ratio is too high or low, or if the atomic number exceeds 83, the nucleus will be radioactive.

       Radioactive: $^{60}_{27}Co$ —odd proton, odd neutron

                   $^{92}_{41}Nb$ —odd proton, odd neutron

                   ^{226}Ra —high atomic number

       Stable: $^{58}_{26}Fe$ —even proton, even neutron, stable neutron/proton ratio

               ^{202}Hg —even proton, even neutron, stable neutron/proton ratio

Nuclear Transmutations (Section 21.3)

21.28   The element is technetium, Tc. $^{96}_{42}Mo + ^{2}_{1}H \rightarrow ^{98}_{43}Tc$

21.30   (a) $^{14}_{7}N + ^{4}_{2}He \rightarrow ^{17}_{8}O + ^{1}_{1}H$ (b) $^{40}_{19}K + ^{0}_{-1}e\,(orbital\,electron) \rightarrow ^{40}_{18}Ar$

        (c) $^{27}_{13}Al + ^{4}_{2}He \rightarrow ^{30}_{14}Si + ^{1}_{1}H$ (d) $^{58}_{26}Fe + 2\,^{1}_{0}n \rightarrow ^{0}_{-1}e + ^{60}_{27}Co$

        (e) $^{235}_{92}U + ^{1}_{0}n \rightarrow ^{135}_{54}Xe + ^{99}_{38}Sr + 2\,^{1}_{0}n$

21.32   (a) $^{238}_{92}U + ^{1}_{0}n \rightarrow ^{239}_{92}U + ^{0}_{0}\gamma$ (b) $^{16}_{8}O + ^{1}_{1}H \rightarrow ^{13}_{7}N + ^{4}_{2}He$

        (c) $^{18}_{8}O + ^{1}_{0}n \rightarrow ^{19}_{9}F + ^{0}_{-1}e$

Rates of Radioactive Decay (Section 21.4)

21.34   (a)   The suggestion is not reasonable. The energies of nuclear states are very large relative to ordinary temperatures. Thus, merely changing the temperature by less than 100 K would not be expected to significantly affect the behavior of nuclei with regard to nuclear decay rates.

        (b)   No. Radioactive decay has no activation energy like a chemical reaction. Activation energy is the minimum amount of energy required to initiate a chemical reaction. Radioactive decay is a spontaneous nuclear transformation from a less stable to a more stable nuclear configuration. Radioisotopes are by definition in a "transition state," prone to nuclear change or decay. Changes in external conditions such as temperature, pressure or chemical state provide insufficient energy to either excite or relax an unstable nucleus.

21.36   Calculate the decay constant, k, and then $t_{1/2}$. $t = 4$ h 39 min = 279 min

$$k = \frac{-1}{t} \ln \frac{N_t}{N_o} = \frac{-1}{279 \text{ min}} \ln \frac{0.25 \text{ mg}}{2.000 \text{ mg}} = 0.007453 = 0.0.0075 \text{ min}^{-1}$$

Using Equation 21.20, $t_{1/2} = 0.693/k = 0.693/0.007453 \text{ min}^{-1} = 92.980 = 93$ min

21.38   Follow the logic in Sample Exercise 21.6. In this case, we are given initial sample mass as well as mass at time t, so we can proceed directly to calculate k (Equation 21.20, and then t (Equation 21.19). *Solve.*

$$k = 0.693 / t_{1/2} = 0.693/27.8 \text{ d} = 0.02493 = 0.0249 \text{ d}^{-1}$$

$$t = \frac{-1}{k} \ln \frac{N_t}{N_o} = \frac{-1}{0.02493 \text{ d}^{-1}} \ln \frac{0.75}{6.25} = 85.06 = 85 \text{ d}$$

21.40   (a)   Proceeding as in Solution 21.39, calculate number of ^{60}Co atoms and k in s^{-1}.

$$3.75 \text{ mg Co} \times \frac{1 \text{ g}}{1000 \text{ mg}} \times \frac{1 \text{ mol Co}}{60 \text{ g Co}} \times \frac{6.022 \times 10^{23} \text{ Co atoms}}{1 \text{ mol Co}} = 3.76375 \times 10^{19} = 3.76 \times 10^{19}$$

$$5.26 \text{ yr} \times \frac{365 \text{ d}}{1 \text{ yr}} \times \frac{24 \text{ h}}{1 \text{ d}} \times \frac{3600 \text{ s}}{1 \text{ h}} = 1.659 \times 10^{8} = 1.66 \times 10^{8} \text{ s}$$

$$k = \frac{0.693}{t_{1/2}} = \frac{0.693}{1.659 \times 10^{8}} = 4.178 \times 10^{-9} = 4.18 \times 10^{-9} \text{ s}^{-1}$$

Rate = kN = $(4.178 \times 10^{-9} \text{ s}^{-1})(3.76375 \times 10^{19} \text{ atoms}) = 1.57 \times 10^{11}$ atoms/s

$(1.57 \times 10^{11} \text{ atoms/s})(600 \text{ s}) = 9.43 \times 10^{13}$ ^{60}Co atoms decay in 600 s

9.43×10^{13} beta particles emitted by a 3.75 mg sample in 600 s

        (b) $\dfrac{1.57 \times 10^{11} \text{ dis}}{\text{s}} \times \dfrac{1 \text{ Bq}}{1 \text{ dis/s}} = 1.57 \times 10^{11}$ Bq

The activity of the sample is 1.57×10^{11} Bq.

21.42   Follow the logic in Sample Exercise 21.6.

$$t = \frac{-1}{k} \ln \frac{N_t}{N_o}; k = 0.693/5715 \text{ yr} = 1.213 \times 10^{-4} \text{ yr}^{-1}$$

$$t = \frac{-1}{1.213 \times 10^{-4} \text{ yr}^{-1}} \ln \frac{38.0}{58.2} = 3.52 \times 10^3 \text{ yr}$$

21.44   Follow the procedure outlined in Sample Exercise 21.6. The original quantity of ^{238}U is 75.0 mg plus the amount that gave rise to 18.0 mg of ^{206}Pb. This amount is $18.0(238/206) = 20.8$ mg.

$$k = 0.693/4.5 \times 10^9 \text{ yr} = 1.54 \times 10^{-10} = 1.5 \times 10^{-10} \text{ yr}^{-1}$$

$$t = \frac{-1}{k} \ln \frac{N_t}{N_o} = \frac{-1}{1.54 \times 10^{-10} \text{ yr}^{-1}} \ln \frac{75.0}{95.8} = 1.59 \times 10^9 \text{ yr}$$

Energy Changes (Section 21.6)

21.46 $\Delta E = c^2 \Delta m = (2.9979246 \times 10^8 \text{ m/s})^2 \times 0.1 \text{ mg} \times \dfrac{1 \text{ g}}{1000 \text{ mg}} \times \dfrac{1 \text{ kg}}{1000 \text{ g}} \times \dfrac{1 \text{ kJ}}{1000 \text{ J}} = 9 \times 10^6 \text{ kJ}$

21.48 Δm = mass of individual protons and neutrons – mass of nucleus

$\Delta m = 10(1.0072765 \text{ amu}) + 11(1.0086649 \text{ amu}) - 20.98846 \text{ amu} = 0.1796189 = 0.17962 \text{ amu}$

$$\Delta E = (2.9979246 \times 10^8 \text{ m/s})^2 \times 0.1796189 \text{ amu} \times \frac{1 \text{ g}}{6.0221421 \times 10^{23} \text{ amu}} \times \frac{1 \text{ kg}}{1000 \text{ g}}$$

$$= 2.680664 \times 10^{-11} = 2.6807 \times 10^{-11} \text{ J}/^{21}\text{Ne nucleus required}$$

$$2.680664 \times 10^{-11} \frac{\text{J}}{\text{nucleus}} \times \frac{6.0221421 \times 10^{23} \text{ nuclei}}{\text{mol}}$$

$$= 1.6143 \times 10^{13} \text{ J/mol } ^{21}\text{Ne binding energy}$$

21.50   Nuclear mass is atomic mass minus mass of electrons. Nuclear binding energy is nuclear mass minus mass of the separate nucleons, converted to energy using Equation 21.22. Divide by the total number of nucleons to find binding energy per nucleon.

     (a)   Nuclear mass

         ^{14}N: $13.999234 - 7(5.485799 \times 10^{-4} \text{ amu}) = 13.995394$

         ^{48}Ti: $47.935878 - 22(5.485799 \times 10^{-4} \text{ amu}) = 47.923809$

         ^{129}Xe: $128.904779 - 54(5.485799 \times 10^{-4} \text{ amu}) = 128.875156 \text{ amu}$

     (b)   Nuclear binding energy

         ^{14}N: $\Delta m = 7(1.0072765) + 7(1.0086649) - 13.995394 = 0.1161959 = 0.116196 \text{ amu}$

$$\Delta E = 0.1161959 \text{ amu} \times \frac{1 \text{ g}}{6.0221421 \times 10^{23} \text{ amu}} \times \frac{1 \text{ kg}}{1000 \text{ g}} \times \frac{8.987551 \times 10^{16} \text{ m}^2}{\text{s}^2}$$

$$= 1.734127 \times 10^{-11} = 1.73413 \times 10^{-11} \text{ J}$$

         ^{48}Ti: $\Delta m = 22(1.0072765) + 26(1.0086649) - 47.923809 = 0.4615614 = 0.461561 \text{ amu}$

$$\Delta E = 0.4615614 \text{ amu} \times \frac{1 \text{ g}}{6.0221421 \times 10^{23} \text{ amu}} \times \frac{1 \text{ kg}}{1000 \text{ g}} \times \frac{8.987551 \times 10^{16} \text{ m}^2}{\text{s}^2}$$

$$= 6.888428 \times 10^{-11} = 6.88843 \times 10^{-11} \text{ J}$$

^{129}Xe: $\Delta m = 54(1.0072765) + 75(1.0086649) - 128.875156 = 1.1676425 = 1.167643$ amu

$$\Delta E = 1.1676425 \text{ amu } \times \frac{1\,g}{6.0221421 \times 10^{23}\,\text{amu}} \times \frac{1\,kg}{1000\,g} \times \frac{8.987551 \times 10^{16}\,m^2}{s^2}$$

$$= 1.742610 \times 10^{-10}\,J$$

(c)   Binding energy per nucleon

^{14}N: 1.73413×10^{-11} J / 14 nucleons $= 1.23866 \times 10^{-12}$ J/nucleon

^{48}Ti: 6.888428×10^{-11} J / 48 nucleons $= 1.43509 \times 10^{-12}$ J/nucleon

^{129}Xe: 1.742610×10^{-10} J/129 nucleons $= 1.350860 \times 10^{-12}$ J/nucleon

21.52   First, calculate nuclear masses from atomic masses. Then, the calculated Δm is for one group of single nuclides involved in a reaction, labeled Δm/'atomic reaction'. Multiplying by Avogadro's number changes the quantity to 'mol of reaction'. Because energy is released, the sign of ΔE is negative.

^{1}H: 1.00782 amu $- 1(5.485799 \times 10^{-4}$ amu$) = 1.0072714 = 1.00727$ amu

^{2}H: 2.01410 amu $- 1(5.485799 \times 10^{-4}$ amu$) = 2.0135514 = 2.01355$ amu

^{3}H: 3.10605 amu $- 1(5.485799 \times 10^{-4}$ amu$) = 3.1055014 = 3.10550$ amu

^{3}He: 3.10603 amu $- 2(5.485799 \times 10^{-4}$ amu$) = 3.1049328 = 3.10493$ amu

^{4}He: 4.00260 amu $- 2(5.485799 \times 10^{-4}$ amu$) = 4.0015028 = 4.00150$ amu

(a) $\Delta m = 4.0015028 + 1.0086649 - 3.1055014 - 2.0135514 = -0.1088851 = -0.10889$ amu

$$\Delta E = \frac{-0.1088851 \text{ amu}}{\text{'atomic reaction'}} \times \frac{1\,g}{6.022 \times 10^{23}\,\text{amu}} \times \frac{6.022 \times 10^{23} \text{ 'atomic reaction'}}{\text{mol of reaction}}$$

$$\times \frac{1\,kg}{10^3\,g} \times (2.99792458 \times 10^8\,m/s)^2 = -9.7861 \times 10^{12}\,J/\text{mol}$$

(b) $\Delta m = 3.1049328 + 1.0086649 - 2(2.0135514) = -0.0864949 = -0.08649$ amu

$\Delta E = -7.774 \times 10^{12}$ J/mol

(c) $\Delta m = 4.0015028 + 1.0072714 - 2.0135514 - 3.1049328 = -0.1097100 = -0.10971$ amu

$\Delta E = -9.8602 \times 10^{12}$ J/mol

21.54   Nuclear mass $= 61.928345$ amu $- 28(5.485799 \times 10^{-4}$ amu$) = 61.912985$

Binding energy $= 28(1.0072765) + 34(1.0086649) - 61.912985 = 0.585363$ amu

$$\Delta E = 0.585363 \text{ amu } \times \frac{1\,g}{6.0221421 \times 10^{23}\,\text{amu}} \times \frac{1\,kg}{1000\,g} \times \frac{8.987551 \times 10^{16}\,m^2}{s^2}$$

$$= 8.73606 \times 10^{-11}\,J$$

Binding energy/nucleon $= 8.73606 \times 10^{-11}$ J / 62 $= 1.40904 \times 10^{-12}$ J/nucleon

The value given for iron-56 in Table 21.7 is 1.41×10^{-12} J/nucleon. These values are the same, to three significant figures.

Effects and Uses of Radioisotopes (Sections 21.7–21.9)

21.56   Radioisotopes used as diagnostic tools are introduced into the body and carried to the point where imaging or some other diagnostic data is needed. We want the decay products of these radioisotopes to leave the body and do as little damage as possible on the way. Gamma rays are penetrating radiation and can escape the body more easily than other radioactive decay products. Also, gamma rays leaving the body can be easily detected using scintillation counters. This is particularly important when imaging is the goal of the procedure.

Alpha emitters are never used as diagnostic tools because alpha particles are ionizing but do not move easily through the body. Trapped inside the body, alpha particles initiate the ionization of water, which ultimately produces free radicals that disrupt the normal operation of cells.

21.58   (a)   Statements (i) and (iv) are true. Natural uranium contains about 0.7% ^{235}U; it must be enriched to about 3-5% for use as a fuel. ^{238}U undergoes neutron-induced fission according to Equation 21.12; no neutrons are produced.

(b)   Statement (iii) explains why ^{238}Pu cannot be used for nuclear power plants or nuclear weapons. Other isotopes of plutonium are used as fuels [(i) and (iv)]. Particles that are the products of nuclear decay processes cannot be detected by the human eye (ii).

21.60   (a)   A moderator slows neutrons, so that they are more easily captured by fissioning nuclei.

(b)   Water is the moderator in a pressurized water generator.

(c)   Graphite is used as a moderator in gas-cooled reactors, and D_2O is used in heavy-water reactors.

21.62   *Analyze/Plan.* Use conservation of A and Z to complete the equations, keeping in mind the symbols and definitions of various decay products.   *Solve.*

(a) $^{235}_{92}U + {}^{1}_{0}n \rightarrow {}^{160}_{62}Sm + {}^{72}_{30}Zn + 4\,{}^{1}_{0}n$

(b) $^{239}_{94}Pu + {}^{1}_{0}n \rightarrow {}^{144}_{58}Ce + {}^{94}_{36}Kr + 2\,{}^{1}_{0}n$

21.64   (a)   If the spent fuel rods are more radioactive than the original rods, the products of fission must lie outside the belt of stability and be radioactive themselves.

(b)   The heavy (Z > 83) nucleus has a high neutron/proton ratio. The lighter radioactive fission products (for example, barium-142 and krypton-91) also have high neutron/proton ratios, because only 2 or 3 free neutrons are produced during fission. The preferred decay mode to reduce the neutron/proton ratio is beta decay, which has the effect of converting a neutron into a proton. Both barium-142 (86 n, 56 p) and krypton-91 (55 n, 36 p) undergo beta decay.

21.66   (a)   A *heavy water reactor* and a *gas-cooled rector* can use natural uranium as a fuel.

(b)   A *fast breeder reactor* does not use a moderator.

(c)   A *high-temperature pebble-bed reactor* can be refueled without shutting down.

21.68   X-rays, alpha particles and gamma rays are classified as ionizing radiation.

21.70   (a) $1\ Ci = 3.7 \times 10^{10}$ dis/s; $1\ Bq = 1$ dis/s

$$15\ \text{mCi} \times \frac{1\ Ci}{1000\ \text{mCi}} \times 3.7 \times 10^{10}\ \text{dis/s} = 5.55 \times 10^{8} = 5.6 \times 10^{8}\ \text{dis/s} = 5.6 \times 10^{8}\ Bq$$

(b) $1\ Gy = 1\ J/kg$; $1\ Gy = 100$ rad

$$5.55 \times 10^{8}\ \text{dis/s} \times 240\ s \times 0.075 \times \frac{8.75 \times 10^{-14}\ J}{\text{dis}} \times \frac{1}{65\ \text{kg}} = 1.345 \times 10^{-5} = 1.3 \times 10^{-5}\ J/kg$$

$$1.3 \times 10^{-5}\ J/kg \times \frac{1\ Gy}{1\ J/kg} = 1.3 \times 10^{-5}\ Gy;\ 1.3 \times 10^{-5}\ Gy \times \frac{100\ \text{rad}}{1\ Gy} = 1.3 \times 10^{-3}\ \text{rad}$$

(c)   rem = rad (RBE); Sv = Gy (RBE)

$$1.3 \times 10^{-3}\ \text{rad}\ (1.0) = 1.3 \times 10^{-3}\ \text{rem} \times \frac{1000\ \text{mrem}}{1\ \text{rem}} = 1.3\ \text{mrem}$$

$$1.3 \times 10^{-5}\ Gy\ (1.0) = 1.3 \times 10^{-5}\ Sv$$

(d)   The radiation dose (1.3 mrem) is much less than that for a typical mammogram.

Additional Exercises

21.71   *Analyze/Plan.* Atomic number is number of protons; mass number is number of (protons + neutrons). The element symbol is determined by atomic number. Check the list of magic numbers, Figure 21.2 and Table 21.4 to determine which isotope is unstable. Use information in Table 21.3 about positron emission to determine which isotope will produce potassium-39.

(a)   (i) $^{38}_{19}K$ (ii) $^{40}_{19}K$ (iii) $^{39}_{20}Ca$ (iv) $^{40}_{20}Ca$

(b) $^{38}_{19}K$ is most likely to be unstable. Both $^{38}_{19}K$ and $^{40}_{19}K$ are odd proton/odd neutron isotopes. $^{38}_{19}K$ is more likely to be unstable because it has a lower neutron/proton ratio. (The two isotopes of Ca have magic numbers of protons and are thus more likely to be stable.)

(c) $^{39}_{20}Ca$ has a magic number of protons. $^{40}_{20}Ca$ has magic numbers of protons and neutrons.

(d) $^{39}_{20}Ca \rightarrow ^{39}_{19}K + ^{0}_{+1}e$

21.73 $^{3}_{2}He + ^{3}_{2}He \rightarrow ^{4}_{2}He + 2^{1}_{1}H$. Use nuclear masses calculated in Solution 21.52.

$\Delta m = 4.00150 + 2(1.00727) - 2(3.10493) = -0.19382$ amu

$$\Delta E = \frac{-0.19382\ \text{amu}}{\text{'atomic reaction'}} \times \frac{1\ g}{6.022 \times 10^{23}\ \text{amu}} \times \frac{6.022 \times 10^{23}\ \text{'atomic reaction'}}{\text{mol of reaction}}$$

$$\times \frac{1\ \text{kg}}{10^{3}\ g} \times (2.99792458 \times 10^{8}\ \text{m/ s})^{2} = -1.7420 \times 10^{13}\ J/\text{mol}$$

21.75   A ^{2}He nucleus has 2 protons and no neutrons. Electrostatic repulsion between two positively charged protons in very close proximity causes ^{2}He to be unstable.

21.77   (a)   Z is $^{297}_{117}$[117]; Q is $^{48}_{20}$Ca

(b)   Isotope Q has 20 protons and 28 neutrons; these are both magic numbers. Even though Q has an unfavorable neutron-to-proton ratio, the special stability associated with magic numbers of protons and neutrons explains its long half-life.

(c)   The target isotope was $^{248}_{96}$Cm. $^{248}_{96}$Cm + $^{48}_{20}$Ca → $^{296}_{116}$Lv

21.78

Time (h)	N_t (dis/min)	ln N_t
0	180	5.193
2.5	130	4.868
5.0	104	4.644
7.5	77	4.34
10.0	59	4.08
12.5	46	3.83
17.5	24	3.18

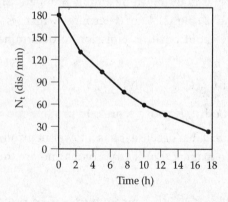
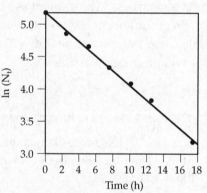

The plot on the left is a graph of activity (disintegrations per minute) vs. time. Choose $t_{1/2}$ at the time where $N_t = 1/2\,N_o = 90$ dis/min. $t_{1/2} \approx 6.0$ h.

Rearrange Equation 21.19 to obtain the linear relationship shown on the right.

$\ln(N_t / N_o) = -kt$; $\ln N_t - \ln N_o = -kt$; $\ln N_t = -kt + \ln N_o$

The slope of this line = $-k = -0.11$; $t_{1/2} = 0.693/0.11 = 6.3$ h.

21.79 1×10^{-6} curie $\times \dfrac{3.7 \times 10^{10} \text{ dis/s}}{\text{curie}} = 3.7 \times 10^4$ dis/s

rate $= 3.7 \times 10^4$ nuclei/s $= kN$

$$k = \frac{0.693}{t_{1/2}} = \frac{0.693}{28.8 \text{ yr}} \times \frac{1 \text{ yr}}{365 \times 24 \times 3600 \text{ s}} = 7.630 \times 10^{-10} = 7.63 \times 10^{-10} \text{ s}^{-1}$$

$$3.7 \times 10^{4} \text{ nuclei/s} = (7.63 \times 10^{-10}/\text{s}) \text{ N}; \text{ N} = 4.849 \times 10^{13} = 4.8 \times 10^{13} \text{ }^{90}\text{Sr nuclei}$$

$$\text{mass } ^{90}\text{Sr} = 4.849 \times 10^{13} \text{ nuclei} \times \frac{89.907738 \text{ g Sr}}{6.022 \times 10^{23} \text{ nuclei}} = 7.2 \times 10^{-9} \text{ g Sr}$$

(mass of ^{90}Sr from webelements.com)

21.81   (a)   The C−OH bond of the acid and the O−H bond of the alcohol break in this reaction. Initially, ^{18}O is present in the C−^{18}OH group of the alcohol. In order for ^{18}O to end up in the ester, the ^{18}O−H bond of the alcohol must break. This requires that the C−OH bond in the acid also breaks. The unlabeled O from the acid ends up in the H_2O product.

(b)   No. When $TOCH_3$ is used to react with CH_3COOH, T will end up in the H_2O product, regardless of whether the C−OT or O−T bond breaks in the reaction.

21.82   Assume that no depletion of iodide from the water because of plant uptake has occurred. Then the activity after 30 days would be:

$$k = 0.693/t_{1/2} = 0.693/8.02 \text{ d} = 0.0864 \text{ d}^{-1}$$

$$\ln\frac{N_t}{N_o} = -(0.0864 \text{ d}^{-1})(30 \text{ d}) = -2.592 = -2.6; \frac{N_t}{N_o} = 0.07485 = 0.07$$

We thus expect $N_t = 0.07485(214) = 16.0 = 2 \times 10^1$ counts/min. The measured value of 15.7 counts/min is within experimental uncertainty of the expected value, 2×10^1 (16.0) counts/minute, for no iodine uptake by the plant. We conclude that the plant absorbs no iodine.

21.83   (a)   (i)   X is $^{11}_6$C. The long-hand reaction is $^{14}_7\text{N} + ^1_1\text{p} \rightarrow ^{11}_6\text{C} + ^4_2\text{He}$.

(ii)   X is $^1_0$n. The long-hand reaction is $^{18}_8\text{O} + ^1_1\text{p} \rightarrow ^{18}_9\text{F} + ^1_0\text{n}$.

(b)   (iii)   d is $^2_1$H. The long-hand reaction is $^{14}_7\text{N} + ^2_1\text{H} \rightarrow ^{15}_8\text{O} + ^1_0\text{n}$.

It makes sense that "d" represents deuterium, $^2_1$H, an isotope of hydrogen.

21.85   First, calculate k in s^{-1}

$$k = \frac{0.693}{12.3 \text{ yr}} \times \frac{1 \text{ yr}}{365 \text{ d}} \times \frac{1 \text{ d}}{24 \text{ h}} \times \frac{1 \text{ h}}{3600 \text{ s}} = 1.7866 \times 10^{-9} = 1.79 \times 10^{-9} \text{ s}^{-1}$$

From Equation 21.18, $1.50 \times 10^3 \text{ s}^{-1} = (1.7866 \times 10^{-9} \text{ s}^{-1})(\text{N})$;

$N = 8.396 \times 10^{11} = 8.40 \times 10^{11}$. In 26.00 g of water, there are

$$26.00 \text{ g H}_2\text{O} \times \frac{1 \text{ mol H}_2\text{O}}{18.02 \text{ g H}_2\text{O}} \times \frac{6.022 \times 10^{23} \text{ H}_2\text{O}}{1 \text{ mol H}_2\text{O}} \times \frac{2 \text{ H}}{1 \text{H}_2\text{O}} = 1.738 \times 10^{24} \text{ H atoms}$$

The mole fraction of $^3_1$H atoms in the sample is thus

$$8.396 \times 10^{11}/1.738 \times 10^{24} = 4.831 \times 10^{-13} = 4.83 \times 10^{-13}$$

21.86   (a) $\Delta m = \Delta E/c^2$; $\Delta m = \dfrac{3.9 \times 10^{26} \text{ J/s}}{(2.9979246 \times 10^8 \text{ m/s})^2} \times \dfrac{1 \text{ kg-m}^2/\text{s}^2}{1 \text{ J}} = 4.3 \times 10^9 \text{ kg/s}$

         The rate of mass loss is 4.3×10^9 kg/s. (Fewer sig figs in the value of c produce the same result.)

         (b)   The mass loss arises from fusion reactions that produce more stable nuclei from less stable ones, e.g., Equations 21.26-21.29.

         (c)   Express the mass lost by the sun in terms of protons per second consumed in fusion reactions like Equations 21.26, 21.27 and 21.29.

$$\dfrac{4.3 \times 10^9 \text{ kg}}{\text{s}} \times \dfrac{1 \text{ proton}}{1.673 \times 10^{-24} \text{ g}} \times \dfrac{1000 \text{ g}}{1 \text{ kg}} = 2.594 \times 10^{36} = 3 \times 10^{36} \text{ protons/s}$$

21.87 $1000 \text{ Mwatts} \times \dfrac{1 \times 10^6 \text{ watts}}{1 \text{ Mwatt}} \times \dfrac{1 \text{ J}}{1 \text{ watt-s}} \times \dfrac{1 \ ^{235}\text{U atom}}{3 \times 10^{-11} \text{ J}} \times \dfrac{1 \text{ mol U}}{6.02214 \times 10^{23} \text{ atoms}}$

         $\times \dfrac{235 \text{ g U}}{1 \text{ mol}} \times \dfrac{3600 \text{ s}}{1 \text{ h}} \times \dfrac{24 \text{ h}}{1 \text{ d}} \times \dfrac{365 \text{ d}}{1 \text{ yr}} \times \dfrac{100}{40} (\text{efficiency}) = 1.03 \times 10^6 = 1 \times 10^6 \text{ g U/ yr}$

21.88 $2 \times 10^{-12} \text{ curies} \times \dfrac{3.7 \times 10^{10} \text{ dis/s}}{1 \text{ curie}} = 7.4 \times 10^{-2} = 7 \times 10^{-2} \text{ dis/s}$

$$\dfrac{7.4 \times 10^{-2} \text{ dis/s}}{75 \text{ kg}} \times \dfrac{8 \times 10^{-13} \text{ J}}{\text{dis}} \times \dfrac{1 \text{ rad}}{1 \times 10^{-2} \text{ J/ g}} \times \dfrac{3600 \text{ s}}{\text{h}} \times \dfrac{24 \text{ h}}{1 \text{ d}}$$

$$\times \dfrac{365 \text{ d}}{1 \text{ yr}} = 2.49 \times 10^{-6} = 2 \times 10^{-6} \text{ rad/yr}$$

         Recall that there are 10 rem/rad for alpha particles.

$$\dfrac{2.49 \times 10^{-6} \text{ rad}}{1 \text{ yr}} \times \dfrac{10 \text{ rem}}{1 \text{ rad}} = 2.49 \times 10^{-5} = 2 \times 10^{-5} \text{ rem/yr}$$

Integrative Exercises

21.89   Calculate the molar mass of $NaClO_4$ that contains 29.6% ^{36}Cl. Atomic mass of the enhanced Cl is 0.296(36.0) + 0.704(35.453) = 35.615 = 35.6. The molar mass of $NaClO_4$ is then (22.99 + 35.615 + 64.00) = 122.605 = 122.6. Calculate N, the number of ^{36}Cl nuclei, the value of k in s^{-1}, and the activity in dis/s.

$$53.8 \text{ mg NaClO}_4 \times \dfrac{1 \text{ g}}{1000 \text{ mg}} \times \dfrac{1 \text{ mol NaClO}_4}{122.605 \text{ g NaClO}_4} \times \dfrac{1 \text{ mol Cl}}{1 \text{ mol NaClO}_4} \times \dfrac{6.022 \times 10^{23} \text{ Cl atoms}}{\text{mol Cl}}$$

$$\times \dfrac{29.6 \ ^{36}\text{Cl atoms}}{100 \text{ Cl atoms}} = 7.822 \times 10^{19} = 7.82 \times 10^{19} \ ^{36}\text{Cl atoms}$$

$$k = 0.693/t_{1/2} = \dfrac{0.693}{3.0 \times 10^5 \text{ yr}} \times \dfrac{1 \text{ yr}}{365 \times 24 \times 3600 \text{ s}} = 7.32 \times 10^{-14} = 7.3 \times 10^{-14} \text{ s}^{-1}$$

         rate = kN = $(7.32 \times 10^{-14} \text{ s}^{-1})(7.822 \times 10^{19} \text{ nuclei}) = 5.729 \times 10^6 = 5.7 \times 10^6 \text{ dis/s}$

21.91   Refer to "Chemistry Put to Work: Gas Separations" in Section 10.8.

(a)   The atomic weight of naturally occurring uranium is 238.02891 g/mol. The molar mass of UF_6 is then [238.02891 + 6(18.998403)] = 352.01933 g/mol.

$$g = \frac{MM \times RT}{VP} = \frac{352.02\ g}{30.0\ L} \times \frac{0.082058\ L\text{-atm}}{mol\text{-}K} \times \frac{350\ K}{(695/760)\ atm} = 368.522 = 369\ g$$

Check. This seems like a large mass, but 30.0 L is more than the molar volume of a gas at STP, so it is reasonable that more than one mole of UF_6 is in the flask.

(b)   Of the 369 g sample, 0.720% is $^{235}UF_6$.
(368.522 × 0.00720) = 2.65336 = 2.65 g $^{235}UF_6$.

The mass of ^{235}U in 2.65 g $^{235}UF_6$ is

$$2.65336\ g\ ^{235}UF_6 \times \frac{235.044\ g\ ^{235}U}{349.034\ g\ ^{235}UF_6} = 1.78681 = 1.79\ g\ ^{235}U$$

(c)   According to information in "Chemistry Put to Work: Gas Separations," the ratio of effusion rates is 1.0043. That is, $^{235}UF_6$ effuses (diffuses) 1.0043 times faster than $^{238}UF_6$. That is, the mass of $^{235}UF_6$ in the diffused sample is 1.0043 times greater than in the initial sample. The mass of $^{235}UF_6$ in the diffused sample is then (2.65336 g $^{235}UF_6$ initial × 1.0043) = 2.66477 = 2.66 g $^{235}UF_6$.

$$2.66477\ g\ ^{235}UF_6 \times \frac{235.044\ g\ ^{235}U}{349.034\ g\ ^{235}UF_6} = 1.79449 = 1.79\ g\ ^{235}U$$

(d)   One more round of enrichment yields (0.266477 g $^{235}UF_6$ × 1.0043) = 2.67623 = 2.68 g $^{235}UF_6$. The mass % of $^{235}UF_6$ in the sample is then

$$\frac{2.67623\ g\ ^{235}UF_6}{368.522\ g\ sample} \times 100 = 0.726206 = 0.726\ \%\ ^{235}UF_6$$

[This is the same result as twice applying the 1.0043 times enrichment to the natural abundance of ^{235}U. (0.720% × 1.0043 × 1.0043) = 0.726205 = 0.726%]

21.92   (a) $0.18\ Ci \times \dfrac{3.7 \times 10^{10}\ dis/s}{Ci} \times \dfrac{3600\ s}{h} \times \dfrac{24\ h}{d} \times 245\ d = 1.41 \times 10^{17} = 1.4 \times 10^{17}$ alpha particles

(b) $P = nRT/V = 1.41 \times 10^{17}$ He atoms $\times \dfrac{1\ mol\ He}{6.022 \times 10^{23}\ atoms} \times \dfrac{295\ K}{0.0250\ L} \times \dfrac{0.08206\ L\text{-}atm}{mol\text{-}K}$

$= 2.27 \times 10^{-4} = 2.3 \times 10^{-4}$ atm = 0.17 torr

21.93   Calculate N_t in dis/min/g C from 1.5×10^{-2} dis/0.788 g $CaCO_3$. N_o = 15.3 dis/min/g C. Calculate k from $t_{1/2}$, calculate t from ln (N_t / N_o) = –kt.

$C(s) + O_2(g) \rightarrow CO_2(g) + Ca(OH_2)(aq) \rightarrow CaCO_3(s) + H_2O(l)$

1 C atom → 1 $CaCO_3$ molecule

$$\frac{1.5 \times 10^{-2}\ Bq}{0.788\ g\ CaCO_3} \times \frac{1\ dis/s}{1\ Bq} \times \frac{60\ s}{1\ min} \times \frac{100.1\ g\ CaCO_3}{12.01\ g\ C} = 9.52 = 9.5\ dis/min/g\ C$$

$$k = 0.693/t_{1/2} = 0.693/5.700 \times 10^3 \text{ yr} = 1.216 \times 10^{-4} = 1.22 \times 10^{-4} \text{ yr}^{-1}$$

$$t = -\frac{1}{k} \ln \frac{N_t}{N_0} = \frac{-1}{1.216 \times 10^{-4} \text{ yr}^{-1}} \ln \frac{9.52 \text{ dis/min/g C}}{15.3 \text{ dis/min/g C}} = 3.90 \times 10^3 \text{ yr}$$

21.94   (a) $Ba(NO_3)_2(aq) + Na_2SO_4(aq) \rightarrow BaSO_4(s) + 2NaNO_3(aq)$

(b) $1.25 \text{ mmol } Ba^{2+} + 1.25 \text{ mmol } SO_4^{2-} \rightarrow 1.25 \text{ mmol } BaSO_4$

Neither reactant is in excess, so the activity of the filtrate is due entirely to $[SO_4^{2-}]$ from dissociation of $BaSO_4(s)$. Calculate $[SO_4^{2-}]$ in the filtrate by comparing the activity of the filtrate to the activity of the reactant.

$$\frac{0.050 \ M \ SO_4^{2-}}{1.22 \times 10^6 \ \text{Bq/mL}} = \frac{x \ M \text{ filtrate}}{250 \ \text{Bq/mL}}$$

$[SO_4^{2-}]$ in the filtrate $= 1.0246 \times 10^{-5} = 1.0 \times 10^{-5} \ M$

$K_{sp} = [Ba^{2+}][SO_4^{2-}]; \ [SO_4^{2-}] = [Ba^{2+}]$

$K_{sp} = (1.0246 \times 10^{-5})^2 = 1.0498 \times 10^{-10} = 1.0 \times 10^{-10}$

22 Chemistry of the Nonmetals

Visualizing Concepts

22.2 (a) Acid–base (Brønsted)

(b) Charges on species from left to right in the reaction: 0, 0, 1+, 1–

(c) $NH_3(aq) + H_2O(l) \rightleftharpoons NH_4^+(aq) + OH^-(aq)$

22.4 Both gases are colorless and odorless, so they cannot be distinguished by inspection. (In practice, for safety reasons, one should never work in a lab with unlabeled bottles.) One big difference in the chemical properties of the two gases is their ability to support combustion. Oxygen supports combustion, whereas nitrogen does not. Heat a small piece of steel wool, place it at the mouth of one of the bottles. If the metal flares or glows more strongly, the gas is oxygen. If not, the gas is nitrogen.

22.5 *Analyze.* Given: space-filling models of molecules containing nitrogen and oxygen atoms. Find: molecular formulas and Lewis structures.

Plan. Nitrogen atoms are blue, and oxygen atoms are red. Count the number of spheres of each color to determine the molecular formula. From each molecular formula, count the valence electrons (N = 5, O = 6) and draw a correct Lewis structure. Resonance structures are likely.

Solve. (We list the formulas and Lewis structures from left to right across each row of space-filling models.)

N_2O_5 40 valence electrons, 20 e⁻ pairs

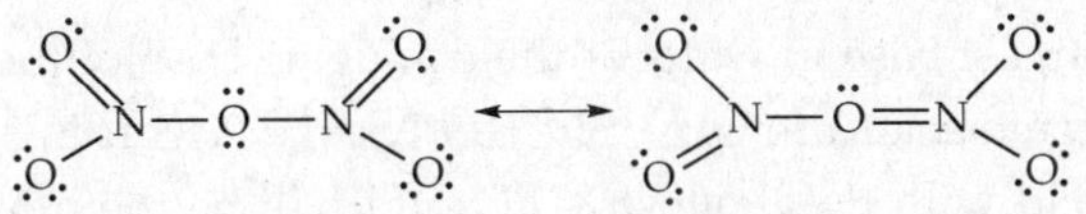

Many other resonance structures are possible. Those with double bonds to the central oxygen (like the right-hand structure above) do not minimize formal charge and are less significant in the net bonding model.

N_2O_3 28 e⁻, 14 e⁻ pairs

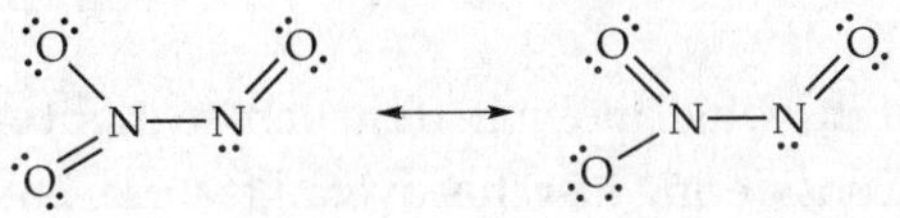

N_2O_4 34 e⁻, 17 e⁻ pairs

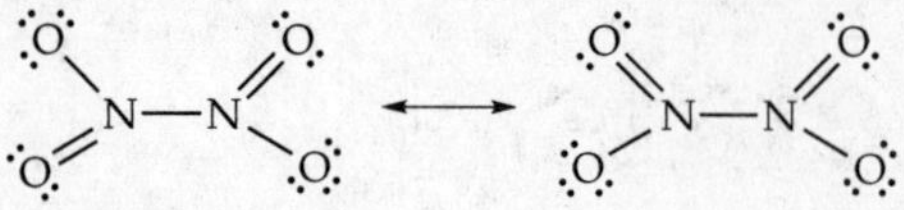

Other equivalent resonance structures with different arrangement of the double bonds are possible.

NO   11 e⁻, 5.5 e⁻ pairs

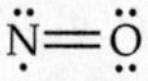

We place the odd electron on N because of electronegativity arguments.

NO_2 17 e⁻, 8.5 e⁻ pairs

We place the odd electron on N because of electronegativity arguments.

N_2O 16 e⁻, 8 e⁻ pairs

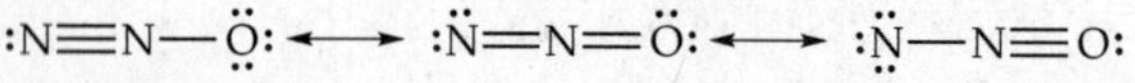

The right-most structure above does not minimize formal charge and makes smaller contribution to the net bonding model.

22.7   (a)   Atomic radius increases moving downward in a group because the principal quantum number (n) of the valence electrons increases. As n increases, the average distance of an electron from the nucleus increases and so does atomic radius.

       (b)   Anionic radii are greater than atomic radii because of increased electrostatic repulsions among electrons. Additional electrons in the same principle quantum level lead to additional electrostatic repulsion. This increases the energy of the electrons, and their average distance from the nucleus; the anionic radii are thus greater than the atomic radii.

       (c)   The anion that is the strongest base in water is the conjugate base of the weakest conjugate acid. The conjugate acids are OH⁻, SH⁻, and SeH⁻. According to trends in binary hydrides, the acid with the longest X–H bond will be the most readily ionized and the strongest acid. SeH⁻ is thus the strongest acid and OH⁻ the weakest. Therefore, O^{2-} is the strongest base in water.

22.8   *Analyze/Plan.* Evaluate the graph, describe the trend in data, recall the general trend for each of the properties listed, and use details of the data to discriminate between possibilities.

       Solve. The general trend is an increase in value moving from left to right across the period, with a small discontinuity at S. Considering just this overall feature, both (a), first ionization, energy and (c), electronegativity, increase moving from left to right, so these are possibilities. (b), Atomic radius, decreases, and can be eliminated. Because Si is a solid and Cl and Ar are gases at room temperature, melting points must decrease across the row; (d), melting point, can be eliminated. According to data in Tables 22.2, 22.5, 22.7, and

22.8, (e), X–X single bond enthalpies, show no consistent trend. Furthermore, there is no known Ar–Ar single bond, so no value for this property can be known; (e) can be eliminated.

Now let's examine trends in (a), first ionization energy, and (c), electronegativity, more closely. From electronegativity values in Chapter 8, we see a continuous increase with no discontinuity at S, and no value for Ar. Values for (a), first ionization energy, from Chapter 7 do match the pattern in the figure. The slightly lower value of I_1 for S is the result of a decrease in repulsion when an electron is removed from a fully occupied orbital. In summary, only (a), first ionization energy, fits the property depicted in the graph.

22.10   *Analyze/Plan.* The structure shown is a diatomic molecule or ion, depending on the value of n. Each species has 10 valence electrons and 5 electron pairs.   *Solve.*

(a)   Only second row elements are possible, because of the small covalent radius required for multiple bonding. Likely candidates are CO, N_2, NO^+, CN^-, and C_2^{2-}.

:C≡O: :N≡N: $[:N≡O:]^+$ $[:C≡N:]^-$ $[:C≡C:]^{2-}$

(b)   Because C_2^{2-} has the highest negative charge, it is likely to be the strongest H^+ acceptor and strongest Brønsted base. This is confirmed in Section 22.9 under "Carbides."

Periodic Trends and Chemical Reactions (Section 22.1)

22.12   Metals: (a) Ga, (b) Mo, (f) Ru   nonmetals: (e) Xe   metalloid: (c) Te, (d) As

22.14   (a)   Cl   (b)   K

(c)   K in the gas phase (lowest ionization energy), Li in aqueous solution (most positive E° value)

(d)   Ne; Ne and Ar are difficult to compare to the other elements because they do not form compounds and their radii are not measured in the same way as other elements. However, Ne is several rows to the right of C and surely has a smaller atomic radius. The next smallest is C.

(e)   C

(f)   C (graphite, diamond, fullerenes, carbon nanotubes, and graphene)

22.16   (a)   Nitrogen is a highly electronegative element. In HNO_3 it is in its highest oxidation state, +5, and thus is more readily reduced than phosphorus, which forms stable P–O bonds.

(b)   The difference between the third row element and the second lies in the smaller size of C as compared with Si, and the fact that Si has additional orbitals available to accommodate more than an octet of electrons.

(c)   Two of the carbon compounds, C_2H_4 and C_2H_2, contain C–C π bonds. Si does not readily form π bonds (to itself or other atoms), so Si_2H_4 and Si_2H_2 are not known as stable compounds.

22.18   (a) $Mg_3N_2(s) + 6\,H_2O(l) \rightarrow 2\,NH_3(g) + 3\,Mg(OH)_2(s)$

            Because $H_2O(l)$ is a reactant, the state of NH_3 in the products could be expressed as $NH_3(aq)$.

        (b) $2\,C_3H_7OH(l) + 9\,O_2(g) \rightarrow 6\,CO_2(g) + 8\,H_2O(l)$

        (c) $MnO_2(s) + C(s) \xrightarrow{\Delta} CO(g) + MnO(s)$ or

            $MnO_2(s) + 2\,C(s) \xrightarrow{\Delta} 2\,CO(g) + Mn(s)$ or

            $MnO_2(s) + C(s) \xrightarrow{\Delta} CO_2(g) + Mn(s)$

        (d) $AlP(s) + 3\,H_2O(l) \rightarrow PH_3(g) + Al(OH)_3(s)$

        (e) $Na_2S(s) + 2\,HCl(aq) \rightarrow H_2S(g) + 2\,NaCl(aq)$

Hydrogen, the Noble Gases, and the Halogens (Sections 22.2–22.4)

22.20   Yes, the physical properties of H_2O are different than those of D_2O. Deuterium atoms have a neutron, whereas protium atoms do not, so deuterium is almost twice as heavy as protium. We expect physical properties influenced by mass to be different for the two isotopes of hydrogen. In fact, D_2O has higher melting and boiling points, and a greater density than H_2O.

22.22   In its standard state, hydrogen is a gas, and thus a nonmetal, like the halogens. Hydrogen can gain an electron to form an anion with a 1– charge. Chemically, hydrogen can combine with group 1A metals to form ionic compounds, where H⁻ is the anion.

22.24   (a)   Electrolysis of brine; reaction of carbon with steam; reaction of methane with steam; by-product in petroleum refining

        (b)   Synthesis of ammonia; synthesis of methanol; reducing agent; hydrogenation of unsaturated vegetable oils

22.26   (a) $2\,Al(s) + 6\,H^+(aq) \rightarrow 2\,Al^{3+}(aq) + 3\,H_2(g)$

        (b) $Mg(s) + H_2O(g) \rightarrow MgO(s) + H_2(g)$

        (c) $MnO_2(s) + H_2(g) \rightarrow MnO(s) + H_2O(g)$

        (d) $CaH_2(s) + 2\,H_2O(l) \rightarrow Ca(OH)_2(aq) + 2\,H_2(g)$

22.28   (a)   molecular   (b)   ionic   (c)   metallic

22.30   Electrolysis of water is the cleanest way to produce hydrogen, but it is energy intensive. To make this process sustainable, the energy must come from renewable sources, such as hydroelectric or nuclear power plants, wind generators, or solar cells. The biomass production of fuel for electric power generation would also be a sustainable energy source.

22.32   Your friend cannot be correct. In general, noble gas elements have very stable electron configurations with complete s and p subshells. They have very large positive ionization energies; they do not lose electrons easily. They have positive electron affinities; they do not attract electrons to themselves. They do not easily gain, lose or share electrons, so they do not readily form the chemical bonds required to create compounds. To date, the only known compounds of noble gases involve Xe and Kr bound to other nonmetals. Specifically, there are no known compounds of Ne.

22.34   (a) ClO_3^-, Cl +5   (b)   HI, I, –1   (c) ICl_3; I, +3; Cl, –1

       (d)   NaOCl, Cl, +1   (e) $HClO_4$, Cl, +7   (f) XeF_4; Xe, +4; F, –1

22.36   (a)   potassium chlorate, Cl, +5   (b)   calcium iodate, I, +5

       (c)   aluminum chloride, Cl, –1   (d)   bromic acid, Br, +5

       (e)   paraperiodic acid, I, +7   (f)   xenon tetrafluoride, F, –1

22.38   (a)   The more electronegative the central atom, the greater the extent to which it withdraws charge from oxygen, in turn making the O–H bond more polar, and enhancing ionization of H^+.

       (b)   HF reacts with the silica which is a major component of glass:

$$6HF(aq) + SiO_2(s) \rightarrow SiF_6^{2-}(aq) + 2H_2O(l) + 2H^+(aq)$$

       (c)   Iodide is oxidized by sulfuric acid, as shown in Figure 22.10.

       (d)   The major factor is size; there is not room about Br for the three chlorides plus the two unshared electron pairs that would occupy the bromine valence shell orbitals.

Oxygen and the Other Group 6A Elements (Sections 22.5 and 22.6)

22.40   (a) $CaO(s) + H_2O(l) \rightarrow Ca^{2+}(aq) + 2\,OH^-(aq)$

       (b) $Al_2O_3(s) + 6\,H^+(aq) \rightarrow 2\,Al^{3+}(aq) + 3\,H_2O(l)$

       (c) $Na_2O_2(s) + 2\,H_2O(l) \rightarrow 2\,Na^+(aq) + 2\,OH^-(aq) + H_2O_2(aq)$

       (d) $N_2O_3(g) + H_2O(l) \rightarrow 2\,HNO_2(aq)$

       (e) $2\,KO_2(s) + 2\,H_2O(l) \rightarrow 2\,K^+(aq) + 2\,OH^-(aq) + O_2(g) + H_2O_2(aq)$

       (f) $NO(g) + O_3(g) \rightarrow NO_2(g) + O_2(g)$

22.42   (a) Mn_2O_7 (higher oxidation state of Mn)

       (b) SnO_2 (higher oxidation state of Sn)

       (c) SO_3 (higher oxidation state of S)

       (d) SO_2 (more nonmetallic character of S)

       (e) Ga_2O_3 (more nonmetallic character of Ga)

       (f) SO_2 (more nonmetallic character of S)

22.44   (a) SCl_4, +4   (b) SeO_3, +6   (c) $Na_2S_2O_3$, +2   (d) H_2S, –2

       (e) H_2SO_4, +6   (f) SO_2, +4   (g)   HgTe, –2

Oxygen (a group 6A element) is in the –2 oxidation state in compounds (b), (c), (e) and (f).

22.46   An aqueous solution of SO_2 contains H_2SO_3 and is acidic. Use H_2SO_3 as the reducing agent and balance assuming acid conditions.

22 Chemistry of the Nonmetals   **Solutions to Exercises**

(a) $2[MnO_4^-(aq) + 8H^+(aq) + 5e^- \rightarrow Mn^{2+}(aq) + 4H_2O(l)]$

$$\underline{5[H_2SO_3(aq) + H_2O(l) \rightarrow SO_4^{2-}(aq) + 4H^+(aq) + 2e^-]}$$

$$2MnO_4^-(aq) + 5H_2SO_3(aq) \rightarrow 2MnSO_4(aq) + 3SO_4^{2-}(aq) + 3H_2O(l) + 4H^+(aq)$$

(b) $Cr_2O_7^{2-}(aq) + 14H^+(aq) + 6e^- \rightarrow 2Cr^{3+}(aq) + 7H_2O(l)$

$$\underline{3[H_2SO_3(aq) + H_2O(l) \rightarrow SO_4^{2-}(aq) + 4H^+(aq) + 2e^-]}$$

$$Cr_2O_7^{2-}(aq) + 3H_2SO_3(aq) + 2H^+(aq) \rightarrow 2Cr^{3+}(aq) + 3SO_4^{2-}(aq) + 4H_2O(l)$$

(c) $Hg_2^{2+}(aq) + 2e^- \rightarrow 2Hg(l)$

$$\underline{H_2SO_3(aq) + H_2O(l) \rightarrow SO_4^{2-}(aq) + 4H^+(aq) + 2e^-}$$

$$Hg_2^{2+}(aq) + H_2SO_3(aq) + H_2O(l) \rightarrow 2Hg(l) + SO_4^{2-}(aq) + 4H^+(aq)$$

22.48 SF_4, 34 e$^-$ SF_5^-, 42 e$^-$

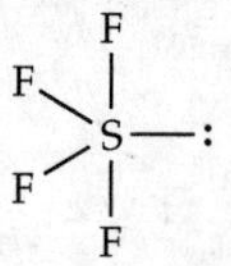

trigonal bipyramidal   octahedral
electron pair geometry   electron pair geometry

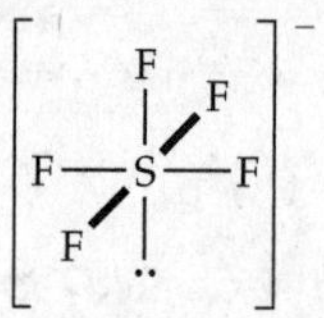

see-saw   square pyramidal
molecular geometry   molecular geometry

22.50   (a) $Al_2Se_3(s) + 6\,H^+(aq) \rightarrow 2\,Al^{3+}(aq) + 3\,H_2Se(g)$

        (b) $Cl_2(aq) + S_2O_3^{2-}(aq) + H_2O(l) \rightarrow 2\,Cl^-(aq) + S(s) + SO_4^{2-}(aq) + 2\,H^+(aq)$

Nitrogen and the Other Group 5A Elements (Sections 22.7 and 22.8)

22.52   (a)   NO, +2   (b) N_2H_4, −2   (c)   KCN, −3

        (d) $NaNO_2$, +3   (e) NH_4Cl, −3   (f) Li_3N, −3

22.54   (a)

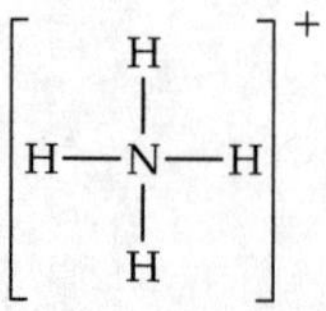

The ion is tetrahedral. The oxidation state of N is −3.

(b)

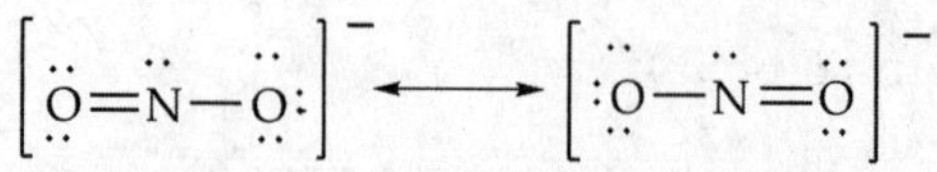

The ion is bent with a 120° O–N–O angle. The oxidation state of N is +3.

(c)

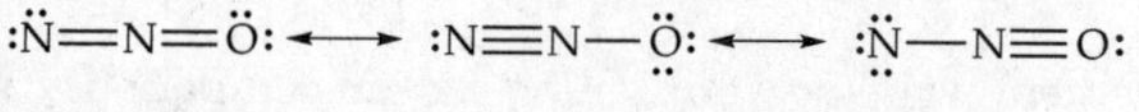

The molecule is linear. Again, the third resonance form makes less contribution to the structure because of the high formal charges involved. The oxidation state of N is +1.

400
Copyright © 2015 Pearson Education, Inc.

(d) $\ddot{O}{=}\dot{N}{-}\ddot{O}\!: \longleftrightarrow :\ddot{O}{-}\dot{N}{=}\ddot{O}:$

The molecule is bent (nonlinear). The odd electron resides on N because it is less electronegative than O. The oxidation state of N is +4.

22.56   (a) $4\,Zn(s) + 2\,NO_3^-(aq) + 10\,H^+(aq) \rightarrow 4\,Zn^{2+}(aq) + N_2O(g) + 5\,H_2O(l)$

         (b) $4\,NO_3^-(aq) + S(s) + 4\,H^+(aq) \rightarrow 4\,NO_2(g) + SO_2(g) + 2\,H_2O(l)$

             (or $6\,NO_3^-(aq) + S(s) + 4\,H^+(aq) \rightarrow 6\,NO_2(g) + SO_4^{2-}(aq) + 2\,H_2O(l)$

         (c) $2\,NO_3^-(aq) + 3\,SO_2(g) + 2\,H_2O(l) \rightarrow 2\,NO(g) + 3\,SO_4^{2-}(aq) + 4\,H^+(aq)$

         (d) $N_2H_4(g) + 5\,F_2(g) \rightarrow 2\,NF_3(g) + 4\,HF(g)$

         (e) $4\,CrO_4^{2-}(aq) + 3\,N_2H_4(aq) + 4\,H_2O(l) \rightarrow 4\,Cr(OH)_4^-(aq) + 4\,OH^-(aq) + 3\,N_2(g)$

22.58   (a) $NO_3^-(aq) + 4H^+(aq) + 3e^- \rightarrow NO(g) + 2H_2O(l)$

         (b) $HNO_2(aq) \rightarrow NO_2(g) + H^+(aq) + 1e^-$

22.60   (a) PO_4^{3-}, +5   (b) H_3AsO_3, +3   (c) Sb_2S_3, +3

         (d) $Ca(H_2PO_4)_2$, +5   (e) K_3P, –3   (f) $GaAs$, –3

22.62   (a)   Only two of the hydrogens in H_3PO_3 are bound to oxygen. The third is attached directly to phosphorus, and not readily ionized, because the H–P bond is not very polar.

         (b)   The smaller, more electronegative nitrogen withdraws more electron density from the O–H bond, making it more polar and more likely to ionize.

         (c)   Phosphate rock consists of $Ca_3(PO_4)_2$, which is only slightly soluble in water. The phosphorus is unavailable for plant use.

         (d) N_2 can form stable π bonds to complete the octet of both N atoms. Because phosphorus atoms are larger than nitrogen atoms, they do not form stable π bonds with themselves and must form σ bonds with several other phosphorus atoms (producing P_4 tetrahedral or chain structures) to complete their octets.

         (e)   In solution Na_3PO_4 is completely dissociated into Na^+ and PO_4^{3-}. PO_4^{3-}, the conjugate base of the very weak acid HPO_4^{2-}, has a K_b of 2.4×10^{-2} and produces a considerable amount of OH^- by hydrolysis of H_2O.

22.64   (a) $PCl_5(l) + 4\,H_2O(l) \rightarrow H_3PO_4(aq) + 5\,HCl(aq)$

         (b) $2\,H_3PO_4(aq) \xrightarrow{\Delta} H_4P_2O_7(aq) + H_2O(l)$

         (c) $P_4O_{10}(s) + 6\,H_2O(l) \rightarrow 4\,H_3PO_4(aq)$

Carbon, the Other Group 4A Elements, and Boron (Sections 22.9–22.11)

22.66   (a) H_2CO_3 (b)   NaCN   (c) $KHCO_3$ (d) C_2H_2 (e) $Fe(CO)_5$

22.68   (a) $CO_2(g) + OH^-(aq) \rightarrow HCO_3^-(aq)$

         (b) $NaHCO_3(s) + H^+(aq) \rightarrow Na^+(aq) + H_2O(l) + CO_2(g)$

(c) $2\,CaO(s) + 5\,C(s) \xrightarrow{\Delta} 2\,CaC_2(s) + CO_2(g)$

(d) $C(s) + H_2O(g) \xrightarrow{\Delta} H_2(g) + CO(g)$

(e) $CuO(s) + CO(g) \rightarrow Cu(s) + CO_2(g)$

22.70   (a) $2\,Mg(s) + CO_2(g) \rightarrow 2\,MgO(s) + C(s)$

(b) $6\,CO_2(g) + 6\,H_2O(l) \xrightarrow{h\nu} C_6H_{12}O_6(aq) + 6\,O_2(g)$

(c) $CO_3^{2-}(aq) + H_2O(l) \rightarrow HCO_3^{-}(aq) + OH^{-}(aq)$

22.72   (a) SiO_2, +4   (b) $GeCl_4$, +4   (c) $NaBH_4$, +3

(d) $SnCl_2$, +2   (e) B_2H_6, +3   (f) BCl_3, +3

22.74   (a)   carbon   (b)   lead   (c)   germanium

22.76   Carbon forms carbonates rather than silicates to take advantage of its ability to form π bonds. Because of its relatively compact 2p valence orbitals, carbon can engage in effective π-type overlap and form multiple bonds with itself and other elements. Carbonate, CO_3^{2-}, the anion present in carbonates, takes advantage of this ability to form stable π bonds, and is additionally stabilized by resonance. In silicates, silicon atoms form only single bonds and are tetrahedral.

22.78   (a) $x = 1$. The charge on $Si_3O_8^{4-}$ anion is 4–, that on Al^{3+} cation is 3+. One Na^+ cation is required to balance charge.

(b) $x = 2$. The charge on $Si_4O_{11}^{6-}$ anion is 6–, that on the cations is 2+. Two OH^- anions are required to balance charge.

22.80   (a) $B_2H_6(g) + 6\,H_2O(l) \rightarrow 2\,H_3BO_3(aq) + 6\,H_2(g)$

(b) $4\,H_3BO_3(s) \xrightarrow{\Delta} H_2B_4O_7(s) + 5\,H_2O(g)$

(c) $B_2O_3(s) + 3\,H_2O(l) \rightarrow 2\,H_3BO_3(aq)$

Additional Exercises

22.82   (a) $BrO_3^{-}(aq) + XeF_2(aq) + H_2O(l) \rightarrow Xe(g) + 2HF(aq) + BrO_4^{-}(aq)$

(b) BrO_3^{-}, +5; BrO_4^{-}, +7

22.83   (a) $SO_2(g) + H_2O(l) \rightleftharpoons H_2SO_3(aq)$

(b) $Cl_2O_7(g) + H_2O(l) \rightleftharpoons 2\,HClO_4(aq)$

(c) $Na_2O_2(s) + 2\,H_2O(l) \rightarrow H_2O_2(aq) + 2\,NaOH(aq)$

(d) $BaC_2(s) + 2\,H_2O(l) \rightarrow Ba^{2+}(aq) + 2\,OH^{-}(aq) + C_2H_2(g)$

(e) $2\,RbO_2(s) + 2\,H_2O(l) \rightarrow 2\,Rb^+(aq) + 2\,OH^{-}(aq) + O_2(g) + H_2O_2(aq)$

(f) $Mg_3N_2(s) + 6\,H_2O(l) \rightarrow 3\,Mg(OH)_2(s) + 2\,NH_3(g)$

(g) $NaH(s) + H_2O(l) \rightarrow NaOH(aq) + H_2(g)$

22.85 Assume that the reactions occur in acidic solution. The half-reaction for reduction of H_2O_2 is in all cases $H_2O_2(aq) + 2H^+(aq) + 2e^- \rightarrow 2H_2O(aq)$.

(a) $N_2H_4(aq) + 2H_2O_2(aq) \rightarrow N_2(g) + 4H_2O(l)$

(b) $SO_2(g) + H_2O_2(aq) \rightarrow SO_4^{2-}(aq) + 2H^+(aq)$

(c) $NO_2^-(aq) + H_2O_2(aq) \rightarrow NO_3^-(aq) + H_2O(l)$

(d) $H_2S(g) + H_2O_2(aq) \rightarrow S(s) + 2H_2O(l)$

(e)
$$2H^+(aq) + H_2O_2(aq) + 2e^- \rightarrow 2H_2O(l)$$
$$2[Fe^{2+}(aq) \rightarrow Fe^{3+}(aq) + e^-]$$
$$\overline{2Fe^{2+}(aq) + H_2O_2(aq) + 2H^+(aq) \rightarrow 2Fe^{3+}(aq) + 2H_2O(l)}$$

22.86 Sulfur has a total of six valence electrons; its possible oxidation states range from –2 to +6. SO_3 has sulfur in its maximum +6 oxidation state; it cannot lose electrons and serve as a reducing agent. SO_2 has sulfur in the +4 oxidation state. It can lose electrons, be oxidized and serve as a reducing agent.

22.87
$$S(g) + O_2(g) \rightarrow SO_2(g) \qquad \Delta H = -296.9 \text{ kJ} \qquad (a)$$
$$SO_2(g) + 1/2\, O_2(g) \rightarrow SO_3(g) \qquad \Delta H = -98.3 \text{ kJ} \qquad (b)$$
$$SO_3(g) + H_2O(l) \rightarrow H_2SO_4(aq) \qquad \Delta H = -130 \text{ kJ} \qquad (c)$$
$$\overline{S(g) + 3/2\, O_2(g) + H_2O(l) \rightarrow H_2SO_4(aq) \qquad \Delta H = -525 \text{ kJ}}$$

$$5000 \text{ lb } H_2SO_4 \times \frac{453.6 \text{ g}}{1 \text{ lb}} \times \frac{1 \text{ mol } H_2SO_4}{98.09 \text{ g}} \times \frac{-525 \text{ kJ}}{\text{mol } H_2SO_4} = -1.21 \times 10^7 \text{ kJ}$$

One mole of H_2SO_4 produces 525 kJ of heat, 5000 lb of H_2SO_4 produces 1.21×10^7 kJ.

22.89 (a) Although P_4, P_4O_6 and P_4O_{10} all have four P atoms in a tetrahedral arrangement, the bonding *between* P atoms and *by* P atoms is not the same in the three molecules. In P_4, the 4 P atoms are bound only to each other by P–P single bonds with strained bond angles of approximately 60°. In the two oxides, the 4 P atoms are directly bound to oxygen atoms, not to each other. Bonding by P atoms in P_4O_6 and P_4O_{10} is very similar. Each contains the P_6O_6 cage, formed by four P_3O_3 rings that share a P–O–P edge. Phosphorus bonding to oxygen maintains the overall P_4 tetrahedron but allows the P atoms to move away from each other so that the angle strain is relieved relative to molecular P_4. The P–O–P and O–P–O angles in both oxides are near the ideal 109°. In P_4O_6, each P is bound to 3 O atoms and has a lone pair completing its octet. In P_4O_{10}, the lone pair is replaced by a terminal O atom and each P is bound to 3 bridging and 1 terminal O atom.

(b)

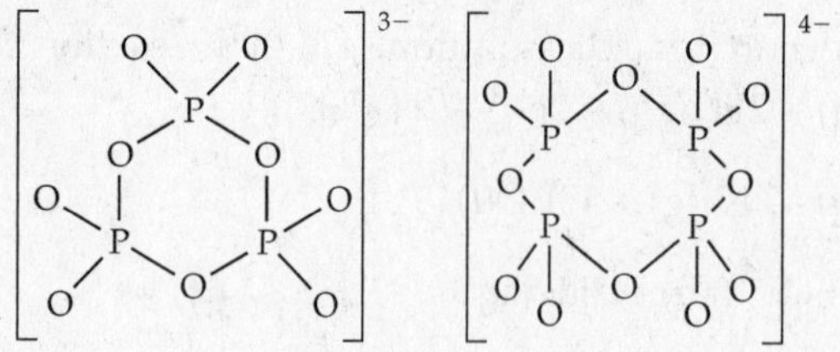

In both structures there are unshared pairs on all oxygens to give octets and the geometry around each P is approximately tetrahedral.

22.90 $GeO_2(s) + C(s) \xrightarrow{\Delta} Ge(l) + CO_2(g)$

$Ge(l) + 2\,Cl_2(g) \rightarrow GeCl_4(l)$

$GeCl_4(l) + 2\,H_2O(l) \rightarrow GeO_2(s) + 4\,HCl(g)$

$GeO_2(s) + 2\,H_2(g) \rightarrow Ge(s) + 2\,H_2O(l)$

22.91   (a)   The charge on the aluminosilicate ion shown ($AlSi_3O_{10}$) is 5– (+3 from Al, +12 from Si, –20 from O).

       (b)   In the silicate ions pictured in Figure 22.33, Si is central and O is either bridging or terminal. We want to select a structure from Figure 22.33 that has the same ratio of central atoms and O atoms as $AlSi_3O_{10}^{5-}$. The ratio of central to O atoms in $AlSi_3O_{10}^{5-}$ is 4 to 10, or 2 to 5. The analogous ion in Figure 22.33 is $Si_2O_5^{2-}$. The $AlSi_3O_{10}^{5-}$ ion will have a sheet structure similar to the one shown in Figure 22.33(c), with ¼ of the Si central atoms replaced by Al central atoms.

Integrative Exercises

22.93   From Appendix C, we need only ΔH_f° for F(g), so that we can estimate ΔH for the process:

$$F_2(g) \rightarrow F(g) + F(g); \qquad \Delta H^\circ = 160\ kJ$$
$$XeF_2(g) \rightarrow Xe(g) + F_2(g) \qquad -\Delta H_f^\circ = 109\ kJ$$
$$\overline{XeF_2(g) \rightarrow Xe(g) + 2F(g) \qquad \Delta H^\circ = 269\ kJ}$$

The average Xe–F bond enthalpy is thus 269/2 = 134 kJ. Similarly,

$$XeF_4(g) \rightarrow Xe(g) + 2F_2(g) \qquad -\Delta H_f^\circ = 218\ kJ$$
$$2F_2(g) \rightarrow 4F(g) \qquad\qquad \Delta H^\circ = 320\ kJ$$
$$\overline{XeF_4(g) \rightarrow Xe(g) + 4F(g) \qquad \Delta H^\circ = 538\ kJ}$$

Average Xe–F bond energy = 538/4 = 134 kJ

$$XeF_6(g) \rightarrow Xe(g) + 3F_2(g) \qquad -\Delta H_f^\circ = 298\ kJ$$
$$3F_2(g) \rightarrow 6F(g) \qquad\qquad \Delta H^\circ = 480\ kJ$$
$$\overline{XeF_6(g) \rightarrow Xe(g) + 6F(g) \qquad \Delta H^\circ = 778\ kJ}$$

Average Xe–F bond energy = 778/6 = 130 kJ

The average bond enthalpies are: XeF_2, 134 kJ; XeF_4, 134 kJ; XeF_6, 130 kJ. They are remarkably constant in the series.

22.95   *Analyze/Plan.* $\Delta G° = -RT \ln K$. Use $\Delta G°$ for ozone from Appendix C to calculate K for the reaction at 290.0 K, assuming no electrical input. *Solve.*

The reaction under consideration is: $3 O_2(g) \rightarrow 2 O_3(g)$

$\Delta G°$ for ozone at 298 K = 163.4 kJ/mol

$\Delta G°$ for the reaction is then 2(163.4 kJ) – 3(0 kJ) = 326.8 kJ.

$$\ln K = \frac{-\Delta G°}{RT} = \frac{-(326.8 \times 10^3)J}{8.314 \text{ J/ K} \times 298.0 \text{ K}} = -131.903 = -131.9; \quad K = 5 \times 10^{-58}$$

22.97   (a) $2NH_4ClO_4(s) \overset{\Delta}{\rightarrow} N_2(g) + 2HCl(g) + 3H_2O(g) + 5/2\, O_2(g)$

           $NH_4ClO_4(s) \overset{\Delta}{\rightarrow} 1/2\, N_2(g) + HCl(g) + 3/2\, H_2O(g) + 5/4\, O_2(g)$

   (b) $\Delta H° = \Sigma \Delta H_f° \text{ prod} - \Sigma \Delta H \text{ react}$

         $\Delta H° = \Delta H_f° HCl(g) + 3/2\, \Delta H_f° H_2O(g) + 1/2\, \Delta H_f° N_2(g) + 5/4\, \Delta H_f° O_2(g) - \Delta H_f° NH_4ClO_4 \Delta H°$

           $= -92.30 \text{ kJ} + 3/2(-241.82 \text{ kJ}) + 1/2\,(0 \text{ kJ}) + 5/4\,(0 \text{ kJ}) - (-295.8 \text{ kJ})$

           $= -159.2 \text{ kJ/ mol } NH_4ClO_4$

   (c)   The aluminum reacts exothermically with $O_2(g)$ and HCl(g) produced in the decomposition, providing additional heat and thrust.

   (d)   There are (1/2 + 1 + 3/2 + 5/4) = 4.25 mol gas per mol NH_4ClO_4 decomposed

$$1 \text{ lb } NH_4ClO_4 \times \frac{453.6 \text{ g}}{1 \text{ lb}} \frac{1 \text{ mol } NH_4ClO_4}{117.49 \text{ g } NH_4ClO_4} \times \frac{4.25 \text{ mol gas}}{1 \text{ mol } NH_4ClO_4} = 16.408 = 16.4 \text{ mol gas}$$

$$V = \frac{nRT}{P} = 16.408 \text{ mol gas} \times \frac{0.08206 \text{ L - atm}}{\text{mol - K}} \times \frac{273 \text{ K}}{1 \text{ atm}} = 367.57 = 368 \text{ L}$$

22.98   (a) $N_2H_4(g) + O_2(g) \rightarrow N_2(g) + 2H_2O(l)$

   (b) $\Delta H° = \Delta H_f° N_2(g) + 2\Delta H_f° H_2O(l) - \Delta H_f° N_2H_4(aq) - \Delta H_f° O_2(g)$

         $= 0 + 2(-285.83) - 95.40 - 0 = -667.06 \text{ kJ}$

   (c) $\dfrac{9.1 \text{ g } O_2}{1 \times 10^6 \text{ g } H_2O} \times \dfrac{1.0 \text{ g } H_2O}{1 \text{ mL } H_2O} \times \dfrac{1000 \text{ mL}}{1 \text{ L}} \times 3.0 \times 10^4 \text{ L} = 273 = 2.7 \times 10^2 \text{ g } O_2$

         $2.73 \times 10^2 \text{ g } O_2 \times \dfrac{1 \text{ mol } O_2}{32.00 \text{ g } O_2} \times \dfrac{1 \text{ mol } N_2H_4}{1 \text{ mol } O_2} \times \dfrac{32.05 \text{ g } N_2H_4}{1 \text{ mol } N_2H_4} = 2.7 \times 10^2 \text{ g } N_2H_4$

22.100   *Plan.* Vol air → kg air → g H_2S → g FeS. Use the ideal-gas equation to change volume of air to mass of air, (assuming 1.00 atm, 298 K and an average molar mass (MM) for air of 29.0 g/mol. Use (20 mg H_2S/kg air) to find the mass of H_2S in the given mass of air.

     Solve.

$$V_{air} = 12 \text{ ft} \times 20 \text{ ft} \times 8 \text{ ft} \times \frac{12^3 \text{ in}^3}{\text{ft}^3} \times \frac{2.54^3 \text{ cm}^3}{1^3 \text{ in}^3} \times \frac{1 \text{ L}}{1000 \text{ cm}^3} = 5.4368 \times 10^4 = 5 \times 10^4 \text{ L}$$

$$g_{air} = \frac{PV \text{ MM}}{RT}; \text{ assume } P = 1.00 \text{ atm, } T = 298 \text{ K, MM}_{air} = 29.0 \text{ g/mol}$$

$$g_{air} = \frac{1.00 \text{ atm} \times 5.4368 \times 10^4 \text{ L} \times 29.0 \text{ g/mol}}{298 \text{ K}} \times \frac{\text{mol - K}}{0.08206 \text{ L - atm}} = 64,476 = 6 \times 10^4 \text{ g air}$$

$$6.4476 \times 10^4 \text{ g air} \times \frac{1 \text{ kg}}{1000 \text{ g}} \times \frac{20 \text{ mg H}_2\text{S}}{1 \text{ kg air}} \times \frac{1 \text{ g}}{1000 \text{ mg}} = 1.2895 = 1 \text{ g H}_2\text{S}$$

$$FeS(s) + 2HCl(aq) \rightarrow FeCl_2(aq) + H_2S(g)$$

$$1.2895 \text{ g H}_2 \times \frac{1 \text{ mol H}_2}{34.08 \text{ g H}_2\text{S}} \times \frac{1 \text{ mol FeS}}{1 \text{ mol H}_2\text{S}} \times \frac{87.91 \text{ g FeS}}{1 \text{ mol FeS}} = 3.3263 = 3 \text{ g FeS}$$

22.102   (a)   MnSi: more than one element, so not metallic; high melting, so not molecular; insoluble in water, so not ionic; therefore covalent network.

         (b) $MnSi(s) + HF(aq) \rightarrow SiH_4(g) + MnF_4(s)$

         Reduction of Mn(IV) to Mn(II) is unlikely, because F^- is an extremely weak reducing agent. E°_{red} for $F_2(g) + 2 e^- \rightarrow 2F^-(aq) = 2.87 \text{ V}$

22.103   The most definitive experiment is X-ray crystallography to determine the Si–Si bond length. Compare the experimentally determined bond length to the Si–Si lengths in elemental Si, which is a covalent-network solid with only single bonds. If the length from the new compound is shorter than the bonds in elemental Si, the bond in the new compound has at least some double bond character. The difference in bond lengths must be greater than the uncertainty in the experimentally determined Si–Si bond length in the new compound. (We assume the structure of elemental Si has been determined with high precission.)

Visible absorption spectroscopy might be useful if the compound is colored. Otherwise, infrared absorption spectroscopy might give some useful information, but the intensity of double bond stretching frequencies is often small, and the peaks can be difficult to assign.

22.104 $N_2H_5^+(aq) \rightarrow N_2(g) + 5H^+(aq) + 4e^-$ $E^{\circ}_{red} = -0.23 \text{ V}$

Reduction of the metal should occur when E°_{red} of the metal ion is more positive than about –0.15 V. This is the case for (b) Sn^{2+} (marginal), (c) Cu^{2+}, (d) Ag^+ and (f) Co^{3+}

22.106   (a) $HOOC-CH_2-COOH \xrightarrow{P_2O_5} C_3O_2 + 2H_2O$

         (b) $20.0 \text{ g } C_3H_4O_4 \times \frac{1 \text{ mol } C_3H_4O_4}{104.06 \text{ g } C_3H_4O_4} \times \frac{1 \text{ mol } C_3O_2}{1 \text{ mol } C_3H_4O_4} \times \frac{68.03 \text{ g } C_3O_2}{1 \text{ mol } C_3O_2} = 13.1 \text{ g } C_3O_2$

         (c)   24 valence e^-, 12 e^- pair $\ddot{\text{O}}{=}\text{C}{=}\text{C}{=}\text{C}{=}\ddot{\text{O}}$

(d)   C=O, about 1.23 Å; C=C 1.34 Å or less. Because consecutive C=C bonds require sp hybrid orbitals on C (as in allene, C_3H_4), we might expect the orbital overlap requirements of this bonding arrangement to require smaller than usual C=C distances.

(e)   The product has the formula $C_3H_4O_2$.

28 valence e^-, 14 e^- pr

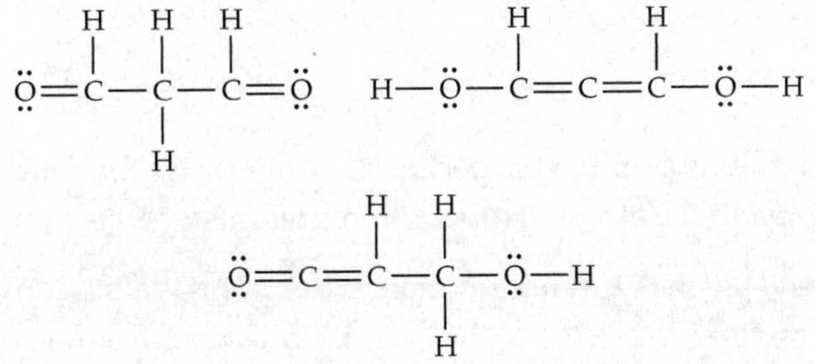

Three possibilities are shown above. The O=C=C group on the lower structure is uncommon and less likely than the two symmetrical structures.

23 Transition Metals and Coordination Chemistry

Visualizing Concepts

23.1   *Analyze/Plan.* Given graphs of three properties moving across the fourth period of the chart, match each graph to the atomic property it represents. *Solve.*

Graph (a) is the trend in maximum oxidation state. This value corresponds to the maximum number of (4s+3d) electrons that can be removed from a neutral atom. It increases up to a maximum of +7 for Mn, then decreases. The decrease is partly because of an increase in the attraction of 3d electrons for the nucleus as Z_{eff} increases.

Graph (b) is the trend in effective nuclear charge, Z_{eff}. Moving from left to right in a period, Z_{eff} increases because the increase in Z is not offset by a significant increase in shielding.

Graph (c) is the trend in radius. Increasing Z_{eff} leads to decreasing atomic radius.

23.3   *Analyze.* Given a ball-and-stick figure of a ligand, write the Lewis structure and answer questions about the ligand.

Plan. Assume that each atom in the Lewis structure obeys the octet rule. Complete each octet with unshared electron pairs or multiple bonds, depending on the bond angles in the ball-and-stick model. Black = C, blue = N, red = O, gray = H.

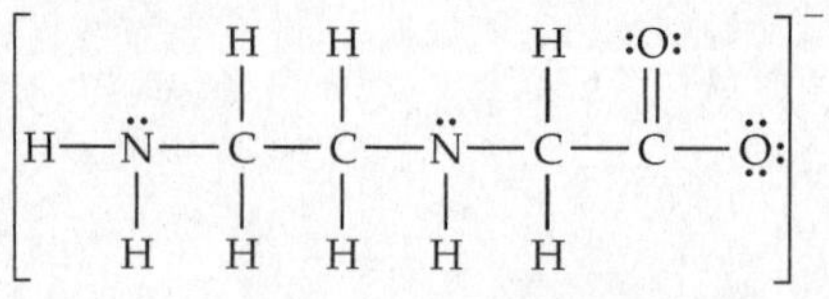

There is a second resonance structure with the double bond drawn to the second O atom.

Check. Write the molecular formula, count the valence electron pairs and see if it matches your structure. $[C_4H_9N_2O_2]^-$ $(16 + 9 + 10 + 12 + 1) = 48$ valence e^-, 24 e^- pair. Our Lewis structure also has 24 e^- pairs.

(a)   Donor atoms have unshared electron pairs. The potential donors in this structure are the two N and two O atoms.

The ligand is tridentate. (Even though there are four possible donor atoms, the structure would be strained if all four were bound to one metal center. It is likely that only one of the two O atoms binds to the same metal as the two N atoms.)

(b)   An octahedral complex has 6 coordination sites. A single ligand has 3 likely donors, so two ligands are needed. From a steric perspective, the likely donors would be the 2 N atoms and 1 of the carbonyl oxygen atoms. The chelate bite of a carboxyl group is relatively small and would require an $O-M-O$ angle of less than 90°.

23.4   *Analyze.* Given a ball-and-stick structure, name the complex ion, which has a 1– charge.

Plan. Write the chemical formula of the complex ion, determine the oxidation state of the metal, and name the complex.

Solve. $[Pt(NH_3)Cl_3]^-$. Oxidation numbers: $[Pt + 0 + 3(-1) = -1, Pt = +2, Pt(II)$

Arrange the ligands alphabetically, followed y the metal. Because the complex is an anion, add the suffix -ate, then the oxidation state of the metal: aminotrichloroplatinate(II)

23.5   *Analyze.* Given 5 structures, visualize which are identical to (1) and which are geometric isomers of (1).

Plan. There are two possible ways to arrange MA_3X_3. The first has bond angles of 90° between all similar ligands; this is structure (1). The second has one 180° angle between similar ligands. Visualize which description fits each of the five structures.

Solve. (1) has all 90° angles between similar ligands.

(2) has a 180° angle between similar ligands (see the blue ligands in the equatorial plane of the octahedron)

(3) has all 90° angles between similar ligands

(4) has all 90° angles between similar ligands

(5) has a 180° angle between similar ligands (see the blue axial ligands)

Structures (3) and (4) are identical to (1); (2) and (5) are geometric isomers.

23.7   *Analyze.* Given the visible colors of two solutions, determine the colors of light absorbed by each solution.

Plan. Apparent color is transmitted or reflected light, absorbed color is basically the complement of apparent color. Use the color wheel in Figure 23.25 to obtain the complementary absorbed color for the solutions.

Solve. Moving from left to right, the solutions appear blue-green (cyan), yellow, green, and red. The solutions absorb red-orange, violet, red, and green.

23.9   *Analyze/Plan.* Given the linear diagram and axial labels, answer the questions and predict crystal field splitting. Orbitals with lobes nearest ligand charges (or partial charges) will be highest in energy; orbitals with lobes away from charges are lowest in energy.

Solve. Diagram (c) is the best choice. The d_{z^2} orbital has lobes nearest the charges and is at the highest energy. The $d_{x^2-y^2}$ and d_{xy} orbitals have lobes in the xy-plane farthest from the charges and are lowest in energy. The d_{xz} and d_{yz} orbitals point between the respective axes and are intermediate in energy.

23 Transition Metals and Coordination Chemistry

Solutions to Exercises

$$d_{z^2}$$

$$d_{xz}, d_{yz}$$

$$d_{xz}, d_{x^2-y^2}$$

23.10   *Analyze.* Given the colors of two low spin Fe(II) complexes, determine which complex contains the stronger field ligand. *Plan.* We can make this direct comparison because both solutions contain low spin d^6 ions. A solution that appears one color absorbs visible light of the complementary color. Use the color wheel in Figure 23.25 to decide which color and approximate wavelength of visible light is absorbed by the two solutions. The stronger-field ligand causes a larger d-orbital splitting and absorbs light with the shorter wavelength.

Solve. The green solution absorbs red light in the 650 to 750 nm range. The red solution absorbs green light in the 490 to 560 nm range. The complex that produces the red solution has the larger d-orbital splitting and the stronger-field ligand.

The Transition Metals (Section 23.1)

23.12   Trend (b) explains the peak in maximum oxidation state of the transition-metal elements near groups 7B and 8B. As effective nuclear charge increases, d-electrons are more strongly attracted to the nucleus and more difficult to remove from the atom.

23.14   Refer to Figure 23.5. Among the period 4 transition metals, only Sc and Zn do not have at least one oxidation state with partially filled 3d orbitals. Sc^{3+} has zero 3d electrons and Zn^{2+} has ten 3d electrons. In Sc^{3+} the 3d orbitals are unoccupied; in Zn^{2+} the 3d orbitals are filled.

23.16   (a) Co^{3+}, $[Ar]3d^6$; 6 valence d-electrons   (b) Cu^+; $[Ar]3d^{10}$; 10 valence d-electrons

  (c) Cd^{2+}, $[Kr]4d^{10}$; 10 valence d-electrons   (d) Os^{3+}: $[Xe]4f^{14}5d^5$; 5 valence d-electrons

23.18   Antiferromagnetic materials cannot be used to make permanent magnets. In an antiferromagnetic material, coupled spins are aligned in opposite directions and the opposing spins exactly cancel.

23.20   (a) Fe_2O_3 has all Fe atoms in the +3 oxidation state, whereas Fe_3O_4 contains Fe atoms in both the +2 and +3 states. Each Fe_3O_4 formula unit has one Fe(II) and two Fe(III).

  (b)   In an antiferromagnetic material, spins on coupled atoms are oppositely aligned, producing a net spin of zero. This is only possible for Fe_2O_3, where all Fe atoms have the same oxidation state, d-electron configuration and number of unpaired electrons. In Fe_3O_4, Fe(II) and Fe(III) atoms have different d-electron configurations and different numbers of unpaired electrons. Assuming an Fe(II) is coupled to an Fe(III), even if spins on coupled centers are oppositely aligned, their spins do not fully cancel and the material is ferrimagnetic.

410

Copyright © 2015 Pearson Education, Inc.

Transition-Metal Complexes (Section 23.2)

23.22 (a) *Coordination number* is the number of atoms bound directly to the metal in a metal complex.

(b) Ligands in a metal complex usually have a lone (unshared, nonbonding) pair of electrons. NH_3 is neutral, CN^- is negatively charged.

(c) No, ligands with positive charges are not common. Metal atoms in a complex usually have a positive charge, so a positively charged ligand would neither be electrostatically attracted to the metal nor provide the electron density required to form a metal-ligand bond.

(d) In $Co(NH_3)_6Cl_3$, there are two types of chemical bonds. The bonds formed between Co^{3+} and $:NH_3$ are covalent. They are the result of Lewis acid-Lewis base interactions, where $:NH_3$ is the electron pair donor (Lewis base) and Co^{3+} is the electron pair acceptor (Lewis acid). The interactions between the $[Co(NH_3)_6]^{3+}$ complex cation and the three Cl^- anions are ionic bonds.

(e) The most common coordination numbers for metal complexes are six and four. (Coordination numbers of two, three, and five are less common, but not unknown.)

23.24 (a) $[Cr(H_2O)_6]^{3+}$

(b) Three. Because the Cl^- ions are not in the coordination sphere, all 3 anions react with Ag^+ to form 3 moles of AgCl(s).

(c) If the empirical formula of the compound is $CrCl_3$ and Cr^{3+} has a coordination number of six, some or all of the Cl^- ions must be shared by more than one Cr^{3+} ion. This produces a network solid that is very difficult to dissolve or break down.

23.26 (a) Coordination number = 6, oxidation number = +3

(b) 4, +2

(c) 6, +4

(d) 6, +3

(e) 6, +3

(f) 5, +2

Common Ligands in Coordination Chemistry (Section 23.3)

23.28 (a) 2 coordination sites, 2 N donor atoms

(b) 2 coordination sites, 2 N donor atoms

(c) 2 coordination sites, 2 O donor atoms (Although there are four potential O donor atoms in $C_2O_4^{2-}$, it is geometrically impossible for more than two of these to be bound to a single metal ion.)

(d) 4 coordination sites, 4 N donor atoms

(e) 6 coordination sites, 2 N and 4 O donor atoms

23.30   (a)   6   (b)   6   (c)   6   (d)   6

23.32   (a)   Pyridine is a **mono**dentate ligand because it has one N donor atom and therefore occupies one coordination site in a metal complex.

(b)   K for this reaction will be less than one. Two free pyridine molecules are replaced by one free bipy molecule. There are more moles of particles in the reactants than products, so ΔS is predicted to be negative. Processes with a net decrease in entropy are usually nonspontaneous, have positive ΔG, and values of K less than one. This equilibrium is likely to be spontaneous in the reverse direction.

23.34   (a)   The complex in the figure has tetrahedral geometry about the silver.

(b)   The ligands are neutral molecules and the metal is Ag(I), so the complex will have a 1+ charge.

(c)   Yes, one nitrate ion, NO_3^-, will be present in the crystal to provide charge balance for the complex cation.

(d)   [Ag(*o*-phen)$_2$]NO$_3$

(e)   bis(*ortho*-phenanthroline)silver(I) nitrate

Nomenclature and Isomerism in Coordination Chemistry (Section 23.4)

23.36   (a)   [Mn(H$_2$O)$_4$Br$_2$]ClO$_4$ (b)   [Cd(bipy)$_2$]Cl$_2$

(c)   K[Co(*o*-phen)Br$_4$]   (d)   Cs[Cr(NH$_3$)$_2$(CN)$_4$]

(e)   [Rh(en)$_3$][Co(ox)$_3$]

23.38   (a)   dichloroethylenediamminecadmium(II)

(b)   potassium hexacyanomanganate(II)

(c)   pentaamminecarbonatochromium(III) chloride

(d)   tetraamminediaquairidium(III) nitrate

23.40   Complex 1 has a coordination number of 6 and octahedral geometry about the metal. There are 5 monodentate ligands of one kind and one of another.

Complex 2 has a coordination number of 6 and octahedral geometry. There are three monodentate ligands of one kind and three of another.

Complex 3 has a coordination number of 6 and octahedral geometry about the metal. There are 5 monodentate ligands of one kind and one of another.

(a)   Only complex 2 has geometric isomers. It is the only complex with different possible arrangements of the same ligands. There is only one unique way to arrange 5 ligands of one kind and one of another, as in complex 1 and 3.

(b)   Only complex 1 can have linkage isomers. Thiocyanate ion, SCN$^-$, can coordinate through either N or S. It is the only ligand in the three complexes that has this ability.

(c)   None of the complexes has optical isomers. None of the complexes has bidentate ligands, which are usually present in optically active octahedral complexes.

(d)   Only complex 3 can have coordination sphere isomers. This is where an anion can either be a ligand or a counterion. Complex 3 is the only example with anions that are not specifically written as ligands.

23.42   Two geometric isomers are possible for an octahedral MA_3B_3 complex (see below). All other arrangements, including mirror images, can be rotated into these two structures. Neither isomer is optically active.

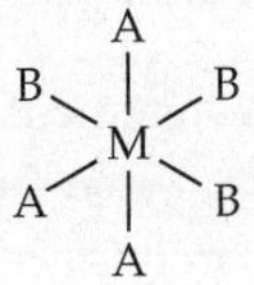

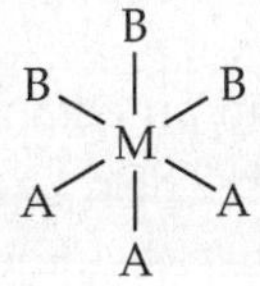

23.44

(a)

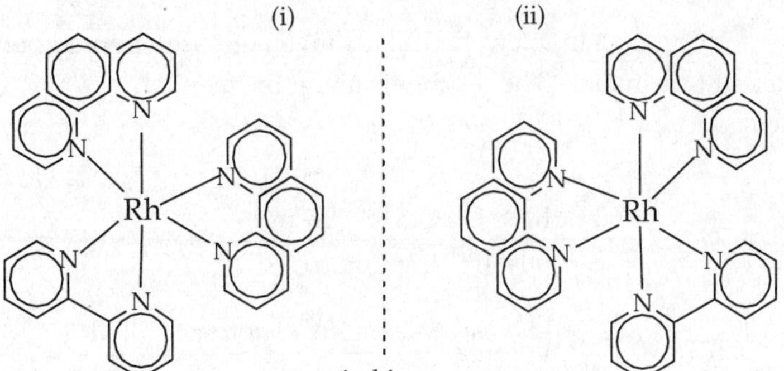

optical isomers

(b)

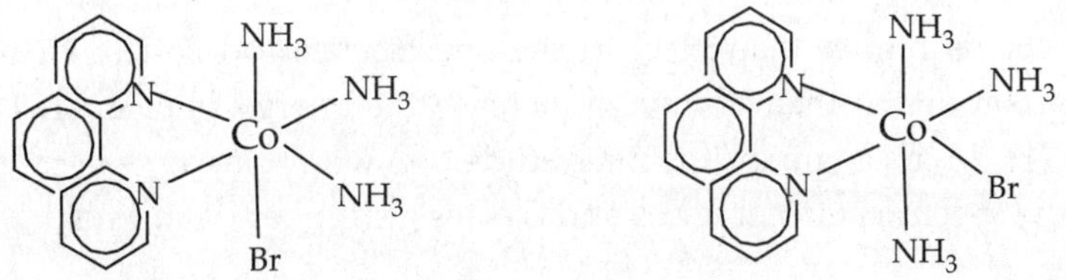

(c)

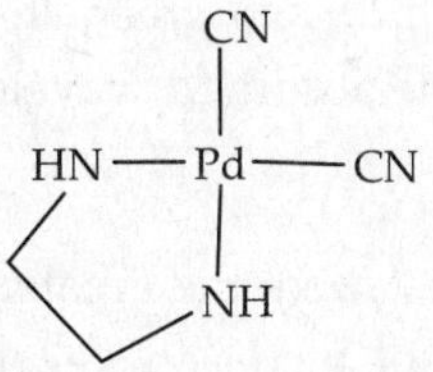

Color and Magnetism in Coordination Chemistry; Crystal-Field Theory
(Sections 23.5 and 23.6)

23.46   (a) $E(J/photon) = hc/\lambda$. $\lambda = hc/E$.

$$\lambda = \frac{6.626 \times 10^{-34} \text{ J-s}}{4.51 \times 10^{-19} \text{ J/ photon}} \times \frac{2.998 \times 10^8 \text{ m}}{s} \times \frac{1 \text{nm}}{1 \times 10^{-9} \text{m}} = 440.46 = 440 \text{ nm}$$

(b)   If the complex absorbs only 441 nm visible light, it absorbs violet and appears yellow.

23.48   (a) Ag^+, $[Kr]3d^{10}$. The complex is diamagnetic; all electrons are paired.

(b) Cu^{2+}, $[Ar]3d^9$. The complex is paramagnetic. The geometry is square planar, but there would be one unpaired electron in any of the d-orbital energy level diagrams.

(c) Ru^{2+}, $[Kr]3d^6$. The complex is octahedral and bipy is a strong field ligand. The 6 d-electrons are paired in the 3 lower energy d orbitals. The complex is diamagnetic.

(d) Co^{2+}, $[Ar]3d^7$. The complex is paramagnetic. The complex is either tetrahedral or square planar, but in either diagram there will be at least one unpaired electron.

23.50   (a)   The six ligands in an octahedral arrangement are oriented along the x, y, and z axes of the metal. The d_{xy}, d_{xz}, and d_{yz} metal orbitals point between the x, y, and z axes, and also between the ligands in this arrangement.

(b)   The four ligands in a tetrahedral arrangement are oriented between the x, y, and z axes of the metal. The d orbitals along the axes, $d_{x^2-y^2}$ and d_{z^2}, point between the ligands.

23.52   (a) $\Delta E = hc/\lambda = \dfrac{6.626 \times 10^{-34}\text{ J-s} \times 2.998 \times 10^8\text{ m/ s}}{500 \times 10^{-9}\text{ m photon}} = 3.973 \times 10^{-19} = 3.97 \times 10^{-19}\text{ J/ photon}$

$\Delta = 3.973 \times 10^{-19}\text{ J/photon} \times \dfrac{6.022 \times 10^{23}\text{ photons}}{1\text{ mol}} \times \dfrac{1\text{ kJ}}{1000\text{ J}} = 239.25 = 239\text{ kJ/mol}$

(b) NH_3 is higher than H_2O in the spectrochemical series, an ordering of ligands according to their ability to increase the energy gap, Δ. If H_2O is replaced by NH_3 in the complex, the magnitude of Δ would increase because NH_3 is higher in the spectrochemical series and creates a stronger ligand field.

23.54   (a)   Red. The $[Ni(bipy)_2]^{2+}$ ion absorbs 520 nm green light and appears as the complementary color, red. This fits the absorbed wavelength vs observed color trend shown in Figure 23.30.

(b)   The shorter the wavelength of light absorbed, the greater the value of Δ, and the stronger the ligand field. The order of increasing ligand field strength is the order of decreasing wavelength absorbed.

$H_2O < NH_3 < en < bipy$

23.56   (a) Fe^{3+}, d^5 (b) Mn^{2+}, d^5 (c) Ag^+, d^{10}

(d) Cr^{3+}, d^3 (e) Sr^{2+}, d^0

23.58   No. A strong-field ligand is one that interacts strongly with d-electrons of the metal, creating a large Δ splitting. However, the valence d-electrons of the metal are not directly involved in metal-ligand bonding. The field strength of the ligand is not directly related to the strength of metal-ligand bonding.

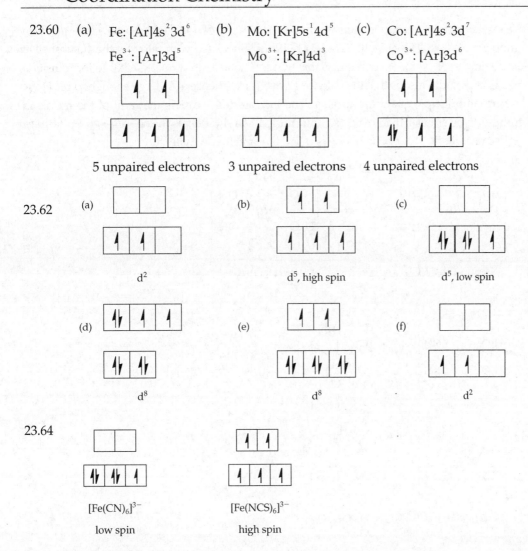

23.60 (a) Fe: $[Ar]4s^2 3d^6$ (b) Mo: $[Kr]5s^1 4d^5$ (c) Co: $[Ar]4s^2 3d^7$

Fe^{3+}: $[Ar]3d^5$ Mo^{3+}: $[Kr]4d^3$ Co^{3+}: $[Ar]3d^6$

5 unpaired electrons   3 unpaired electrons   4 unpaired electrons

23.62 (a)   (b)   (c)

d^2 d^5, high spin d^5, low spin

(d)   (e)   (f)

d^8 d^8 d^2

23.64

$[Fe(CN)_6]^{3-}$ $[Fe(NCS)_6]^{3-}$

low spin   high spin

Both complexes contain Fe^{3+}, a d^5 ion. CN^-, a strong field ligand, produces such a large Δ that the splitting energy is greater than the pairing energy, and the complex is low spin. NCS^- produces a smaller Δ, so it is energetically favorable for d-electrons to be unpaired in the higher energy d-orbitals. NCS^- is a much weaker-field ligand than CN^-. It is probably weaker than NH_3 and near H_2O in the spectrochemical series.

Additional Exercises

23.65 The paper clip must contain a significant amount of Ni, a ferromagnetic metal. At ambient temperature, the paper clip is below its Curie temperature, behaves ferromagnetically, and is strongly attracted to the permanent magnet. The lighter heats the left paperclip above its Curie temperature (354 °C), and it switches from from ferromagnetic to paramagnetic behavior. That is, below its Curie temperature, the spins of the unpaired electrons in Ni are perfectly aligned and the clip is strongly attracted to the permanent magnet. Above the Curie temperature, the unpaired spins become randomly aligned, and the paper clip loses most of its attraction for the permanent magnet.

23.66 We expect radii in a group to increase moving down the periodic table as principle quantum number increases. However, the nuclear build-up associated with filling of the 4f subshell at the beginning of period 6 counteracts this trend. The increased nuclear charge for transition metals of period 6 means that the valence electrons experience a Z_{eff} large enough to offset the increase in principle quantum number. The increased Z_{eff} causes the radii of the metals in group 6 to be smaller than expected, and period 5 and 6 metals in the same group to have very similar radii. This phenomenon is called the lanthanide contraction.

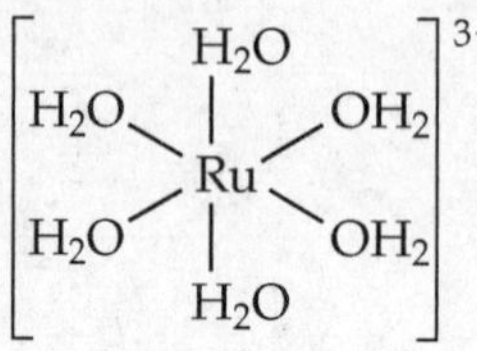

23.68 (a)

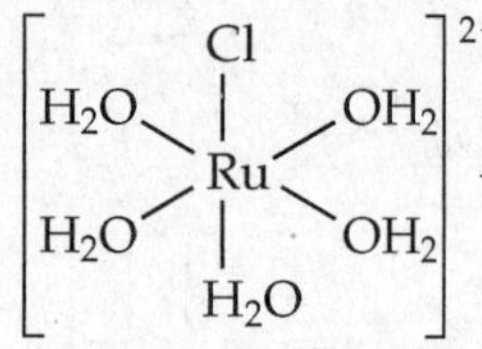

$$[Ru(H_2O)_5Cl]Cl_2 \longrightarrow [Ru(H_2O)_6]Cl_3$$

23.69 (a)

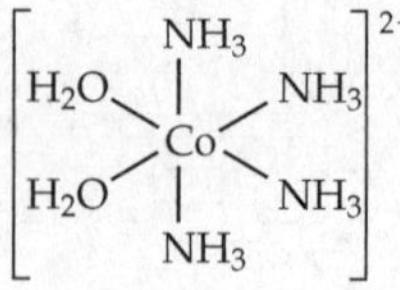

octahedral (a), octahedral (b)

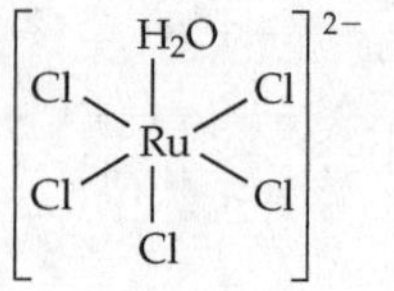

octahedral (c), 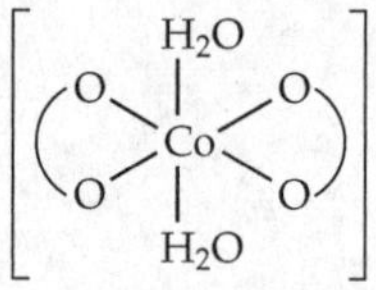 octahedral (d)

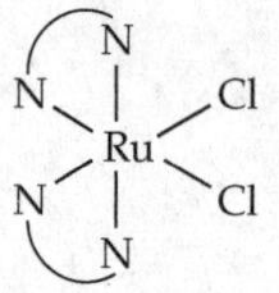

(a)   *cis*-tetraamminediaquacobalt(II) nitrate

(b)   sodium aquapentachlororuthenate(III)

(c)   ammonium *trans*-diaquabisoxalatocobaltate(III)

(d)   *cis*-dichlorobisethylenediammineruthenium(II)

23.70 (a)   In these octahedral complex ions, a mirror plane contains the central metal ion, and reflects the part of the ion on one side of the plane into the part on the other side of the plane. In all cases, assume ligands can be rotated so that H atoms either sit on the mirror plane, or are reflected across the mirror plane into another H atom. The complex ions in (a), (b), and (c) above all have multiple mirror planes. Below we describe one mirror plane in each complex ion; it is possible to visualize other correct planes in each complex.

In complex (a), one mirror contains the Co, two NH_3, and two H_2O ligands; it reflects one of the remaining NH_3 ligands into the other. In complex (b), one plane contains the Ru, H_2O, and 3 Cl⁻ ligands; it reflects the other two Cl⁻ ligands into each other. In complex (c), one plane contains the Co and both oxalate ligands; it reflects the two H_2O ligands into each other. Complex (d) does not have a mirror plane; the orientation of the en ligands precludes the presence of a mirror plane.

(b) Only the complex in 23.69(d) has optical isomers. The mirror images of (a)-(c) can be superimposed on the original structure. The chelating ligands in (d) prevent its mirror images (enantiomers) from being superimposable.

23.72 (a) In a square planar complex such as $[Pt(en)Cl_2]$, if one pair of ligands is trans, the remaining two coordination sites are also trans to each other. Ethylenediamine is a relatively short bidentate ligand that cannot occupy trans coordination sites, so the trans isomer is unknown.

(b) The minimum steric requirement for a bidentate ligand is a medium-length chain between the two coordinating atoms that will occupy the trans positions. In terms of reaction rate theory, it is unlikely that a flexible bidentate ligand will be in exactly the right orientation to coordinate trans. The ligand with the four-carbon chain is more able to occupy trans positions, although four carbons may still not be long enough. And, the flexibility of its backbone means that it is still an unlikely trans bidentate ligand.

23.73 We will represent the end of the bidentate ligand containing the CF_3 group by a shaded oval, the other end by an open oval:

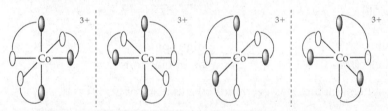

23.75 (a) Zero. The CO ligand is a neutral molecule and the charge on the complex is zero, so nickel must be present as Ni(0).

(b) The electron configuration of a Ni atom in the absence of a ligand field is $[Ar]4s^2 3d^8$. A tetrahedral complex with 8 d-electrons would have two unpaired electrons and be paramagnetic. Because the compound is diamagnetic, the Ni atom must have 10 d-electrons. The electron configuration of Ni in the complex is $[Ar]3d^{10}$.

(c) tetracarbonylnickel(0)

23.77 (a) left shoe

(c) wood screw

(e) a typical golf club

23.79   (a)   Formally, the two Ru centers have different oxidation states; one is +2 and the other is +3.

       (b)

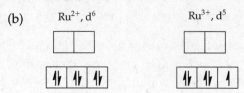

       (c)   There is extensive bonding-electron delocalization in the isolated pyrazine molecule. When pyrazine acts as a bridging ligand, its delocalized molecular orbitals provide a pathway for delocalization of the "odd" d-electron in the Creutz-Taube ion. The two metal ions appear equivalent because the odd d-electron is delocalized across the pyrazine bridge.

23.81   (a)   oxyhemoglobin   deoxyhemoglobin

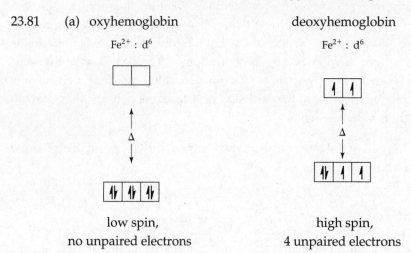

               low spin,   high spin,

     no unpaired electrons   4 unpaired electrons

       (b)   In deoxyhemoglobin, H_2O is bound to Fe in place of O_2.

       (c)   The two forms of hemoglobin have different colors because they absorb different wavelengths of visible light. They differ by just the H_2O or O_2 ligand, which means that the two ligands have slightly different ligand fields. Oxyhemoglobin appears red and absorbs green light, whereas deoxyhemoglobin appears bluish and absorbs longer wavelength yellow-green light. O_2 has a stronger ligand field than H_2O.

       (d)   According to Table 18.1, air is 20.948 mole percent O_2. This translates to 209,480 ppm O_2. This is approximately 500 times the 400 ppm concentration of CO in the experiment. If air with a CO concentration $1/500^{th}$ that of O_2 converts 1/10 of the oxyhemoglobin to carboxyhemoglobin, the equilibrium constant for binding CO is much larger than that for binding O_2.

       (e)   If CO is a stronger field ligand than O_2, carboxyhemoglobin will absorb shorter wavelengths than hemoglobin. It will absorb blue-green light and appear orange-red.

23.82   (a)   The term *isoelectronic* means that the three ions have the same number of electrons.

       (b)   In each ion, the metal is in its maximum oxidation state and has a d^0 electron configuration. That is, the metal ions have no *d*-electrons, so there should be no *d-d* transitions.

       (c)   A *ligand-metal charge transfer* transition occurs when an electron in a filled ligand orbital is excited to an empty d-orbital of the metal.

       (d)   Absorption of 565 nm yellow light by MnO_4^- causes the compound to appear violet, the complementary color. CrO_4^{2-} appears yellow, so it is absorbing violet light of approximately 420 nm. The wavelength of the LMCT transition for chromate, 420 nm, is shorter than the wavelength of LCMT transition in permanganate, 565 nm. This means that there is a larger energy difference between filled ligand and empty metal orbitals in chromate than in permanganate.

       (e)   UV. A white compound indicates that no visible light is absorbed. Going left on the periodic chart from Mn to Cr, the absorbed wavelength got shorter and the energy difference between ligand and metal orbitals increased. The 420 nm absorption by CrO_4^- is at the short wavelength edge of the visible spectrum. It is not surprising that the ion containing V, further left on the chart, absorbs at a still shorter wavelength in the ultraviolet region and that VO_4^{3-} appears white.

23.83   The higher the oxidation state of the metal, the smaller the energy separation between the ligand orbitals and the empty d-orbitals on the metal. The oxidation states of the metals in the tetrahedral oxoanions are: Mn, +7; Cr, +6; V, +5. From Solution 23.82, the energy separation between the ligand orbitals and the empty d-orbitals on the metals increases in the order Mn < Cr < V.

23.84   Application of pressure would result in shorter metal ionoxide distances. This would have the effect of increasing the ligand-electron repulsions, and would result in a larger splitting in the d-orbital energies. Thus, application of pressure should result in a shift in the absorption to a higher energy and shorter wavelength.

23.86

23.87 A large part of the metal-ligand interaction is electrostatic attraction between the positively charged metal and the fully or partially negatively charged ligand. For the same ligand, the greater the charge on the metal or the shorter the M–L separation, the stronger the interaction and the more stable the complex. The greater positive charge and smaller ionic radius of a metal in the 3+ oxidation state means that, for the same ligand, complexes with metals in the 3+ state are more stable than those with metals in the 2+ state.

23.88 (a) Only one (b) Two

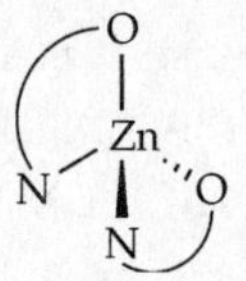

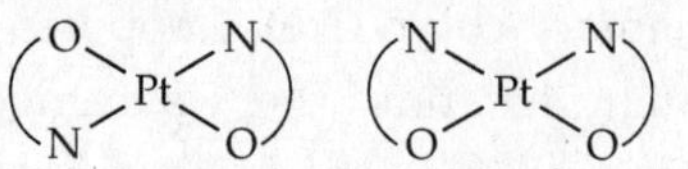

(c) Four; two are geometric, the other two are stereoisomers of each of these.

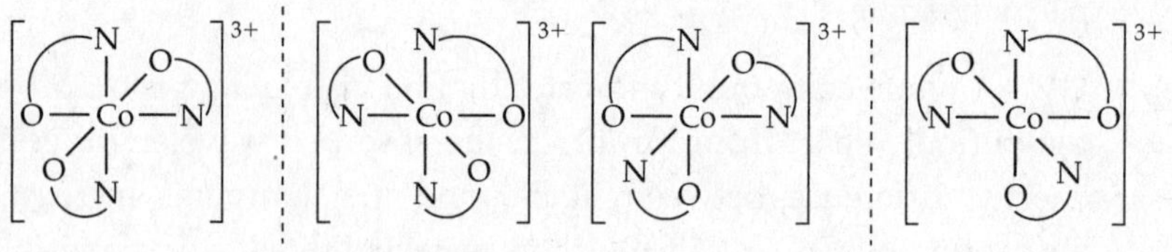

23.89 (a) Zero. The CO ligand is a neutral molecule and the charge on the complex is zero, so chromium must be present as Cr(0).

(b) The electron configuration of a Cr atom in the absence of a ligand field is $[Ar]4s^2 3d^4$. An octahedral complex with 4 d-electrons would have unpaired electrons and be paramagnetic. In order for the complex to be diamagnetic, all valence electrons must be paired. The electron configuration of Cr in the complex is $[Ar]3d^6$. In a strong field complex, the 6 d-electrons will be paired in the three lower energy d orbitals.

(c) A colorless complex has no d-d electron transitions. This indicates a large D for the complex, which means CO is a strong-field ligand.

(d) hexacarbonylchromium(0)

Integrative Exercises

23.90 (a)

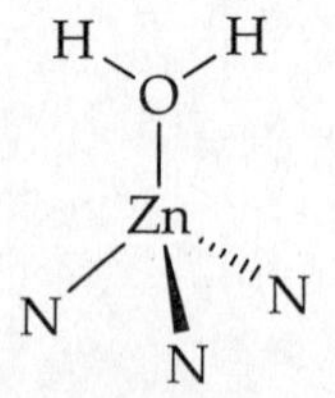

(b) The pK_a of pure water is 14, that of carbonic anhydrase is 7.5. The active site of carbonic anhydrase is much more acidic than the bulk water. In carbonic anhydrase, the Zn^{2+} ion withdraws electron density from the O atom of water. The electronegative oxygen atom compensates by withdrawing electron-density from the O—H bond. The O—H bond is polarized and H becomes more ionizable, more acidic than in the bulk solvent. This is similar to the effect of an electronegative central atom in an oxyacid such as H_2SO_4.

(c) When the water molecule is deprotonated, the ligand coordinated to water becomes hydroxide ion, OH⁻. The three N atoms are unaffected.

(d) In $[Zn(H_2O)_6]^{2+}$, the Zn^{2+} ion has six bound O atoms from which to withdraw electron density. Each O atom donates less electron density than the single O atom in carbonic anhydrase and each O atom withdraws less electron density from its O—H bonds. The O—H bonds in $[Zn(H_2O)_6]^{2+}$ are less polarized and less acidic than those in carbonic anhydrase. $[Zn(H_2O)_6]^{2+}$ is a weaker acid and has a higher pK_a than carbonic anhydrase.

(e) No, we do not expect carbonic anhydrase to have a deep color like hemoglobin. The Zn^{2+} ion is a d^{10} metal center. Its d-orbitals are completely occupied and there is no possibility for the d-d transitions that lead to colored complexes.

23.92 Determine the empirical formula of the complex, assuming a 100 g sample.

$$10.0 \text{ g Mn} \times \frac{1 \text{ mol Mn}}{54.94 \text{ g Mn}} = 0.1820 \text{ mol Mn; } 0.182 / 0.182 = 1$$

$$28.6 \text{ g K} \times \frac{1 \text{ mol K}}{39.10 \text{ g K}} = 0.7315 \text{ mol K; } 0.732 / 0.182 = 4$$

$$8.8 \text{ g C} \times \frac{1 \text{ mol C}}{12.0 \text{ g C}} = 0.7327 \text{ mol C; } 0.733 / 0.182 = 4$$

$$29.2 \text{ g Br} \times \frac{1 \text{ mol Br}}{79.904 \text{ g Br}} = 0.3654 \text{ mol Br; } 0.365 / 0.182 = 2$$

$$23.4 \text{ g O} \times \frac{1 \text{ mol O}}{16.00 \text{ g O}} = 1.463 \text{ mol O; } 1.46 / 0.182 = 8$$

There are 2 C and 4 O per oxalate ion, for a total of two oxalate ligands in the complex. To match the conductivity of $K_4[Fe(CN)_6]$, the oxalate and bromide ions must be in the coordination sphere of the complex anion. Thus, the compound is $K_4[Mn(ox)_2Br_2]$.

23.93 (a) $\Delta G° = -nFE°$. The positive E° values for both sets of complexes correspond to $-\Delta G°$ values. Negative values of $\Delta G°$ mean that both processes are spontaneous. For both o-phen and CN⁻ ligands, the Fe(II) complex is more thermodynamically favorable than the Fe(III) complex.

(b) The CN⁻ complex, with the smaller positive E° value, is more difficult to reduce.

(c) That both the Fe(II) complexes are low spin means that both CN⁻ and o-phen are strong-field ligands. The negatively charged CN⁻ has a stronger electrostatic interaction with Fe^{3+} than the neutral o-phen has. This stabilizes the Fe(III) complex of CN⁻ relative to the Fe(III) complex of o-phen, which reduces the driving force for reduction of $[Fe(CN)_6]^{3-}$ relative to reduction of $[Fe(o\text{-phen})_3]^{3+}$. The E° value and magnitude of $\Delta G°$ for the reduction of the CN⁻ complex are thus smaller than those values for the o-phen complex.

23.95   (a)   The reaction that occurs increases the conductivity of the solution by producing a greater number of charged particles, particles with higher charges, or both. It is likely that H_2O from the bulk solvent exchanges with a coordinated Br^- according to the reaction below. This reaction would convert the 1:1 electrolyte, $[Co(NH_3)_4Br_2]Br$, to a 1:2 electrolyte, $[Co(NH_3)_3(H_2O)Br]Br_2$.

         (b) $[Co(NH_3)_4Br_2]^+(aq) + H_2O(l) \rightarrow [Co(NH_3)_4(H_2O)Br]^{2+}(aq) + Br^-(aq)$

         (c)   Before the exchange reaction, there is one mole of free Br^- per mole of complex. mol Br^- = mol Ag^+

M = mol/L; L $AgNO_3$ = mol $AgNO_3$/M $AgNO_3$

$$\frac{3.87 \text{ g complex}}{0.500 \text{ L soln}} \times \frac{1 \text{ mol complex}}{366.77 \text{ g complex}} \times 0.02500 \text{ L soln used} =$$

$$5.276 \times 10^{-4} = 5.28 \times 10^{-4} \text{ mol complex}$$

$$5.276 \times 10^{-4} \text{ mol complex} \times \frac{1 \text{ mol } Br^-}{1 \text{ mol complex}} \times \frac{1 \text{ mol } Ag^+}{1 \text{ mol } Br^-} \times \frac{1 \text{ L } Ag^+(aq)}{0.0100 \text{ mol } Ag^+(aq)}$$

$$= 0.05276 \text{ L} = 52.8 \text{ mL } AgNO_3(aq)$$

         (d)   After the exchange reaction, there are 2 mol free Br^- per mol of complex. Because M $AgNO_3$(aq) and volume of complex solution are the same for the second experiment, the titration after conductivity changes will require twice the volume calculated in part (c), 105.52 = 106 mL of 0.0100 M $AgNO_3$(aq).

23.97   Use Hess' law to calculate $\Delta G°$ for the desired equilibrium. Then $\Delta G°$ = –RTlnK to calculate K.

$$\begin{array}{ll} Hb + CO \rightarrow HbCO & \Delta G° = -80 \text{ kJ} \\ \underline{HbO_2 \rightarrow Hb + O_2} & \underline{\Delta G° = 70 \text{ kJ}} \\ HbO_2 + Hb + CO \rightarrow HbCO + Hb + O_2 & \\ HbO_2 + CO \rightarrow HbCO + O_2 & \Delta G° = -10 \text{ kJ} \end{array}$$

$$\Delta G° = -RT\ln K, \ln K = \frac{-\Delta G°}{RT} = \frac{-(-10 \text{ kJ})}{8.314 \text{ J/K-mol} \times 298 \text{ K}} \times \frac{1000 \text{ J}}{\text{kJ}} = 4.036 = 4.04$$

$$K = e^{4.04} = 56.61 = 57$$

23.98   (a)

$$\begin{array}{ll} [Cd(CH_3NH_2)_4]^{2+} \rightleftharpoons Cd^{2+}(aq) + 4 CH_3NH_2(aq) & \Delta G° = 37.2 \text{ kJ} \\ \underline{Cd^{2+}(aq) + 2 \text{ en}(aq) \rightleftharpoons [Cd(en)_2]^{2+}(aq)} & \underline{\Delta G° = -60.7 \text{ kJ}} \\ Cd(CH_3NH_2)_4]^{2+} + 2 \text{ en}(aq) \rightleftharpoons [Cd(en)_2]^{2+}(aq) + 4 CH_3NH_2(aq) & \Delta G° = -23.5 \text{ kJ} \end{array}$$

$\Delta G°$ = –RTlnK; –23.5 kJ = -2.35×10^4 J

$$-2.35 \times 10^4 \text{ J} = \frac{-8.314 \text{ J}}{\text{K-mol}} \times 298 \text{ K} \times \ln K; \ln K = 9.485, K = 1.32 \times 10^4$$

The magnitude of K is large, so the reaction favors products. The bidentate chelating ligand en will spontaneously replace the monodentate ligand CH_3NH_2. This is an illustration of the chelate effect.

(b) Using the stepwise construction from part (a),

$\Delta H° = 57.3 \text{ kJ} - 56.5 \text{ kJ} = 0.8 \text{ kJ}$

$\Delta S° = 67.3 \text{ J/K} + 14.1 \text{ J/K} = 81.4 \text{ J/K}$

$-T\Delta S = -298 \text{ K} \times 81.4 \text{ J/K} = -2.43 \times 10^4 \text{ J} = -24.3 \text{ kJ}$

The chelate effect is mainly the result of entropy. The reaction is spontaneous because of the increase in the number of free particles and corresponding increase in entropy going from reactants to products. The enthalpic contribution is essentially zero because the bonding interactions of the two ligands are very similar and the reaction is not "downhill" in enthalpy.

(c) $\Delta H°$ will be very small and negative. When NH_3 replaces H_2O in a complex (Closer Look Box), the tighter bonding of the NH_3 ligand causes a substantial negative $\Delta H°$ for the substitution reaction. When a bidentate amine ligand replaces a monodentate amine ligand of similar bond strength, $\Delta H°$ is very small and either positive (part (c)) or negative (Closer Look Box). In the case of NH_3 replacing CH_3NH_2, the bonding characteristics are very similar. The presence of CH_3 groups in CH_3NH_2 produces some steric hindrance in $[Cd(CH_3NH_2)_4]^{2+}$. This complex is at a slightly higher energy than $[Cd(NH_3)_4]^{2+}$, which experiences no steric hindrance, so $\Delta H°$ will have a negative sign but a very small magnitude. Relief of steric hindrance leads to a very small negative $\Delta H°$ for the substitution reaction.

23.100 The process can be written:

$$H_2(g) + 2 e^- \rightarrow 2 H^+(aq) \qquad\qquad E°_{red} = 0.0 \text{ V}$$

$$Cu(s) \rightarrow Cu^{2+} + 2 e^- \qquad\qquad E°_{red} = 0.337 \text{ V}$$

$$\underline{Cu^{2+}(aq) + 4 NH_3(aq) \rightarrow [Cu(NH_3)_4]^{2+}(aq) \qquad\qquad "E°_f" = ?}$$

$$H_2(g) + Cu(s) + 4 NH_3(aq) \rightarrow 2 H^+(aq) + [Cu(NH_3)_4]^{2+}(aq) \qquad E = 0.08 \text{ V}$$

$$E = E° - RT \ln K; \quad K = \frac{[H^+]^2[Cu(NH_3)_4^{2+}]}{P_{H_2}[NH_3]^4}$$

$P_{H_2} = 1 \text{ atm}, \; [H^+] = 1 \, M, \; [NH_3] = 1 \, M, \; [Cu(NH_3)_4]^{2+} = 1 \, M, \; Q = 1$

$E = E° - RT \ln(1); \quad E = E° - RT(0); \quad E = E° = 0.08 \text{ V}$

Because we know $E°$ values for two steps and the overall reaction, we can calculate "$E°$" for the formation reaction and then K_f, using $E° = \dfrac{0.0592}{n} \log K_f$ for the step.

$E_{cell} = 0.08 \text{ V} = 0.0 \text{ V} - 0.337 \text{ V} + "E°_f" \qquad "E°_f" = 0.08 \text{ V} + 0.337 \text{ V} = 0.417 \text{ V} = 0.42 \text{ V}$

$"E°_f" = \dfrac{0.0592}{n} \log K_f; \quad \log K_f = \dfrac{n(E°_f)}{0.0592} = \dfrac{2(0.417)}{0.0592} = 14.0878 = 14$

$K_f = 10^{14.0878} = 1.2 \times 10^{14} = 10^{14}$

23.101   (a)   The units of the rate constant and the rate dependence on the identity of the second ligand show that the reaction is second order. Therefore, the rate-determining step cannot be the dissociation of water, because that would be independent of the concentration and identity of the incoming ligand. The alternative mechanism, a bimolecular association of the incoming ligand with the complex, is indicated.

       (b)   The larger the value of the rate constant, the faster the exchange reaction, the stronger the donor ability of the nitrogen-containing ligand toward Ru(III). The rate constants indicate that the order of nitrogen-donor strength is:

          pyridine $>$ SCN$^-$ $>$ CH$_3$CN.

       (c)   Ru(III) is a d^5 ion. In a low-spin d^5 octahedral complex, there is one unpaired electron.

24 The Chemistry of Life: Organic and Biological Chemistry

Visualizing Concepts

24.2 *Analyze/Plan*. Given structural formulas, specify which molecules are unsaturated. Consider the definition of *unsaturated* and apply it to the molecules in the exercise. *Solve*.

Unsaturated molecules contain one or more multiple bonds. Saturated molecules contain only single bonds. Molecules (c) and (d) are unsaturated.

24.3 *Analyze/Plan*. Given structural formulas, decide which molecule will undergo addition. Consider which functional groups are present in the molecules, and which are most susceptible to addition. *Solve*.

(a) Molecule (iii), an alkene, will readily undergo addition. Addition reactions are characteristic of alkenes. [Molecule (i) will not typically undergo addition, because its delocalized electron cloud is too difficult to disrupt. Molecules (b) and (d) contain carbonyl groups (actually carboxylic acid groups) that do not typically undergo addition, except under special conditions.]

(b) Molecule (i) is an aromatic hydrocarbon.

(c) Molecule (i) most readily undergoes a substitution reaction.

24.4 *Analyze/Plan*. Given condensed structural formulas, predict which molecule will have the highest boiling point. Boiling point is determined by strength of intermolecular forces; for neutral molecules with similar molar masses, the strongest intermolecular force is hydrogen bonding.

We are also asked to identify various functional groups, given condensed structural formulas. Refer to Table 24.6 for information on functional groups. *Solve*.

(a) Molecule (ii), an alcohol, forms hydrogen bonds with like molecules; it has the highest boiling point. Only F–H, O–H, and N–H bonds fit the strict definition of hydrogen bonding.

(b) Molecule (iv) is most oxidized; it has the most oxygen atoms.

(c) None of the compounds is an ether. Compound (iv) is an ester, which has an ether linkage.

(d) Compound (iv) is an ester.

(e)   None of the compounds is a ketone. Ketones are characterized by a carbonyl group bonded to two alkyl groups. Compound (i) is an aldehyde; the carbonyl group is bonded to one alkyl group and one hydrogen atom.

24.5   *Analyze.* Given structural formulas, decide which molecules are capable of isomerism, and what type. *Plan.* Analyze each molecule for possible structural, geometric, and optical isomers/enantiomers.   *Solve.*

(a) $C_5H_{11}NO_2$, structural, geometric, and optical. There are many ways to arrange the atoms in molecules with this empirical formula, so there are many structural isomers. There is one point of unsaturation in the given molecule, the C=O group; structural isomers with their point of unsaturation at a C=C group could have geometric isomers as well. The C atom to which the $-NH_3^+$ group is bound is a chiral center, so there are enantiomers. All amino acids except glycine have two possible enantiomers.

(b) $C_7H_5O_2Cl$, structural, and geometric. The most obvious isomers for this aromatic compound are ortho, meta, and para geometric isomers. Because the molecule has several points of unsaturation, the number of structural isomers is limited, but there are a few possibilities with two triple bonds. Switching the –OH and –Cl groups also generates a structural isomer. There are no chiral centers, so no optical isomers.

(c) C_5H_{10}, structural, and geometric (cis-trans). There are many structural isomers for this empirical formula. The straight-chain alkene shown also has geometric (cis-trans) isomers.

(d) C_3H_8. There are no other structural, geometric, or optical isomers for this molecule.

24.6   *Analyze/Plan.* Given ball-and-stick models, select the molecule that fits the description given. From the models, decide the type of molecule or functional group represented.

Solve. Molecule (i) is a sugar, (ii) is an ester with a long hydrocarbon chain, (iii) is an organic base and a component of nucleic acids, (iv) is an amino acid, and (v) is an alcohol.

(a)   Molecule (i) is a disaccharide composed of galactose (left) and glucose (right); it can be hydrolyzed to form a solution containing glucose. Because it is the only sugar molecule depicted, it was not necessary to know the exact structure of glucose to answer the question.

(b)   Amino acids form zwitterions, so the choice is molecule (iv).

(c)   Molecule (iii) is an organic base present in DNA (again, the only possible choice).

(d)   Molecule (v) because alcohols react with carboxylic acids to form esters.

(e)   Molecule (ii), because it has a long hydrocarbon chain and an ester functional group.

Introduction to Organic Compounds; Hydrocarbons
(Sections 24.1 and 24.2)

24.8   (a)   False. Hexane contains six carbon atoms and pentane contains only five.

        (b)   True. For molecules with similar structures, the larger the molecule, the stronger ther dispersion forces and the higher the boiling point.

        (c)   True. The carbon atoms involved in an alkyne group are sp hybridized.

        (d)   False. There is only one way to arrange the three carbon atoms in propane.

24.10

$$N\equiv\underset{7}{C}-\underset{6}{\overset{H}{\underset{|}{C}}}-\underset{5}{\overset{H}{\underset{|}{C}}}-\underset{4}{\overset{H}{C}}=\underset{3}{\overset{H}{C}}-\underset{2}{\overset{OH}{\underset{|}{C}}}-\underset{1}{\overset{O}{\overset{||}{C}}}-H$$

(with H below C6, H below C5, H below C3)

        (a)   C2, C5, and C6 have sp^3 hybridization (4 e⁻ domains around C)

        (b)   C7 has sp hybridization (2 e⁻ domains around C)

        (c)   C1, C3, and C4 have sp^2 hybridization (3 e⁻ domains around C)

24.12   From Table 8.4, the bond enthalpies in kJ/mol are: C–H, 413; C–C, 348; C–O, 358; C–Cl, 328. Based on bond enthalpies, compounds that contain C–Cl bonds are definitely more reactive than simple alkane hydrocarbons. We might expect the reactivity of molecules containing C–O bonds to be similar to that of simple alkane hydrocarbons.

        The bond enthalpies indicate that C–H bonds are most difficult to break, and C–Cl bonds least difficult. However, they do not explain the relative reactivity of C–O bonds, or stability of C–C bonds. The reactivity of molecules containing C–O and C–Cl bonds is enhanced because of their unequal charge distribution, which attracts reactants that are either electron deficient (electrophilic) or electron rich (nucleophilic).

24.14   All the classifications listed are hydrocarbons; they contain only the elements hydrogen and carbon.

        (a)   *Alkanes* are hydrocarbons that contain only single bonds.

        (b)   *Cycloalkanes* contain at least one ring of three or more carbon atoms joined by single bonds. Because it is a type of alkane, all bonds in a cycloalkane are single bonds.

        (c)   *Alkenes* contain at least one C=C double bond.

        (d)   *Alkynes* contain at least one C≡C triple bond.

        (e)   A *saturated hydrocarbon* contains only single bonds. Alkanes and cycloalkanes fit this definition.

        (f)   An *aromatic hydrocarbon* contains one or more planar, six-membered rings of carbon atoms with delocalized π-bonding throughout the ring.

24.16   (a)   3,3,5-trimethylheptane

        (b)   3,4,4-trimethylheptane

        (c)

$$CH_3CH_2CH_2-\overset{\overset{\displaystyle CH_3}{|}}{CH}-\overset{\overset{\displaystyle CH_3}{|}}{CH}-CH_2-CH_2-\overset{\overset{\displaystyle CH_3}{|}}{CH}-CH_3$$

(d)

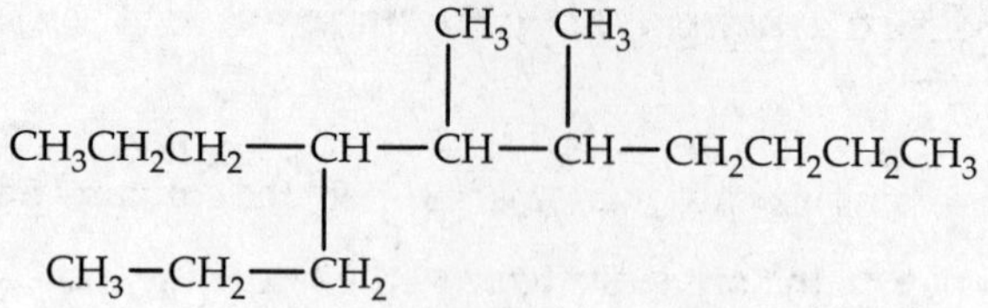

$$CH_3 \quad CH_3$$

$$CH_3CH_2CH_2-CH-CH-CH-CH_2CH_2CH_2CH_3$$

$$CH_3-CH_2-CH_2$$

(e)

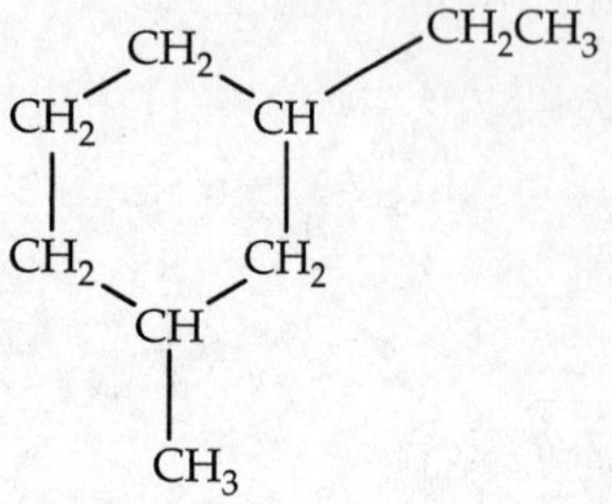

$$CH_2 \qquad CH_2CH_3$$

$$CH_2 \qquad CH$$

$$CH_2 \qquad CH_2$$

$$CH$$

$$CH_3$$

24.18   (a)

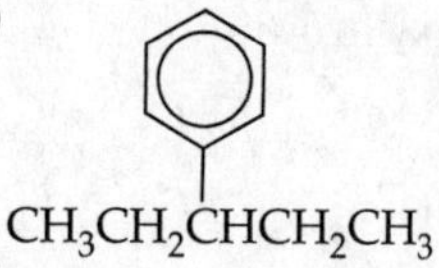

$$CH_3CH_2CHCH_2CH_3$$

(b)

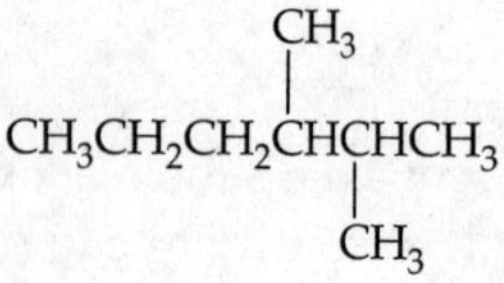

$$CH_3$$

$$CH_3CH_2CH_2CHCHCH_3$$

$$CH_3$$

(c)

$$CH_3$$

$$CH_3CH_2CH_2CH_2CH_2CCH_3$$

$$CH_2CH_3$$

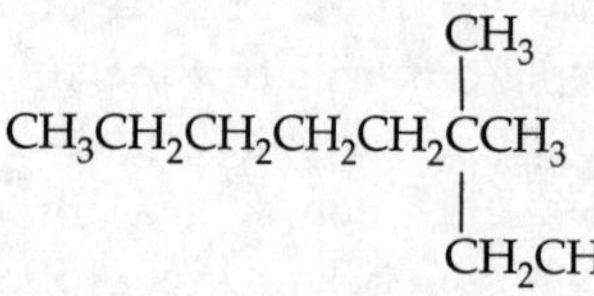

(A more correct name for this compound is 3,3-dimethyloctane.)

(d)   2,4-dimethylhexane

(e)   methylcyclobutane

24.20   Octane number can be increased by increasing the fraction of branched-chain alkanes or aromatics, because these have high octane numbers. This can be done by cracking. The octane number also can be increased by adding an anti-knock agent such as tetraethyl lead, $Pb(C_2H_5)_4$ (no longer legal); methyl t-butyl ether (MTBE); or an alcohol, methanol, or ethanol.

Alkenes, Alkynes, and Aromatic Hydrocarbons (Section 24.3)

24.22   (a)   The molecule $CH_3CH=CH_2$ is unsaturated because it contains a double bond. It is possible to add more hydrogen to the molecule.

(b)   The formula $CH_3CH_2CH=CH_3$ has too many H atoms bound to the right-most C atom; the formula as it stands implies 5 bonds to this atom. A correct formula is $CH_3CH_2CH=CH_2$, with 2 H atoms on the right-most C atom.

24.24   cyclic alkane, H_2C—CH_2 , C_6H_{12}

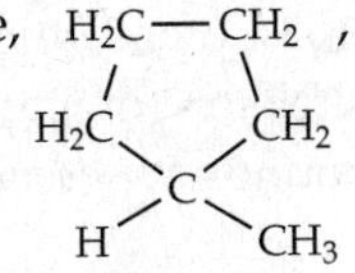

$$\begin{array}{c}
H_2C\!-\!CH_2 \\
H_2C \quad\; CH_2 \\
C \\
H \quad\; CH_3
\end{array}$$

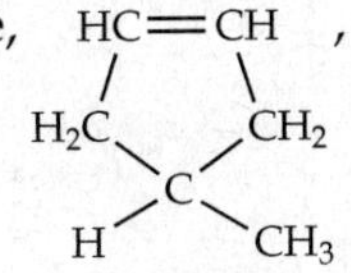

cyclic alkene, HC＝CH , C_6H_{10}

$$\begin{array}{c}
HC\!=\!CH \\
H_2C \quad\; CH_2 \\
C \\
H \quad\; CH_3
\end{array}$$

alkyne, CH_3–CH_2–C≡C–CH_2–CH_3, C_6H_{10}

aromatic hydrocarbon, C_6H_6

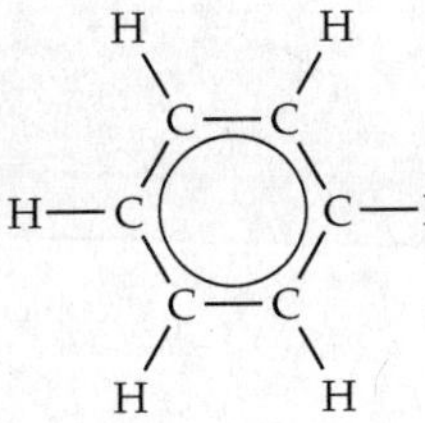

24.26 C_nH_{2n-2}

24.28 CH_3—CH_2—CH_2—CH＝CH_2
pentene

CH_3—CH_2—CH＝CH—CH_3
2-pentene

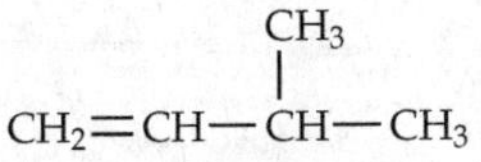

CH_2＝CH—$\overset{\displaystyle CH_3}{\underset{\displaystyle |}{CH}}$—$CH_3$
3-methyl-1-butene

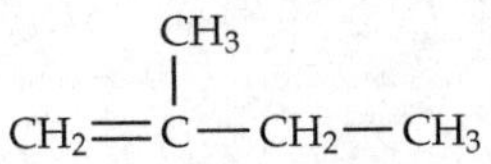

CH_2＝$\overset{\displaystyle CH_3}{\underset{\displaystyle |}{C}}$—$CH_2$—$CH_3$
2-methyl-1-butene

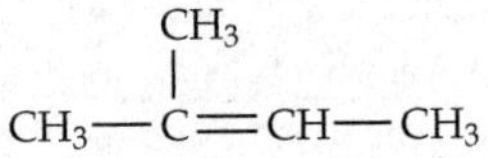

CH_3—$\overset{\displaystyle CH_3}{\underset{\displaystyle |}{C}}$＝$CH$—$CH_3$
2-methyl-2-butene

24.30   (a) 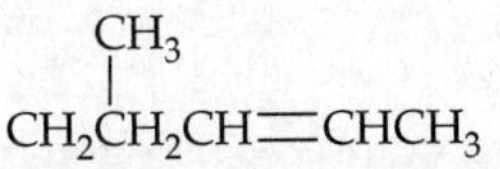 $\overset{\displaystyle CH_3}{\underset{\displaystyle |}{}}$
CH_2CH_2CH＝$CHCH_3$

(b) 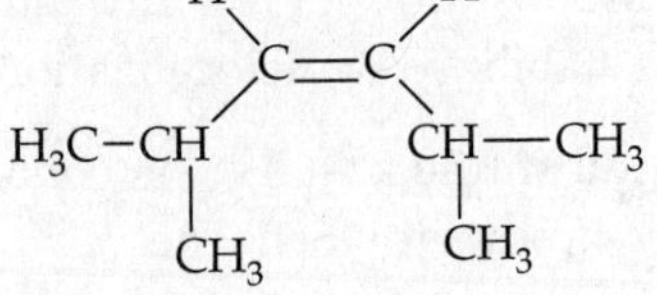 H_3C—$\overset{\displaystyle H}{\underset{\displaystyle CH}{\underset{\displaystyle |}{C}}}$＝$\overset{\displaystyle H}{\underset{\displaystyle CH}{\underset{\displaystyle |}{C}}}$—$CH_3$ (with CH_3 groups below)

(c)   benzene ring with CH_3, CH_3 substituents

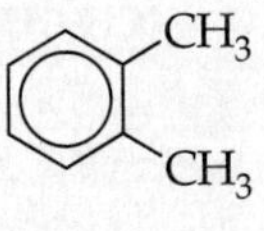

(d)   1-butyne

(e)   *trans*-2-heptene

24.32   Butene is an alkene, C_4H_8. There are two possible placements for the double bond:

CH_2＝$CHCH_2CH_3$ or CH_3CH＝$CHCH_3$
1-butene   2-butene

These two compounds are *structural isomers*. For 2-butene, there are two different, noninterchangeable ways to construct the carbon skeleton (owing to the absence of free rotation around the double bond). These two compounds are *geometric isomers*.

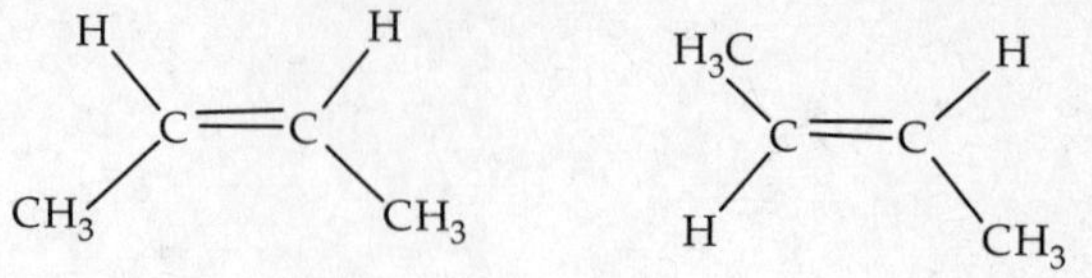

cis-2-butene   *trans*-2-butene

24.34

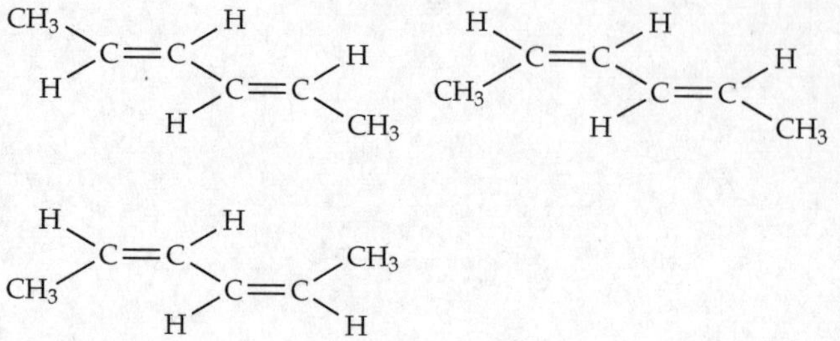

24.36   (a)

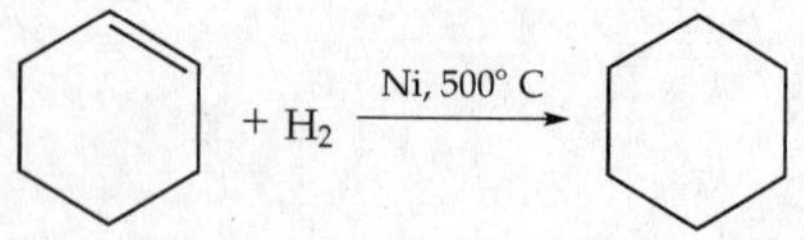

(b)

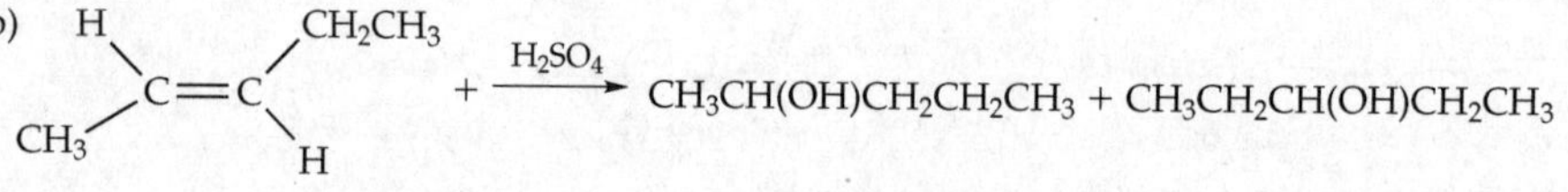

$CH_3CH(OH)CH_2CH_2CH_3 + CH_3CH_2CH(OH)CH_2CH_3$

(c)

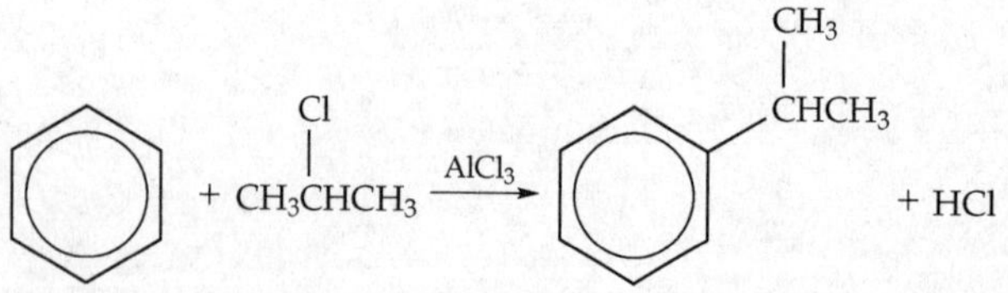

24.38   (a)   The reaction of Br_2 with an alkene to form a colorless halogenated alkane is an addition reaction. Aromatic hydrocarbons do not readily undergo addition reactions, because their π-electrons are stabilized by delocalization.

(b)   *Plan.* Use a Friedel-Crafts reaction to substitute a $-CH_2CH_3$ onto benzene. Do a second substitution reaction to get *para*-bromoethylbenzene.   *Solve.*

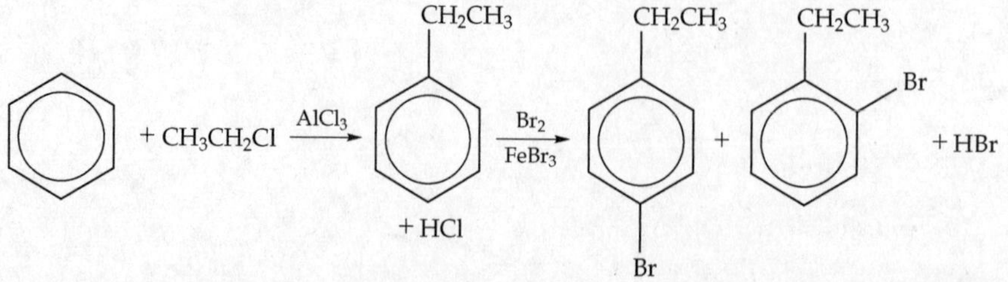

It appears that ortho, meta, and para geometric isomers of bromoethylbenzene would be possible. However, because of electronic effects beyond the scope of this chapter, the ethyl group favors formation of ortho and para isomers, but not the meta. The ortho and para products must be separated by distillation or some other technique.

24.40 The partially positive end of the hydrogen halide, $\overset{\delta^+}{H} - \overset{\delta^-}{X}$, is attached to the π electron cloud of the alkene cyclohexene. The electrons that formed the π bond in cyclohexene form a sigma bond to the H atom of HX, leaving a halide ion, X^-. The intermediate is a carbocation; one of the C atoms formerly involved in the π bond is now bound to a second H atom. The other C atom formerly involved in the π bond carries a full positive charge and forms only three sigma bonds, two to adjacent C atoms and one to H.

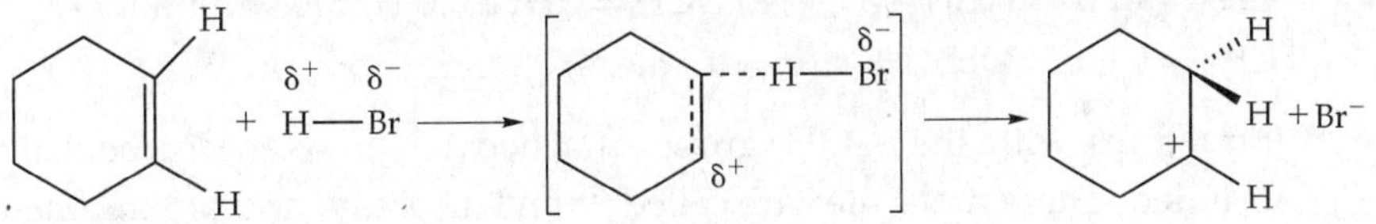

24.42

$$
\begin{array}{lr}
 & \underline{\Delta H} \\
C_{10}H_8(l) + 12\,O_2(g) \rightarrow 10\,CO_2(g) + 4\,H_2O(l) & -5157\text{ kJ} \\
-[C_{10}H_{18}(l) + 29/2\,O_2(g) \rightarrow 10\,CO_2(g) + 9\,H_2O(l) & -(-6286)\text{ kJ} \\
\hline
C_{10}H_8(l) + 5\,H_2O(l) \rightarrow C_{10}H_{18}(l) + 5/2\,O_2(g) & +1129\text{ kJ} \\
5/2\,O_2(g) + 5\,H_2(g) \rightarrow 5\,H_2O(l) & 5(-285.8)\text{ kJ} \\
\hline
C_{10}H_8(l) + 5\,H_2(g) \rightarrow C_{10}H_{18}(l) & -300\text{ kJ}
\end{array}
$$

Compare this with the heat of hydrogenation of ethylene:

$C_2H_4(g) + H_2(g) \rightarrow C_2H_6(g)$; $\Delta H = -84.7 - (52.3) = -137$ kJ. This value applies to just one double bond. For five double bonds, we would expect about –685 kJ. The fact that hydrogenation of napthalene yields only –300 kJ indicates that the overall energy of the napthalene molecule is lower than expected for five isolated double bonds. The resonance energy is then [–685 kJ – (–300 kJ)] or –385 kJ. Resonance has stabilized or lowered the overall energy of naphthalene by 385 kJ.

Functional Groups and Chirality (Sections 24.4 and 24.5)

24.44 (a) $-\overset{\displaystyle O}{\overset{\|}{C}}-O$, ester

(b) –Cl, halocarbon; –OH, alcohol (aromatic alcohols are phenols)

(c) $-\overset{\displaystyle O}{\overset{\|}{C}}-NH-$, amide

(d) alkane

(e) $-C{=}C-$, alkene; $-\overset{\displaystyle O}{\overset{\|}{C}}H$, aldehyde

(f) $-\overset{\displaystyle O}{\overset{\|}{C}}-$, ketone

3333

24.46 (a) C_4H_8O, cyclic structure (tetrahydrofuran): ring of H_2C–O–CH_2 / H_2C—CH_2

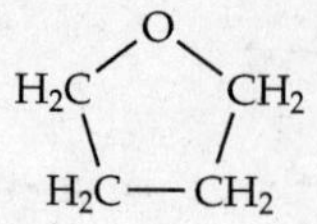

(b)

$CH_3CH_2CCH_3$ (with $=O$), $CH_3CH_2CH_2C$—H (with $=O$)

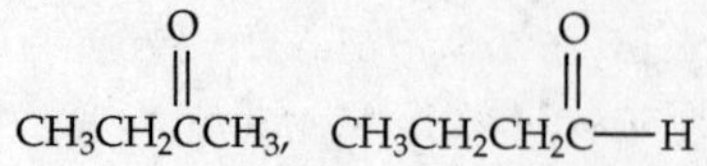

$CH_2=CH_2CH_2CH_2OH$, $CH_3CH=CHCH_2OH$, (cis and trans)
$CH_2=CHCH(OH)CH_3$ (enantiomers)

(Structures with the —OH group attached to an alkene carbon atom are not included. These molecules are called "vinyl alcohols" and are not the major form at equilibrium.)

24.48 (a) 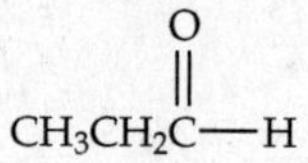 CH_3CH_2C—H (with $=O$) (b) 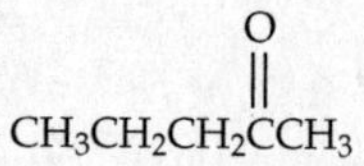 $CH_3CH_2CH_2CCH_3$ (with $=O$)

(c) 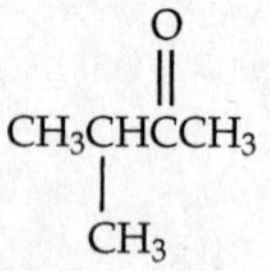 CH_3CHCCH_3 (with $=O$), with CH_3 below (d) 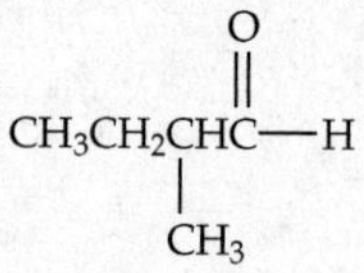 CH_3CH_2CHC—H (with $=O$), with CH_3 below

24.50 (a) 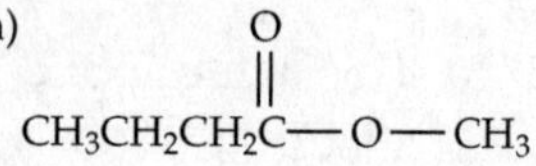 $CH_3CH_2CH_2C$—O—CH_3 (with $=O$)
methylbutanoate

(b) 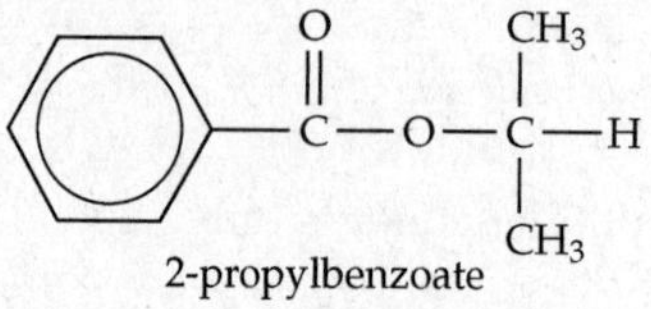 benzene ring—C(=O)—O—C(H)(CH_3)(CH_3)
2-propylbenzoate

(c) 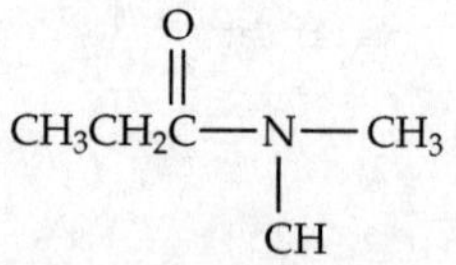 CH_3CH_2C—N—CH_3 (with $=O$), with CH below
N, N-dimethylpropanamide

24.52 (a)

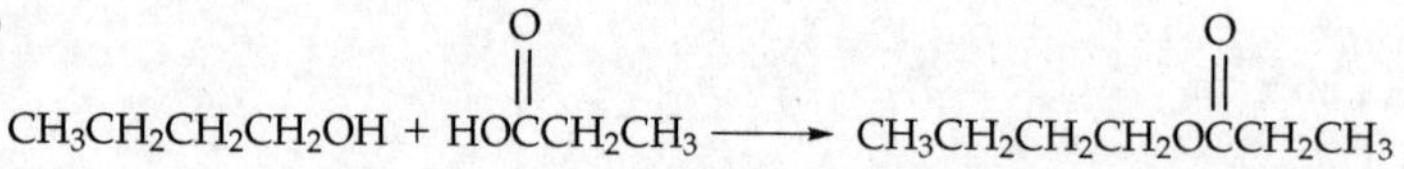

$CH_3CH_2CH_2CH_2OH + HOCCH_2CH_3$ (with $=O$) ⟶ $CH_3CH_2CH_2CH_2OCCH_2CH_3$ (with $=O$)

1-butanol   propionic acid   butyl proprionate
(propanoic acid)

(b)

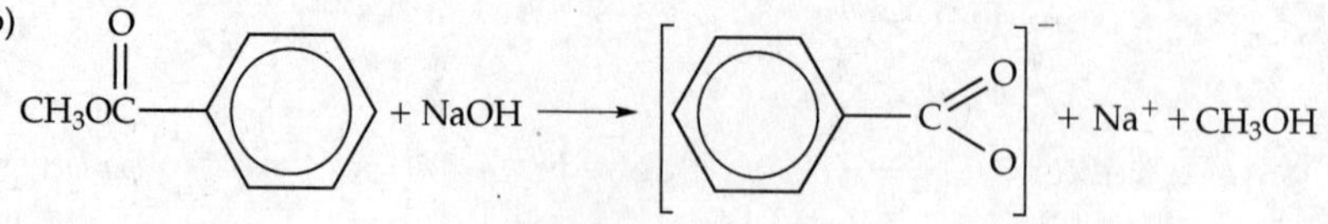

CH_3OC—(benzene ring) (with $=O$) + NaOH ⟶ [(benzene ring)—C(=O)O^- O] + Na^+ + CH_3OH

24.54 $2\ CH_3COOH(l) \longrightarrow CH_3COCH_3(l) + H_2O(l)$ (with two $=O$)

CH_3C—[OH + H]—O—CCH_3 (with $=O$) ⟶ CH_3—C(=O)—O—C(=O)—CH_3 + H_2O

32
Copyright © 2015 Pearson Education, Inc.

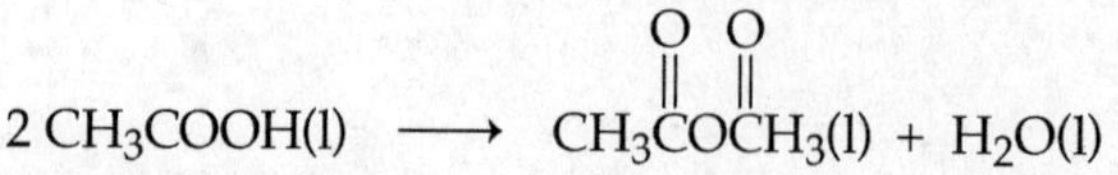

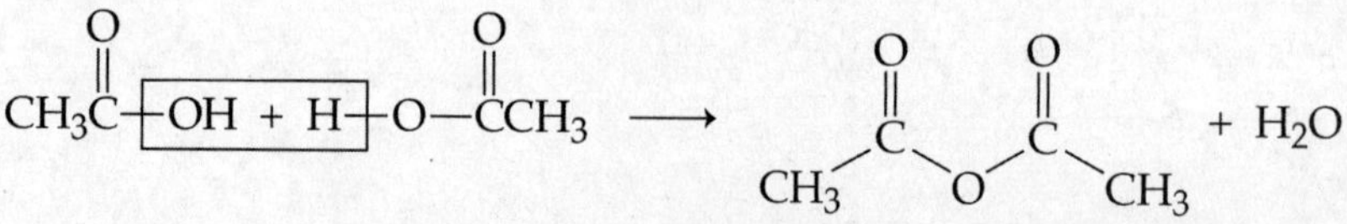

24.56   (a) $CH_3CH_2CH_2CH_2CH(C_2H_5)CH_2OH$

(b)

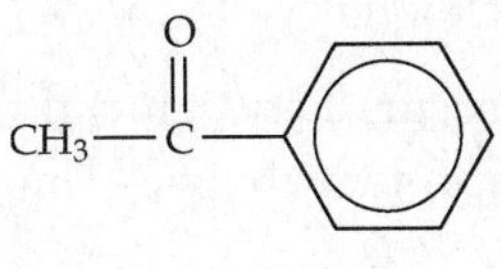

(c)

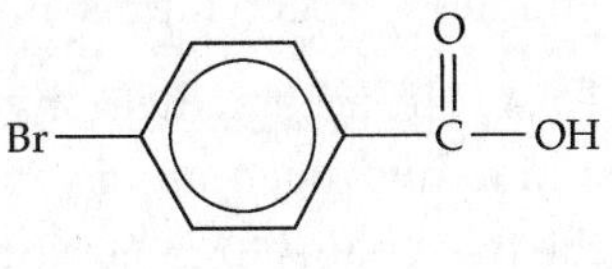

(d) $CH_3CH_2OCH_2CH_2CH_2CH_3$

(e)

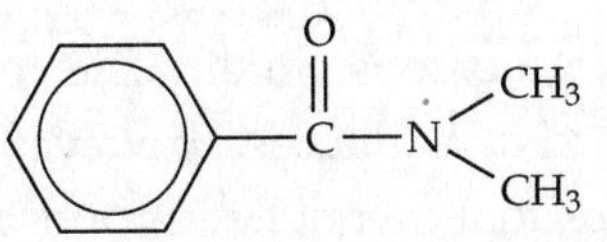

24.58

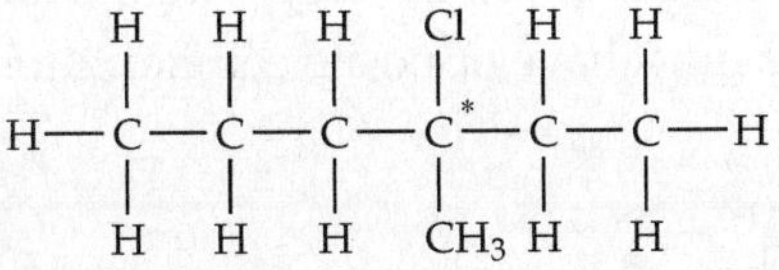

Yes, the molecule has optical isomers. The chiral carbon atom is attached to chloro, methyl, ethyl, and propyl groups. (If the root was a 5-carbon chain, the molecule would not have optical isomers because two of the groups would be ethyl groups.)

Proteins (Section 24.7)

24.60   (a)   True

(b)   True. Lysine is in the group of *basic* amino acids. Near pH 7, these amino acids have more positively charged functional groups than negatively charged groups.

(c)   False. There is one amide group, but other N-containing group is an amine; it is separated from the carbonyl group by one C atom.

(d)   False. Leucine and isoleucine are structural isomers, but not enantiomers.

(e)   False. The R group of valine is hydrophobic; it is a nonpolar hydrocarbon. The R group of arginine is hydrophilic; although it is larger than that of valine, it has one imine and two amine groups capable of hydrogen-bonding interactions with water molecules.

24.62

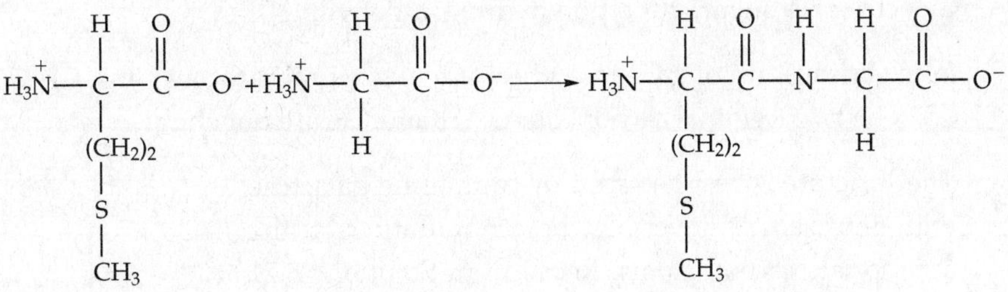

methionine   glycine   methionylglycine

24.64   (a)   Valine, serine, glutamic acid

(b)   Six (assuming the tripeptide contains all three amino acids):

Gly-Ser-Glu, GSE; Gly-Glu-Ser, GES; Ser-Gly-Glu, SGE; Ser-Glu-Gly, SEG; Glu-Ser-Gly, ESG; Glu-Gly-Ser, EGS

24.66   (a)   False. In an alpha helix, hydrogen bonding occurs between carbonyl groups and amide hydrogen atoms along the protein backbone.

(b)   False. In a beta sheet, hydrogen bonding occurs between carbonyl groups and amide hydrogen atoms along two different protein chains. This interaction zips the two chains together into a beta sheet.

Carbohydrates and Lipids (Sections 24.8 and 24.9)

24.68   (a)   No. Carbon 1 in the cyclic form of glucose is chiral. The configuration at C1 in alpha glucose is the opposite of that in beta glucose. However, there are four other chiral centers in each of the two forms that do not have opposite configurations. In order for molecules to be enantiomers, they must be mirror images of each other; all analogous chiral centers must have opposite configurations.

(b)

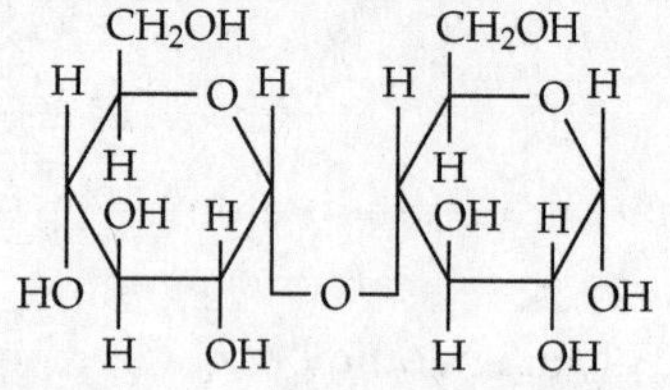

α-linkage

(c)

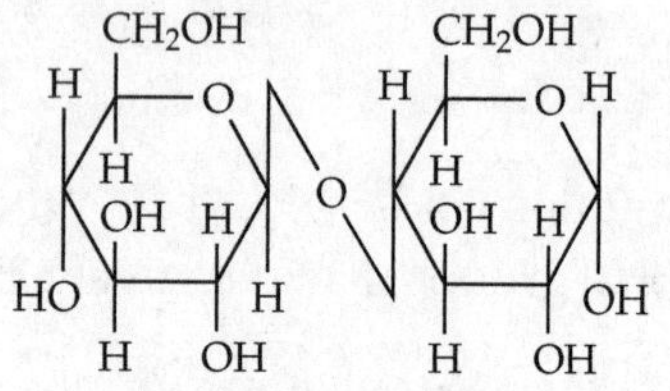

β-linkage

24.70   (a)   The empirical formula of starch is $C_6H_{10}O_5$.

(b)   The six-membered ring form of glucose is the unit that forms the basis of starch. The glucose monomer units are joined by α linkages.

(c)   Ether linkages connect the glucose monomer units in starch.

24.72   (a)   Galactose is a sugar; it is a polyhydroxy aldehyde.

(b)   In the linear form of galactose, there are four chiral carbon atoms, C2, C3, C4, and C5. The two terminal carbon atoms, C1 and C6, are not chiral.

(c)   The structure is best deduced by comparing galactose with glucose, and inverting the configurations at the appropriate carbon atoms. Recall from Solution 25.71 that both the β-form (shown here) and the α-form (OH on carbon 1 on the opposite side of ring as the CH_2OH on carbon 5) are possible.

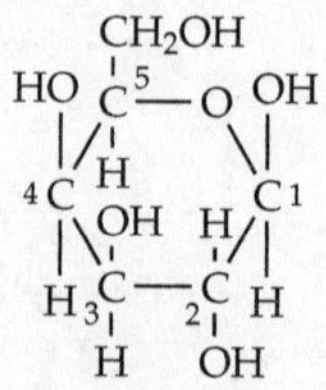

galactose

24.74   (a)   True. Consider the fuels ethane, C_2H_6, and ethanol, C_2H_5OH, where one C–H bond has been replaced by C–O–H, a C–O and an O–H bond. Combustion reactions for the two fuels follow.

$$C_2H_6 + \frac{7}{2}\,O_2 \rightarrow 2\,CO_2 + 3\,H_2O; \quad C_2H_5OH + 3\,O_2 \rightarrow 2\,CO_2 + 3\,H_2O$$

More energy is required to break bonds in the combustion of one mole of ethanol. The reaction is less exothermic overall than the combustion of ethane. As ethane has the more exothermic combustion reaction, we say that more energy is "stored" in C_2H_6 than in C_2H_5OH.

(b)   False. Trans fats contain at least one double bond, with substituents in the trans orientation.

(c)   True.

(d)   False. Monounsaturated fatty acids have one carbon-carbon double bond in the chain, whereas the rest are single bonds.

Nucleic Acids (Section 24.10)

24.76

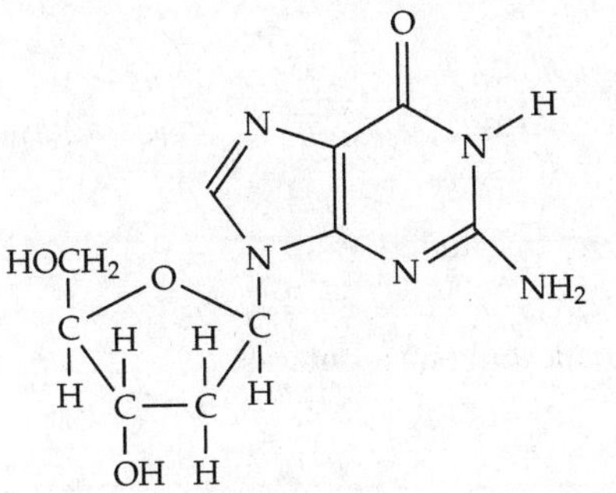

24.78   In the helical structure for DNA, the strands of the polynucleotides are held together by hydrogen-bonding interactions between particular pairs of bases. It happens that adenine and thymine form an especially effective base pair, and that guanine and cytosine are similarly related. Thus, each adenine has a thymine as its opposite number in the other strand, and each guanine has a cytosine as its opposite number. In the overall analysis of the double strand, total adenine must then equal total thymine, and total guanine equals total cytosine.

24.80   Statement (d) best explains the chemical differences between DNA and RNA. Statement (b) is true, but it does not prevent hydrogen bonding between complimentary base pairs.

Additional Exercises

24.82   *Analyze/Plan.* We are asked the number of structural isomers for two specified carbon chain lengths and a certain number of double bonds. Structural isomers have different connectivity. Since the chain length is specified, we can ignore structural isomers created by branching. We are not asked about geometrical isomers, so we ignore those as well. The resulting question is: How many ways are there to place the specified number of double bonds along the specified C chain? *Solve.*

5 C chain with one double bond: 2 structural isomers

C=C–C–C–C   C–C=C–C–C

6 C chain with two double bonds: 6 structural isomers

C=C–C=C–C–C   C=C–C–C=C–C   C=C–C–C–C=C

C–C=C–C=C–C   C=C=C–C–C–C   C–C=C=C–C–C

24.83 (a) 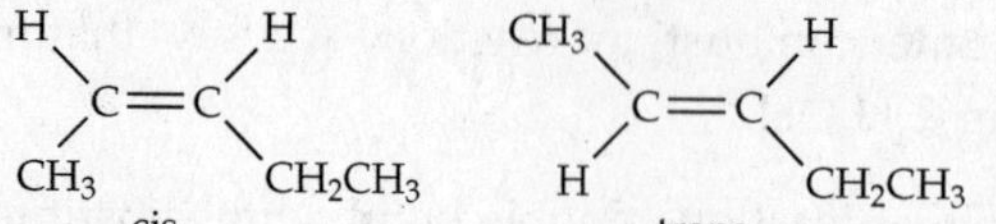

cis   trans

(b) Cyclopentene does not show cis-trans isomerism because the existence of the ring demands that the C–C bonds be cis to one another.

(c) 1-pentyne does not have enantiomers because the geometry about the alkyne group is linear.

24.84 The suffix –ene signifies an alkene, –one a ketone. The molecule has alkene and ketone functional groups.

24.85 (a) 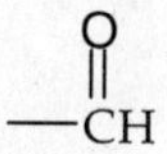 , aldehyde; 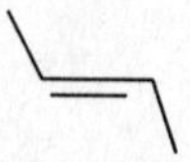 , *trans*-alkene; 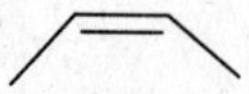 , *cis*-alkene

(b) —O— , ether; —OH , alcohol; C=C , alkene;

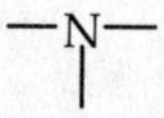 , amine (two of these, one aliphatic and one aromatic);

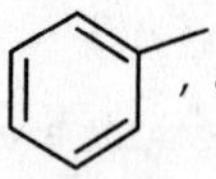 , aromatic (phenyl) ring

(c) 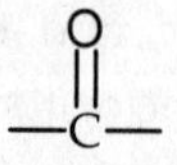 , ketone (2 of these); 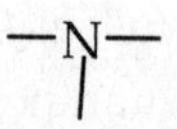 , amine (2 of these)

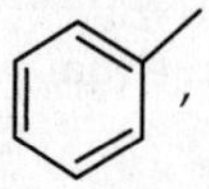 , aromatic (phenyl) ring (2 of these)

(d) 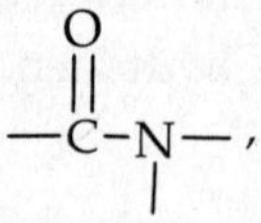 , amide; 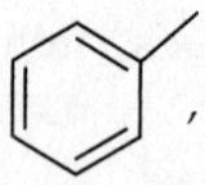 , aromatic (phenyl) ring;

—OH , alcohol (aromatic alcohol, phenol)

24.86 (a) Quinine, molecule (b), and indigo, molecule (c), both contain amine functional groups and would produce basic solutions if dissolved in water.

(b) Acetaminophen, molecule (d), would produce an acidic solution if dissolved in water. None of the four molecules is a carboxylic acid. Acetaminophen is both an amide and a phenol. Both of these functional groups have ionizable H atoms and produce very weakly acidic aqueous solutions.

(c)   Acetaminophen, molecule (d), is probably most water soluble. Both (c) and (d) can form hydrogen bonds with water, but acetaminophen has a lower molecular weight and smaller nonpolar portion to interfere with interactions between solute and solvent.

24.87   (a)

$$CH_3CH_2CH_2COH \quad or \quad (CH_3)_2CHCOH$$

(b)

(c)

$$CH_3{-}CH{-}CH_2 \quad or \quad CH_2{-}CH_2{-}CH_2$$
with OH OH and OH OH

(d)

24.88   (a) $ROH(aq) \rightleftharpoons H^+(aq) + RO^-(aq)$

(b)   The stronger the acid, the larger the value of K_a and the smaller the pK_a. Carboxylic acids, with pK_a values ~5, are much stronger acids than alcohols, with pK_a values ~16. Aqueous solutions of acids are significantly more acidic aqueous solutions of alcohols.

(c)   Carboxylic acids are stronger acids with smaller pK_a values than alcohols for two reasons. First, the electronegative carbonyl oxygen in a carboxylic acid withdraws electron density from the O–H bond, rendering the bond more polar and the H more ionizable. Second, the conjugate base of a carboxylic acid, carboxylate anion, exhibits resonance. This stabilizes the conjugate base and encourages ionization of the carboxylic acid. In an alcohol no electronegative atoms are bound to the carbon that holds the –OH group, and the H is tightly bound to the O.

24.89   In order for indole to be planar, the N atom must be sp^2 hybridized. The nonbonded electron pair on N is in a pure p orbital perpendicular to the plane of the molecule. The electrons that form the π bonds in the molecule are also in pure p orbitals perpendicular to the plane of the molecule. Thus, each of these p orbitals is in the correct orientation for π overlap; the delocalized π system extends over the entire molecule and includes the "nonbonded" electron pair on N. The reason that indole is such a weak base (H^+ acceptor) is that the nonbonded electron pair is delocalized and a H^+ ion does not feel the attraction of a full localized electron pair.

24.91   In the zwitterion form of a tripeptide present in aqueous solution near pH 7, the terminal carboxyl group is deprotonated and the terminal amino group is protonated, resulting in a net zero charge. The molecule has a net charge only if a side (R) group contains a charged (protonated or deprotonated) group. The tripeptide is positively charged if a side group contains a protonated amine. According to Figure 24.18, the only amino acids with protonated amines in their side groups are histidine (His), lysine (Lys), and arginine (Arg). Of the tripeptides listed, only (a) Gly-Ser-Lys will have a net positive charge at pH 7. [Note that aspartic acid (Asp) has a deprotonated carboxyl in its side group, so (c) Phe-Tyr-Asp will have a net negative charge at pH 7.]

24.92   Glu-Cys-Gly is the only possible order. Glutamic acid has two carboxyl groups that can form a peptide bond with cysteine, so there are two possible structures.

$$\underset{\substack{| \\ (CH_2)_2 \\ | \\ COO^-}}{\overset{\substack{O \\ \| }}{H_3\overset{+}{N}CHC}}-\underset{\substack{| \\ CH_2 \\ | \\ SH}}{\overset{\substack{H \quad O \\ | \quad \| }}{NCHC}}-\overset{\substack{H \quad O \\ | \quad \| }}{NCH_2CO^-} \quad or \quad \underset{\substack{| \\ COO^-}}{\overset{\substack{O \\ \| }}{H_3\overset{+}{N}CHCH_2CH_2C}}-\underset{\substack{| \\ CH_2 \\ | \\ SH}}{\overset{\substack{H \quad O \\ | \quad \| }}{NCHC}}-\overset{\substack{H \quad O \\ | \quad \| }}{NCH_2CO^-}$$

24.93   Both glucose and fructose contain six C atoms, so both are hexoses. Glucose contains an aldehyde group at C1, so it is an aldohexose. Fructose has a ketone at C2, so it is a ketohexose.

24.94   DNA and RNA have the bases guanine, cytosine and adenine in common, but DNA contains thymine, whereas RNA contains uracil. Thymine and uracil differ by a single methyl group, so both have a similar hydrogen bonding pattern with the complementary base adenine. This means that there is the possibility of a DNA strand binding to a "complementary" RNA strand.

Given this possibility, RNA is not involved in DNA replication. However, during the process of transcription and in the presence of the enzyme RNA polymerase, a complementary strand of mRNA is assembled along a segment of the backbone of a single DNA strand. During transcription, when there is adenine (A) in the DNA strand, uracil (U) is added to the RNA strand.

Integrative Exercises

24.96   Determine the empirical formula of the unknown compound and its oxidation product. Use chemical properties to propose possible structures.

$$68.1\,g\,C \times \frac{1\,mol\,C}{12.01\,g\,C} = 5.6703; \; 5.6703/1.1375 = 4.98 \approx 5$$

$$13.7\,g\,H \times \frac{1\,mol\,H}{1.008\,g\,H} = 13.5913; \; 13.5913/1.1375 = 11.95 \approx 12$$

$$18.2\,g\,P \times \frac{1\,mol\,O}{16.00\,g\,O} = 1.1375; \; 1.1375/1.1375 = 1$$

The empirical formula of the unknown is $C_5H_{12}O$.

$$69.7\,g\,C \times \frac{1\,mol\,C}{12.01\,g\,C} = 5.8035; \; 5.8035/1.1625 = 4.99 \approx 5$$

$$11.7\,g\,H \times \frac{1\,mol\,H}{1.008\,g\,H} = 11.6071; \; 11.6071/1.1625 = 9.99 \approx 10$$

$$18.6 \, g \, O \times \frac{1 \, mol \, O}{16.00 \, g \, O} = 1.1625; \, 1.1625/1.1625 = 1$$

The empirical formula of the oxidation product is $C_5H_{10}O$.

The compound is clearly an alcohol. Its slight solubility in water is consistent with the properties expected of a secondary alcohol with a five-carbon chain. The fact that oxidation results in a ketone, rather than an aldehyde or a carboxylic acid, tells us that it is a secondary alcohol. Some reasonable structures for the unknown secondary alcohol are:

$$\underset{\underset{OH}{|}}{CH_3CHCH_2CH_2CH_3} \quad \underset{\underset{OH}{|}}{CH_3CHCHCH_2CH_3} \quad \underset{\underset{OH}{|}}{CH_3CHCH(CH_3)_2}$$

24.97 Determine the empirical formula, molar mass, and thus molecular formula of the compound. Confirm with physical data.

$$66.7 \, g \, C \times \frac{1 \, mol \, C}{12.01 \, g \, C} = 5.554 \, \, mol \, C; \, 5.554/1.381 = 4.021 = 4$$

$$11.2 \, g \, H \times \frac{1 \, mol \, H}{1.008 \, g \, H} = 11.11 \, \, mol \, H; \, 11.11/1.381 = 8.043 = 8$$

$$22.1 \, g \, O \times \frac{1 \, mol \, O}{16.00 \, g \, O} = 1.381 \, \, mol \, O; \, 1.381/1.381 = 1$$

The empirical formula is C_4H_8O. Using Equation 10.11 (MM = molar mass):

$$MM = \frac{(2.28 \, g/L)(0.08206 \, L\text{-}atm/mol\text{-}K)(373K)}{0.970\text{-}atm} = 71.9 \, g/mol$$

The formula weight of C_4H_8O is 72, so the molecular formula is also C_4H_8O. Because the compound has a carbonyl group and cannot be oxidized to an acid, the only possibility is 2-butanone.

$$\underset{CH_3CCH_2CH_3}{\overset{\overset{\displaystyle O}{\|}}{}}$$

The boiling point of 2-butanone is 79.6°C, confirming the identification.

24.98 Determine the empirical formula, molar mass, and thus molecular formula of the compound. Confirm with physical data.

$$85.7 \, g \, C \times \frac{1 \, mol \, C}{12.01 \, g \, C} = 7.136 \, mol \, C; \, 7.136/7.136 = 1$$

$$14.3 \, g \, H \times \frac{1 \, mol \, H}{1.008 \, g \, H} = 14.19 \, mol \, H; \, 14.19/7.136 \approx 2$$

Empirical formula is CH_2. Using Equation 10.11 (MM = molar mass):

$$MM = \frac{(2.21 \, g/L)(0.08206 \, L\text{-}atm/mol\text{-}K)(373K)}{(735/760) \, atm} = 69.9 \, g/mol$$

The molecular formula is thus C_5H_{10}. The absence of reaction with aqueous Br_2 indicates that the compound is not an alkene, so the compound is probably the cycloalkane cyclopentane. According to the *Handbook of Chemistry and Physics*, the boiling point of cyclopentane is 49 °C at 760 torr. This confirms the identity of the unknown.

24.100 (a)   At low pH, the amine and carboxyl groups are protonated.   At high pH, the amine and carboxyl groups are deprotonated.

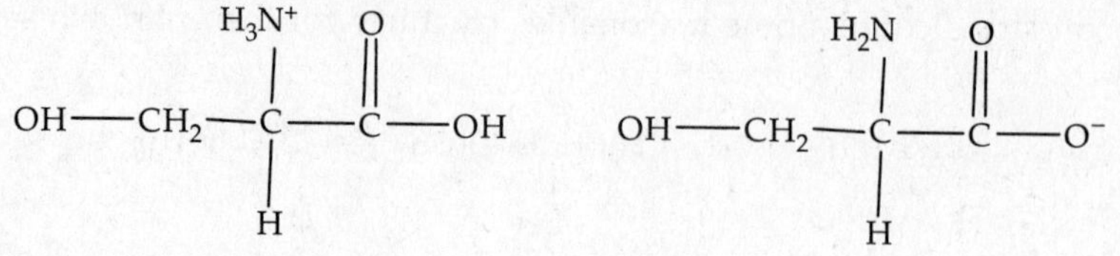

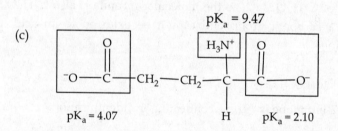

(b)

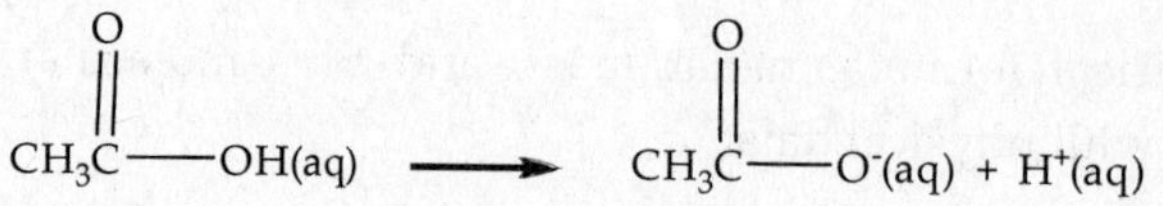

$$K_a = 1.8 \times 10^{-5}, \quad pK_a = -\log(1.8 \times 10^{-5}) = 4.74$$

The conjugate acid of NH_3 is NH_4^+.

$$NH_4^+(aq) \rightleftharpoons NH_3(aq) + H^+(aq)$$

$$K_a = K_w/K_b = 1.0 \times 10^{-14} / 1.8 \times 10^{-5} = 5.55 \times 10^{-10} = 5.6 \times 10^{-10}$$

$$pK_a = -\log(5.55 \times 10^{-10}) = 9.26$$

In general, a –COOH group is stronger acid than a $-NH_3^+$ group. The lower pK_a value for amino acids is for the ionization (deprotonation) of the –COOH group and the higher pK_a is for the deprotonation of the $-NH_3^+$ group.

(c)

By analogy to serine, the carboxyl group near the amine will have pK_a ~2 and the amino group will have pK_a ~9. By elimination, the carboxyl group in the side chain has pK_a ~4.

(d)   From the titration curve, the unknown amino acid has three pK_a values: 2.0, 4.0 and 10.0. The unknown will be an acidic amino acid, because two of the pK_a values are significantly less than 7. The two acidic amino acids are glutamic acid and aspartic acid. From part (c), the pK_a values of glutamic acid are close to but not an exact match to the ones on the titration curve. The most likely candidate is aspartic acid.

Asparagine and glutamine are the amide forms of aspartic and glutamic acid. The "middle" pKa value for these two amino acids represents ionization of the amide hydrogen and will be large than 4.0.

24.101 (a) Because the native form is most stable, it has a lower, more negative free energy than the denatured form. Another way to say this is that ΔG for the process of denaturing the protein is positive.

(b) ΔS is negative in going from the denatured form to the folded (native) form; the native protein is more ordered.

(c) The four S–S linkages are strong covalent links holding the chain in place in the folded structure. A folded structure without these links would be less stable (higher G) and have more motional freedom (more positive entropy).

(d) After reduction, the eight S–S groups will form hydrogen-bond-like interactions with acceptors along the protein backbone, but these will be weaker and less specifically located than the S–S covalent bonds of the native protein. Overall, the tertiary structure of the reduced protein will be looser and less compact because of the loss of the S–S linkages; the entropy will be higher.

(e) The amino acid cysteine must be present in order for –SH bonds to be found in ribonuclease A. (Methionine contains S, but no –SH functional group.)

24.102 $AMPOH^-(aq) \rightleftharpoons AMPO^{2-}(aq) + H^+(aq)$

$pK_a = 7.21; K_a = 10^{-pK_a} = 6.17 \times 10^{-8} = 6.2 \times 10^{-8}$

$K_a = \dfrac{[AMPO^{2-}][H^+]}{[AMPOH^-]} = 6.2 \times 10^{-8}$. When $pH = 7.40$, $[H^+] = 3.98 \times 10^{-8} = 4 \times 10^{-8}$.

Then $\dfrac{[AMPOH^-]}{[AMPO^{2-}]} = 3.98 \times 10^{-8} / 6.17 \times 10^{-8} = 0.6457 = 0.6$